# STUDY GUIDE AND WORKBOOK
# AN INTERACTIVE APPROACH

for
Starr and Taggart's

# BIOLOGY

*The Unity and Diversity of Life*

EIGHTH EDITION

### JANE B. TAYLOR

*Northern Virginia Community College*

### JOHN D. JACKSON

*North Hennepin Community College*

**Wadsworth Publishing Company**
I(T)P® An International Thomson Publishing Company

Belmont, CA • Albany, NY • Bonn • Boston • Cincinnati • Detroit • Johannesburg • London
Madrid • Melbourne • Mexico City • New York • Paris • Singapore • Tokyo • Toronto • Washington

**Biology Editor:** Jack Carey
**Project Development Editor:** Kristin Milotich
**Editorial Assistant:** Michael Burgreen
**Marketing Manager:** Halee Dinsey
**Project Editors:** Dianne Jensis, Howard Severson
**Print Buyer:** Stacey Weinberger
**Permissions Editor:** Veronica Olivia
**Copy Editor:** Publishers' Design and Production Services, Inc.
**Art Editor:** Roberta Broyer
**Compositor:** Publishers' Design and Production Services, Inc.
**Cover Designer:** Gary Head
**Cover Image:** *Minnehaha Falls*, © Richard Hamilton Smith
**Printer:** Courier

For more information, contact Wadsworth Publishing Company, 10 Davis Drive, Belmont, CA 94002, or electronically at
http://www.thomson.com/wadsworth.html

International Thomson Publishing Europe
Berkshire House 168-173
High Holborn
London, WC1V 7AA, England

International Thomson Editores
Campos Eliseos 385, Piso 7
Col. Polanco
11560 México D.F. México

Thomas Nelson Australia
102 Dodds Street
South Melbourne 3205
Victoria, Australia

International Thomson Publishing Asia
221 Henderson Road
#05-10 Henderson Building
Singapore 0315

Nelson Canada
1120 Birchmount Road
Scarborough, Ontario
Canada M1K 5G4

International Thomson Publishing Japan
Hirakawacho Kyowa Building, 3F
2-2-1 Hirakawacho
Chiyoda-ku, Tokyo 102, Japan

International Thomson Publishing GmbH
Königswinterer Strasse 418
53227 Bonn, Germany
Halfway House, 1685 South Africa

International Thomson Publishing Southern Africa
Building 18, Constantia Park
240 Old Pretoria Road

ISBN 0-534-53010-9

# CONTENTS

# Photo Credits:

**Chapter 7**
p. 75 (36): David Fisher.
p. 75 (37): David Fisher.

**Chapter 22**
p. 232 (10): Gary Grimes and Steven L'Hernault.
p. 232 (12): Tony Brain/SPL/Photo Researchers.
p. 232 (14): T.J. Beveridge, University of Guelph/BPS.
p. 232 (15): Tony Brain/SPL/Photo Researchers.

**Chapter 23**
p. 240 (20): M. Abbey/Visuals Unlimited.
p. 240 (21): T.E. Adams/Visuals Unlimited.
p. 240 (22): John Clegg/Ardea, London.
p. 243: D.J. Patterson/Seaphot Limited: Planet Earth Pictures.
p. 246 (20): Jan Hinsch/SPL/Photo Researchers.
p. 246 (22): John Clegg/Ardea, London.
p. 246 (23): M. Abbey/Visuals Unlimited.

**Chapter 24**
p. 250: G. T. Cole, University of Texas, Austin/BPS.
p. 252 (inset left): Ed Reschke.
p. 252 (inset right): Ed Reschke.
p. 254 (1): Victor Duran.
p. 254 (2): John E. Hodgin.
p. 254(3): Jane Burton/Bruce Coleman Ltd.
p. 254 (4): M. Eichelberger/Visuals Unlimited.
p. 255 (6): Ken Davis/Tom Stack & Associates.
p. 255 (7): Robert C. Simpson/Nature Stock.
p. 255 (8): G.L. Barron, University of Guelph.

**Chapter 25**
p. 262: Jane Burton/Bruce Coleman Ltd.
p. 264: A. & E. Bomford/Ardea, London.
p. 267: Edward S. Ross.

**Chapter 26**
p. 293: Jane Burton/Bruce Coleman Ltd.
p. 294 (a): Ian Took/Biofotos.
p. 294 (b): John Mason/Ardea, London.
p. 294 (c): Kjell B. Sandved.
p. 294 (d): Douglas Faulkner/Sally Faulkner Collection.

**Chapter 27**
p. 303: Bill Wood/Bruce Coleman Ltd.
p. 310: Christopher Crowley.
p. 311 (8): Bill Wood/Bruce Coleman Ltd.
p. 311 (9): Reinhard/ZEFA.
p. 311 (10): Peter Scoones/Seaphot Ltd.: Planet Earth Pictures.
p. 312 (11): Erwin Christian/ZEFA.
p. 312 (12): Peter Scoones/Seaphot Ltd.: Planet Earth Pictures.
p. 312 (13): Allan Power/Bruce Coleman Ltd.
p. 312 (14): Rick M. Harbo.
p. 312 (15): Hervé Chaumeton/Agence Nature.

**Chapter 29**
p. 328 (1): D.E. Akin and I.L. Risgby, Richard B. Russel Agricultural Research Center, Agricultural Research Service, U.S., Department of Agriculture, Athens, GA.
p. 328 (2): Biophoto Associates.
p. 328 (3): Biophoto Associates.
p. 328 (4): Kingsley R. Stern.
p. 328 (5): Biophoto Associates.
p. 328 (6): Jan Robert Factor/Photo Researchers.
p. 331: Robert and Linda Mitchell.
p. 332 (center): Carolina Biological Supply Company.
p. 332 (right): James W. Perry.
p. 333 (center): Ray F. Evert.
p. 333 (right): James W. Perry.
p. 335 (above): C.E. Jeffree et al., *Planta*, 172 (1):20–37, 1987; reprinted by permission of C.E. Jeffree and Springer-Verlag.
p. 335 (below): Jeremy Burgess/SPL Photos/Photo Researchers.
p. 337: Chuck Brown.
p. 340: H.A. Core, W.A. Cote, and A. C. Day, *Wood Structure and Identification*, Second edition, Syracuse University Press.

**Chapter 31**
p. 360 (a): Patricia Schulz.
p. 360 (b): Patricia Schulz.
p. 360 (c): Ray F. Evert.
p. 360 (c): Ray F. Evert.
p. 360 (e): Ripon Microslides.
p. 360 (f): Ripon Microslides.

**Chapter 32**
p. 368 (top): Hervé Chaumeton/Agence Nature.
p. 368 (middle): Barry L. Runk/Grant Heilman Photography
p. 368 (bottom): Mauseth.

**Chapter 35**
p. 404: Manfred Kage/Peter Arnold, Inc.
p. 408: C. Yokochi and J. Rohen, *Photographic Anatomy of the Human Body*, Second Edition, Igaku-Shoin Ltd., 1979.

**Chapter 36**
p. 419: Ed Reschke.

**Chapter 38**
p. 443: Ed Reschke.
p. 449: D. Fawcett, *The Cell*, Philadelphia: W.B. Saunders Co., 1966.

**Chapter 45**
p. 524: Ed Reschke.
p. 531: Lennart Nilsson, *A Child Is Born*, © 1966, 1977 Dell Publishing Company, Inc.
p. 531: Lennart Nilsson, *A Child Is Born*, © 1966, 1977 Dell Publishing Company, Inc.

# PREFACE

*Tell me and I will forget, show me and I might
remember, involve me and I will understand.*
—Chinese Proverb

The proverb outlines three levels of learning, each successively more effective than the method preceding it. The writer of the proverb understood that humans learn most efficiently when they involve themselves in the material to be learned. This study guide is like a tutor; when properly used it increases the efficiency of your study periods. The interactive exercises actively involve you in the most important terms and central ideas of your text. Specific tasks ask you to recall key concepts and terms and apply them to life; they test your understanding of the facts and indicate items to reexamine or clarify. Your performance on these tasks provides an estimate of your next test score based on specific material. Most important, though, this biology study guide and text together help you make informed decisions about matters that affect your own well-being and that of your environment. In the years to come, human survival on planet Earth will require administrative and managerial decisions based on an informed biological background.

## HOW TO USE THIS STUDY GUIDE

Following this preface, you will find an outline that shows you how the study guide is organized and that will help you use it efficiently. Each chapter begins with a title and an outline list of the 1- and 2-level headings in that chapter. The Interactive Exercises follow, wherein each chapter is divided into sections of one or more of the main (1-level) headings. These main-level headings are labeled 10.1, 10.2, and so on. This system exactly matches the section numbers used in the textbook. The Interactive Exercises begin with a list of Selected Words (other than boldfaced terms) selected by the authors as those that are most likely to enhance understanding. In the text chapters, these words appear in italics, quotation marks, or roman type. This is followed by a list of Boldfaced, Page-Referenced Terms, which appear bold-

faced in the text. These terms are essential to understanding each study guide section of a particular chapter. Space is provided by each term for you to formulate a definition in your own words. Next is a series of different types of exercises that includes completion, short answer, true/false, fill-in-the-blanks, matching, choice, dichotomous choice, label and match, crossword puzzles, problems, labeling, sequencing, multiple-choice, and completion of tables.

A Self-Quiz immediately follows the Interactive Exercises. This quiz is composed primarily of multiple-choice questions, although sometimes we present another examination device or some combination of devices. Any wrong answers in the quiz indicate portions of the text you need to reexamine. A series of Chapter Objectives/Review Questions follows each Self-Quiz. These are tasks that you should be able to accomplish if you have understood the assigned reading in the text. Some objectives require you to compose a short answer or long essay, while others require drawing a sketch or supplying correct words.

The final part of each chapter is named Integrating and Applying Key Concepts. It invites you to try your hand at applying major concepts to situations in which there is not necessarily a single pat answer, and so none is provided in the chapter answer section (except for a problem in Chapter 11). Your text generally will provide enough clues to get you started on an answer, but this part is intended to stimulate your thought and provoke group discussions. A separate publication titled *Critical Thinking Exercises for Starr and Taggart's Biology: The Unity and Diversity of Life,* 8th Edition, is available. Corresponding with the text chapters, these exercises present problem situations that concentrate on the critical and higher-level thinking skills used by scientists. Solving these problems requires you to apply chapter information to form new perspectives, analyze data, draw conclusions, make predictions, and identify basic assumptions.

*A person's mind, once stretched by a new idea, can
never return to its original dimension.*
—Oliver Wendell Holmes

# STRUCTURE OF THIS STUDY GUIDE

The outline below shows how each chapter in this study guide is organized.

Chapter Number   ———————————————▶    **10**

Chapter Title   ———————————————▶    **MEIOSIS**

Chapter Vignette   ———————————————▶    *Octopus Sex and Other Stories*

Chapter Outline   ———————————————▶

**10.1 COMPARISON OF ASEXUAL AND SEXUAL REPRODUCTION**

**10.2 HOW MEIOSIS HALVES THE CHROMOSOME NUMBER**
Think "Homologues"
Two Divisions, Not One

**10.3 A VISUAL TOUR OF THE STAGES OF MEIOSIS**

**10.4 KEY EVENTS OF MEIOSIS I**
Prophase I Activities
Metaphase I Alignments

**10.5 FROM GAMETES TO OFFSPRING**
Gamete Formation in Plants
Gamete Formation in Animals
More Shufflings at Fertilization

**10.6 MEIOSIS AND MITOSIS COMPARED**

Interactive Exercises   ———————————————▶ The interactive exercises are divided into numbered sections (that match the numbered sections of the text) by titles of main headings and page references. Each section begins with a list of author-selected words that appear in the text chapter in italics, quotation marks, or roman type. This is followed by a list of important boldfaced, page-referenced terms from each section of the chapter. Each section ends with interactive exercises that vary in type and require constant interaction with the important chapter information.

Self-Quiz   ———————————————▶ Usually a set of multiple-choice questions that sample important blocks of text information.

Chapter Objectives/Review Questions   ———————▶ Combinations of relative objectives to be met and questions to be answered.

Integrating and Applying Key Concepts   ———————▶ Applications of text material to questions for which there may be more than one correct answer.

Answers to Interactive Exercises and Self-Quiz   ———————▶ Answers for all interactive exercises can be found at the end of this study guide by chapter and title, and the main headings with their page references, followed by answers for the Self-Quiz.

# 1

---

# METHODS AND CONCEPTS IN BIOLOGY

---

## Interactive Exercises

Note: In the answer sections of this book, a specific molecule is most often indicated by its abbreviation. For example, adenosine triphosphate is ATP.

---

*Biology Revisited* (pp. 2–3)

## 1.1. DNA, ENERGY, AND LIFE (pp. 4–5)

*Selected Words:* from *DNA to RNA to protein, transfer* of energy, *internal* environment

In addition to the boldfaced terms, the text features other important terms essential to understanding the assigned material. "Selected Words" is a list of these terms, which appear in the text in italics, in quotation marks, and occasionally in roman type. Latin binomials found in this section are underlined and in roman type to distinguish them from other italicized words.

## Boldfaced, Page-Referenced Terms

The page-referenced terms are important; they are in boldface type in the chapter. Write a definition for each term in your own words without looking at the text. Next, compare your definition with that given in the chapter or in the text glossary. If your definition seems inaccurate, allow some time to pass and repeat this procedure until you can define each term rather quickly (how fast you answer is a gauge of your learning effectiveness).

(4) cell _____

_____

(4) DNA _____

_____

(4) RNAs _____

_____

(4) reproduction _____

_____

(4) energy _____

_____

(5) metabolism _____

_____

(5) photosynthesis _____

_____

(5) aerobic respiration _____

_____

(5) receptors _____

_____

(5) homeostasis _____

_____

## Fill-in-the-Blanks

All living things consist of one or more (1) _____ , the smallest units of matter having the capacity for life. The most important molecule in a cell is a nucleic acid known as (2) _____ . Encoded in this molecule's structure are the instructions for assembling a dazzling array of (3) _____ from a limited number of smaller building blocks, the (4) _____ acids. (5) _____ are a class of worker proteins that build, juggle, and split all of the complex molecules of life when they receive an energy boost. (6) _____ carry out DNAs instructions by working as partners with some enzymes. Think about the information encoded in DNA flowing to (7) _____ and then to (8) _____ . One of life's defining features is (9) _____ , the process by which parents transmit DNA instructions for duplicating their traits to offspring. (10) _____ is most simply defined as the capacity to do work. Nothing in the universe happens without a complete or partial (11) _____ of energy. (12) _____ refers to the cell's capacity to extract and transform energy from its surroundings and use energy to maintain itself, grow, and make more cells. Leaves contain cells that carry on the process of (13) _____ by intercepting energy from the sun and using it to produce molecules of the energy carrier called (14) _____ . These molecules transfer energy to metabolic workers inside the cell where enzyme molecules assemble sugars with potential (stored) energy. In most organisms, some of the potential energy stored in sugars is released to form more ATP by way of a metabolic process known as aerobic (15) _____ . Organisms have certain molecules and structures called (16) _____ that can detect specific stimuli related to changes in their environment. Following a snack, simple sugars and other molecules leave the gut and enter the blood, which is part of the body's (17) _____ environment. Blood sugar level then rises and stimulates secretion of the hormone insulin by the pancreas. Insulin stimulates cells to take up sugar molecules from the internal environment and return blood sugar concentration levels to normal. This is an example of (18) _____ , the capacity to maintain rather constant physical and chemical conditions inside an organism within some tolerable range, even when external conditions vary.

## 1.2. ENERGY AND LIFE'S ORGANIZATION (pp. 6–7)

### Boldfaced, Page-Referenced Terms

(6) cell _____

(6) multicelled organisms _____
_____

(6) population _____
_____

(6) community _____
_____

(6) ecosystem _____
_____

(7) biosphere _____
_____

(7) producers _____
_____

(7) consumers _____
_____

(7) decomposers _____
_____

## Matching

Choose the most appropriate answer to match with each of the following terms.

1. ___organ system
2. ___cell
3. ___community
4. ___ecosystem
5. ___molecule
6. ___DNA
7. ___organelle
8. ___population
9. ___subatomic particle
10. ___tissue
11. ___biosphere
12. ___energy
13. ___multicelled organism
14. ___organ
15. ___atom

A. One or more tissues interacting as a unit
B. A proton, neutron, or electron
C. A well-defined structure within a cell, performing a particular function
D. All of the regions of Earth where organisms can live
E. A capacity to make things happen, to do work
F. The smallest unit of life
G. Two or more organs whose separate functions are integrated to perform a specific task
H. Two or more atoms bonded together
I. All of the populations interacting in a given area
J. The smallest unit of a pure substance that has the properties of that substance
K. A special molecule; sets living things apart from the nonliving world
L. A community interacting with its nonliving environment
M. An individual composed of cells arranged in tissues, organs, and often organ systems
N. A group of individuals of the same species in a particular place at a particular time
O. A group of cells that work together to carry out a particular function

## Sequence

This exercise provides further practice with the levels of organization in nature. Arrange in hierarchical order with the largest, most inclusive category first and the smallest, most exclusive category last.

16. ___     A. Tissue
17. ___     B. Community
18. ___     C. Molecule
19. ___     D. Biosphere
20. ___     E. Organ system
21. ___     F. Organelle
22. ___     G. Ecosystem
23. ___     H. Atom
24. ___     I. Cell
25. ___     J. Population
26. ___     K. Subatomic particle
27. ___     L. Multicelled organism
28. ___     M. Organ

*Fill-in-the-Blanks*

Plants and other organisms that produce their own food are (29) _____ and serve as an energy entry point for the world of life. Animals feed directly or indirectly on energy stored in tissues of the producers; they are known as (30) _____ . (31) _____ are bacteria and fungi that feed on tissues or remains of other organisms and break down biological molecules to simple materials that may be cycled back to producers. Thus, there is a one-way flow of (32) _____ through organisms and a (33) _____ of materials among them that organizes life in the biosphere.

## 1.3. SO MUCH UNITY, YET SO MANY SPECIES (pp. 8–9)

*Selected Words:* Quercus alba, Q. rubra, prokaryotic, eukaryotic

*Boldfaced, Page-Referenced Terms*

(8) species _____

_____

(8) genus _____

_____

(8) Archaebacteria _____

_____

(8) Eubacteria _____

_____

(8) Protista _____

_____

(8) Fungi _____

_____

(8) Plantae _____

_____

(8) Animalia _____

_____

*Fill-in-the-Blanks*

Different "kinds" of organisms are referred to as (1) _____. A (2) _____ is the first of a two-part name of each organism that encompasses all the species having perceived similarities to one another. For example, the pronghorn antelope is known by the two-part name *Antilocapra americana; Antilocapra* is the (3) _____ name and *americana* is the (4) _____ name. In the classification of organisms, genera that share common ancestry are grouped in the same (5) _____. Related families are grouped in the same (6) _____ and related orders are grouped in the same (7) _____. Related classes are grouped into a (8) _____ that is then assigned to one of the five kingdoms of life. The adjective (9) _____ describes single-celled organisms that lack a nucleus while the adjective (10) _____ describes single-celled and multicelled organisms whose DNA is enclosed within a nucleus.

## Complete the Table

11. Fill in the table below by entering the correct name of each kingdom of life described.

| Kingdom | Description |
|---|---|
| a. | Eukaryotic, multicelled, photosynthetic producers |
| b. | Prokaryotic, very successful in distribution living nearly everywhere, single cells, producers, consumers, or decomposers |
| c. | Eukaryotic, mostly multicelled, decomposers and consumers, food is digested outside their cells and bodies |
| d. | Prokaryotic, live only in extreme habitats, such as the ones that prevailed when life originated |
| e. | Eukaryotic, single-celled species and some multicelled forms, larger than bacteria but internally more complex |
| f. | Eukaryotic, diverse multicelled consumers, actively move at least during some stage of their life. |

## Sequence

Arrange in correct hierarchical order with the largest, most inclusive category first and the smallest, most exclusive category last. This exercise classifies an animal with the common name of "beaver." Refer to p. 8 in the text and Appendix I, pp. A1-A3.

12. ___     A. Class: Mammalia

13. ___     B. Family: Castoridae

14. ___     C. Genus: *Castor*

15. ___     D. Kingdom: Animalia

16. ___     E. Order: Rodentia

17. ___     F. Phylum: Chordata

18. ___     G. Species: *canadensis*

## 1.4. AN EVOLUTIONARY VIEW OF LIFE'S DIVERSITY (pp. 10–11)

**Selected Words:** *hemophilia A,* Staphylococcus

**Boldfaced, Page-Referenced Terms**

(10) mutation _____

_____

(10) adaptive trait _____

_____

(10) evolution _____

_____

(10)  artificial selection _____

_____

(11)  natural selection _____

_____


## Choice

For questions 1–10, choose from the following:

a. evolution through artificial selection          b. evolution through natural selection

1. ____ Pigeon breeding

2. ____ Antibiotics are powerful agents of this process

3. ____ A favoring of adaptive traits in nature

4. ____ The selection of one form of a trait over another taking place under contrived, manipulated conditions

5. ____ Refers to change that is occurring within a line of descent over time

6. ____ The outcome of differences in survival and reproduction among individuals that differ in one or more trait

7. ____ Darwin viewed this as a simple model for natural selection

8. ____ Breeders are the "selective agents"

9. ____ The mechanism whereby antibiotic resistance evolves

10. ____ Changing of the relative frequencies of different forms of moths through successive generations


## 1.5. THE NATURE OF BIOLOGICAL INQUIRY (pp. 12–13)

## 1.6. *Focus on Science:* THE POWER AND PITFALLS OF EXPERIMENTAL TESTS (pp. 14–15)

## 1.7. THE LIMITS OF SCIENCE (p. 16)

*Selected Words:* *inductive* logic, *deductive* logic, *prediction*, *potentially falsifiable* hypotheses, *biological therapy,* Escherichia coli, *one variable, quantitative* terms

### Boldfaced, Page-Referenced Terms

(12)  hypotheses _____

_____

(12)  prediction _____

_____

(12)  test _____

_____

(12)  logic _____

_____

(12) experiments _____

_____

(12) control groups _____

_____

(12) variables _____

_____

(12) scientific theory _____

_____

## Sequence

Arrange the following steps of the scientific method in correct chronological sequence. Write the letter of the first step next to 1, the letter of the second step next to 2, and so on.

1. ___    A. Develop hypotheses about what the solution or answer to a problem might be.

2. ___    B. Devise ways to test the accuracy of predictions drawn from the hypothesis (use of observations, models, and experiments).

3. ___    C. Repeat or devise new tests (different tests might support the same hypothesis).

4. ___    D. Make a prediction, using hypotheses as a guide; the "if-then" process.

5. ___    E. If the tests do not provide the expected results, check to see what might have gone wrong.

6. ___    F. Objectively analyze and report the results from tests and the conclusions drawn.

7. ___    G. Identify a problem or ask a question about nature.

## Labeling

Assume that you have to identify what object is hidden inside a sealed, opaque box. Your only tools to test the contents are a bar magnet and a triple-beam balance. Label each of the following with an O (for observation) or a C (for conclusion).

8. ___ The object has two flat surfaces.

9. ___ The object is composed of nonmagnetic metal.

10. ___ The object is not a quarter, a half-dollar, or a silver dollar.

11. ___ The object weighs $x$ grams.

12. ___ The object is a penny.

## Complete the Table

13. Complete the following table of concepts important to understanding the scientific method of problem solving. Choose from scientific experiment, variable, prediction, inductive logic, control group, hypothesis, deductive logic, and theory.

| Concept | Definition |
|---|---|
| a. | An educated guess about what the answer (or solution) to a scientific problem might be |
| b. | A statement of what one should be able to observe in nature if one looks; the "if-then" process |
| c. | A related set of hypotheses that, taken together, form a broad explanation of a fundamental aspect of the natural world |
| d. | A carefully designed test that manipulates nature into revealing one of its secrets |
| e. | Used in scientific experiments to evaluate possible side effects of a test being performed on an experimental group |
| f. | The control group is identical to the experimental group except for the *key factor* under study |
| g. | An individual makes inferences about specific consequences or specific predictions that must follow from a hypothesis |
| h. | An individual derives a general statement from specific observations |

## Dichotomous Choice

An Italian physician, Francisco Redi, published a paper in 1688 in which he challenged the doctrine of spontaneous generation, the proposition that living things could arise from dead material. Although many examples of spontaneous generation were described in his day, Redi's work dealt particularly with disproving the notion that decaying meat could be transformed into flies. He tested his ideas in a laboratory.

Circle one of two possible answers given between parentheses in each statement; questions 14–18 deal with spontaneous generation.

14. "Two sets of jars are filled with meat or fish. One set is sealed; the other is left open so that flies can enter the jars." This description deals with a(n) (hypothesis/experiment).
15. The description in the last question also includes a (prediction/control).
16. Prior to his test, Redi suggested that "Worms are derived directly from the droppings of flies." This statement represents a(n) (theory/hypothesis).
17. "Worms (maggots) will appear only in the second set of jars" represent a(n) (prediction/hypothesis).
18. The statement, "mice arise from a dirty shirt and a few grains of wheat placed in a dark corner" is best called a(n) (belief/test).
19. From a multitude of individual observations he made of the natural world, Charles Darwin proposed the theory of organic evolution. This was an example of (deductive logic/inductive logic).
20. Since the time that Darwin proposed the theory of organic evolution, countless numbers of biologists have discovered evidence of various kinds that conform to the general theory. This is an example of (deductive logic/inductive logic).

21. The control group is identical to the experimental group except for the (hypothesis/variable) being studied.
22. Systematic observations, model development, and conducting experiments are all methods employed to (make predictions/test predictions).
23. Science is distinguished from faith in the supernatural by (cause and effect/experimental design).
24. Through their failure to use large-enough samples in their experiments, scientists encounter (bias in reporting results/sampling error).
25. Science emphasizes reporting test results in (quantitative/qualitative) terms.

## Completion

26. Questions whose answers are _____ in nature do not readily lend themselves to scientific analysis and experimentation.
27. Scientists often stir up controversy when they explain a part of the world that was considered beyond natural explanation—that is, belonging to the "_____."
28. The external world, not internal _____, must be the testing ground for scientific beliefs.

---

# Self-Quiz

___ 1. About 12 to 24 hours after a meal, a person's blood-sugar level normally varies from about 60 to 90 mg per 100 ml of blood, through it may attain 130 mg/100 ml after meals high in carbohydrates. That the blood-sugar level is maintained within a fairly narrow range despite uneven intake of sugar is due to the body's ability to carry out _____.
   a. predictions
   b. inheritance
   c. metabolism
   d. homeostasis

___ 2. Different species of Galapagos Island finches have different beak types that enable them to obtain different kinds of food. One species removes tree bark with a sharp beak to forage for insect larvae and pupae whereas another species has a large, powerful beak capable of crushing and eating large, heavy coated seeds. These statements illustrate _____.
   a. adaptation
   b. metabolism
   c. puberty
   d. homeostasis

___ 3. A boy is color-blind just as his grandfather was, even though his mother had normal vision. This situation is the result of _____.
   a. adaptation
   b. inheritance
   c. metabolism
   d. homeostasis

___ 4. The digestion of food, the production of ATP by respiration, the construction of the body's proteins, cellular reproduction by cell division, and the contraction of a muscle are all part of _____.
   a. adaptation
   b. inheritance
   c. metabolism
   d. homeostasis

___ 5. Which of the following does *not* involve using energy to do work?
   a. atoms bonding together to form molecules
   b. the division of one cell into two cells
   c. the digestion of food
   d. none of these

___ 6. The experimental group and control group are identical except for _____.
   a. the number of variables studied
   b. the variable under study
   c. the two variables under study
   d. the number of experiments performed on each group

___ 7. A hypothesis should *not* be accepted as valid if _____.
   a. the sample studied is determined to be representative of the entire group
   b. a variety of different tools and experimental designs yield similar observations and results
   c. other investigators can obtain similar results when they conduct the experiment under similar conditions
   d. several different experiments, each without a control group, systematically eliminate each of the variables except one

___ 8. The principal point of evolution by natural selection is that _____.
   a. it measures the difference in survival and reproduction that has occurred among individuals who differ from one another in one or more traits
   b. even bad mutations can improve survival and reproduction of organisms in a population
   c. evolution does not occur when some forms of traits increase in frequency and others decrease or disappear with time
   d. individuals lacking adaptive traits make up more of the reproductive base for each new generation

___ 9. Which match is incorrect?
   a. Kingdom Animalia—diverse multicelled consumers, actively move at least during some stage of their life
   b. Kingdom Plantae—multicelled photosynthetic producers
   c. Kingdom Archaebacteria—relatively simple cells, successful in distribution living nearly everywhere
   d. Kingdom Fungi—mostly multicelled decomposers and consumers, food is digested outside their cells and bodies
   e. Kingdom Protista—internally complex single cells, some multicellular

___10. The least inclusive of the taxonomic categories listed is _____.
   a. family
   b. phylum
   c. class
   d. order
   e. genus

# Chapter Objectives/Review Questions

This section lists general and detailed chapter objectives that can be used as review questions. You can make maximum use of these items by writing answers on a separate sheet of paper. Fill in answers where blanks are provided. To check for accuracy, compare your answers with information given in the chapter or glossary.

*Page*    *Objectives/Questions*

(4)    1.  A special molecule called _____ acid, or DNA, sets living things apart from the nonliving world.

(4)    2.  The _____ is an organized unit that can survive and reproduce on its own, given DNA, raw materials, and inputs of energy.

(4)    3.  To carry out the instructions in DNA, some enzymes work as partners with another class of nucleic acids, the _____ .

(4)    4.  Describe the process of reproduction in terms of DNA instructions.

(4–5)  5.  Define *energy* and explain its role in life processes.

(5)    6.  Explain the function of the process known as metabolism.

(5)    7.  By a process called _____ , some leaf cells intercept energy from the sun and uses it to form _____ , an energy carrier.

(5)    8.  By the process of aerobic _____ , cells can release stored energy from sugars and other molecules and form more ATP.

(5)    9.  _____ are certain molecules and structures that can detect specific kinds of information about the environment.

(5)    10. A fairly constant level of physical and chemical conditions inside an organism represents a state of _____ .

(6)    11.   Arrange in order, from smallest to largest, the levels of organization that occur in nature. Define each as you list it.

(7)    12.   Explain how the actions of producers, consumers, and decomposers create an interdependency among organisms.

(7)    13.   Describe the general pattern of energy flow through Earth's life forms and explain how Earth's resources are used again and again (cycled).

(8)    14.   Explain the use of genus and species names by considering your Latin name, *Homo sapiens.*

(8)    15.   List the six kingdoms of life; briefly describe organisms placed in each.

(8)    16.   Arrange in order, from greater to fewer organisms included, the following categories of classification: class, family, genus, kingdom, order, phylum, and species.

(8–9)  17.   Distinguish between these terms: *prokaryotic, eukaryotic.*

(10)   18.   _____ , molecular changes in DNA, are the original source of variations in heritable traits.

(10)   19.   An _____ trait is any trait that helps an organism survive and reproduce under a given set of environmental conditions.

(10)   20.   As organisms move through time in successive generations, the character of populations change; this is called _____ .

(10)   21.   Darwin used _____ selection as a model for natural selection.

(10–11) 22.  Define *natural selection* and briefly describe what is occurring when a population is said to evolve.

(11)   23.   Explain what is meant by the term *diversity* and speculate about what caused the great diversity of life forms on Earth.

(12)   24.   Compose sentences using each of the following terms to demonstrate a good understanding of them: hypotheses, prediction, test, logic, inductive logic, deductive logic, and experiment.

(13)   25.   Generally, members of a control group should be identical to those of the experimental group except for the key factor under study, the _____ .

(13)   26.   Explain what is meant by scientific theory; cite an actual example.

(14)   27.   Experiments deal with potentially _____ hypotheses that can be tested in the natural world in ways that might disprove them.

(15)   28.   If experimenters fail to use large-enough samples, they run the risk of _____ error.

(15)   29.   Describe the meaning of "bias" in relation to data interpretation by experimenters.

(15)   30.   Science emphasizes presenting test results in _____ terms.

(16)   31.   Explain how the methods of science differ from answering questions by using subjective thinking and systems of belief.

---

## Interpreting and Applying Key Concepts

1. Humans have the ability to maintain body temperature very close to 37°C.
   a. What conditions would tend to make the body temperature drop?
   b. What measures do you think your body takes to raise body temperature when it drops?
   c. What conditions would cause body temperature to rise?
   d. What measures do you think your body takes to lower body temperature when it rises?
2. Do you think that all humans living on Earth today should be grouped in the same species?
3. What sorts of topics are usually regarded by scientists as untestable by the kinds of methods that scientists generally use?

# 2

# CHEMICAL FOUNDATIONS FOR CELLS

## Interactive Exercises

Note: In the answer sections of this book, a specific molecule is most often indicated by its abbreviation. For example, deoxyribonucleic acid is DNA.

*Leafy Clean-Up Crews* (pp. 20–21)

## 2.1. REGARDING THE ATOMS (pp. 22–23)

## 2.2. *Focus on Sciences:* USING RADIOISOTOPES TO DATE THE PAST, TRACK CHEMICALS, AND SAVE LIVES (pp. 23–24)

***Selected Words:*** Spartina, *formulas, chemical equations, reactants, products, mole, phytoremediation, unstable, PET, radiation therapy*

In addition to the boldfaced terms, the text features other important terms essential to understanding the assigned material. "Selected Words" is a list of these terms, which appear in the text in italics, in quotation marks, and occasionally in roman type. Latin binomials found in this section are underlined and in roman type to distinguish them from other italicized words.

## Boldfaced, Page-Referenced Terms

The page-referenced terms are important; they are in boldface type in the chapter. Write a definition for each term in your own words without looking at the text. Next, compare your definition with that given in the chapter or in the text glossary. If your definition seems inaccurate, allow some time to pass and repeat this procedure until you can define each term rather quickly (how fast you can answer is a gauge of your learning effectiveness).

(22) elements _____

_____

(22) trace element _____

_____

(22) atoms _____

_____

(22) protons _____

_____

(22) electrons _____

_____

(22) neutrons _____

_____

(22) atomic number _____

_____

(23) mass number _____

_____

(23) isotope _____

_____

(23) radioisotopes _____

_____

(23) radioactive decay _____

_____

(23) chemical bonds _____

_____

(24) radiometric dating _____

_____

(24) half-life _____

_____

(25) tracer _____

_____

## Matching

Choose the most appropriate answer for each term.

1. ___ radiometric dating
2. ___ atoms
3. ___ protons
4. ___ trace element
5. ___ neutrons
6. ___ electrons
7. ___ radioactive decay
8. ___ atomic number
9. ___ mass number
10. ___ elements
11. ___ isotope
12. ___ radioisotopes
13. ___ tracer
14. ___ half-life
15. ___ chemical bonds
16. ___ radiation therapy

A. Unions between the electron structures of atoms
B. Radioisotopes used with scintillation counters to reveal the pathway or destination of a substance
C. Subatomic particles with a negative charge
D. Positively charged subatomic particles within the nucleus
E. The time it takes for half of the nuclei in a given amount of a radioactive element to decay into a different element
F. Method used to measure proportions of isotopes in a rock-trapped mineral and the daughter isotopes formed in the same rock
G. Atoms of a given element that differ in the number of neutrons
H. Spontaneous process that transforms an unstable atom into a different isotope
I. Refers to the number of protons in an atom
J. Radioactive isotopes
K. Chemical elements representing less than 0.01 percent of body weight
L. Destroys or impairs living cancer cells
M. The number of protons and neutrons in the nucleus of one atom nucleus
N. Fundamental forms of matter that occupy space, have mass, and cannot be broken down into something else
O. Smallest units that retain the properties of a given element
P. Subatomic particles within the nucleus carrying no charge

## Fill-in-the-Blanks

The expression $12H_2O$ plus $6 CO_2 \longrightarrow 6O_2$ plus $C_6H_{12}O_6$ plus $H_2O$ is known as the chemical

(17) _____ for photosynthesis. $H_2O$ is the (18) _____ for water. The arrow in the expression above

means (19) _____ . The (20) _____ are to the left of the arrow and the (21) _____ are to the

right of the arrow. In the expression one can count 12 hydrogen atoms on the left side of the arrow;

therefore one should be able to count (22) _____ hydrogen atoms on the right side of the arrow. By

reference to Table 2.1, p. 22, one can determine the weight of one mole of $H_2SO_4$ to be: (23) _____ .

## 2.3. THE NATURE OF CHEMICAL BONDS (pp. 26–27)

**Selected Words:** *volumes of space, lowest available energy level, higher energy levels, inert, bonds*

**Boldfaced, Page-Referenced Terms**

(26) orbitals _____

_____

(26) shell model _____

_____

(27) molecule _____

_____

(27) compounds _____

_____

(27) mixture _____

_____

## Matching

Choose the one best answer for each.

1. ___ mixture
2. ___ shells
3. ___ lowest energy level
4. ___ inert
5. ___ orbitals
6. ___ compounds
7. ___ molecule
8. ___ higher energy levels

A. Regions of space around an atom's nucleus where electrons are likely to be at any one instant
B. Results when two or more atoms bond together
C. Two or more elements are simply intermingling in proportions that usually vary
D. A series of orbitals arranged around the nucleus
E. Types of molecules composed of two or more different elements in proportions that never vary
F. Energy of electrons farther from the nucleus than the first orbital
G. Refers to atoms with no vacancies in their shells, hence showing little tendency to enter chemical reactions
H. Energy of electrons in the orbital closest to the nucleus

## Complete the Table

9. Complete the table (refer to Figure 2.12, p. 27, in the text) by entering the name of the element and its symbol in the appropriate spaces.

| Element | Symbol | Atomic Number | First Shell | Electron Distribution Second Shell | Third Shell | Fourth Shell |
|---------|--------|---------------|-------------|--------------|-------------|--------------|
| a. | | 20 | 2 | 8 | 8 | 2 |
| b. | | 6 | 2 | 4 | | |
| c. | | 17 | 2 | 8 | 7 | |
| d. | | 1 | 1 | | | |
| e. | | 11 | 2 | 8 | 1 | |
| f. | | 7 | 2 | 5 | | |
| g. | | 8 | 2 | 6 | | |

## Identification

10. Following the model (number of protons and neutrons shown in the nucleus), identify the indicated atoms of the elements illustrated by entering appropriate electrons in this form: (2e⁻).

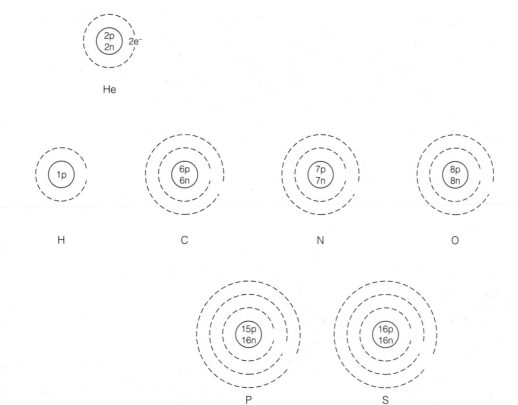

## 2.4. IMPORTANT BONDS IN BIOLOGICAL MOLECULES (pp. 28–29)

**Selected Words:** *sharing, single* covalent bond, *double* covalent bond, *triple* covalent bond, *nonpolar* covalent bond, *polar* covalent bond, no net charge

### Boldfaced, Page-Referenced Terms

(28) ion _____

_____

(28) ionic bond _____

_____

(28) covalent bond _____

_____

(29) hydrogen bond _____

_____

## Identification

1. Following the model, complete the diagram by adding arrows to identify the transfer of electron(s), showing how positive magnesium and negative chlorine ions form ionic bonds to create a molecule of $MgCl_2$ (magnesium chloride).

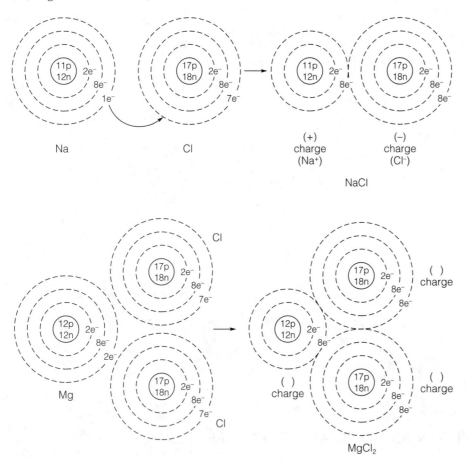

2. Following the model of hydrogen gas shown, complete the diagram by placing electrons (as dots) in the outer shells to identify the nonpolar covalent bonding that forms oxygen gas; similarly, identify polar covalent bonds by completing electron structures to form a water molecule.

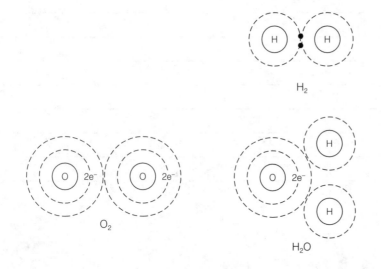

*Short Answer*

3. Distinguish between a nonpolar covalent bond and a polar covalent bond. Cite an example of each (pp. 28–29).

_____

_____

_____

_____

4. Describe one example of a large biological molecule within which hydrogen bonds exist (p. 29).

_____

_____

_____

## 2.5. PROPERTIES OF WATER (pp. 30–31)

*Selected Words:* *no net charge, attracts, repels, liquid water, spheres of hydration, dissolved*

### Boldfaced, Page-Referenced Terms

(30) hydrophilic substances _____

_____

(30) hydrophobic substances _____

_____

(30) temperature _____

_____

(30) evaporation _____

_____

(31) cohesion _____

_____

(31) solutes _____

_____

## Fill-in-the-Blanks

The (1) _____ of water molecules allows them to hydrogen-bond with each other. Water molecules hydrogen-bond with polar molecules, which are (2) _____ (water-loving). Polarity causes water to repel oil and other nonpolar substances, which are (3) _____ (water-dreading). (4) _____ is a measure of the constant motion of molecules. Liquid water changes its temperature more slowly than air because of the great amount of heat required to break the high number of (5) _____ bonds between water molecules; this property helps stabilize temperature in aquatic habitats and cells. Large energy inputs increase molecular motion to the point where hydrogen bonds stay broken, releasing individual molecules from the water surface. This process is known as (6) _____ . Below 0°C, water molecules become locked in the less dense lattlicelike bonding pattern of (7) _____ , which is less dense than water. Collective hydrogen bonding creates a high tension on surface water molecules resulting in (8) _____ , the property of water that explains how long, narrow water columns rise to the tops of tall trees. Water is an excellent (9) _____ in which ions and polar molecules readily dissolve. Substances dissolved in water are known as (10) _____ . A substance is (11) _____ in water when spheres of (12) _____ form around its individual ions or molecules.

## 2.6. ACIDS, BASES, AND BUFFERS (pp. 32–33)

*Selected Words:* *donate H+, accept H+, acidic solutions, basic solutions, acid stomach, chemical burns, coma, tetany, acidosis, alkalosis*

### Boldfaced, Page-Referenced Terms

(32) hydrogen ions _____

_____

(32) hydroxide ions _____

_____

(32) pH scale _____

_____

(32) acids _____

_____

(32) bases _____

_____

(33) buffer system _____

_____

(33) salts _____

_____

## Matching

Choose the most appropriate answer for each.

1. ___ acid stomach
2. ___ acids
3. ___ pH scale
4. ___ chemical burns
5. ___ H+
6. ___ bases
7. ___ examples of basic solutions
8. ___ coma
9. ___ acidosis
10. ___ OH−
11. ___ tetany
12. ___ examples of acid solutions
13. ___ alkalosis
14. ___ buffer system

A. A sometimes irreversible state of unconsciousness
B. $CO_2$ builds up in the blood, too much $H_2CO_3$ forms and blood pH severely decreases
C. Hydroxide ion
D. Substances that accept $H^+$ when dissolved in water
E. An uncorrected increase in blood pH
F. Used to measure $H^+$ concentration in various fluids
G. A partnership between a weak acid and the base that forms when it dissolves in water; counters slight pH shifts
H. Hydrogen ion or proton
I. Baking soda, seawater, egg white
J. Can be caused by ammonia, drain cleaner, and sulfuric acid in car batteries
K. Substances that donate $H^+$ when dissolved in water
L. Lemon juice, gastric fluid, coffee
M. Can be caused by eating too much fried chicken or certain other foods
N. A potentially lethal pH stage in which the body's skeletal muscles enter a state of uncontrollable contraction

## Complete the Table

15. Complete the following table by consulting Figure 2.19, page 32 in the text.

| Fluid | pH Value | Acid/Base |
|---|---|---|
| a. Blood | | |
| b. Saliva | | |
| c. Urine | | |
| d. Stomach Acid | | |
| e. Seawater | | |
| f. Some acid rain | | |

---

# Self-Quiz

___ 1. Each element has a unique _____ , which refers to the number of protons present in its atoms.
   a. isotope
   b. mass number
   c. atomic number
   d. radioisotope

___ 2. A molecule is _____ .
   a. a combination of two or more atoms
   b. less stable than its constituent atoms separated
   c. electrically charged
   d. a carrier of one or more extra neutrons

___ 3. If lithium has an atomic number of 3 and an atomic mass of 7, it has _____ neutron(s) in its nucleus.
   a. one
   b. two
   c. three
   d. four
   e. seven

___ 4. Substances that are nonpolar and repelled by water are _____ .
   a. hydrolyzed
   b. nonpolar
   c. hydrophilic
   d. hydrophobic

___ 5. A hydrogen bond is _____ .
   a. a sharing of a pair of electrons between a hydrogen nucleus and an oxygen nucleus
   b. a sharing of a pair of electrons between a hydrogen nucleus and either an oxygen or a nitrogen nucleus
   c. formed when a small electronegative atom of a molecule weakly interacts with a hydrogen atom that is already participating in a polar covalent bond
   d. none of the above

___ 6. An ionic bond is one in which _____ .
   a. electrons are shared equally
   b. electrically neutral atoms have a mutual attraction
   c. two charged atoms have a mutual attraction due to electron transfer
   d. electrons are shared unequally

___ 7. A covalent bond is one in which _____ .
   a. electrons are shared
   b. electrically neutral atoms have a mutual attraction
   c. two charged atoms have a mutual attraction due to electron transfer
   d. electrons are lost

___ 8. A nonpolar covalent bond implies that _____ .
   a. one negative atom bonds with a hydrogen atom
   b. the bond is double
   c. there is no difference in charge at the ends (the two poles) of the bond
   d. atoms of different elements do not exert the same pull on shared electrons

___ 9. The shapes of large molecules are controlled by _____ bonds.
   a. hydrogen
   b. ionic
   c. covalent
   d. inert
   e. single

___10. A solution with a pH of 10 is _____ times as basic as one with a pH of 8.
   a. 2
   b. 3
   c. 10
   d. 100
   e. 1,000

___11. A control that minimizes unsuitable pH shifts is a(n) _____ .
   a. neutral molecule
   b. salt
   c. base
   d. acid
   e. buffer

## Chapter Objectives/Review Questions

*Page*      *Objectives/Questions*

(21)      1. The use of living plants to withdraw harmful substances from the environment is known as _____ .

(22)      2. Be able to explain what an element is.

(22)      3. Define *trace element* and list some examples.

(22–23)   4. List and describe the three types of subatomic particles and describe the reason that hydrogen is an exception.

(22–23)  5.  The number of protons in an atom is referred to as the _____ number of that element; the combined number of protons and neutrons in the atomic nucleus is referred to as the _____ number of that element.

(23)  6.  Distinguish between isotopes, radioisotopes, and radioactive decay.

(23)  7.  Be able to read a chemical equation ( Figure 2.5, p. 23).

(24)  8.  _____-_____ is the time it takes for half of a given quantity of any radioisotope to decay into a different, less unstable daughter isotope.

(25)  9.  Describe what tracers are and how they are used.

(25)  10.  _____ uses radioactively labeled compounds to yield images of metabolically active and inactive tissues in a patient.

(25)  11.  Describe the use of radiation therapy.

(26)  12.  Explain the relationship of orbitals to shells.

(27)  13.  Be able to sketch shell models of the atoms described in the text, Figure 2.12, page 27.

(27)  14.  Explain why helium and neon are known as inert atoms.

(27)  15.  A _____ is formed when two or more atoms bond together.

(27)  16.  A _____ contains atoms of two or more elements whose proportions never vary.

(27)  17.  How does a mixture differ from a compound?

(28)  18.  An _____ is an atom that becomes positively or negatively charged.

(28)  19.  An association of two oppositely charged ions is a(n) _____ bond.

(28)  20.  In a(n) _____ bond, two atoms share electrons.

(29)  21.  Explain why $H_2O$ is an example of a nonpolar covalent bond and $H_2O$ has two polar covalent bonds.

(29)  22.  In a(n) _____ bond, a small, highly electronegative atom of a molecule weakly interacts with a hydrogen atom that is already participating in a polar covalent bond.

(30)  23.  Polar molecules attracted to water are _____ ; all nonpolar molecules are _____ and are repelled by water.

(30)  24.  _____ is a measure of molecular motion; during _____ , an input of heat energy converts liquid water to the gaseous state.

(30)  25.  Describe the formation of ice in terms of hydrogen bonding.

(31)  26.  Completely describe the properties of water that move water from the roots to the tops of the tallest trees.

(31)  27.  Distinguish a solvent from a solute.

(31)  28.  Describe what happens when a substance is dissolved in water.

(32)  29.  The ionization of water is the basis of the _____ scale.

(32)  30.  _____ are substances that donate $H^+$ ions when they dissolve in water, and _____ are substances that accept $H^+$ with when dissolved in water.

(32)  31.  Black coffee with a pH of 5 is a(n) _____ solution while baking soda with a pH of 9 is a(n) _____ solution.

(32).  32.  Define *buffer*; cite an example and describe how buffers operate.

(32–33) 33.  Be able to define the following terms: *acid stomach, chemical burns, coma, tetany, acidosis,* and *alkalosis.*

(33)  34.  A _____ is produced by a chemical reaction between an acid and a base.

## Integrating and Applying Key Concepts

1. Explain what would happen if water were a nonpolar molecule instead of a polar molecule. Would water be a good solvent for the same kinds of substances? Would the nonpolar molecule's specific heat likely be higher or lower than that of water? Would surface tension be affected? Cohesive nature? Ability to form hydrogen bonds? Is it likely that the nonpolar molecules could form unbroken columns of liquid? What implications would that hold for trees?

# 3

# CARBON COMPOUNDS IN CELLS

## Interactive Exercises

*Carbon, Carbon, In the Sky—Are You Swinging Low and High?* (pp. 36–37)

### 3.1. PROPERTIES OF ORGANIC COMPOUNDS (pp. 38–39)

### 3.2. HOW CELLS USE ORGANIC COMPOUNDS (p. 40)

### 3.3. *Focus on the Environment:* FOOD PRODUCTION AND A CHEMICAL ARMS RACE (p. 41)

*Selected Words: single* covalent bond, *double* covalent bond, *carbohydrates, lipids, proteins, nucleic acids, herbicides, insecticides, fungicides*

## Boldfaced, Page-Referenced Terms

(38) organic compounds _____

_____

(39) functional groups _____

_____

(39) alcohols _____

_____

(40) enzymes _____

_____

(40) functional-group transfer _____

_____

(40) electron transfer _____

_____

(40) rearrangement _____

_____

(40) condensation _____

_____

(40) cleavage _____

_____

(40) polymer _____

_____

(40) monomers _____

_____

(40) hydrolysis _____

_____

(41) toxin _____

_____

## Labeling

Study the structural formulas of organic compounds. By referencing Figure 3.2, page 38, and Figure 3.3, page 39 in the text, identify the circled functional groups (sometimes repeated) by entering the correct name in the blanks with matching numbers.

1. _____

2. _____

3. _____

4. _____

5. _____

6. _____

7. _____

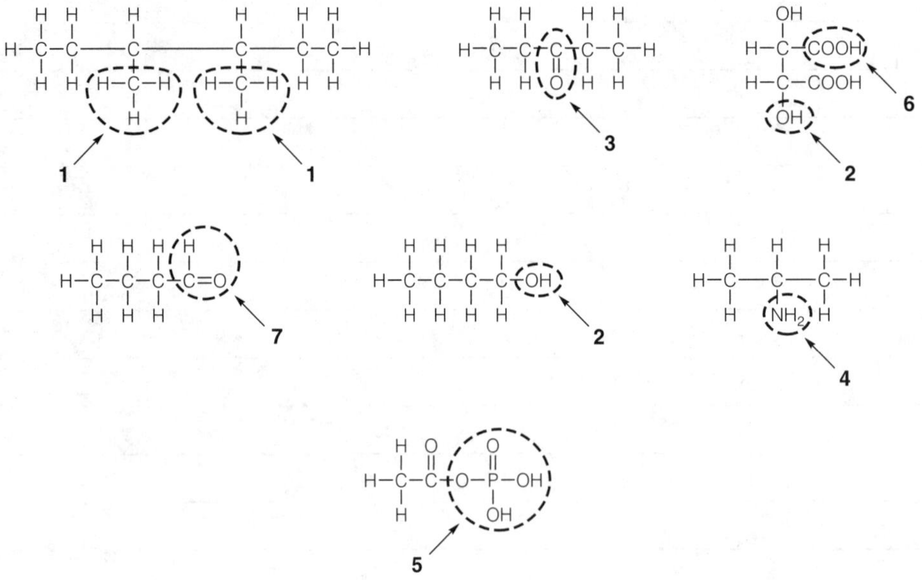

## Short Answer

8. State the general role of enzymes as they relate to the chemistry of life (p. 40).

_____

_____

_____

## Complete the Table

9. Complete the following table summarizing five categories of reactions that are mediated by enzymes.

| Reaction Category | Reaction Description |
|---|---|
| a. Rearrangement | |
| b. Cleavage | |
| c. Condensation | |
| d. Electron transfer | |
| e. Functional-group transfer | |

## Identification

10. Study the structural formulas of the two adjacent amino acids shown below. Identify the enzyme action that causes formation of a covalent bond and a water molecule (through a condensation reaction) by circling an H atom from one amino acid and an –OH group from the other amino acid. Also circle the covalent bond that formed the dipeptide.

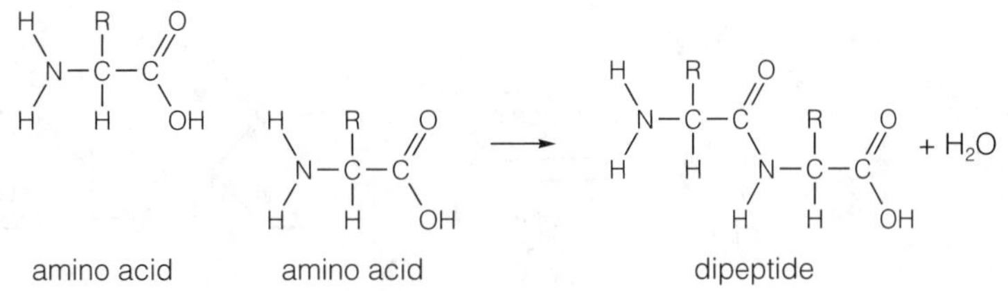

amino acid        amino acid                              dipeptide

## Short Answer

11. Describe hydrolysis through enzyme action for the molecules in exercise 10 (p. 40).

_____

_____

_____

_____

_____

# 3.4. CARBOHYDRATES (pp. 42–43)

**Selected Words:** *mono-, oligo-, di-, poly-* saccharides, bonds *angle*

## Boldfaced, Page-Referenced Terms

(42) carbohydrate _____

_____

(42) monosaccharides _____

_____

(42) oligosaccharides _____

_____

(42) polysaccharides _____

_____

## Identification

1. In the diagram below, identify condensation reaction sites between the two glucose molecules by circling the components of the water removed that allow a covalent bond to form between the glucose molecules. Note that the reverse reaction is hydrolysis and that both condensation and hydrolysis reactions require enzymes in order to proceed.

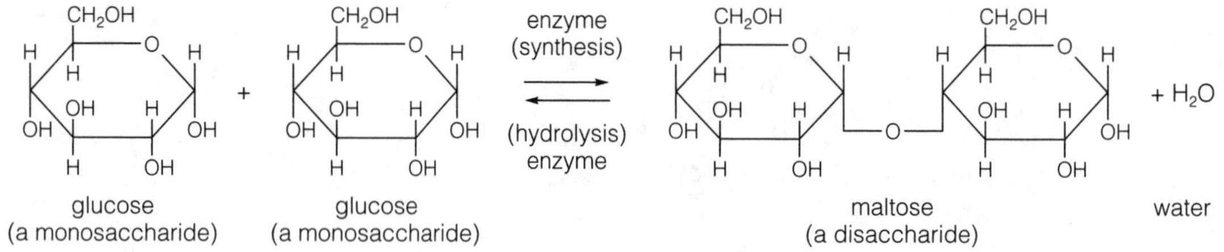

glucose
(a monosaccharide)      glucose
(a monosaccharide)                    maltose
(a disaccharide)            water

*Complete the Table*

2. In the table below, enter the name of the carbohydrate described by its carbohydrate class and functions.

| Carbohydrate | Carbohydrate Class | Function |
|---|---|---|
| a. | | Most plentiful sugar in nature; transport form of carbohydrates in plants; table sugar; formed from glucose and fructose |
| b. | | Five-carbon sugar occurring in DNA |
| c. | | Main energy source for most organisms; precursor of many organic organisms; serve as building blocks for larger carbohydrates |
| d. | | Structural material of plant cell walls; formed from glucose chains |
| e. | | Five-carbon sugar occurring in RNA |
| f. | | Sugar present in milk; formed from glucose and galactose |
| g. | | A branched starch that invites rapid mobilization of glucose; formed from glucose chains |
| h. | | Main structural material in some external skeletons and other hard body parts of some animals and fungi |
| i. | | Animal starch; stored especially in liver and muscle tissue; formed from glucose chains |
| j. | | Long-term glucose storage form in plants; a starch formed from glucose chains |

## 3.5. LIPIDS (pp. 44–45)

*Selected Words: unsaturated, saturated*

*Boldfaced, Page-Referenced Terms*

(44) lipids _____

_____

(44) fatty acids _____

_____

(44) triglycerides _____

_____

(45) phospholipid _____

_____

(45) sterols _____

(45) waxes _____

## Labeling

1.  In the appropriate blanks, label the molecules shown as saturated or unsaturated.

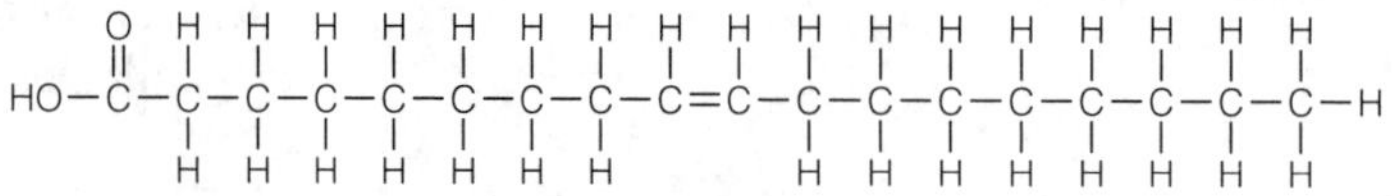

a                 oleic acid

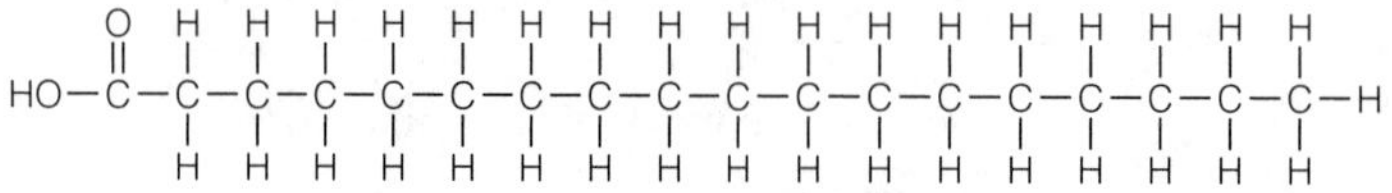

b                 stearic acid

       a. _____               b. _____

## Identification

2.  Combine glycerol with three fatty acids to form a triglyceride by circling the participating atoms that will identify three covalent bonds. Also circle the covalent bonds in the triglyceride.

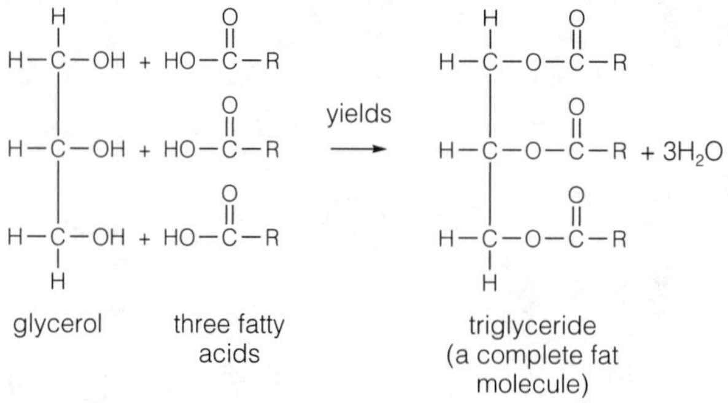

glycerol      three fatty          triglyceride
             acids            (a complete fat
                                    molecule)

## Short Answer

3.  Describe the structure and biological functions of phospholipid molecules (p. 45).

_____

_____

_____

_____

## Matching

Choose the most appropriate answer for each item. Some letters may be used more than once.

4. ___atherosclerotic plaques

5. ___richest source of body energy

6. ___honeycomb material

7. ___cholesterol

8. ___saturated tails

9. ___butter and lard

10. ___lack fatty acid tails

11. ___main cell membrane component

12. ___all possess a rigid backbone of four fused carbon rings

13. ___plant cuticles

14. ___triglycerides

15. ___precursors of testosterone, estrogen, and bile salts

16. ___unsaturated tails

17. ___vegetable oil

18. ___vertebrate insulation

19. ___furnishes lubrication for skin and feathers

A. Fatty acids
B. Neutral fats
C. Phospholipids
D. Waxes
E. Sterols

## 3.6. AMINO ACIDS AND THE PRIMARY STRUCTURE OF PROTEINS (pp. 46–47)

## 3.7. EMERGENCE OF THE THREE-DIMENSIONAL STRUCTURE OF PROTEINS
(pp. 48–49)

*Selected Words:* peptide bonds, primary structure, secondary structure, tertiary structure, quaternary structure

*Boldfaced, Page-Referenced Terms*

(46) proteins _____

_____

(46) amino acid _____

_____

(46) dipeptide _____

_____

(46) polypeptide chain _____

_____

(49) lipoproteins _____

_____

(49) glycoproteins _____

_____

(49) denaturation _____

_____

## Matching

For exercises 1, 2, and 3, match the major parts of *every* amino acid by entering the letter of the part in the blank corresponding to the number on the molecule.

1. ___

2. ___

3. ___

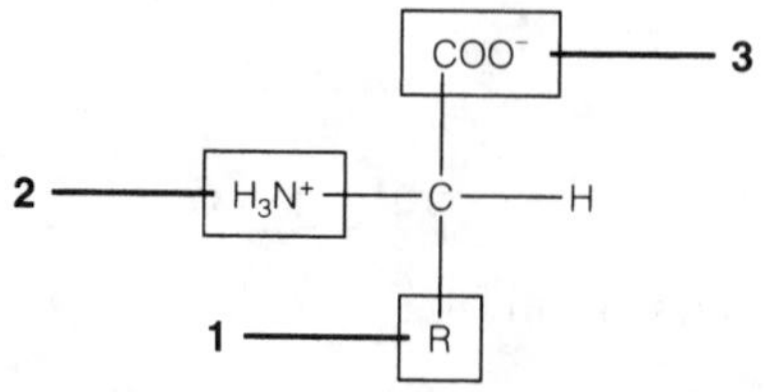

A. R group (a symbol for a characteristic group of atoms that differ in number and arrangement from one amino acid to another)
B. Carboxyl group (ionized)
C. Amino group (ionized)

## Identification

4. Four ionized amino acids (in cellular solution) are shown in the upper illustration below; form the proper covalent bonds by circling the acids and ions involved in polypeptide formation. Circle the newly formed peptide bonds of the completed polypeptide on the lower illustration.

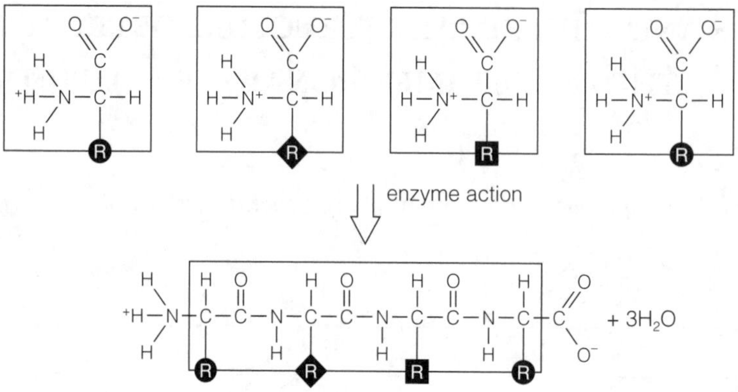

enzyme action

$+ 3H_2O$

## Matching

Choose the most appropriate answer for each term.

5. ___amino acid

6. ___peptide bond

7. ___polypeptide chain

8. ___primary structure

9. ___proteins

10. ___secondary structure

11. ___tertiary structure

12. ___dipeptide

13. ___quaternary structure

14. ___lipoproteins

15. ___glycoproteins

16. ___denaturation

A. A coiled or extended pattern of protein structure caused by regular intervals of H bonds
B. Three or more amino acids joined in a linear chain
C. Proteins with linear or branched oligosaccharides covalently bonded to them; found on animal cell surfaces, in cell secretion, or on blood proteins
D. Folding of a protein through interactions among R groups of a polypeptide chain
E. Form when freely circulating blood proteins encounter and combine with cholesterol, or phospholipids
F. The type of covalent bond linking one amino acid to another
G. Hemoglobin, a globular protein of four chains, is an example
H. Breaking weak bonds in large molecules (such as protein) to change its shape so it no longer functions
I. Formed when two amino acids join together
J. Lowest level of protein structure; has a linear, unique sequence of amino acids an acid group, a hydrogen atom, and an R group
K. A small organic compound having an amino group, an acid group, a hydrogen atom, and an R group
L. The most diverse of all the large biological molecules; constructed from pools of only twenty kinds of amino acids

## 3.8. NUCLEOTIDES AND THE NUCLEIC ACIDS (pp. 50–51)

### Boldfaced, Page-Referenced Terms

(50) nucleotides _____

_____

(50) ATP _____

_____

(50) coenzymes _____

_____

(50) nucleic acids _____

_____

(51) DNA _____

_____

(51) RNAs _____

_____

## Matching

For exercises 1, 2, and 3, match the following answers to the parts of a nucleotide shown in the diagram.

1. ___
2. ___
3. ___

A. A five-carbon sugar (ribose or deoxyribose)
B. Phosphate group
C. A nitrogen-containing base that has either a single-ring or double-ring structure

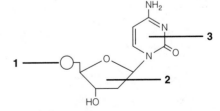

## Identification

4. In the following diagram of a single-stranded nucleic acid molecule, encircle as many complete nucleotides as possible. How many complete nucleotides are present?

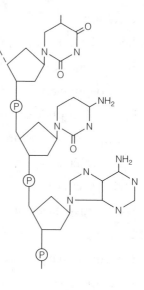

## Matching

Choose the most appropriate answer for each term.

5. ___adenosine triphosphate

6. ___RNA

7. ___DNA

8. ___NAD⁺ and FAD

9. ___cAMP

A. Single nucleotide strand; function in processes by which genetic instructions are used to build proteins
B. ATP, a cellular energy carrier
C. Nucleotide chemical messenger
D. Single nucleotide units; coenzymes; transport hydrogen ions and their associated electrons from one cell reaction site to another
E. Double nucleotide strand; encodes genetic instructions with nucleotide sequences

## Complete the Table

10. Complete the table below by entering the correct name of the major cellular organic compounds suggested in the "types" column (choose from carbohydrates, lipids, proteins, and nucleic acids).

| Cellular Organic Compounds | Types |
|---|---|
| a. | Phospholipids |
| b. | Antibodies |
| c. | Enzymes |
| d. | Genes |
| e. | Glycogen, starch, cellulose, and chitin |
| f. | Glycerides |
| g. | Saturated and unsaturated fats |
| h. | Coenzymes |
| i. | Sterols, oils, and waxes |
| j. | Glucose and sucrose |

# Self-Quiz

___ 1. Amino acids are linked by _____ bonds to form the primary structure of a protein.
   a. disulfide
   b. hydrogen
   c. ionic
   d. peptide

___ 2. Proteins _____ .
   a. are weapons against disease-causing bacteria and other invaders
   b. are composed of nucleotide subunits
   c. translate protein-building instructions into actual protein structures
   d. lack diversity of structure and function

___ 3. Lipids _____ .
   a. include fats that are broken down into one fatty acid molecule and three glycerol molecules
   b. are composed of monosaccharides
   c. serve as food reserves in many organisms
   d. include cartilage and chitin

___ 4. DNA _____ .
   a. is one of the adenosine phosphates
   b. is one of the nucleotide coenzymes
   c. contains protein-building instructions
   d. are composed of monosaccharides

___ 5. Most of the chemical reactions in cells must have _____ present before they proceed.
   a. RNA
   b. salt
   c. enzymes
   d. fats

___ 6. Carbon is part of so many different substances because _____ .
   a. carbon generally forms two covalent bonds with a variety of other atoms
   b. carbon generally forms four covalent bonds with a variety of atoms
   c. carbon ionizes easily
   d. carbon is a polar compound

___ 7. The information built into a protein's amino acid sequence plus a coiled pattern of that chain and the addition of more folding yields the _____ level of protein structure.
   a. quaternary
   b. primary
   c. secondary
   d. tertiary

___ 8. _____ are compounds used by cells as transportable packets of quick energy, storage forms of energy, and structural materials.
   a. Lipids
   b. Nucleic acids
   c. Carbohydrates
   d. Proteins

___ 9. Hydrolysis could be correctly described as the _____ .
   a. heating of a compound in order to drive off its excess water and concentrate its volume
   b. breaking of a long-chain compound into its subunits by adding water molecules to its structure between the subunits
   c. linking of two or more molecules by the removal of one or more water molecules
   d. constant removal of hydrogen atoms from the surface of a carbohydrate

___10. Genetic instructions are encoded in the bases of _____ ; molecules of _____ function in processes using genetic instructions to construct proteins.
   a. DNA; DNA
   b. DNA; RNA
   c. RNA; DNA
   d. RNA; RNA

# Chapter Objectives/Review Questions

*Page*    *Objectives/Questions*

(38)     1.  The molecules of life are _____ compounds.
(38)     2.  Each carbon atom can form as many as _____ covalent bonds with other carbon atoms as well as with atoms of other elements.
(39)     3.  A _____ has only hydrogen atoms attached to a carbon backbone.
(39)     4.  Hydroxyl, amino, and carboxyl are examples of _____ groups.
(40)     5.  Describe the role of enzymes in the metabolism of life.
(40)     6.  Only living _____ can synthesize carbohydrates, lipids, proteins, and nucleic acids.
(41)     7.  A _____ is an organic compound that is a normal metabolic product of one species that can harm or kill members of other species.
(41)     8.  Distinguish between *herbicides* and *insecticides*.
(42)     9.  Define *carbohydrates*; be able to list their general functions.
(42)    10.  Most carbohydrates consist of _____ , _____ , and _____ in a 1:2:1 ratio.
(42)    11.  Name the three classes of carbohydrates.
(42)    12.  A _____ means "one monomer of sugar"; be able to give common examples and their functions.
(42)    13.  An _____ is a short chain of two or more covalently bonded sugar monomers; be able to give examples of well-known disaccharides and their functions.
(42–43) 14.  A _____ is a straight or branched chain of hundreds or thousands of sugar monomers, of the same or different kinds; able to give common examples and their functions.
(44)    38.  Define *lipids*; list their general functions.
(44)    15.  Describe a "fatty acid"; a _____ molecule has three fatty acid tails attached to a backbone of glycerol; distinguish a saturated fatty acid from an unsaturated fatty acid.
(44)    16.  Describe the structure of triglycerides; list examples and their functions.
(45)    17.  A _____ has two fatty acid tails and a hydrophilic head attached to a glycerol backbone; list examples and their cellular functions.
(45)    18.  Define *sterols* and describe their chemical structure; cite the importance of the sterols known as cholesterol and steroid hormones.
(45)    19.  Be able to list the names of some sterols and their derivatives; describe their functions.
(45)    20.  Lipids called _____ have long-chain fatty acids, tightly packed and linked to long-chain alcohols or to carbon rings; list examples and their functions.
(46)    21.  Describe protein structures and cite their general functions; be able to sketch the three general parts of any amino acid.
(46–49) 22.  Describe how the primary, secondary, tertiary, and quaternary structure of proteins results in complex three-dimensional structures.
(49)    23.  Distinguish lipoproteins from glycoproteins.
(49)    24.  _____ refers to the loss of a molecule's three-dimensional shape through disruption of the weak bonds responsible for it.
(50)    25.  List the three parts of every nucleotide.
(50)    26.  Name the nucleotide type that transfers a phosphate group to an acceptor molecule to energize it.
(50–51) 27.  Describe the structure of DNA and RNA; list the functions of each.

---

# Integrating and Applying Key Concepts

1.  Humans can obtain energy from many different food sources. Do you think this ability is an advantage or a disadvantage in terms of long-term survival? Why?
2.  If the ways that atoms bond affect molecular shapes, do the ways that molecules behave toward one another influence the shapes of organelles? Do the ways that organelles behave toward one another influence the structure and function of the cells?

# 4

# CELL STRUCTURE AND FUNCTION

## Interactive Exercises

**Selected Words:** *cellulae, compound light microscope, transmission electron microscope, scanning electron microscope, scanning tunneling microscope*

## Boldfaced, Page-Referenced Terms

(55) cell theory _____

_____

(56) cell _____

_____

(56) plasma membrane _____

_____

(56) cytoplasm _____

_____

(56) ribosomes _____

_____

(56) eukaryotic cells _____

_____

(56) prokaryotic cells _____

_____

(56) lipid bilayer _____

_____

(57) surface-to-volume ratio _____

_____

(58) micrograph _____

_____

(58) wavelength _____

_____

## Label-Match

Although cells vary in many specific ways, they are all alike in a few basic respects. Identify each part of the illustration. Choose from cytoplasm, DNA-containing region (nucleus in eukaryotes), and cytoplasm. Complete the exercise by matching and entering the letter of the proper description in the parentheses following each label.

1. _____ (     )

2. _____ (     )

3. _____ (     )

A. Includes everything enclosed by the plasma membrane; a semifluid substance in which particles, filaments, and often compartments are organized
B. Thin, outermost membrane that maintains the cell as a distinct entity with metabolism occurring within
C. Molecules of heredity are here along with molecules that can read or copy hereditary instructions

*Short Answer*

4. Explain why most cells are, of necessity, very small.

_____

_____

_____

*Matching*

Select the single best answer.

5. ___electron wavelength

6. ___micrograph

7. ___transmission electron microscope

8. ___scanning tunneling microscope

9. ___wavelength

10. ___compound light microscope

11. ___scanning electron microscope

A. Glass lenses bend incoming light rays to form an enlarged image of a cell or some other specimen
B. The distance from one wave's peak to the peak of the wave behind it
C. A computer analyzes a tunnel formed in electron orbitals that is produced between the tip of a needlelike probe and an atom
D. A narrow beam of electrons moves back and forth across the surface of a specimen coated with a thin metal layer
E. A photograph of an image formed with a microscope
F. About 100,000 times shorter than those of visible light
G. Electrons pass through a thin section of cells to form an image

## 4.3. THE DEFINING FEATURES OF EUKARYOTIC CELLS (pp. 60–63)

*Selected Words:* compartmentalization

*Boldfaced, Page-Referenced Terms*

(60) organelle _____

_____

## Complete the Table

1. Complete the following table to identify eukaryotic organelles and their functions.

| Organelle or Structure | Main Function |
|---|---|
| a. | Contains most of the cell's DNA |
| b. | Synthesis of polypeptide chains |
| c. | Route and modify new polypeptide chains; lipid synthesis |
| d. | Further modify polypeptide chains into mature proteins; sort and ship proteins and lipids |
| e. | Different types function in transport or storage of substances; digestion inside cells and other functions |
| f. | Efficient ATP production |
| g. | Cell shape, internal organization, and movements |

## 4.4. THE NUCLEUS (pp. 64–65)

*Selected Words:* *two* lipid bilayers

### Boldfaced, Page-Referenced Terms

(64) nucleus _____

_____

(64) nuclear envelope _____

_____

(64) nucleolus _____

_____

(65) chromatin _____

_____

(65) chromosome _____

_____

### Short Answer

1. List the two major functions of the nucleus.

_____

_____

_____

*Complete the Table*

2. Complete the following table about the eukaryotic nucleus. Enter the name of each nuclear component described.

| Nuclear Component | Description |
|---|---|
| a. | One or more masses in the nucleus; sites where the protein and RNA subunits of ribosomes are assembled |
| b. | Two lipid bilayer membranes thick; proteins and protein pores span the bilayers |
| c. | A cell's total collection of DNA and its associated proteins |
| d. | Fluid interior portion of the nucleus |
| e. | An individual DNA molecule and its associated proteins in the nucleus |

*Fill-in-the-Blanks*

(3) _____ assemble polypeptide chains on ribosomes in the (4) _____ . Many of the newly formed chains are used or stockpiled in the cytoplasm but many others pass through the (5) _____ system. This system is a series of (6) _____ , including the endoplasmic reticulum, Golgi bodies, and vesicles. Many (7) _____ take on final form and get packaged in (8) _____ in the (9) _____ system. In addition, (10) _____ are assembled and packaged here. Some vesicles deliver proteins and lipids to regions of the cell where new (11) _____ are to be built whereas others store these compounds for specific uses. Still other vesicles move to the (12) _____ membrane and release their contents outside the cell.

## 4.5. THE CYTOMEMBRANE SYSTEM (pp. 66–67)

*Selected Words: rough ER, smooth ER*

*Boldfaced, Page-Referenced Terms*

(66) cytomembrane system _____

_____

(66) endoplasmic reticulum or ER _____

_____

(66) Golgi bodies _____

_____

(67) lysosome _____

_____

(67) peroxisomes _____

_____

## Matching

Study the illustration below and match each component of the cytomembrane system with the most correct description of function. Some components may be used more than once.

1. ___Assembly of polypeptide chains

2. ___Lipid assembly

3. ___DNA instructions for building polypeptide chains

4. ___Initiate protein modification following assembly

5. ___Proteins and lipids take on final form

6. ___Sort and package lipids and proteins for transport to final destinations following modification

7. ___Vesicles formed at plasma membrane transport substances into cytoplasm

8. ___Sacs of enzymes that break down fatty acids and amino acids, forming hydrogen peroxide

9. ___Special vesicles budding from Golgi bodies that become organelles of intracellular digestion

10. ___Transport unfinished proteins to a Golgi body

11. ___Transport finished Golgi products to the plasma membrane

12. ___Release Golgi products at the plasma membrane

13. ___Transport unfinished lipids to a Golgi body

A. spaces within smooth membranes of ER
B. nucleus
C. Golgi body
D. vesicles from Golgi
E. vesicles budding from rough ER
F. endocytosis with vesicles
G. exocytosis with vesicles
H. spaces within rough ER
I. ribosomes in the cytoplasm
J. vesicles budding from smooth ER
K. lysosomes
L. peroxisomes

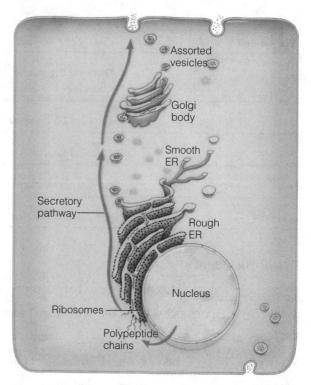

## 4.6. MITOCHONDRIA (p. 68)

## 4.7. SPECIALIZED PLANT ORGANELLES (p. 69)

*Selected Words:* chromoplasts, amyloplasts

*Boldfaced, Page-Referenced Terms*

(68) mitochondria _____

_____

(69) chloroplasts _____

_____

(69) central vacuole _____

_____

*Choice*

For questions 1–20, choose from the following:

a. mitochondria     b. chloroplasts     c. amyloplasts     d. central vacuole     e. chromoplasts

1.____Occur only in photosynthetic eukaryotic cells

2.____ATP molecules form when organic compounds are degraded

3.____Plastids that lack pigments

4.____A muscle cell might have thousands

5.____Plastids that have an abundance of carotenoids but no chlorophylls

6.____ATP-forming reactions require oxygen

7.____Causes fluid pressure to build up inside a living plant cell

8.____Organelles that absorb sunlight energy and produce ATP

9.____The source of the yellow-to-red colors of many flowers, leaves, and fruits

10.____Inner folds are called cristae

11.____May increase so much in volume that it takes up 50 to 90 percent of the cell's interior

12.____Plastid with internal areas known as *grana* and *stroma*

13.____Plastids that resemble certain photosynthetic bacteria

14.____Two distinct compartments are created by a double-membrane system

15.____Store starch grains and are abundant in cells of stems, tubers, and seeds

16.____Resemble nonphotosynthetic bacteria in terms of size and biochemistry

17.____The site of photosynthesis in plant cells

18.____Fluid-filled; stores amino acids, sugars, ions, and toxic wastes

19.____All photosynthetic and nonphotosynthetic eukaryotic cells have one or more

20.____Possess two outer membrane layers that surround a semifluid interior, one membrane wrapped over the other

## 4.8. COMPONENTS OF THE CYTOSKELETON (pp. 70–71)
## 4.9. THE STRUCTURAL BASIS OF CELL MOTILITY (pp. 72–73)

*Selected Words:* *plus* end, *minus* end, <u>Colchicum</u>, *colchicine, taxol,* <u>Taxus</u> <u>brevifolia</u>, <u>Amoeba</u> <u>proteus</u>

### Boldfaced, Page-Referenced Terms

(70) cytoskeleton _____

_____

(70) microtubules _____

_____

(70) microfilaments _____

_____

(70) intermediate filaments _____

_____

(72) pseudopods _____

_____

(72) flagellum _____

_____

(72) cilium _____

_____

### Dichotomous Choice

Circle one of two possible answers given between parentheses in each statement.

1. The cytoskeleton gives (prokaryotic/eukaryotic) cells their shape, internal organization, and movement.
2. (Protein/Carbohydrate) subunits form the basic components of microtubules, microfilaments, and intermediate filaments.
3. Microtubules and microfilaments, acting singly or collectively, underlie nearly all the (chemistry/movements) of eukaryotic cells.
4. Hollow microtubule cylinders consist of (tubulin/actin) subunits arranged in uniform orientation.
5. Growing microtubules elongate much faster at their (minus/plus) ends.
6. Typically, the (minus/plus) end of a microtubule becomes anchored in loose material in the cytoplasm or at an MTOC.
7. An MTOC is a (microfilament organizing center/microtubule organizing center).
8. MTOCs include (microtubules, microfilaments, and intermediate filaments/centrioles, centrosomes, and kinetochores).
9. Colchicine (inhibits assembly and promotes disassembly of microtubules/stabilizes existing microtubules and prevents formation of new ones).
10. Taxol (inhibits assembly and promotes disassembly of microtubules/stabilizes existing microtubules and prevents formation of new ones).
11. (Taxol/Colchicine) has been used to reduced the uncontrolled cell division underlying benign and malignant tumor growth.
12. The thinnest of cytoskeletal elements are (microtubules/microfilaments).
13. Microfilaments consist of two helically twisted polypeptide chains assembled from (tubulin/actin) monomers.

14. Microfilaments participate in many cell movements, particularly those (deep within cells/at cell surfaces).
15. (Intermediate filaments/Accessory proteins) allow microfilaments and microtubules to serve a variety of functions in different regions of the cell.
16. The most stable elements of the cytoskeleton are (microfilaments/intermediate filaments).
17. The major function of the known groups of intermediate filaments is (movement/strength and support).
18. The pseudopods of *Amoeba proteus* are extended by controlled (assembly/disassembly) of microfilaments beneath its plasma membrane.
19. Within muscle cells, ATP-activated myosin polymers repeatedly bind and release adjacent microfilaments; this action causes the microfilaments to slide on past them and (lengthen/shorten) the cell.
20. In response to the sun's position, chloroplasts flowing in plant cells due to myosin monomers "walking" over bundles of microfilaments is known as (amoeboid motion/cytoplasmic streaming).
21. (Flagella/Cilia) are long, whiplike motile structures on cells.
22. (Flagella/Cilia) are short but very numerous motile structures on cells.
23. The cross-sectional array of 9 + 2 microtubules is found in (cilia/centrioles).
24. The human respiratory tract is lined with beating (flagella/cilia).
25. 9 + 2 arrays of microtubules arise from a (centriole/pseudopod), one type of MTOC.

## 4.10. CELL SURFACE SPECIALIZATIONS (pp. 74–75)

*Selected Words: tight* junctions, *adhering* junctions, *gap* junctions

### Boldfaced, Page-Referenced Terms

(74) cell wall _____

_____

(74) primary wall _____

_____

(74) secondary wall _____

_____

(75) plasmodesma _____

_____

### Matching

Choose the most appropriate answer to match with each of the following terms.

1. ___ primary wall
2. ___ secondary wall
3. ___ plasmodesmata
4. ___ tight junctions
5. ___ adhering junctions
6. ___ gap junctions

A. Link the cells of epithelial tissues lining the body's outer surface, inner cavities, and organs
B. Numerous tiny channels crossing the adjacent primary walls of living plant cells and connecting their cytoplasm
C. Formed of cementing secretions of pectin, hemicellulose, glycoproteins, and cellulose
D. Link the cytoplasm of neighboring animal cells and are open channels for the rapid flow of signals and substances
E. Formed of rigid layers; reinforces plant cell shape
F. Join cells in tissues of the skin, heart, and other organs subject to stretching

## 4.11. PROKARYOTIC CELLS—THE BACTERIA (pp. 76–77)

*Selected Words:* *prokaryotic,* Escherichia coli, Pseudomonas marginalis

### Boldfaced, Page-Referenced Terms

(76) nucleoid _____

_____

(76) bacterial flagella _____

_____

### Fill-in-the-Blanks

With one exception, (1) _____ are the smallest and most structurally simple cells. The word *prokaryotic* means "before the (2) _____ ," which implies that bacteria evolved before cells possessing nuclei existed. Most bacteria have a cell (3) _____ that surrounds the plasma membrane; this membrane controls movement of substances to and from the (4) _____ . Some bacteria have one or more long, threadlike motile structures, the (5) _____ , that extend from the cell surface; they permit rapid movements through fluid environments. Bacterial cells have many (6) _____ upon which polypeptide chains are assembled. Bacterial cell cytoplasm is continuous with an irregularly shaped region of DNA, the (7) _____ that lacks surrounding membranes.

## Self-Quiz

### Label-Match

Identify each indicated part of the accompanying illustrations. Complete the exercise by matching and entering the letter of the proper function description in the parentheses following each label. Some letter choices must be used more than once.

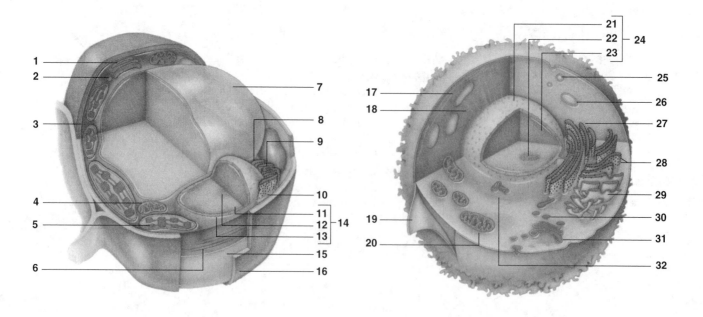

1. _____ _____ ( )

2. _____ ( )

3. _____ (cytoskeletal component) ( )

4. _____ ( )

5. _____ ( )

6. _____ (cytoskeletal component) ( )

7. _____ _____ ( )

8. _____ _____ _____ ( )

9. _____ ( )

10. _____ _____ _____ ( )

11. _____ + _____ ( )

12. _____ ( )

13. _____ _____ ( )

14. _____ ( )

15. _____ _____ ( )

16. _____ _____ ( )

17. _____ (cytoskeletal component) ( )

18. _____ (cytoskeletal component) ( )

19. _____ _____ ( )

20. _____ ( )

21. _____ _____ ( )

22. _____ ( )

23. _____ + _____ ( )

24. _____ ( )

25. _____ ( )

26. _____ ( )

27. _____ _____ _____ ( )

28. _____ ( )

29. _____ _____ _____ ( )

30. _____ ( )

31. _____ _____ ( )

32. _____ ( )

A. Pore-riddled two-membrane structure; outermost part of the nucleus

B. Porous, but provides protection and structural support for plant cells

C. Its growth increases the cell surface area available for absorption

D. Free of ribosomes; the main site of lipid synthesis in many cells

E. Formed as buds from Golgi membranes of animal cells and some fungal cells

F. A pair of small cylinders composed of triplet microtubules; produce microtubules of cilia and flagella and act as basal bodies

G. Different types function in transport or storage of various substances; digestion in cell; other functions

H. The largest element of its type; its minus ends anchor the cylinder made of tubulin subunits

I. A membrane-bound compartment that houses DNA in eukaryotic cells

J. Modification of polypeptide chains into mature proteins; sorting, and shipping of proteins and lipids for secretion or for use in the cell

K. Site of assembly of protein and RNA into ribosomal subunits

L. Thinnest element of its type; takes part in diverse movements, and in the formation and maintenance of cell shape

M. Function of food production and some starch storage

N. Thin, outermost membrane that maintains the cell as a distinct entity; substances and signals continually move across it, in highly controlled ways

O. Site of aerobic respiration

P. Initial modification of protein structure after formation on ribosomes

Q. Genetic material and the other substances within the nucleus

R. Assembly of polypeptide chains

## Multiple Choice

____33. Membranes consist of _____ .
   a. a lipid bilayer
   b. a protein bilayer
   c. phospholipids and proteins
   d. both a and c are correct

____34. Which of the following is *not* found as a part of prokaryotic cells?
   a. Ribosomes
   b. DNA
   c. Nucleus
   d. Cytoplasm
   e Cell wall

____35. Which of the following statements most correctly describes the relationship between cell surface area and cell volume?
   a. As a cell expands in volume, its diameter increases at a rate faster than its surface area does.
   b. Volume increases with the square of the diameter, but surface area increases only with the cube.
   c. If a cell were to grow four times in diameter, its volume of cytoplasm increases sixteen times and its surface area increases sixty-four times.
   d. Volume increases with the cube of the diameter, but surface area increases only with the square.

____36. Most cell membrane functions are carried out by _____ .
   a. carbohydrates
   b. nucleic acids
   c. lipids
   d. proteins

____37. Animal cells dismantle and dispose of waste materials by _____ .
   a. using centrally located vacuoles
   b. several lysosomes fusing with a vesicle that encloses the wastes
   c. microvilli packaging and exporting the wastes
   d. mitochondrial breakdown of the wastes

____38. The nucleolus is the site where _____ .
   a. the protein and RNA subunits of ribosomes are assembled
   b. the chromatin is formed
   c. chromosomes are bound to the inside of the nuclear envelope
   d. chromosomes duplicate themselves

____39. The _____ is free of ribosomes, curves through the cytoplasm, and is the main site of lipid synthesis.
   a. lysosome
   b. Golgi body
   c. smooth ER
   d. rough ER

____40. Which of the following is *not* present in all cells?
   a. Cell wall
   b. Plasma membrane
   c. Ribosomes
   d. DNA molecules

____41. As a part of the cytomembrane system, the _____ modify lipids and proteins to permit sorting and packaging for specific locations.
   a. endoplasmic reticulum
   b. Golgi bodies
   c. peroxisomes
   d. lysosomes

____42. Chloroplasts _____ .
   a. are specialists in oxygen-requiring reactions
   b. function as part of the cytoskeleton
   c. trap sunlight energy and produce organic compounds
   d. assist in carrying out cell membrane functions

____43. Mitochondria convert energy stored in _____ to forms that the cell can use, principally ATP.
   a. water
   b. carbon compounds
   c. $NADPH_2$
   d. carbon dioxide

____44. _____ are sacs of enzymes that bud from ER; they produce potentially harmful hydrogen peroxide while breaking down fatty acids and amino acids.
   a. Lysosomes
   b. Glyoxysomes
   c. Golgi bodies
   d. Peroxisomes

____45. Two classes of cytoskeletal elements underlie nearly all movements of eukaryotic cells; they are _____ .
   a. desmins and vimentins
   b. actin and microfilaments
   c. microtubules and microfilaments
   d. microtubules and myosin

## Choice

Cells of the organisms in the six kingdoms of life share the following characteristics: plasma membrane, DNA, RNA, and ribosomes. For questions 46–57, choose the kingdom(s) possessing the characteristics listed below. Some questions will require more than one kingdom as the answer.

a. Archaebacteria, Eubacteria     b. Protistans     c. Fungi     d. Plants     e. Animals

46.___cytoskeleton

47.___cell wall

48.___nucleus

49.___nucleolus

50.___central vacuole

51.___simple flagellum

52.___photosynthetic pigment

53.___chloroplast

54.___endoplasmic reticulum

55.___Golgi body

56.___lysosome

57.___complex flagella and cilia

---

# Chapter Objectives/Review Questions

*Page*     *Objectives/Questions*

(55)    1.   Be able to list the three generalizations that together constitute the cell theory.

(56)    2.   List and describe the three major regions that all cells have in common.

(56)    3.   _____ cells contain distinct arrays of organelles.

(56)    4.   _____ cells, by contrast, have no nucleus or organelles.

(56–57)   5.   Describe the lipid bilayer arrangement for the plasma membrane.

(57)    6.   Cell size is necessarily limited because its volume increases with the _____ but surface area increases only with the _____ .

(58–59)   7.   Briefly describe the operating principles of light microscopes, phase-contrast microscopes, scanning tunneling microscopes, transmission electron microscopes, and scanning electron microscopes.

(60)    8.   Briefly describe the function and cellular location of the organelles typical of most eukaryotic cells: nucleus, ribosomes, endoplasmic reticulum, Golgi body, various vesicles, mitochondria, and the cytoskeleton.

(60)    9.   In eukaryotic cells, _____ separate different incompatible chemical reactions in space and time.

(64)    10.   Describe the nature of the nuclear envelope and relate its function to its structure.

(64–65) 11.   _____ are sites where the protein and RNA subunits of ribosomes are assembled.

(65)    12.   _____ is the cell's collection of DNA molecules and associated proteins; a _____ is an individual DNA molecule and associated proteins.

(66)    13.   Explain how the endoplasmic reticulum (rough and smooth types), peroxisomes, Golgi bodies, lysosomes, and a variety of vesicles function together as the cytomembrane system.

(66)    14.   Describe the function of exocytosis and exocytic vesicles; also describe the function of endocytosis and endocytic vesicles.

(67)    15.   _____ are organelles of intracellular digestion that bud from Golgi membranes of animal cells and some fungal cells.

(67)    16.   Define and describe the function of peroxisomes.

(68)    17.   Within _____ , energy stored in organic molecules is released by enzymes and used to form many ATP molecules in the presence of oxygen.

(69)    18.   Describe the detailed structure of the chloroplast, the site of photosynthesis (include grana and stroma).

(69)    19.   Give the function of the following plant organelles: chloroplasts, chromoplasts, amyloplasts, and the central vacuole.

(70) 20. Elements of the _____ gives eukaryotic cells their internal organization, overall shape, and capacity to move

(70) 21. List the three major structural elements of the cytoskeleton and give the general function of each.

(71) 22. Describe the general function of accessory proteins.

(72) 23. *Amoeba proteus*, a soft-bodied protistan, crawls on _____ .

(72) 24. Both cilia and flagella have an internal microtubule arrangement called the "_____" array.

(72) 25. Microtubules of flagella and cilia arise from _____ that remain at the base of those completed structures as basal bodies.

(74) 26. Distinguish a primary cell wall from a secondary cell wall in leafy plants.

(74–75) 27. Describe the location and function of plasmodesmata.

(75) 28. What is meant by *ground substance*? _____ junctions link the cells of epithelial cells; _____ junctions keep cells together in tissues; _____ junctions link the cytoplasm of neighboring cells.

(76) 29. Describe the structure of a generalized prokaryotic cell. Include the bacterial flagellum, nucleoid, pili, capsule, cell wall, plasma membrane, cytoplasm, and ribosomes.

---

# Integrating and Applying Key Concepts

1. Which parts of a cell constitute the minimum necessary for keeping the simplest of living cells alive?

2. How did the existence of a nucleus, compartments, and extensive internal membranes confer selective advantages on cells that developed these features?

# 5

# A CLOSER LOOK AT CELL MEMBRANES

## Interactive Exercise

*It isn't Easy Being Single* (pp. 80–81)

## 5.1. MEMBRANE STRUCTURE AND FUNCTION (pp. 82–83)

## 5.2. *Focus on Science:* TESTING IDEAS ABOUT CELL MEMBRANES (pp. 84–85)

*Selected Words: observed* ratio of proteins, ratio *predicted, freeze-fracturing, freeze-etching*

*Boldfaced, Page-Referenced Terms*

(82) phospholipid _____

_____

(82) lipid bilayer _____

_____

(82) fluid mosaic model _____

_____

(83) transport proteins _____

_____

(83) receptor proteins _____

_____

(83) recognition proteins _____

_____

(83) adhesion proteins _____

_____

(84) centrifuge _____

_____

## *Labeling*

Identify each numbered membrane component in the illustration.

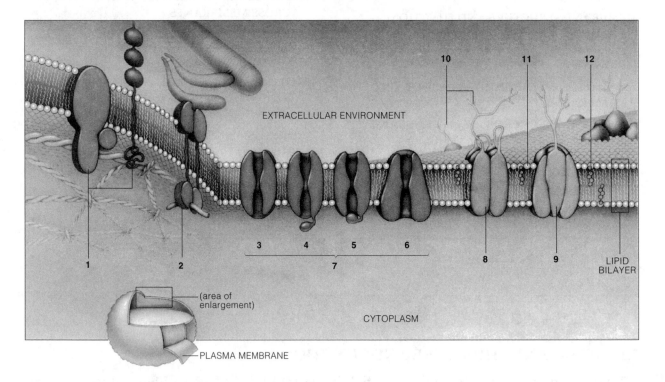

1. _____ proteins

2. _____ protein

3. _____ _____ protein

4. _____ _____ protein (open)

5. _____ _____ protein (closed)

6. _____ _____ protein

7. _____ proteins

8. _____ protein

9. _____ protein

10. _____ groups

11. _____

12. _____

## Matching

Choose the most appropriate answer for each.

13. ___fluid mosaic model

14. ___transport proteins

15. ___freeze-fracturing and freeze-etching

16. ___recognition proteins

17. ___cholesterol and phytosterol

18. ___phospholipid

19. ___lipid bilayer

20. ___adhesion proteins

21. ___centrifuge

22. ___receptor proteins

A. Oligosaccharide chains identify specific cell types
B. Two layers of phospholipids, the structural basis of cell membranes
C. Bind extracellular substances such as hormones that trigger changes in cell activities
D. Separates cell components according to their relative densities
E. Water-soluble substances move through their interiors; also bind and release molecules or ions to move them through membranes
F. Phosphate-containing head and two fatty acid tails attached to a glycerol backbone
G. Helps cells of the same type locate and stick to one another
H. Determination of the nature of cell membranes by observation
I. The most abundant sterols in animal and plant membranes, respectively
J. A mixture of phospholipids, glycolipids, sterols, and proteins; motions and interaction of component parts

## 5.3. HOW SUBSTANCES CROSS CELL MEMBRANES (pp. 86–87)

## 5.4. THE DIRECTIONAL MOVEMENT OF WATER ACROSS MEMBRANES (pp. 88–89)

*Selected Words: net* movement, *negative* charge, *positively* charged substance, facilitated diffusion, *against* the concentration gradient, *tonicity*

### Boldfaced, Page-Referenced Terms

(86) selective permeability _____

_____

(86) concentration gradient _____

_____

(86) diffusion _____

_____

(87) electric gradient _____

_____

(87) pressure gradient _____

_____

(87) passive transport _____

_____

(87) active transport _____

_____

(87) exocytosis _____

_____

(87) endocytosis _____

_____

(88) bulk flow _____

_____

(88) osmosis _____

_____

(88) hypotonic solution _____

_____

(88) hypertonic solution _____

_____

(88) isotonic solution _____

_____

(89) hydrostatic pressure _____

_____

(89) osmotic pressure _____

_____

(89) plasmolysis _____

_____

*Fill-in-the-Blanks*

Membranes show selective (1) _____ in that they allow some substances but not others to cross them in certain ways, at certain times. (2) _____ refers to the number of molecules of a substance in a specified region, as in volume of fluid or air.  A (3) _____ gradient is a difference in the number of molecules or ions of a given substance in two different regions.  When like molecules move *down* their concentration gradient, it is called (4) _____ ; this is a key factor in moving substances across cell membranes and through fluid portions of cytoplasm.  When the concentration gradient is steep, diffusion is (5) [choose one] ( ) slower ( ) faster.  As the gradient decreases and the number of molecules moving down the gradient decreases, diffusion is (6) [choose one] ( ) slower ( ) faster.  When the net distribution of molecules in a diffusion system is nearly uniform in the two adjoining regions, it is called dynamic (7) _____ . Diffusion is (8) [choose one] ( ) slower ( ) faster at higher temperature; this is due to more rapidly moving molecules that collide more often.  If a difference in charge exists between two adjoining regions, the rate and direction of diffusion may be changed by an (9) _____ gradient.  Within a cell, the side of the membrane that is negatively charged will exert a(n) (10) [choose one] ( ) attraction ( ) repulsion for positive sodium ions and enhance the flow of sodium across the membrane.  A difference in pressure, or the (11) _____ gradient between two adjoining diffusion regions, can also influence the rate and direction of diffusion.

## Matching

Choose the most appropriate answer for each.

12. ___bulk flow

13. ___osmosis

14. ___tonicity

15. ___hypotonic solution

16. ___hypertonic solution

17. ___isotonic solutions

18. ___hydrostatic pressure

19. ___osmotic pressure

20. ___plasmolysis

A. Refers to the relative solute concentrations of two fluids
B. Having the same solute concentrations
C. Mass movement of one or more substances in response to pressure, gravity, or another external force
D. The amount of force preventing further inward diffusion of water
E. The fluid on one side of a membrane that contains more solutes than the fluid on the other side of the membrane
F. The diffusion of water in response to a water concentration gradient between two regions separated by a selectively permeable membrane
G. Osmotically induced shrinkage of cytoplasm
H. The fluid on one side of a membrane that contains fewer solutes than the fluid on the other side of the membrane
I. A fluid force exerted against a cell wall and/or membrane enclosing the fluid

## Complete the Table

21. Complete the table below listing substances by entering the name of the correct membrane transport mechanism that will move that substance across a membrane. Choose from diffusion, osmosis, facilitated diffusion (passive transport), and active transport.

| Substance | Transport Mechanism |
|---|---|
| a. $H_2O$ | |
| b. $CO_2$ | |
| c. $Na^+$ | |
| d. Glucose | |
| e. $O_2$ | |
| f. $K^+$ | |
| g. Amino acids | |
| h. Substances pumped through interior of a transport protein; energy input required | |
| i. Substances move through interior of a transport protein; no energy input required | |
| j. Lipid soluble substances | |

## True/False

If the statement is true, write a T in the blank. If the statement is false, make it correct by changing the under-lined word(s) and writing the correct word(s) in the answer blank.

_____22. Because membranes exhibit selective permeability, concentrations of dissolved substances can <u>increase</u> on one side of the membrane or the other.

_____23. A water concentration gradient is influenced by the number of <u>solute</u> molecules present on both sides of the membrane.

_____24. The relative concentrations of solutes in two fluids is referred to as <u>isotonic</u>.

_____25. An animal cell placed in a <u>hypertonic</u> solution would swell and perhaps burst.

_____26. Water tends to move from <u>hypotonic</u> solutions to areas with more solutes.

_____27. Physiological saline is 0.9 percent NaCl; red blood cells placed in such a solution will not gain or lose water; therefore, one could state that the fluid in red blood cells is <u>hypertonic</u>.

_____28. A solution of 80 percent water, 20 percent solute is <u>more</u> concentrated than a solution of 70 percent water, 30 percent solute.

_____29. The movement of water molecules and solutes in the same direction in response to a pressure gradient is called <u>osmosis</u>.

_____30. Plant cells placed in a <u>hypotonic</u> solution will swell.


## 5.5. PROTEIN-MEDIATED TRANSPORT (pp. 90–91)

### Boldfaced, Page-Referenced Terms

(91) calcium pump _____

_____

(91) sodium-potassium pump _____

_____


### Choice

For questions 1–14, choose from the following:

a. passive transport          b. active transport          c. applies to both active and passive transport

1. ____The calcium pump

2. ____Part of the transport protein closes in behind the bound solute—and part opens up to the opposite side of the membrane

3. ____Involves a carrier protein that has not been energized

4. ____A carrier protein has a specific site that weakly binds a substance

5. ____At any given time, the *net* direction of movement depends on how many solute molecules make random contact with vacant binding sites in the interior of proteins

6. ____The transport protein must receive an energy boost, usually from ATP

7. ____Solute binding to a carrier protein leads to changes in protein shape

8. ____The sodium-potassium pump

9.___Transport proteins span the bilayer, and their interior is able to open on both sides of it

10.___A solute is pumped across the cell membrane *against* its concentration gradient

11.___The solute binding site becomes altered when ATP donates energy to the carrier protein to allow easier solute binding

12.___Passive two-way transport would continue until solute concentrations became equal on both sides of the membrane; other processes such as cellular solute use may influence the outcome

13.___Net movement will be down the solute's concentration gradient

14.___Think of the solute as hopping onto the transport protein on one side of the membrane, then hopping off on the other side

## 5.6. EXOCYTOSIS AND ENDOCYTOSIS (pp. 92–93)

*Selected Words:* *receptor-mediated* endocytosis, *bulk-phase* endocytosis

### Boldfaced, Page-Referenced Terms

(93) phagocytosis _____

_____

### Matching

Choose the most appropriate answer for each.

1. ___exocytosis
2. ___receptor-mediated endocytosis
3. ___bulk-phase endocytosis
4. ___phagocytosis
5. ___membrane cycling

A. A cell engulfs microorganisms, large edible particles, and cellular debris
B. Membrane initially used for endocytic vesicles returns receptor proteins and lipids back to the plasma membrane
C. Vesicles form around small volumes of extracellular fluid of various content
D. A cytoplasmic vesicle moves to the cell surface, its own membrane fuses with the plasma membrane while its contents are released to the environment
E. Chemical recognition and binding of specific substances; pits of clathrin baskets sink into the cytoplasm and close on themselves

## Self-Quiz

___ 1. White blood cells use _____ to devour disease agents invading your body.
   a. diffusion
   b. bulk flow
   c. osmosis
   d. phagocytosis

___ 2. _____ cells depend on the calcium-pump mechanism.
   a. Intestine
   b. Nerve
   c. Muscle
   d. Amoeba

___ 3. _____ proteins bind extracellular substances, such as hormones, and bring about changes in cell activities.
a. Receptor
b. Adhesion
c. Transport
d. Recognition

___ 4. In a lipid bilayer, tails point inward and form a(n) _____ region that excludes water.
a. acidic
b. basic
c. hydrophilic
d. hydrophobic

___ 5. A protistan adapted to life in a freshwater pond is collected in a bottle and transferred to a saltwater bay. Which of the following is likely to happen?
a. The cell bursts.
b. Salts flow out of the protistan cell.
c. The cell shrinks.
d. Enzymes flow out of the protistan cell.

___ 6. Which of the following is *not* a form of active transport?
a. Sodium-potassium pump
b. Endocytosis
c. Exocytosis
d. Bulk flow

___ 7. Which of the following is *not* a form of passive transport?
a. Osmosis
b. Passive transport
c. Bulk flow
d. Exocytosis

___ 8. $O_2$, $CO_2$, $H_2O$, and other small, electrically neutral molecules move across the cell membrane by _____ .
a. electric gradients
b. receptor-mediated endocytosis
c. passive transport
d. active transport

___ 9. Ions such as $H^+$, $Na^+$, $K^+$, and $Ca^{++}$ move across cell membranes by _____ .
a. receptor-mediated endocytosis
b. pressure gradients
c. passive transport
d. active transport

___10. Receptors, pits, and clathrin baskets participate in _____ .
a. passive transport
b. receptor-mediated endocytosis
c. bulk-phase endocytosis
d. active transport

---

# Chapter Objectives/Review Questions

*Page    Objective/Review Questions*

(82)  1.  _____ molecules are the most abundant component of cell membranes.
(82)  2.  Being a composite of phospholipids, glycolipids, sterols, and proteins, the membrane is said to have a _____ quality.
(82)  3.  A lipid bilayer has a _____ behavior as a result of motions and interactions of its component parts.
(83)  4.  Most membrane functions are carried out by _____ embedded in or attached to one of the surfaces of the membrane.
(83)  5.  Be able to state the general functions of transport proteins, receptor proteins, recognition proteins, and adhesion proteins.
(84)  6.  What is the value of a centrifuge to cellular studies? Freeze-fracturing and freeze-etching?
(86)  7.  Explain the concept of selective permeability as it applies to cell membrane function.
(86)  8.  The difference in the number of molecules or ions of a given substance in two regions is defined as _____ _____ .
(86)  9.  The net movement of like molecules down a concentration gradient is called _____ .
(87) 10.  In addition to concentration gradients, diffusion rates can be influenced by temperature, _____ gradients, and _____ gradients.
(87) 11.  Distinguish passive transport from active transport.

(87) 12. _____ involves fusion of a vesicle formed in the cytoplasm with the cell's plasma membrane; _____ involves an inward sinking of a small patch of plasma membrane, which then seals back on itself to form a cytoplasmic vesicle.

(88) 13. Define and cite an example of *bulk flow*.

(88) 14. The movement of water across membranes in response to solute concentration gradients, fluid pressure, or both, is called _____ .

(88) 15. Define *tonicity*.

(88) 16. Water tends to move from a _____ solution (less solutes) to a _____ solution (more solutes).

(88) 17. Water shows no net osmotic movement in an _____ solution.

(89) 18. A fluid force directed against a wall or membrane is _____ pressure.

(89) 19. Define *osmotic pressure*.

(89) 20. _____ is responsible for wilting lettuce.

(90) 21. Describe the mechanism and result of solute binding to a transport protein.

(90) 22. With _____ transport, a solute simply diffuses through the interior of a _____ protein.

(91) 23. The calcium and sodium-potassium pump are examples of _____ transport wherein net solute movement is against the concentration gradient.

(92) 24. Describe mechanisms involved in receptor-mediated endocytosis.

(92) 25. How does bulk-phase endocytosis differ from phagocytosis?

(93) 26. Explain the role of endocytosis and exocytosis in membrane recycling.

---

## Integrating and Applying Key Concepts

If there were no such thing as active transport, how would the lives of organisms be affected?

# 6

# GROUND RULES OF METABOLISM

## Interactive Exercises

*Growing Old with Molecular Mayhem* (pp. 96–97)

## 6.1. ENERGY AND THE UNDERLYING ORGANIZATION OF LIFE (pp. 98–99)

*Selected Words: thermal* energy, *"system", "surroundings"*

*Boldfaced, Page-Referenced Terms*

(96) free radical _____

_____

(98) metabolism _____

_____

(98) potential energy _____

_____

(98) kinetic energy _____

_____

(98) heat _____

_____

(98) chemical energy _____

_____

(98) kilocalorie _____

_____

(98) first law of thermodynamics _____

_____

(99) second law of thermodynamics _____

_____

(99) entropy _____

_____

## True/False

If the statement is true, write a T in the blank. If the statement is false, make it correct by changing the underlined word(s) and writing the correct word(s) in the answer blank.

_____1. The _first_ law of thermodynamics states that entropy is constantly increasing in the universe.

_____2. At rest your body steadily gives off heat equal to that from a _100-watt_ light bulb.

_____3. When you eat a potato, some of the stored chemical energy of the food is converted into _kinetic_ energy that moves your muscles.

_____4. _Energy_ is the capacity to accomplish work.

_____5. The amount of low-quality energy in the universe is _decreasing_.

## Matching

In the blank preceding each item, indicate if the first law of thermodynamics (I) or the second law of thermodynamics (II) is best described.

6. ___Cooling of a cup of coffee.

7. ___Evaporation of gasoline into the atmosphere.

8. ___A hydroelectric plant at a waterfall, producing electricity.

9. ___The creation of a snowman by children.

10. ___The glow of an incandescent bulb following the flow of electrons through a wire.

11. ___A typesetter arranging the type for a page of a biology textbook.

12. ___The movement of a gas-powered automobile.

13. ___The glow of a firefly.

14. ___Humans running the 100-meter dash following their usual food intake.

15. ___The death and decay of an organism.

## Fill-in-the-Blanks

For questions 16–20, choose from these possibilities:

(a) chemical energy      (b) entropy      (c) heat      (d) kilocalorie      (e) metabolism

16. _____ is a measure of the amount of disorder in a system.
17. _____ includes all of the activities by which a cell acquires energy and materials and uses them to build, break apart, store, and release substances in controlled processes that are typical for that cell.
18. The measure of energy that can heat 1,000 grams of water from 14.5°C to 15.5°C at standard pressure is _____.
19. _____ is the potential energy stored in the attractive forces (bonds) that cause atoms to group together into molecules.
20. _____ that results from collisions among molecules and their surroundings is a kind of kinetic energy also called thermal energy.

## 6.2. THE DIRECTIONAL NATURE OF METABOLISM (pp. 100–101)

## 6.3. ENERGY TRANSFERS AND CELLULAR WORK (pp. 102–103)

*Selected Words:* *phenylketonuria (PKU); biosynthetic* pathway, *degradative* pathway

### Boldfaced, Page-Referenced Terms

(100) chemical reaction _____

_____

(100) chemical equilibrium _____

_____

(101) law of conservation of mass _____

_____

(101) exergonic reaction _____

_____

(101) endergonic reaction _____

_____

(102) ATP, adenosine triphosphate _____

_____

(102) phosphorylation _____

_____

(102) ADP, adenosine diphosphate _____

_____

(102) ATP/ADP cycle _____

_____

(103) metabolic pathways _____

_____

(103) substrates _____

_____

(103) intermediates _____

_____

(103) end products _____

_____

(103) energy carriers _____

_____

(103) enzymes _____

_____

(103) cofactors _____

_____

(103) transport proteins _____

_____

## Fill-in-the-Blanks

The substances present at the end of a chemical reaction, the (1) _____ , may have less or more energy than did the starting substances, the (2) _____ . reactions are (3) _____ in that they can proceed in forward and reverse directions. Such reactions tend to approach chemical (4) _____ , a state in which the reactions are proceeding at about the same (5) _____ in both directions.

## Labeling

Classify each of the following reactions as *endergonic* or *exergonic*.

_____ 6. Burning wood at a campfire.

_____ 7. The products of a chemical reaction have more energy than the reactants.

_____ 8. Glucose + oxygen → carbon dioxide + water plus energy

_____ 9. The reactants of a chemical reaction have more energy than the product.

_____ 10. The reaction releases energy.

## Labeling

Study the diagram at the right; choose the most appropriate answer for each: A or B.

11.____a degradative reaction

12.____an endergonic reaction

13.____a biosynthetic reaction

14.____an exergonic reaction

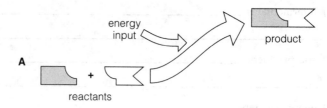

## Matching

Match the most appropriate letter from the set below to its number.

15. ____intermediates

16. ____degradative pathways

17. ____chemical equilibrium

18. ____cofactors

19. ____transport proteins

20. ____metabolic pathway

21. ____reactants (substrates)

22. ____energy carriers

23. ____biosynthetic pathways

24. ____enzymes

A. An orderly series of reactions catalyzed by enzymes
B. Small organic molecules are assembled into larger organic molecules
C. Mainly ATP; donate(s) energy to reactions
D. Small molecules and metal ions that assist enzymes or serve as carriers
E. Substances able to enter into a reaction
F. Compounds formed between the beginning and end of a metabolic pathway
G. Organic compounds are broken down in stepwise reactions
H. Proteins (usually) that catalyze reactions
I. Rate of forward reaction = rate of reverse reaction
J. Membrane-bound substances that adjust concentration gradients in ways that influence the direction of metabolic reactions

## Matching

Study the sequence of reactions below. Identify the components of the reactions by selecting items from the following list and entering the correct letter in the appropriate blank.

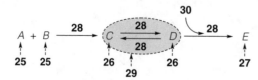

| 25. ___ | 28. ___ | A. cofactor | D. end product |
|---------|---------|-------------|----------------|
| 26. ___ | 29. ___ | B. intermediates | E. reversible reaction |
| 27. ___ | 30. ___ | C. reactants | F. enzymes |

## Fill-in-the-Blanks

Photosynthetic cells possess the chemical mechanisms to convert sunlight energy to the chemical energy of molecules known as (31) _____ , a type of nucleotide. It is important to recognize that all cells lack the ability to extract energy *directly* from food molecules such as glucose. The energy in the food molecules must first be harnessed and converted to the energy of (32) _____ _____ . Cells must obtain their energy from these molecules. An ATP molecule consists of (33) ____ (a nitrogen-containing compound), the five-carbon sugar (34) _____ , and a string of three (35) _____ groups. These components are held together by (36) _____ bonds, some of which are rather unstable. Many different enzymes carry out (37) _____ reactions to release the terminal phosphate group. This releases (38) _____ that can be used by cells to carry out many types of biological tasks. ATP molecules serve as the cell's (39) _____ currency. In the ATP/ADP cycle, an energy input drives the binding of ADP to a phosphate group or to unbound phosphate ($P_i$), forming (40) _____ . Attaching a phosphate group to a molecule is called (41) _____ ; this energizes that molecule and prepares it to enter a specific reaction in a metabolic pathway.

## Labeling

Identify the molecule to the right and label its parts.

42. _____

43. _____

44. _____

45. The name of this molecule is _____

    _____ .

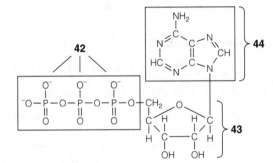

Name the types of pathways in the diagram of chemical reaction sequences below:

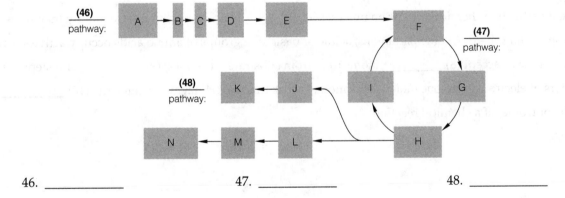

46. _____          47. _____          48. _____

## 6.4. ENZYME STRUCTURE AND FUNCTION (pp. 104–105)
## 6.5. FACTORS INFLUENCING ENZYME ACTIVITY (pp. 106–107)

*Selected Words: catalytic* molecules, "energy hill", *transition* state, <u>Thermoanobacter</u>, hormone

### *Boldfaced, Page-Referenced Terms*

(104) activation energy _____

_____

(104) active sites _____

_____

(104) induced-fit model _____

_____

(106) allosteric control _____

_____

(107) feedback inhibition _____

_____

(107) coenzymes _____

_____

### *Fill-in-the-Blanks*

(1) _____ are highly selective proteins that act as catalysts, which means they greatly enhance the rate at which specific reactions approach (2) _____ . The specific substance upon which a particular enzyme acts is called its (3) _____ ; this substance fits into the enzyme's crevice, which is called its (4) _____ _____ . The (5) _____ –_____ model describes how a substrate contacts the site without a perfect fit. Enzymes increase reaction rates by lowering the required (6) _____ _____ . Sometimes (6) is lowered because the reactant molecule(s) are oriented by the enzyme into positions that put their mutually attractive chemical groups on precise (7) _____ courses much more frequently than if the enzyme weren't there. In other situations, acidic or basic side groups of amino acids occupy active sites and easily donate or accept (8) _____ atoms to or from substrate molecules; (9) _____ and stepwise transfers of electrons from one molecule to another work this way. Enzymes change the (10) _____ , not the outcome, of a chemical reaction.

### *Identification*

Below is an "energy hill" reaction diagram. Identify the different amounts of energy in numbers 11, 12, and 15. Identify the two reaction pathways in numbers 13 and 14.

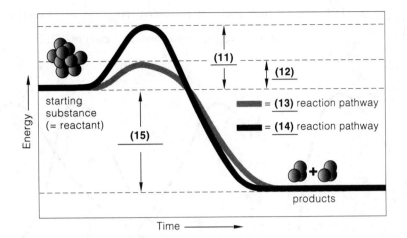

11. _____  _____  _____  _____
12. _____  _____  _____  _____
13. _____
14. _____
15. _____  _____  _____  _____  _____

## Matching

Match the items on the sketch below with the list of descriptions. Some answers may require more than one letter.

16. ___

17. ___

18. ___

19. ___

20. ___

21. ___

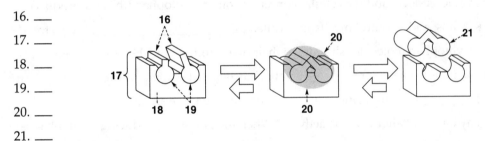

A. Transition state, the time of the most precise fit between enzyme and substrate
B. Complementary active site of the enzyme
C. Enzyme, a protein with catalytic power
D. Product or reactant molecules that an enzyme can specifically recognize
E. Product or reactant molecule
F. Bound enzyme-substrate complex

## Fill-in-the-Blanks

In the graph (below left) maximum enzyme activity occurs at (22) _____ °C, and minimum activity occurs at (23) _____ °C. In the graph (below right), enzyme (24) _____ functions best in basic solutions, enzyme (25) _____ works best in neutral solutions, and enzyme (26) _____ functions best in acidic solutions.

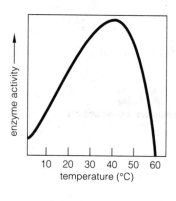
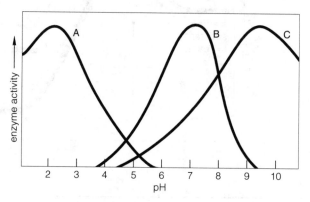

(27) _____ and (28) _____ are two factors that influence the rates of enzyme activity. Extremely high fevers can destroy the three-dimensional shape of an enzyme, which may adversely affect (29) _____ and cause death. The increased heat energy disrupts weak (30) _____ holding the enzyme in its three-dimensional shape. Most enzymes function best at about pH (31) _____ . One example of a control governing enzyme activity is the (32) _____ pathway in bacteria. The bacterium synthesizes tryptophan and other (33) _____ _____ to construct its proteins. When no more tryptophan is needed, the rate of protein synthesis slows and the cellular concentration of tryptophan (the end-product) (34) [choose one] (□) rises (□) falls, and a control mechanism called (35) _____ _____ begins to operate. In this case, excess tryptophan molecules shut down their own production when the unused molecules of tryptophan act to inhibit a key (36) _____ in the biochemical pathway. The inhibited enzyme is governed by (37) _____ control. Such enzymes have an active site and a (38) _____ site where specific substances may bind and alter enzyme activity. When the pathway producing tryptophan is blocked, fewer tryptophan molecules are present to inhibit the key enzyme and production (39) [choose one] (□) rises (□) falls. Another type of biochemical control in humans and other multicelled organisms operates through signaling agents called (40) _____ .

## Matching

Choose the one most appropriate answer for each.

41. ___metal ions

42. ___FAD

43. ___coenzymes

44. ___NADP$^+$

45. ___NAD$^+$

A. Large organic molecules; derived in part from vitamins; transfer hydrogens and electrons

B. A coenzyme, nicotinamide adenine dinucleotid phosphate; in reduced form carries H$^+$ and electrons to other reaction sites

C. A coenzyme, nicotinamide adenine transfers hydrogens and electrons to other reaction sites

D. Example: Fe$^{++}$ serves as a cofactor and is a component of enzymatic cytochromes

E. A coenzyme, flavin adenine dinucleotide; transfers hydrogens and electrons to other reaction sites

## 6.6. ELECTRON TRANSFERS THROUGH TRANSPORT SYSTEMS (p. 108)

## 6.7. *Focus on Science:* YOU LIGHT UP MY LIFE—VISIBLE SIGNS OF METABOLIC ACTIVITY (p. 109)

*Selected Words:* electron donor molecule, "oxidized", electron-acceptor molecule, "reduced", bioluminescent gene transfers, *Mycobacterium tuberculosis, Salmonella, gene therapy*

### Boldfaced, Page-Referenced Terms

(108) oxidation-reduction reaction _____

_____

(108) electron transport system _____

_____

(109) bioluminescence _____

_____

### Fill-in-the-Blanks

In cells, the release of energy from glucose proceeds in controlled steps of a degradative pathway, so that (1) _____ molecules form along the route from glucose to carbon dioxide and water. At each step in a metabolic pathway, a specific (2) _____ lowers the activation energy for the formation of an intermediate compound. At each step in the pathway, only some of the bond energy is released. In the internal membrane systems of chloroplasts and mitochondria, the liberated electrons released from the breaking of chemical bonds are sent through (3) _____ _____ systems; these organized systems consist of enzymes and (4) _____ , bound in a cell membrane, that transfer electrons in a highly organized sequence. Oxidation-reduction refers to (5) _____ transfers. In terms of oxidation-reduction reactions, a molecule in the sequence that donates electrons is said to be (6) _____ , while molecules accepting electrons are said to be (7) _____ . Electron transport systems "intercept" excited electrons and make use of the (8) _____ they release. If we think of the electron transport system as a staircase, excited electrons at the top of the staircase have the (9) [choose one] ( ) most ( ) least energy. As the electrons are transferred from one electron carrier to another, some (10) _____ can be harnessed to do biological (11) _____ . One type of biological work occurs when energy released during electron transfers (down the steps) is used to move ions in ways that set up ion concentration and electric gradients across a membrane. These gradients are essential for the formation of (12) _____ .

## Matching

Match the lettered statements to the numbered items on the sketch.

13. ___

14. ___

15. ___

16. ___

17. ___

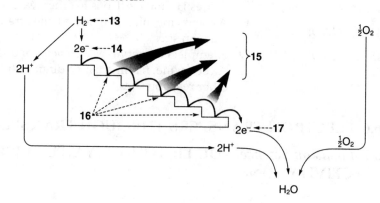

A. Represent the cytochrome molecules in an electron transport system
B. Electrons at their highest energy level
C. Released energy harnessed and used to produce ATP
D. Electrons at their lowest level
E. The separation of hydrogen atoms into protons and electron

## Fill-in-the-Blanks

Some organisms, such as fireflies and kitty boos emit light, a phenomenon known as (18) _____ .

(18) occurs as (19) _____ is first phosphorylated by (20) _____ and then oxidized by (21) _____ acting together with the enzyme (22) _____ . Light is emitted as excited (23) _____ return to their lower energy levels.

# Self-Quiz

____ 1. An important principle of the second law of thermodynamics states that _____ .
   a. energy can be transformed into matter, and therefore we can get something for nothing
   b. energy can only be destroyed during nuclear reactions, such as those that occur inside the sun
   c. if energy is gained by one region of the universe, another place in the universe also must gain energy in order to maintain the balance of nature
   d. matter tends to become increasingly more disorganized

____ 2. Essentially, the first law of thermodynamics states that _____ .
   a. one form of energy cannot be converted into another
   b. entropy is increasing in the universe
   c. energy cannot be created or destroyed
   d. energy cannot be converted into matter or matter into energy

____ 3. An enzyme is best described as _____ .
   a. an acid
   b. protein
   c. a catalyst
   d. a fat
   e. both b and c

____ 4. Which is *not* true of enzyme behavior?
   a. Enzyme shape may change during catalysis.
   b. The active site of an enzyme orients its substrate molecules, thereby promoting interaction of their reactive parts.
   c. All enzymes have an active site where substrates are temporarily bound.
   d. An individual enzyme can catalyze a wide variety of different reactions.

____ 5. When $NAD^+$ combines with hydrogen, the $NAD^+$ is _____ .
   a. reduced
   b. oxidized
   c. phosphorylated
   d. denatured

____ 6. A substance that gains electrons is _____ .
   a. oxidized
   b. a catalyst
   c. reduced
   d. a substrate

____ 7. In _____ pathways, carbohydrates, lipids, and proteins are broken down in stepwise reactions that lead to products of lower energy.
   a. intermediate
   b. biosynthetic
   c. induced
   d. degradative

____ 8. As to major function, $NAD^+$, FAD, and $NADP^+$ are classified as _____ .
   a. enzymes
   b. phosphate carriers
   c. cofactors that function as coenzymes
   d. end products of metabolic pathways

____ 9. When a phosphate bond is linked to ADP, the bond _____ .
   a. absorbs a large amount of free energy when the phosphate group is attached during hydrolysis
   b. is formed when ATP is hydrolyzed to ADP and one hosphate group
   c. is usually found in each glucose molecule; that is why glucose is chosen as the starting point for glycolysis
   d. releases a large amount of usable energy when the phosphate group is split off during hydrolysis

____ 10. An allosteric enzyme _____ .
   a. has an active site where substrate molecules bind and another site that binds with intermediate or end-product molecules
   b. is an important energy-carrying nucleotide
   c. carries out either oxidation reactions or reduction reactions but not both
   d. raises the activation energy of the chemical reaction it catalyzes

# Chapter Objectives/Review Questions

This section lists general and detailed chapter objectives that can be used as review questions. You can make maximum use of these items by writing answers on a separate sheet of paper. Fill in answers where blanks are provided. To check for accuracy, compare your answers with information given in the chapter or glossary.

| Page | | Objectives/Questions |
|---|---|---|
| (98) | 1. | _____ is the controlled capacity to acquire and use energy for stockpiling, breaking apart, building, and eliminating substances in ways that contribute to survival and reproduction. |
| (98–99) | 2. | Define *energy*; be able to state the first and second laws of thermodynamics. |
| (98) | 3. | Explain what is meant by a "system" as related to the laws of thermodynamics. |
| (99) | 4. | _____ is a measure of the degree of randomness or disorder of systems. |
| (99) | 5. | Explain how the world of life maintains a high degree of organization. |
| (100) | 6. | What is meant by a reversible reaction? |
| (101) | 7. | Describe the condition known as chemical equilibrium. |
| (102) | 8. | Adding a phosphate to a molecule is called _____ . |
| (102–103) | 9. | Explain the functioning of the ATP/ADP cycle. |
| (103) | 10. | _____ molecules are the cell's main, renewable energy carriers between sites of metabolic reactions in cells. |
| (103) | 11. | A _____ pathway is an orderly sequence of reactions with specific enzymes acting at each step; the pathways are always linear or circular. |
| (103) | 12. | Give the function of each of the following participants in metabolic pathways: substrates, intermediates, enzymes, cofactors, energy carriers, and end products. |
| (104) | 13. | What are enzymes? Explain their importance. |
| (104) | 14. | The location on the enzyme where specific reactions are catalyzed is the active _____ . |
| (104) | 15. | According to Koshland's _____ - _____ model, each substrate has a surface region that almost but not quite matches chemical groups in an active site. |
| (104) | 16. | When the "energy hill" is made smaller by enzymes in order that particular reactions may proceed, it may be said that the enzyme has lowered the _____ energy. |
| (106) | 17. | Explain what happens to enzymes in the presence of extreme temperatures and pH. |
| (107) | 18. | An enzyme control called _____ inhibition operates in the tryptophan pathway when excess tryptophan molecules inhibit a key allosteric enzyme in the pathway. |
| (107) | 19. | Cofactors called _____ are organic molecules that may be vitamins or are derived in part from vitamins; _____ ions may also serve as cofactors. |
| (107) | 20. | _____ are small molecules or metal ions that assist enzymes or carry atoms or electrons from one reaction site to another. |
| (108) | 21. | _____ -_____ reactions refer to the electron transfers occurring in an electron transport system. |
| (107) | 22. | Explain the differences between $NAD^+$ and $NADH$; $NADP^+$ and $NADPH$; $FAD$ and $FADH_2$. |
| (109) | 23. | Fireflies, various beetles, and some other organisms display _____ when luciferases excite the electrons of luciferins. |

# Integrating and Applying Key Concepts

A piece of dry ice left sitting on a table at room temperature vaporizes. As the dry ice vaporizes into $CO_2$ gas, does its entropy increase or decrease? Tell why you answered as you did.

# 7

# ENERGY-ACQUIRING PATHWAYS

---

## Interactive Exercises

---

*Sun, Rain, and Survival* (pp. 112–113)

## 7.1. PHOTOSYNTHESIS—AN OVERVIEW (pp. 114–115)

*Selected Words:* *photoautotrophs, chemoautotrophs, aerobic respiration, light dependent reactions, light-independent reactions, granum*

### Boldfaced, Page-Referenced Terms

(113) autotrophs _____

_____

(113) heterotrophs _____

_____

(115) chloroplast _____

(118) stroma _____

_____

(115) thylakoid membrane system _____

_____

## Fill-in-the-Blanks

(1) _____ obtain carbon and energy from the physical environment; their carbon source is (2) _____ _____ . (3) _____ autotrophs obtain energy from sunlight. (4) _____ autotrophs are represented by a few kinds of bacteria; they obtain energy by stripping (5) _____ from sulfur or other inorganic substances. (6) _____ feed on autotrophs, each other, and organic wastes; representatives include (7) _____ , fungi, many protistans, and most bacteria. Although energy stored in organic compounds such as glucose may be released by several pathways, the pathway known as (8) _____ _____ releases the most energy.

9. In the space below, supply the missing information to complete the summary equation for photosynthesis:

$$12 \underline{\quad} + \underline{\quad} CO_2 \rightarrow O_2 + C_6H_{12}O_6 + 6 \underline{\quad}$$

10. Supply the appropriate information to state the equation (above) for photosynthesis in words:

(a) _____ molecules of water plus six molecules of (b) _____ _____ (in the presence of pigments, enzymes, and ultraviolet light) yield six molecules of (c) _____ plus one molecule of (d) _____ plus (e) _____ molecules of water.

The two major sets of reactions of photosynthesis are the (11) _____ -_____ reactions and the (12) _____ -_____ reactions. (13) _____ _____ and (14) _____ are the reactants of photosynthesis, and the end product is usually given as (15) _____ . The internal membranes and channels of the chloroplast are the (16) _____ membrane system and are organized into stacks, called (17) _____ . Spaces inside the thylakoid disks and channels form a continuous compartment where (18) _____ ions accumulate to be used to produce ATP. The semifluid interior area surrounding the grana is known as the (19) _____ and is the area where the products of photosynthesis are produced.

## Choice

For questions 20–31, choose the area of the chloroplast that correctly relates to the listed structures and events.

a. thylakoid membrane system

b. stroma

20.____light-independent reactions

21.____the coenzyme NADP+ picks up liberated hydrogen and electrons

22.____granum

23.____sugars are assembled

24.____light-dependent reactions

25.____ATP production

26.____carbon dioxide provides the carbon

27.____sunlight energy is absorbed

28.____where the first stage of photosynthesis proceeds

29.____water molecules are split

30.____NADPH delivers the hydrogen received from water

31.____oxygen is formed

## Identify the Following Structures

32. _____
33. _____
34. _____
35. _____

36. _____
37. _____ _____
38. _____

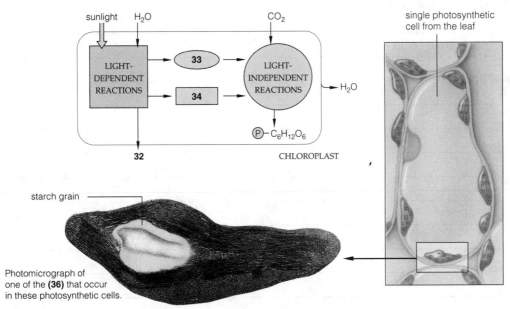

sunlight  H₂O                    CO₂

LIGHT-
DEPENDENT
REACTIONS

**33**

**34**

LIGHT-
INDEPENDENT
REACTIONS

→ H₂O

Ⓟ—C₆H₁₂O₆

**32**

CHLOROPLAST

single photosynthetic
cell from the leaf

starch grain

Photomicrograph of
one of the **(36)** that occur
in these photosynthetic cells.

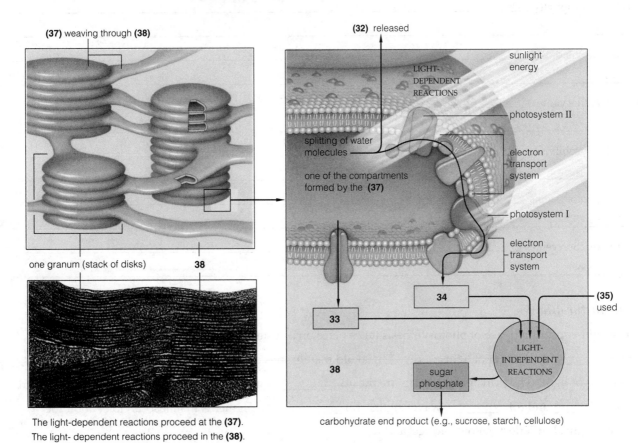

**(37)** weaving through **(38)**

one granum (stack of disks)        **38**

**(32)** released

LIGHT-
DEPENDENT
REACTIONS

sunlight
energy

photosystem II

electron
transport
system

photosystem I

electron
transport
system

splitting of water
molecules

one of the compartments
formed by the  **(37)**

**34**

**(35)**
used

**33**

**38**

sugar
phosphate

LIGHT-
INDEPENDENT
REACTIONS

The light-dependent reactions proceed at the **(37)**.
The light- dependent reactions proceed in the **(38)**.

carbohydrate end product (e.g., sucrose, starch, cellulose)

Overview of the sites where the key steps of both stages of
reactions proceed.  The sections to follow will fill in the details.

## 7.2. SUNLIGHT AS AN ENERGY SOURCE (pp. 116–117)
## 7.3. THE RAINBOW CATCHERS (pp. 118–119)

*Selected Words:* <u>Spirogyra</u>, accessory pigments, Englemann

### Boldfaced, Page-Referenced Terms

(116) wavelength _____

_____

(116) electromagnetic spectrum _____

_____

(116) visible light _____

_____

(116) photons _____

_____

(117) pigments _____

_____

(117) absorption spectrum _____

_____

(118) chlorophylls _____

_____

(118) carotenoids _____

_____

(119) anthocyanins _____

_____

(119) phycobilins _____

_____

(119) photosystems _____

_____

(119) reaction center _____

_____

### Fill-in-the-Blanks

The light-capturing phase of photosynthesis takes place on a system of (1) _____ membranes. A(n) (2) _____ is a packet of light energy. Thylakoid membranes contain (3) _____ , which absorb photons of light. The principal pigments are the (4) _____ , which reflect green wavelengths but absorb (5) _____ and (6) _____ wavelengths. (7) _____ are pigments that absorb violet and blue wavelengths but reflect yellow, orange, and red.

A cluster of 200 to 300 of these pigment proteins is a(n) (8) _____ . When pigments absorb (9) _____ energy, an (10) _____ is transferred from a photosystem to a(n) (11) _____ molecule. (12) _____ refers to the attachment of phosphate to ADP or other organic molecules. Due to the input of light energy, electrons flow through a transport system that causes protons ($H^+$) simultaneously to be pumped into the thylakoid compartments. Electrons then end up in (13) _____ chlorophyll at the end of this transport chain. The flow of protons from the thylakoid compartment through (14) _____ _____ drives the enzyme machinery that phosphorylates (15) _____ , a sequence of events known as the (16) _____ theory of ATP formation.

## Matching

Choose the most appropriate answer.

17. ___chlorophylls
18. ___chlorophyll *b* and carotenoids
19. ___carotenoids
20. ___violet-blue-green-yellow-red
21. ___photons
22. ___chlorophyll *a*
23. ___chloroplast
24. ___phycobilins
25. ___grana
26. ___pigments

A. The main pigment of photosynthesis
B. Packets of energy that have an undulating motion through space
C. The two stages of photosynthesis occur here
D. Molecules that can absorb light
E. Absorb violet and blue wavelengths but transmit red, orange, and yellow
F. Visible light portion of the electromagnetic spectrum
G. Pigments that transfer energy to chlorophyll a
H. Absorb violet-to-blue and red wavelengths; the reason leaves appear green
I. Red and blue pigments
J. The site of the first stage of photosynthesis

## 7.4. THE LIGHT-DEPENDENT REACTIONS (pp. 120–121)
## 7.5. A CLOSER LOOK AT ATP FORMATION IN CHLOROPLASTS (p. 122)

*Selected Words:* P700, type I photosystem, P680, type II photosystem, photosystem I, chemiosmotic theory, NADPH

*Boldfaced, Page-Referenced Terms*

(120) light-dependent reactions _____

_____

(120) electron transport systems _____

_____

(120) cyclic pathway of ATP formation _____

_____

(120) noncyclic pathway of ATP formation _____

_____

(121) photolysis _____

_____

## Complete the Table

1. Complete the following table on elements of the cyclic pathway of the light-dependent reactions.

| Component | Function |
|---|---|
| a. Photosystem I | |
| b. Electrons | |
| c. P700 | |
| d. Electron acceptor | |
| e. Electron transport system | |
| f. ADP | |

## Labeling

The diagram below illustrates noncyclic photophosphorylation. Identify each numbered part of the illustration.

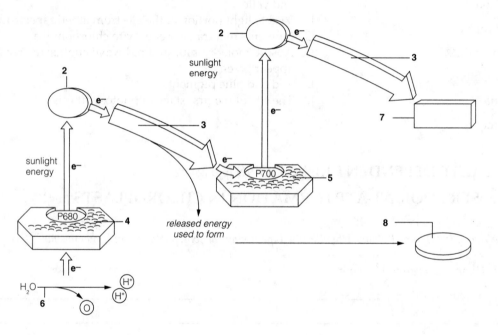

2. _____ _____

3. _____ _____ _____

4. _____ _____

5. _____ _____

6. _____

7. _____

8. _____

## Fill-in-the-Blanks

ATP forms in both the cyclic and noncyclic pathways. When (9) _____ flow through the membrane-bound transport systems, they pick up hydrogen ions ($H^+$) outside the membrane and dump them into the (10) _____ compartment. This sets up $H^+$ concentration and electric (11) _____ across the membrane. Hydrogen ions that were split away from (12) _____ molecules increase the gradients. The ions respond by flowing out through the interior of (13) _____ _____ proteins that span the membrane. Energy associated with the flow drives the binding of unbound phosphate to ADP, the result being (14) _____ . The above description is known as the (15) _____ theory of ATP formation.

The noncyclic pathway also produces (16) _____ by using (17) _____ from water and $H^+$ ions from the thylakoid compartment to reduce $NADP^+$.

## Complete the Table Below

With a check mark (√) indicate for each phase of the light-dependent reactions all items from the left-hand column that are applicable.

| Light-Dependent Reactions | Cyclic Pathway | Noncyclic Pathway | Photolysis Alone |
|---|---|---|---|
| Uses $H_2O$ as a reactant | (18) | (30) | (42) |
| Produces $H_2O$ as a product | (19) | (31) | (43) |
| Photosystem I involved (P 700) | (20) | (32) | (44) |
| Photosystem II involved (P 680) | (21) | (33) | (45) |
| ATP produced | (22) | (34) | (46) |
| NADPH produced | (23) | (35) | (47) |
| Uses $CO_2$ as a reactant | (24) | (36) | (48) |
| Causes $H^+$ to be pumped into the thylakoid compartments from the stroma | (25) | (37) | (49) |
| Produces $O_2$ as a product | (26) | (38) | (50) |
| Produces $H^+$ by breaking apart $H_2O$ | (27) | (39) | (51) |
| Uses ADP and $P_i$ as reactants | (28) | (40) | (52) |
| Uses $NADP^+$ as a reactant | (29) | (41) | (53) |

## 7.6. LIGHT-INDEPENDENT REACTIONS (p. 123)

## 7.7. FIXING CARBON—SO NEAR, YET SO FAR (pp. 124–125)

## 7.8. *Focus on the Environment:* AUTOTROPHS, HUMANS, AND THE BIOSPHERE (p. 126)

*Selected Words:* *"synthesis"*, *photorespiration, mesophyll cells, bundle-sheath cells*

### Boldfaced, Page-Referenced Terms

(123) light-independent reactions _____

_____

(123) RuBP (ribulose bisphosphate) _____

_____

(123) PGA (phosphoglycerate) _____

_____

(123) carbon fixation _____

_____

(123) Calvin-Benson cycle _____

_____

(123) PGAL (phosphoglyceraldehyde) _____

_____

(124) stomata (singular, stoma) _____

_____

(124) C3 plants _____

_____

(124) C4 plants _____

_____

(125) CAM plants _____

_____

(126) chemoautotrophs _____

_____

## Label and Match

Identify each part of the illustration below. Complete the exercise by matching and entering the letter of the proper function description in the parentheses following each label.

1. _____ _____ ( )

2. _____ _____ _____ ( )

3. _____ ( )

4. _____ _____ ( )

5. _____ ( )

6. _____ ( )

7. _____ _____ ( )

8. _____ _____ _____ ( )

9. _____ _____ ( )

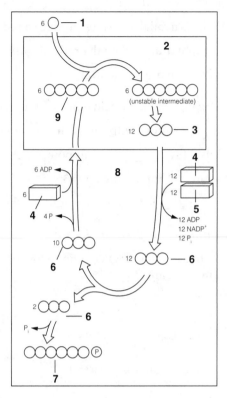

A. A three-carbon sugar, the first sugar produced; goes on to form sugar phosphate and RuBP

B. Typically used at once to form carbohydrate end products of photosynthesis

C. A five-carbon compound produced from PGALs; attaches to incoming $CO_2$

D. A compound that diffuses into leaves; attached to RuBP by enzymes in photosynthetic cells

E. Includes all the chemical reactions that "fix" carbon into an organic compound

F. Three-carbon compounds formed from the splitting of the six-carbon intermediate compound

G. A molecule that was reduced in the noncyclic pathway; furnishes hydrogen atoms to construct sugar molecules

H. A product of the light-dependent reactions; necessary in the light-independent reactions to energize molecules in metabolic pathways.

I. Includes all the chemistry that fixes $CO_2$; converts PGA to PGAL and PGAL to RuBP and sugar phosphates

## Fill-in-the-Blanks

The light-independent reactions can proceed without sunlight as long as (10) _____ and (11) _____ are available. The reactions begin when an enzyme links (12) _____ _____ to (13) _____ _____ , a five-carbon compound. The resulting six-carbon compound is highly unstable and breaks apart at once into two molecules of a three-carbon compound, (14) _____ . This entire reaction sequence is called carbon dioxide (15) _____ . ATP gives a phosphate group to each (16) _____ . This intermediate compound takes on $H^+$ and electrons from NADPH to form (17) _____ . It takes (18) _____ carbon dioxide molecules to produce twelve PGAL. Most of the PGAL becomes rearranged into new (19) _____ molecules—which can be used to fix more (20) _____ _____ . Two (21) _____ are joined together to form a (22) _____ _____ , primed for further reactions. The Calvin-Benson cycle yields enough RuBP to replace those used in carbon dioxide (23) _____ . ADP, NADP$^+$, and phosphate leftovers are sent back to the (24) _____ - _____ reaction sites, where they are again converted to (25) _____ and (26) _____ . (27) _____ _____ formed in the cycle serves as a building block for the plant's main carbohydrates. When RuBP attaches to oxygen instead of carbon dioxide, (28) _____ results; this is typical of (29) _____ plants in hot, dry conditions. If less PGA is available, leaves produce a reduced amount of (30) _____ . C4 plants can still construct carbohydrates when the ratio of carbon dioxide to (31) _____ is unfavorable because of the attachment of carbon dioxide to (32) _____ in certain leaf cells.

Organisms that obtain energy from oxidation of inorganic substances such as ammonium compounds, and iron or sulfur compounds, are known as (33) _____ autotrophs. Such organisms use this energy to build (34) _____ compounds. As an example, some soil bacteria use ammonia molecules as an energy source, stripping them of (35) _____ and (36) _____ ; this leaves (37) _____ and (38) _____ ions that are readily washed out of the soil, thus lowering its (39) _____ .

## Complete the Table Below

With a check mark (√) indicate for each phase of the light-dependent reaction all items from the left-hand column that are applicable.

| Light-Independent Reactions | $CO_2$ Fixation Alone | Conversion of PGA to PGAL | Regeneration of RuBP | Formation of Glucose and other Organic Compounds |
|---|---|---|---|---|
| Requires RuBP as a reactant | (40) | (51) | (62) | (73) |
| Requires ATP as a reactant | (41) | (52) | (63) | (74) |
| Produces ADP as a product | (42) | (53) | (64) | (75) |
| Requires NADPH as a reactant | (43) | (54) | (65) | (76) |
| Produces NADP+ as a reactant | (44) | (55) | (66) | (77) |
| Produces PGA | (45) | (56) | (67) | (78) |
| Produces PGAL | (46) | (57) | (68) | (79) |
| Requires PGAL as a reactant | (47) | (58) | (69) | (80) |
| Produces $P_i$ as a product | (48) | (59) | (70) | (81) |
| Produces $H_2O$ as a product | (49) | (60) | (71) | (82) |
| Requires $CO_2$ as a reactant | (50) | (61) | (72) | (83) |

# Self-Quiz

___ 1. The electrons that are passed to NADPH during noncyclic photophosphorylation were obtained from _____ .
a. water
b. $CO_2$
c. glucose
d. sunlight

___ 2. The cyclic pathway of the light-dependent reactions functions mainly to _____ .
a. fix $CO_2$
b. make ATP
c. produce PGAL
d. regenerate ribulose bisphosphate

___ 3. Chemosynthetic autotrophs obtain energy by oxidizing such inorganic substances as _____ .
a. PGA
b. PGAL
c. sulfur
d. water

___ 4. The ultimate electron and hydrogen acceptor in noncyclic photophosphorylation is _____ .
a. $NADP^+$
b. ADP
c. $O_2$
d. $H_2O$

___ 5. C4 plants have an advantage in hot, dry conditions because _____ .
a. their leaves are covered with thicker wax layers than those of C3 plants
b. their stomates open wider than those of C3 plants, thus cooling their surfaces
c. special leaf cells possess a means of capturing $CO_2$ even in stress conditions
d. they also are capable of carrying on photorespiration

___ 6. Chlorophyll is _____ .
a. on the outer chloroplast membrane
b. inside the mitochondria
c. in the stroma
d. in the thylakoid membrane system

___ 7. Which of the following is applicable to C3 plants?
a. At the end of carbon fixation, the intermediate compound is PGA
b. At the end of carbon fixation, the intermediate compound is oxaloacetate
c. They are more sensitive to cold temperatures than are C4 plants
d. Corn, crabgrass, and sugarcane are examples of C3 plants.

___ 8. Plant cells produce $O_2$ during photosynthesis by _____ .
a. splitting $CO_2$
b. splitting water
c. degradation of the stroma
d. breaking up sugar molecules

___ 9. Plants need _____ and _____ to carry on photosynthesis.
a. oxygen; water
b. oxygen; $CO_2$
c. $CO_2$; $H_2O$
d. sugar; water

___10. The two products of the light-dependent reactions that are required for the light-independent chemistry are _____ and _____ .
a. $CO_2$; $H_2O$
b. $O_2$; NADPH
c. $O_2$; ATP
d. ATP; NADPH

# Chapter Objectives/Review Questions

*Page*      *Objectives/Questions*

(112)      1. When carbon fixation occurs, sunlight energy, which excited electrons during the light-dependent reactions, becomes stored as _____ energy in an organic compound.
(113)      2. Distinguish between organisms known as autotrophs and those known as heterotrophs.
(113)      3. _____ get energy by stripping electrons from sulfur or some other inorganic substance.

(113)    4.  Study the general equation for photosynthesis until you can remember the reactants and products. Reproduce the equation from memory on another piece of paper.

(114–115) 5.  List the two major stages of photosynthesis as well as where in the cell they occur and what reactions occur there.

(114–115) 6.  Describe the details of a familiar site of photosynthesis, the green leaf. Begin with the layers of a leaf cross-section and complete your description with the minute structural sites within the chloroplast where the major sets of photosynthetic reactions occur.

(114–115) 7.  The flattened channels and disklike compartments inside the chloroplast are organized into stacks, the _____ , which are surrounded by a semifluid interior, the _____ ; this is the _____ membrane system.

(115)    8.  The energy-poor molecules that act as raw materials in the photosynthetic equation are _____ and _____ .

(116)    9.  _____ are packets of light energy.

(117)    10.  State what T. Englemann's 1882 experiment with *Spirogyra* revealed.

(118)    11.  _____ absorb violet-to-blue as well as red wavelengths.

(118–119) 12.  The main pigment of photosynthesis is _____ .

(118)    13.  What pigments are responsible for the red and blue colors of red algae and cyanobacteria?

(118)    14.  Name the wavelengths absorbed and transmitted by the carotenoids.

(119)    15.  Describe how the pigments found on thylakoid membranes are organized into photosystems and how they relate to photon light energy.

(120–121) 16.  Contrast the components and functioning of the cyclic and noncyclic pathways of the light-dependent reactions.

(120–121) 17.  Explain what the water split during photolysis contributes to the noncyclic pathway of the light-dependent reactions.

(120–121) 18.  Two energy-carrying molecules produced in the noncyclic pathways are _____ and _____ ; explain why these molecules are necessary for the light-independent reactions.

(121)    19.  _____ _____ is the metabolic pathway most efficient in releasing the energy stored in organic compounds.

(121)    20.  Following evolution of the noncyclic pathway, _____ accumulated in the atmosphere and made _____ respiration possible.

(122)    21.  Explain how the chemiosmotic theory is related to thylakoid compartments and the production of ATP.

(123)    22.  Explain why the light-independent reactions are called by that name.

(123)    23.  Describe the process of carbon dioxide fixation by stating which reactants are necessary to initiate the process and what stable products result from this process.

(123)    24.  Describe the Calvin-Benson cycle in terms of its reactants and products.

(123)    25.  State the fate of all the sugar phosphates produced by photosynthetic reactions in photoautotrophs.

(124–125) 26.  Describe the mechanism by which C4 plants thrive under hot, dry conditions; distinguish this $CO_2$-capturing mechanism from that of C3 plants.

(125)    27.  Describe the carbon-fixing adaptation of the CAM plants living in arid environments.

(126)    28.  Bacteria that are able to obtain energy from ammonium ions, iron, or sulfur compounds are known as _____ .

---

# Integrating and Applying Key Concepts

Suppose that humans acquired all the enzymes needed to carry out photosynthesis. Speculate about the attendant changes in human anatomy, physiology, and behavior that would be necessary for those enzymes actually to carry out photosynthetic reactions.

# 8

# ENERGY-RELEASING PATHWAYS

## Interactive Exercises

*The Killers Are Coming!* (pp. 130–131)

## 8.1. HOW CELLS MAKE ATP (pp. 132–133)

*Selected Words: anaerobic*

*Boldfaced, Page-Referenced Terms*

(132) aerobic respiration _____

_____

(132) glycolysis _____

_____

(133) Krebs cycle _____

_____

(133) electron transport phosphorylation _____

_____

## Short Answer

1. Although various organisms utilize different energy sources, what is the usual form of chemical energy that will drive metabolic reactions? _____

_____

2. Describe the function of oxygen in the main degradative pathway, aerobic respiration. _____

_____

_____

3. List the most common anaerobic pathways and describe the conditions in which they function.

_____

## Fill-in-the-Blanks

Virtually all forms of life depend on a molecule known as (4) _____ as their primary energy carrier. Plants produce adenosine triphosphate during (5) _____ , but plants and all other organisms also can produce ATP through chemical pathways that degrade (take apart) food molecules. The main degradative pathway requires free oxygen and is called (6) _____ _____ .

There are three stages of aerobic respiration. In the first stage, (7) _____ , glucose is partially degraded to (8) _____ . By the end of the second stage, which includes the (9) _____ cycle, glucose has been completely degraded to carbon dioxide and (10) _____ . Neither of the first two stages produces much (11) _____ . During both stages, protons and (12) _____ are stripped from intermediate compounds and delivered to a (13) _____ system. That system is used in the third stage of reactions, electron transport (14) _____ ; passage of electrons along the transport system drives the enzymatic "machinery" that phosphorylates ADP to produce a high yield of (15) _____ . (16) _____ accepts "spent" electrons from the transport system and keeps the pathway clear for repeated ATP production.

Other degradative pathways are (17) _____ , in that something other than oxygen serves as the final electron acceptor in energy-releasing reactions. (18) _____ and anaerobic (19) _____ _____ are the most common anaerobic pathways.

## Completion

20. Complete the equation below, which summarizes the degradative pathway known as aerobic respiration:

$$\text{_____} + \text{_____} \; O_2 \rightarrow 6 \text{_____} + 6 \text{_____}$$

21. Supply the appropriate information to state the equation (above) for aerobic respiration in words:

One molecule of glucose plus six molecules of _____ (in the presence of appropriate enzymes) yield _____ molecules of carbon dioxide plus _____ molecules of water.

## Matching

Choose the one most appropriate answer for each.

22. ___Krebs cycle

23. ___oxygen

24. ___mitochondrion

25. ___electron transport phosphorylation

26. ___enzymes

27. ___ATP

28. ___glycolysis

29. ___aerobic respiration

30. ___cytoplasm

31. ___fermentation pathways and anaerobic electron transport

A. Starting point for three energy-releasing pathways
B. Main energy-releasing pathway for ATP formation
C. Site of glycolysis
D. Third and final stage of aerobic respiration; high ATP yield
E. Oxygen is not the final electron acceptor
F. Catalyze each reaction step in the energy-releasing pathways
G. Second stage of aerobic respiration; pyruvate is broken down to $CO_2$ and $H_2O$
H. Site of the aerobic pathway
I. The final electron acceptor in aerobic pathways
J. The energy form that drives metabolic reactions

## 8.2. GLYCOLYSIS: FIRST STAGE OF THE ENERGY-RELEASING PATHWAYS
(pp. 134–135)

*Selected Words:* *energy-requiring steps, energy-releasing steps, net energy*

### Boldfaced, Page-Referenced Terms

(134) pyruvate _____

_____

(134) substrate-level phosphorylation _____

_____

(134) NAD+ _____

_____

### Fill-in-the-Blanks

(1) _____ organisms can synthesize and stockpile energy-rich carbohydrates and other food molecules from inorganic raw materials. (2) _____ is partially dismantled by the glycolytic pathway; at the end of this process some of its stored energy remains in two (3) _____ molecules. Some of the energy of glucose is released during the breakdown reactions and used in forming the coenzymes (4) _____ and (5) _____ . These reactions take place in the cytoplasm. Glycolysis begins with two phosphate groups being transferred to (6) _____ from two (7) _____ molecules. The addition of two phosphate groups to (6) energizes it and causes it to become unstable and split apart, forming two molecules of (8) _____ . Each (8) gains one (9) _____ group from the cytoplasm, then (10) _____ atoms and electrons from each PGAL are transferred to NAD+, changing this coenzyme to NADH. Two (11) _____

molecules form by substrate-level phosphorylation; the cell's energy investment is paid off. One (12) molecule is released from each 2-PGA as a waste product. The resulting intermediates are rather unstable; each gives up a(n) (13) _____ group to ADP. Once again, two (14) _____ molecules have formed by (15) _____ - _____ phosphorylation. For each (16) _____ molecule entering glycolysis, the net energy yield is two ATP molecules that the cell can use anytime to do work. The end products of glycolysis are two molecules of (17) _____ , each with a (18) _____ -carbon backbone.

(number)

### Sequence

Arrange the following events of the glycolysis pathway in correct chronological sequence. Write the letter of the first step next to 19, the letter of the second step next to 20, and so on.

19. ___
20. ___
21. ___
22. ___
23. ___
24. ___
25. ___
26. ___

A. Two ATPs form by substrate-level phosphorylation; the cell's energy debt is paid off
B. Diphosphorylated glucose (fructose 1,6-bisphosphate) molecules split to form 2 PGALs; this is the first energy-releasing step
C. Two 3-carbon pyruvate molecules form as the end products of glycolysis
D. Glucose is present in the cytoplasm
E. Two more ATPs form by substrate-level phosphorylation, the cell gains ATP; net yield of ATP from glycolysis is 2 ATPs
F. The cell invests two ATPs; one phosphate group is attached to each end of the glucose molecule (fructose 1,6-bisphosphate)
G. Two PGALs gain two phosphate groups from the cytoplasm
H. Hydrogen atoms and electrons from each PGAL are transferred to $NAD^+$, reducing this carrier to NADH

## 8.3. SECOND STAGE OF THE AEROBIC PATHWAY (pp. 136–137)

*Selected Words:* coenzyme

*Boldfaced, Page-Referenced Terms*

(136) mitochondrion _____

_____

(136) acetyl-CoA _____

_____

(136) oxalacetate _____

_____

(136) FAD _____

_____

## Fill-in-the-Blanks

If sufficient oxygen is present, the end product of glycolysis enters a preparatory step, (1) _____

_____ formation. This step converts pyruvate into acetyl CoA, the molecule that enters the

(2) _____ cycle, which is followed by (3) _____ _____ phosphorylation. During these three

processes, (4) _____ (number) additional (5) _____ molecules are generated. In the preparatory

conversions prior to the Krebs cycle and within the Krebs cycle, the food molecule fragments are further

broken down into (6) _____ _____ . During these reactions, hydrogen atoms (with their

(7) _____ ) are stripped from the fragments and transferred to the energy carriers (8) _____ and

(9) _____ .

## Labeling

In exercises 10–14, identify the structure or location; in exercises 15–18, identify the chemical substance involved. In exercise 19, name the metabolic pathway.

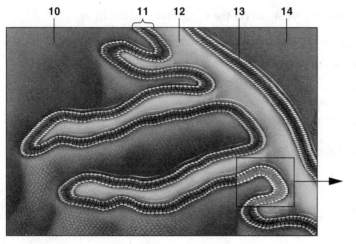

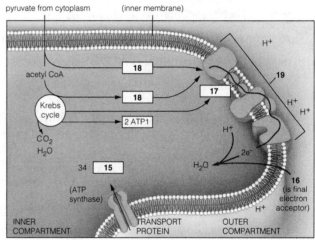

10. _____ _____ of mitochondrion

11. _____ _____ of mitochondrion

12. _____ _____ of mitochondrion

13. _____ _____ of mitochondrion

14. _____

15. _____

16. _____

17. _____

18. _____

19. _____ _____ _____

## 8.4. THIRD STAGE OF THE AEROBIC PATHWAY (pp. 138–139)

*Selected Words:* *electron transport phosphorylation, chemiosmotic theory*

### Fill-in-the-Blanks

NADH delivers its electrons to the highest possible point of entry into a transport system; from each NADH enough $H^+$ is pumped to produce (1) _____ (number) ATP molecules. $FADH_2$ delivers its electrons at a lower point of entry into the transport system; fewer $H^+$ are pumped, and (2) _____ (number) ATPs are produced. The electrons are then sent down highly organized (3) _____ systems located in the inner membrane of the mitochondrion; hydrogen ions are pumped into the outer mitochondrial compartment. According to (4) _____ theory, the hydrogen ions accumulate and then follow a gradient to flow through channel proteins, called ATP (5) _____ , that lead into the inner compartment. The energy of the hydrogen ion flow across the membrane is used to phosphorylate ADP to produce (6) _____ . Electrons leaving the electron transport system combine with hydrogen ions and (7) _____ to form water. These reactions occur only in (8) _____ . From glycolysis (in the cytoplasm) to the final reactions occurring in the mitochondria, the aerobic pathway commonly yields (9) _____ (number) ATP or (10) _____ (number) ATP for every glucose molecule degraded.

### Choice

For questions 11–32, choose from the following. Some correct answers may require more than one letter.

a. preparatory steps     to     Krebs cycle      b. Krebs cycle      c. electron transport phosphorylation

11.____Chemiosmosis occurs to form ATP molecules

12.____Three carbon atoms in pyruvate leave as three $CO_2$ molecules

13.____Chemical reactions occur at transport systems

14.____Coenzyme A picks up a 2-carbon acetyl group

15.____Makes two turns for each glucose molecule entering glycolysis

16.____Two NADH molecules form for each glucose entering glycolysis

17.____Oxaloacetate forms from intermediate molecules

18.____Named for a scientist who worked out its chemical details

19.____Occurs within the mitochondrion

20.____Two $FADH_2$ and six NADH form from one glucose molecule entering glycolysis

21.____Hydrogens collect in the mitochondrion's outer compartment

22.____Hydrogens and electrons are transferred to $NAD^+$ and FAD

23.____Two ATP molecules form by substrate-level phosphorylation

24.____Free oxygen withdraws electrons from the system and then combines with $H^+$ to form water molecules

25.____No ATP is produced

26.____Thirty-two ATPs are produced

27.___Delivery point of NADH and $FADH_2$

28.___Two pyruvates enter for each glucose molecule entering glycolysis

29.___The carbons in the acetyl group leave as $CO_2$

30.___One carbon in pyruvate leaves as $CO_2$

31.___An electron transport system and channel proteins are involved

32.___Two $FADH_2$ and ten NADH are sent to this stage

## 8.5. ANAEROBIC ROUTES OF ATP FORMATION (pp. 140–141)

*Selected Words: pyruvate, "lactic acid", lactate,* Lactobacillus, Saccharomyces cerevisiae, Saccharomyces ellipsoideus

### Boldfaced, Page-Referenced Terms

(140) lactate fermentation _____

_____

(140) alcoholic fermentation _____

_____

(141) anaerobic electron transport _____

_____

### Fill-in-the-Blanks

If (1) _____ is not present in sufficient amounts, the end product of glycolysis enters (2) _____ pathways; in some bacteria and muscle cells, pyruvate is converted into such products as (3) _____ , or in yeast cells it is converted into (4) _____ and (5) _____ _____ .

(6) _____ pathways do not use oxygen as the final (7) _____ acceptor that ultimately drives the ATP-forming machinery. Anaerobic routes must be used by many bacteria and protistans that live in an oxygen-free environment. (8) _____ precedes any of the fermentation pathways. During (8), a glucose molecule is split into two (9) _____ molecules, two energy-rich (10) _____ intermediate molecules form, and the net energy yield from one glucose molecule is two ATP.

In one kind of fermentation pathway, (11) _____ itself accepts hydrogen and electrons from NADH. (11) is then converted to a three-carbon compound, (12) _____ , during this process in a few bacteria species and some animal cells. Human muscle cells can carry on (12) fermentation in times of oxygen depletion; this provides a low yield of ATP.

In yeast cells, each pyruvate molecule from glycolysis forms an intermediate called (13) _____ as a gas (14) _____ _____ is detached from pyruvate with the help of an enzyme. This intermediate accepts hydrogen and electrons from NADH and is then converted to (15) _____ , the end product of alcoholic fermentation.

In both types of fermentation pathways, the net energy yield of two ATPs is formed during (16) _____ . The reactions of the fermentation chemistry regenerate the (17) _____ needed for glycolysis to occur. Anaerobic electron transport is an energy-releasing pathway occurring among the (18) _____ . For example, sulfate-reducing bacteria living in soil or water produce (19) _____ by stripping electrons from a variety of compounds and sending them through membrane transport systems. The inorganic compound (20) _____ ($SO_4^=$) serves as the final electron acceptor and is converted into foul-smelling hydrogen sulfide gas ($H_2S$). Other kinds of bacteria produce ATP by stripping electrons from nitrate ($NO_3^-$), leaving (21) _____ ($NO_2^-$) as the end product. These bacteria are important in the global cycling of (22) _____ .

## Complete the Table

Include a (√) in each box that correctly links an occurrence (left-hand column) with a process (or processes).

| | Glycolysis | Lactate Fermentation | Alcoholic Fermentation | Anaerobic Electron Transport |
|---|---|---|---|---|
| 6-C → 3C / 3C | (23) | (36) | (49) | (62) |
| $NADH \rightarrow NAD^+$ | (24) | (37) | (50) | (63) |
| $NAD^+ \rightarrow NADH$ | (25) | (38) | (51) | (64) |
| $SO_4^= \rightarrow H_2S$ | (26) | (39) | (52) | (65) |
| $NO_3^- \rightarrow NO_2^-$ | (27) | (40) | (53) | (66) |
| $CO_2$ is a waste product | (28) | (41) | (54) | (67) |
| 3-C → 3-C | (29) | (42) | (55) | (68) |
| 3-C → 2-C | (30) | (43) | (56) | (69) |
| ATP is used as a reactant | (31) | (44) | (57) | (70) |
| ATP is produced | (32) | (45) | (58) | (71) |
| pyruvate → ethanol / $CO_2$ | (33) | (46) | (59) | (72) |
| Occurs in animal cells | (34) | (47) | (60) | (73) |
| Occurs in yeast cells | (35) | (48) | (61) | (74) |

## 8.6. ALTERNATIVE ENERGY SOURCES IN THE HUMAN BODY (pp. 142–143)

## 8.7. COMMENTARY: PERSPECTIVE ON LIFE (p. 144)

### True-False

If the statement is true, write a T in the blank. If the statement is false, explain why in the answer blank.

_____1. Glucose is the only carbon-containing molecule that can be fed into the glycolytic pathway.

_____2. Simple sugars, fatty acids, and glycerol that remain after a cell's biosynthetic and storage needs have been met are generally sent to the cell's respiratory pathways for energy extraction.

_____3. Carbon dioxide and water, the products of aerobic respiration, generally get into the blood and are carried to gills or lungs, kidneys, and skin, where they are expelled from the animal's body.

_____4. Energy is recycled along with materials.

_____5. The first forms of life on Earth were most probably photosynthetic eukaryotes.

### Labeling

Identify the process or substance indicated in the illustration below.

6. _____  _____

7. _____

8. _____

9. _____  _____

10. _____  _____

11. _____  _____

12. _____

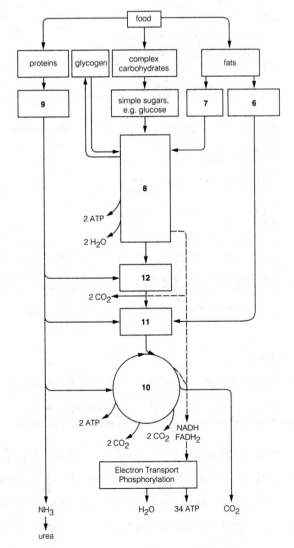

## Complete this Table

Across the top of each column are the principal phases of degradative pathways into which food molecules (in various stages of breakdown) enter or in which specific events occur. Put a check (√) in each box that indicates the phase into which a specific food molecule is fed, or in which a specific event occurs. For example: if simple sugars can enter the glycolytic pathway, put a check in the top left hand box; if not, let the box remain blank.

| | Glycolysis (includes pyruvate) | Acetyl CoA Formation | Krebs Cycle | Electron Transport Phosphorylation | Fermentation Alcohol | Lactate |
|---|---|---|---|---|---|---|
| Complex Carbohydrates → Simple Sugars, which enter | (13) | (28) | (43) | (58) | (73) | (88) |
| Fats → Fatty Acids which enter | (14) | (29) | (44) | (59) | (74) | (89) |
| Glycerol, which enters | (15) | (30) | (45) | (60) | (75) | (90) |
| Proteins → Amino Acids, which enter | (16) | (31) | (46) | (61) | (76) | (91) |
| Intermediate energy carriers (NADH) are produced | (17) | (32) | (47) | (62) | (77) | (92) |
| (FADH$_2$) | (18) | (33) | (48) | (63) | (78) | (93) |
| ATPs produced directly as a result of this process alone | (19) | (34) | (49) | (64) | (79) | (94) |
| NAD⁺ produced (20) | (35) | (50) | (65) | (80) | (95) | |
| FAD produced | (21) | (36) | (51) | (66) | (81) | (96) |
| ADP produced | (22) | (37) | (52) | (67) | (82) | (97) |
| Unbound phosphate (P$_i$) required | (23) | (38) | (53) | (68) | (83) | (98) |
| CO$_2$ produced (waste product) | (24) | (39) | (54) | (69) | (84) | (99) |
| H$_2$O produced (waste product) | (25) | (40) | (55) | (70) | (85) | (100) |
| Atoms here for O$_2$ react | (26) | (41) | (56) | (71) | (86) | (101) |
| NADH required to drive this process | (27) | (42) | (57) | (72) | (87) | (102) |

## Choice

For questions 103–117, refer to the text and Figure 8.12; choose from the following:

a. glucose     b. glucose-6-phosphate     c. glycogen     d. fatty acids     e. triglycerides
f. PGAL     g. acetyl-CoA     h. amino acids     i. glycerol     j. proteins

103. ___Fats that are tapped between meals or during exercise as alternatives to glucose

104. ___Used between meals when free glucose supply dwindles; enters glycolysis after conversion

105. ___Its breakdown yields much more ATP than glucose

106. ___Absorbed in large amounts immediately following a meal

107. ___Represents only 1 percent or so of the total stored energy in the body

108. ___Following removal of amino groups, the carbon backbones may be converted to fats or carbohydrates or they may enter the Krebs cycle

109. ___On the average, represents 78 percent of the body's stored food

110. ___Between meals liver cells can convert it back to free glucose and release it

111. ___Can be stored in cells but not transported across plasma membranes

112. ___Amino groups undergo conversions that produce urea, a nitrogen-containing waste product excreted in urine

113. ___Converted to PGAL in the liver, a key intermediate of glycolysis

114. ___Accumulate inside the fat cells of adipose tissues, at strategic points under the skin

115. ___A storage polysaccharide produced from glucose-6-phosphate following food intake that exceeds cellular energy demand (and increases ATP production to inhibit glycolysis)

116. ___Building blocks of the compounds that represent 21 percent of the body's stored food

117. ___A product resulting from enzymes cleaving circulating fatty acids; enters the Krebs cycle

---

## Self-Quiz

___ 1. Glycolysis would quickly halt if the process ran out of _____ , which serves as the hydrogen and electron acceptor.
   a. NADP$^+$
   b. ADP
   c. NAD$^+$
   d. H$_2$O

___ 2. The ultimate electron acceptor in aerobic respiration is _____ .
   a. NADH
   b. carbon dioxide (CO$_2$)
   c. oxygen (1/2 O$_2$)
   d. ATP

___ 3. When glucose is used as an energy source, the largest amount of ATP is generated by the _____ portion of the entire respiratory process.
   a. glycolytic pathway
   b. acetyl-CoA formation
   c. Krebs cycle
   d. electron transport phosphorylation

___ 4. The process by which about 10 percent of the energy stored in a sugar molecule is released as it is converted into two small organic-acid molecules is _____ .
   a. photolysis
   b. glycolysis
   c. fermentation
   d. the dark reactions

___ 5. During which of the following phases of aerobic respiration is ATP produced directly by substrate-level phosphorylation?
   a. glucose formation
   b. ethanol production
   c. acetyl-CoA formation
   d. glycolysis

___ 6. What is the name of the process by which reduced NADH transfers electrons along a chain of acceptors to oxygen so as to form water and in which the energy released along the way is used to generate ATP?
   a. glycolysis
   b. acetyl-CoA formation
   c. the Krebs cycle
   d. electron transport phosphorylation

___ 7. Pyruvic acid can be regarded as the end product of _____ .
   a. glycolysis
   b. acetyl-CoA formation
   c. fermentation
   d. the Krebs cycle

___ 8. Which of the following is *not* ordinarily capable of being reduced at any time?
   a. NAD$^+$
   b. FAD
   c. oxygen, O$_2$
   d. water

___ 9. ATP production by chemiosmosis involves
_____ .

    a. $H^+$ concentration and electric gradients across a membrane
    b. ATP synthases
    c. formation of ATP in the inner mitochondrial compartment
    d. all of the above

___10. During the fermentation pathways, a net yield of two ATP is produced from _____ ; the $NAD^+$ necessary for _____ is regenerated during the reactions.

    a. the Krebs cycle; glycolysis
    b. glycolysis; electron transport phosphorylation
    c. the Krebs cycle; electron transport phosphorylation
    d. glycolysis; glycolysis

## Matching

Match the following components of respiration to the list of words below. Some components may have more than one answer.

11. ___lactic acid, lactate

12. ___$NAD^+ \rightarrow NADH$

13. ___carbon dioxide is a product

14. ___$NADH \rightarrow NAD^+$

15. ___pyruvic acid (pyruvic acid), used as a reactant

16. ___ATP produced by substrate-level phosphorylation

17. ___glucose

18. ___acetyl-CoA is both a reactant and a product

19. ___oxygen

20. ___water is a product

    A. Glycolysis
    B. Preparatory conversions prior to the Krebs cycle
    C. Fermentation
    D. Krebs cycle
    E. Electron transport

# Chapter Objectives/Review Questions

This section lists general and detailed chapter objectives that can be used as review questions. You can make maximum use of these items by writing answers on a separate sheet of paper. Fill in answers where blanks are provided. To check for accuracy, compare your answers with information given in the chapter or glossary.

Page      Objectives/Questions

(132)    1. No matter what the source of energy may be, organisms must convert it to _____ , a form of chemical energy that can drive metabolic reactions.
(132)    2. The main energy-releasing pathway is _____ respiration.
(132)    3. Give the overall equation for the aerobic respiratory route.
(132)    4. In the first of the three stages of aerobic respiration, one _____ is partially degraded to two pyruvate.
(133)    5. By the end of the second stage of aerobic respiration, which includes the _____ cycle, _____ has been completely degraded to carbon dioxide and water.
(133)    6. Do the first two stages of aerobic respiration yield a high or low quantity of ATP?
(133)    7. The third stage of aerobic respiration is called electron transport _____ , which yields many ATP molecules.
(133)    8. Explain, in general terms, the role of oxygen in aerobic respiration.
(134)    9. Glycolysis occurs in the _____ of the cell.

(134)    10.   Explain the purpose served by the cell investing two ATP molecules into the chemistry of glycolysis.

(134)    11.   State the factors responsible for pyruvic acid entering the acetyl-CoA formation pathway.

(134-135)   12.   Four ATP molecules are produced by _____ - _____ phosphorylation for every two used during glycolysis. (Consult Figure 8.4 in the main text).

(135)    13.   Glycolysis produces _____ (number) NADH, _____ (number) ATP (net) and _____ (number) pyruvate molecules for each glucose molecule entering the reactions.

(136)    14.   Consult Figure 8.5 in the main text. Relate the events that happen during acetyl-coenzyme A formation and explain how the process of acetyl-CoA formation relates glycolysis to the Krebs cycle.

(136)    15.   What happens to the $CO_2$ produced during acetyl-CoA formation and the Krebs cycle?

(138)    16.   Explain how chemiosmosis theory operates in the mitochondrion to account for the production of ATP molecules.

(137)    17.   Calculate the number of ATP molecules produced during the Krebs cycle for each glucose molecule that enters glycolysis.

(138)    18.   Consult Figure 8.8 in the main text and predict what will happen to the NADH produced during acetyl-CoA formation and the Krebs cycle.

(138)    19.   Briefly describe the process of electron transport phosphorylation by stating what reactants are needed and what the products are. State how many ATP molecules are produced through operation of the transport system.

(139)    20.   Be able to account for the total *net yield* of thirty-six ATP molecules produced through aerobic respiration; that is, state how many ATPs are produced in glycolysis, the Krebs cycle, and electron transport system.

(140)    21.   List some environments where there is very little oxygen present and where anaerobic organisms might be found.

(140)    22.   In fermentation chemistry _____ molecules from glycolysis are accepted to construct either lactate or ethyl alcohol; thus, a low yield of _____ molecules continues in the absence of oxygen.

(140-141)   23.   List the main anaerobic energy-releasing pathways and the examples of organisms that use them.

(140-141)   24.   Describe what happens to pyruvate in anaerobic organisms. Then explain the necessity for pyruvate to be converted to a fermentative product.

(140-141)   25.   State which factors determine whether the pyruvate (pyruvic acid) produced at the end of glycolysis will enter into the alcoholic fermentation pathway, the lactate fermentation pathway, or the acetyl-coenzyme A formation pathway.

(142-143)   26.   List some sources of energy (other than glucose) that can be fed into the respiratory pathways.

(142-143)   27.   Predict what your body would do to synthesize its needed carbohydrates and fats if you switched to a diet of 100 percent protein.

(144)    28.   After reading "Perspective on Life" in the main text, outline the supposed evolutionary sequence of energy-extraction processes.

(144)    29.   Scrutinize the sketch in the Commentary in the main text closely; then reproduce the carbon cycle from memory.

# Integrating and Applying Key Concepts

How is the "oxygen debt" experienced by runners and sprinters related to aerobic and anaerobic respiration in humans?

# 9

# CELL DIVISION AND MITOSIS

## Interactive Exercises

Note: In the answer sections of this book, a specific molecule is most often indicated by its abbreviation. For example, adenosine triphosphate is ATP.

*Silver In the Stream of Time* (pp. 148–149)

## 9.1. DIVIDING CELLS: THE BRIDGE BETWEEN GENERATIONS (p. 150)
## 9.2. THE CELL CYCLE (p. 151)

*Selected Words:* *cell division, nuclear* division, *2n, haploid* chromosome number, *n*, G1, S, G2, M, Haemanthus

In addition to the boldfaced terms, the text features other important terms essential to understanding the assigned material. "Selected Words" is a list of these terms, which appear in the text in italics, in quotation marks, and occasionally in roman type. Latin binomials found in this section are underlined and in roman type to distinguish them from other italicized words.

### Boldfaced, Page-Referenced Terms

The page-referenced terms are important; they are in boldface type in the chapter. Write a definition for each term in your own words without looking at the text. Next, compare your definition with that given in the chapter or in the text glossary. If your definition seems inaccurate, allow some time to pass and repeat this procedure until you can define each term rather quickly (how fast you answer is a gauge of your learning effectiveness).

(149) reproduction _____

(150) mitosis _____

_____

(150) meiosis _____

_____

(150) somatic cells _____

_____

(150) germ cells _____

_____

(150) chromosome _____

_____

(150) sister chromatids _____

_____

(150) centromere _____

_____

(150) chromosome number _____

_____

(150) diploid _____

_____

(151) cell cycle _____

_____

(151) interphase _____

_____

## Matching

Choose the most appropriate answer for each.

1. ___centromere
2. ___haploid or *n* cell
3. ___diploid or *2n* cell
4. ___chromosome
5. ___nuclear division
6. ___germ cells
7. ___somatic cells
8. ___sister chromatids
9. ___mitosis and meiosis
10. ___chromosome number

A. Cell lineage set aside for forming gametes and sexual reproduction
B. Necessary for the reproduction of eukaryotic cells
C. Any cell having two of each type of chromosome characteristic of the species
D. The number of each type of chromosome present in a cell (*n* or *2n*)
E. Each DNA molecule with attached proteins
F. Sort out and package parent DNA molecules into new cell nuclei
G. A small chromosome region with attachment sites for microtubules
H. The two attached DNA molecules of a duplicated chromosome
I. Body cells that reproduce by mitosis and cytoplasmic division
J. Possess only one of each type of chromosome characteristic of the species

## Labeling

Identify the stage in the cell cycle indicated by each number.

11. _____

12. _____

13. _____

14. _____

15. _____

16. _____

17. _____

18. _____

19. _____

20. _____

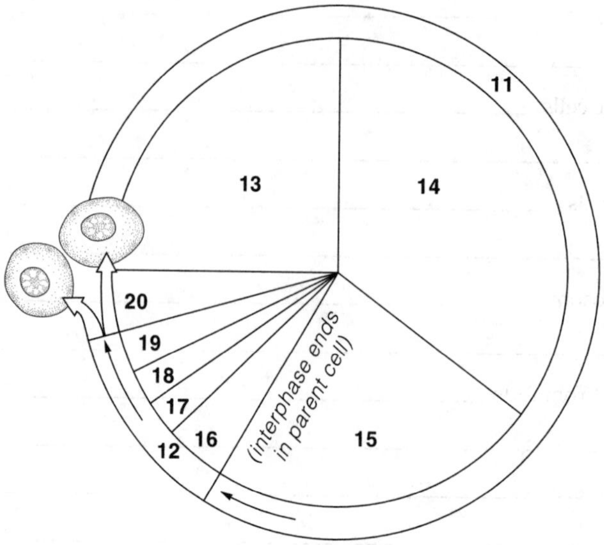

## Matching

Link each time span identified below with the most appropriate number in the preceding labeling section.

21. ____ Period after duplication of DNA during which the cell prepares for division

22. ____ The complete period of nuclear division that is followed by cytoplasmic division (a separate event)

23. ____ DNA duplication occurs now; a time of "synthesis" of DNA and proteins

24. ____ Period of cell growth before DNA duplication; a "gap" of interphase

25. ____ Usually the longest part of a cell cycle

26. ____ Period of cytoplasmic division

27. ____ Period that includes G1, S, G2

## 9.3. THE STAGES OF MITOSIS—AN OVERVIEW (pp. 152–153)

*Selected Words: mitos,* prometaphase, meta-

### Boldfaced, Page-Referenced Terms

(152) prophase _____

_____

(152) metaphase _____

_____

(152) anaphase _____

_____

(152) telophase _____

_____

(152) spindle apparatus _____

_____

(152) centrioles _____

_____

## Label-Match

Identify each of the mitotic stages shown by entering the correct stage in the blank beneath the sketch. Select from late prophase, transition to metaphase (prometaphase), interphase-parent cell, metaphase, early prophase, telophase, interphase-daughter cells, and anaphase. Complete the exercise by matching and entering the letter of the correct phase description in the parentheses following each label.

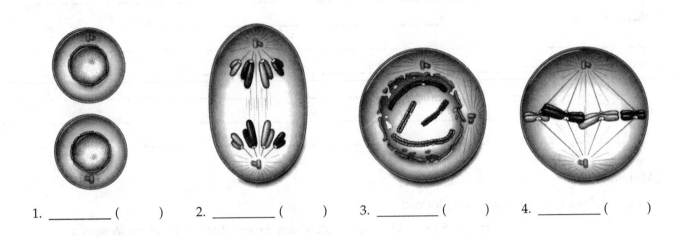

1. _____ (    )    2. _____ (    )    3. _____ (    )    4. _____ (    )

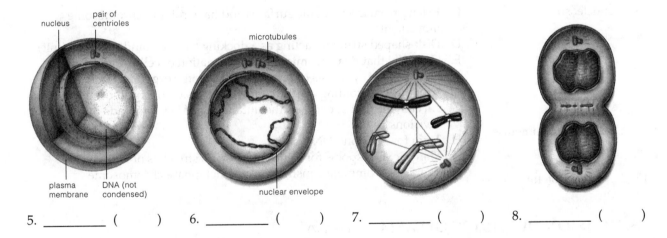

5. _____ (    )    6. _____ (    )    7. _____ (    )    8. _____ (    )

A. Attachments between two sister chromatids of each chromosome break; the two are now separate chromosomes that move to opposite spindle poles
B. Microtubules that form the spindle apparatus penetrate the nuclear region and form the spindle; microtubules become attached to the two sister chromatids of each chromosome
C. The DNA and its associated proteins have started to condense
D. All the chromosomes are now fully condensed and lined up at the equator of the spindle
E. DNA is duplicated and the cell prepares for nuclear division
F. Two daughter cells have formed, each diploid with two of each type of chromosome, just like the parent cell's nucleus

(*continued on next page*)

G. Chromosomes continue to condense. New microtubules are assembled, and they move one of two centriole pairs toward the opposite end of the cell. The nuclear envelope begins to break up

H. New patches of membrane join to form a new nuclear envelope around the decondensing chromosomes. Cytoplasmic division begins before this stage ends

## 9.4. A CLOSER LOOK AT THE CELL CYCLE (pp. 154–155)

*Selected Words:* DNA topoisomerase

### Boldfaced, Page-Referenced Terms

(154) histones _____

_____

(154) nucleosome _____

_____

(154) kinetochore _____

_____

(155) motor proteins _____

_____

### Matching

Choose the most appropriate answer for each.

1. ___S
2. ___histones
3. ___nucleosome
4. ___H1
5. ___kinetochore
6. ___centromere
7. ___metaphase
8. ___DNA topoisomerase
9. ___colchicine
10. ___motor proteins

A. Specific histone that stabilizes the structural array of nucleosomes
B. Period of the cell cycle in which the chromosomes are maximally condensed
C. Extend from microtubule surfaces and have roles in chromosome movements
D. Disk-shaped structure acting as a docking site for spindle microtubules
E. Chemical that disrupts microtubule formation of cell spindles
F. Period of the synthesis of proteins that become organized into structural scaffolding for the condensed version of chromosomes
G. An enzyme that supervises the untangling of DNA threads during cell divisions
H. Term for a histone-DNA spool
I. Many act like spools for winding up small stretches of DNA
J. The most prominent constriction of a metaphase chromosome

## 9.5. DIVISION OF THE CYTOPLASM (pp. 156–157)

## 9.6. *Focus on Science:* HENRIETTA'S IMMORTAL CELLS (p. 158)

### Boldfaced, Page-Referenced Terms

(156) cytoplasmic division _____

_____

(156) cell plate formation _____

_____

(157) cleavage _____

_____

(158) HeLa cells _____

_____

## Choice

For questions 1–10 on cytoplasmic division, choose from the following:

a. plant cells                     b. animal cells

1._____Formation of a cell plate

2._____Contractions continue and in time will divide the cell in two

3._____Cellulose deposits form a crosswall between the two daughter cells

4._____Possess rigid walls that cannot be pinched in two

5._____A cleavage furrow

6._____Deposits form a cementing middle lamella

7._____Through the action of microfilaments the plasma membrane sinks inward

8._____A shallow, ringlike depression forms at the cell surface, above the cell's midsection

9._____As mitosis ends, vesicles converge at the spindle equator

10._____The structure grows at its margins until it fuses with the parent cell's plasma membrane

## Short Answer

11. Describe the origin and the significance of HeLa cells. _____

_____

_____

_____

_____

_____

_____

# Self-Quiz

_____ 1. The replication of DNA occurs _____ .
   a between the growth phases of interphase
   b. immediately before prophase of mitosis
   c. during prophase of mitosis
   d. during prophase of meiosis

_____ 2. In the cell life cycle of a particular cell,
   _____ .
   a. mitosis occurs immediately prior to G1
   b. G2 precedes S
   c. G1 precedes S
   d. mitosis and S precede G1

___ 3. In eukaryotic cells, which of the following can occur during mitosis?
   a. Two mitotic divisions to maintain the parental chromosome number
   b. The replication of DNA
   c. A long growth period
   d. The disappearance of the nuclear envelope and nucleolus

___ 4. Diploid refers to _____ .
   a. having two chromosomes of each type in somatic cells
   b. twice the parental chromosome number
   c. half the parental chromosome number
   d. having one chromosome of each type in somatic cells

___ 5. Somatic cells are _____ cells; germ cells are _____ cells.
   a. meiotic; body
   b. body; body
   c. meiotic; meiotic
   d. body; meiotic

___ 6. If a parent cell has sixteen chromosomes and undergoes mitosis, the resulting cells will have _____ chromosomes.
   a. sixty-four
   b. thirty-two
   c. sixteen
   d. eight
   e. four

___ 7. The correct order of the stages of mitosis is _____ .
   a. prophase, metaphase, telophase, anaphase
   b. telophase, anaphase, metaphase, prophase
   c. telophase, prophase, metaphase, anaphase
   d. anaphase, prophase, telophase, metaphase

   e. prophase, metaphase, anaphase, telophase

___ 8. "The nuclear envelope breaks up completely into vesicles. Microtubules are now free to interact with the chromosomes." These sentences describe the _____ of mitosis.
   a. prophase
   b. metaphase
   c. transition to metaphase
   d. anaphase
   e. telophase

___ 9. During _____ , sister chromatids of each chromosome are separated from each other, and those former partners, now chromosomes, are moved toward opposite poles.
   a. prophase
   b. metaphase
   c. anaphase
   d. telophase

___ 10. Each histone-DNA spool is a single structural unit called a _____ .
   a. kinetochore
   b. motor protein
   c. DNA topoisomerase
   d. nucleosome

___ 11. In the process of cytokinesis, cleavage furrows are associated with _____ cell division, and cell plate formation is associated with _____ cell division.
   a. animal; animal
   b. plant; animal
   c. plant; plant
   d. animal; plant

## Chapter Objectives/Review Questions

| Page | Objectives/Questions |
|---|---|
| (149) | 1. Define the word *reproduction*. |
| (150) | 2. Mitosis and meiosis refer to the division of the cell's _____ . |
| (150) | 3. Distinguish between somatic cells and germ cells as to their location and function. |
| (150) | 4. The eukaryotic chromosome is composed of _____ and _____ . |
| (150) | 5. The two attached threads of a duplicated chromosome are known as sister _____ . |

(150)     6.   The _____ is a small region with attachment sites for the microtubules that move the chromosome during nuclear division

(150)     7.   Any cell having two of each type of chromosome is a _____ cell; a _____ cell, such as an egg or a sperm, have only one type of each chromosome characteristic of a species.

(151)     8.   Be able to list and describe, in order, the various activities occurring in the eukaryotic cell life cycle.

(151)     9.   Interphase of the cell cycle consists of G1, _____ , and G2.

(151)    10.   S is the time in the cell cycle when _____ replication occurs.

(152)    11.   Describe the structure and function of the spindle apparatus.

(152)    12.   Describe the number and movements of centrioles in the cell division of some cells.

(152)    13.   The "_____" is a time of transition when the nuclear envelope breaks up into flattened vesicles prior to metaphase.

(152–153) 14.   Be able to give a detailed description of the cellular events occurring in the prophase, metaphase, anaphase, and telophase of mitosis.

(154)    15.   Describe particular events occurring in G1, S, and G2 of interphase.

(154–155) 16.   Be able to characterize the organization of metaphase chromosomes using the following terms: histones, nucleosome, kinetochore, DNA topoisomerase, and motor proteins.

(156–157) 17.   Compare and contrast cytokinesis as it occurs in plant and animal cell division; use the following concepts: cleavage furrow, microfilaments at the cell's midsection, and cell plate formation.

(158)    18.   Explain how cells from Henrietta Lacks continue to benefit humans everywhere more than forty years after her death.

---

## Integrating and Applying Key Concepts

Runaway cell division is characteristic of cancer. Imagine the various points of the mitotic process that might be sabotaged in cancerous cells in order to halt their multiplication. Then try to imagine how one might discriminate between cancerous and normal cells in order to guide those methods of sabotage most effective in combating cancer.

# 10

# MEIOSIS

## Interactive Exercises

*Octopus Sex and Other Stories* (pp. 160–161)

## 10.1. COMPARISON OF ASEXUAL AND SEXUAL REPRODUCTION (p. 162)
## 10.2. HOW MEIOSIS HALVES THE CHROMOSOME NUMBER (pp. 162–163)

*Selected Words:* *unfertilized* egg cells, *pairs of genes, hom-, homologue to homologue, 2n*

*Boldfaced, Page-Referenced Terms*

(161) germ cells _____

_____

(161) gametes _____

_____

(162) asexual reproduction _____

_____

(162) genes _____

_____

(162) sexual reproduction _____

_____

(162) allele _____

_____

(162) meiosis _____

_____

(162) diploid number _____

_____

(162) homologous chromosomes _____

_____

(162) haploid number _____

_____

## Choice

For questions 1–10, choose from the following:

a. asexual reproduction       b. sexual reproduction

1. _____ Brings together new combinations of alleles in offspring

2. _____ Involves only one parent

3. _____ Each new individual ends up with pairs of genes

4. _____ Commonly involves two parents

5. _____ The production of "clones"

6. _____ Involves meiosis, formation of gametes and fertilization

7. _____ Offspring are genetically identical copies of the parent

8. _____ Produces the variation in traits that forms the basis of evolutionary change

9. _____ The instructions in every pair of genes is identical in all individuals of a species

10. _____ New combinations of alleles lead to variations in physical and behavioral traits

## Dichotomous Choice

Circle one of two possible answers given between parentheses in each statement.

11. _____ Each unique molecular form of the same gene is called a(n) (DNA molecule/allele).

12. _____ (Meiosis/Mitosis) divides chromosomes into separate parcels not once but twice prior to cell division.

13. _____ Sperms and eggs are cells known as (germ/gamete) cells.

14. _____ (Haploid/Diploid) germ cells produce haploid gametes.

15. _____ (Haploid/Diploid) cells possess pairs of homologous chromosomes.

16. _____ (Meiosis/Mitosis) produces cells that have one member of each pair of homologous chromosomes possessed by the species.

17. _____ Identical alleles are found on (homologous chromosomes/sister chromatids).

18. _____ Two attached DNA molecules are known as (sister chromatids/homologous chromosomes).

19. _____ Two attached sister chromatids represent (two/one) chromosome(s).

20. _____ One pair of duplicated chromosomes would be composed of (two/four) chromatids.

21.____With meiosis, chromosomes proceed through (one/two) divisions to yield four haploid nuclei.

22.____DNA is replicated during (prophase I of meiosis I/interphase preceding meiosis I).

23.____During meiosis I, each duplicated (chromosome/chromatid) lines up with its partner, homologue to homologue and then the partners are moved apart from one another.

24.____Cytoplasmic division following Meiosis I results in two (diploid/haploid) daughter cells.

25.____During (meiosis I/meiosis II), the two sister chromatids of each chromosome are separated from each other, and each sister chromatid is now a chromosome in its own right.

26.____If human body cell nuclei contain twenty-three pairs of homologous chromosomes, each resulting gamete will contain (twenty-three/forty-six) chromosomes.

## 10.3. A VISUAL TOUR OF THE STAGES OF MEIOSIS (pp. 164–165)

## 10.4. KEY EVENTS OF MEIOSIS I (pp. 166–167)

*Selected Words:* *non*sister chromatids, *maternal* chromosomes, *paternal* chromosomes

*Boldfaced, Page-Referenced Terms*

(166) crossing over _____

_____

*Matching*

To review the major stages of meiosis, match the following written descriptions with the appropriate sketch. Assume that the cell in this model initially has only one pair of homologous chromosomes (one from a paternal source and one from a maternal source) and crossing over does not occur. Complete the exercise by indicating the diploid (2n = 2) or haploid (n = 1) chromosome number of the cell chosen in the parentheses following each blank.

1. ___( ) A pair of homologous chromosomes prior to S of interphase in a diploid germ cell

2. ___( ) While the germ cell is in S of interphase, chromosomes are duplicated through DNA replication; the two sister chromatids are attached at the centromere

3. ___( ) During meiosis I, each duplicated chromosome lines up with its partner, homologue to homologue

4. ___( ) Also during meiosis I, the chromosome partners separate from each other in anaphase I; cytokinesis occurs, and each chromosome goes to a different cell

5. ___( ) During meiosis II (in two cells), the sister chromatids of each chromosome are separated from each other; four haploid nuclei form; cytokinesis results in four cells (potential gametes)

## Label-Match

Identify each of the meiotic stages shown below by entering the correct stage of either meiosis I or meiosis II in the blank beneath the sketch. Choose from prophase I, metaphase I, anaphase I, telophase I, prophase II, metaphase II, anaphase II, and telophase II. Complete the exercise by matching and entering the letter of the correct stage description in the parentheses following each label.

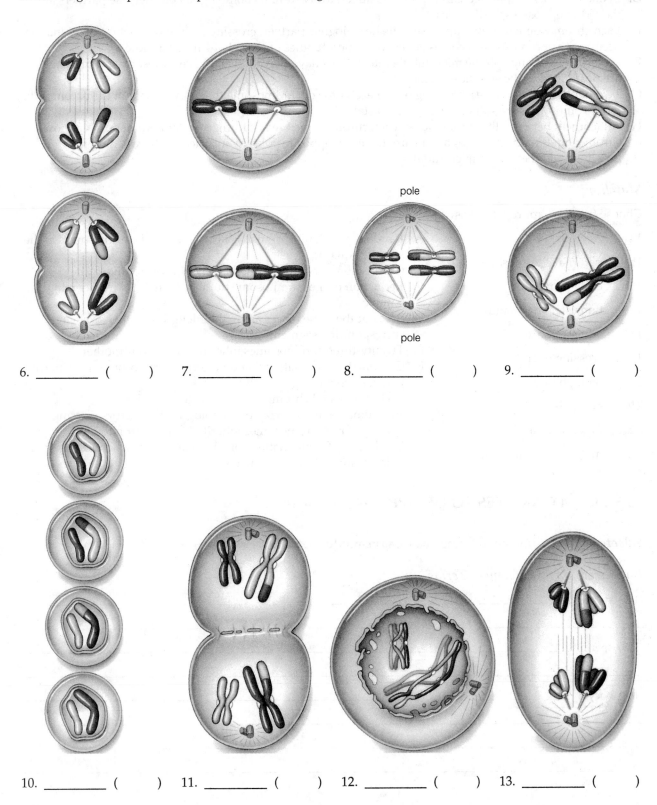

pole

pole

6. _____ (    )    7. _____ (    )    8. _____ (    )    9. _____ (    )

10. _____ (    )    11. _____ (    )    12. _____ (    )    13. _____ (    )

A. The spindle is now fully formed; all chromosomes are positioned midway between the poles of one cell
B. In each of two cells, microtubules attach to the kinetochores of chromosomes, and motor proteins drive the movement of chromosomes toward the spindle's equator
C. Four daughter nuclei form; when the cytoplasm divides, each new cell has a haploid chromosome number, all in the unduplicated state; the cells may develop into gametes
D. In one cell, each duplicated chromosome is pulled away from its homologous partner; the partners are moved to opposite spindle poles
E. Each chromosome is drawn up close to its homologous partner; crossing over and genetic recombination occur; each chromosome becomes attached to microtubules of a newly forming spindle
F. Motor proteins and spindle microtubule interactions have moved all of the chromosomes so that they are positioned in the spindle equator
G. Two haploid cells form, each having one of each type of chromosome that was present in the parent cell; the chromosomes are still in the duplicated state
H. Attachment between the chromatids of each chromosome breaks; former "sister chromatids" are now chromosomes in their own right and are moved to opposite poles by interactions between motor proteins and kinetochore microtubules

## Matching

Choose the most appropriate answer for each.

14. ___ $2^{23}$
15. ___ paternal chromosomes
16. ___ early prophase I
17. ___ nonsister chromatids
18. ___ late prophase I
19. ___ crossing over
20. ___ metaphase I
21. ___ synapsis
22. ___ function of meiosis
23. ___ maternal chromosomes

A. Random positioning of maternal and paternal chromosomes at the spindle equator
B. Chromosomes condense and become thicker rodlike forms
C. Reduction of the chromosome number by half for forthcoming gametes
D. Break at the same places along their length and then exchange corresponding segments
E. Twenty-three chromosomes inherited from your mother
F. Combinations of maternal and paternal chromosomes possible in gametes from one germ cell
G. Each duplicated chromosome is in threadlike form
H. An intimate parallel array of homologues that favors crossing over
I. Twenty-three chromosomes inherited from your father
J. Breaks up old combinations of alleles and puts new ones together in pairs of homologous chromosomes

## 10.5. FROM GAMETES TO OFFSPRING (pp. 168–169)

*Selected Words: gamete*-producing bodies, *spore*-producing bodies

*Boldfaced, Page-Referenced Terms*

(168) spores _____

_____

(168) sperm _____

_____

(168) oocyte _____

_____

(168) egg _____

_____

## Choice

For questions 1–10, choose from the following. Refer to Figure 10.7.

a. animal life cycle        b. plant life cycle        c. both animal and plant life cycles

1. ____ Meiosis results in the production of haploid spores

2. ____ A zygote divides by mitosis

3. ____ Meiosis results in the production of haploid gametes

4. ____ Haploid gametes fuse in fertilization to form a diploid zygote

5. ____ A zygote divides by mitosis to form a diploid sporophyte

6. ____ A spore divides by mitosis to produce a haploid gametophyte

7. ____ A haploid gametophyte divides by mitosis to produce haploid gametes

8. ____ A haploid spore divides by mitosis to produce a gametophyte

9. ____ A diploid body forms from mitosis of a zygote

10. ____ A gamete-producing body and a spore-producing body develop during the life cycle

## Sequence

Arrange the following entities in correct order of development, entering a 1 by the stage that appears first and a 5 by the stage that completes the process of spermatogenesis. Complete the exercise by indicating if each cell is *n* or *2n* in the parentheses following each blank. Refer to Figure 10.8 in the text.

11. ____ (      ) primary spermatocyte

12. ____ (      ) sperm

13. ____ (      ) spermatid

14. ____ (      ) spermatogonium

15. ____ (      ) secondary spermatocyte

## Matching

Choose the most appropriate answer to match with each oogenesis concept. Refer to Figure 10.9 in the text.

16. ____ primary oocyte

17. ____ oogonium

18. ____ secondary oocyte

19. ____ ovum and three polar bodies

20. ____ first polar body

A. The cell in which synapsis, crossing over, and recombination occur

B. A cell that is equivalent to a diploid germ cell

C. A haploid cell formed after division of the primary oocyte that does not form an ovum at second division

D. Haploid cells, but only one of which functions as an egg

E. A haploid cell formed after division of the primary oocyte, the division of which forms a functional ovum

## Short Answer

21. List the various mechanisms that contribute to the huge number of new gene combinations that may

result from fertilization. _____

_____

_____

## 10.6. MEIOSIS AND MITOSIS COMPARED (pp. 170–171)

**Selected Words:** *clones, variations in traits, somatic cell*

### Complete the Table

1. Complete the table below by entering the word "mitosis" or "meiosis" in the blank adjacent to the statement describing one of these processes.

| Description | Mitosis/Meiosis |
|---|---|
| a. Involves one division cycle | |
| b. Functions in growth and tissue repair | |
| c. Daughter cells are haploid | |
| d. Initiated in germ cells | |
| e. Involves two division cycles | |
| f. Daughter cells have one chromosome from each homologous pair | |
| g. Produces spores in plant life cycles | |
| h. Daughter cells have the diploid chromosome number | |
| i. Completed when four daughter cells are formed | |

### Matching

The cell model used in this exercise has two pairs of homologous chromosomes, one long pair and one short pair. Match the descriptions to the numbers of chromosomes shown in the sketches below.

2. ___one cell at the beginning of meiosis II

3. ___a daughter cell at the end of meiosis II

4. ___metaphase I of meiosis

5. ___metaphase of mitosis

6. ___G1 in a daughter cell following mitosis

7. ___prophase of mitosis

A

B

C

D

E

F

The following questions refer to the sketches on the previous page; enter answers in the blanks following each question.

8. How many chromosomes are present in cell E? ____
9. How many chromatids are present in cell E? ____
10. How many chromatids are present in cell C? ____
11. How many chromatids are present in cell D? ____
12. How many chromosomes are present in cell F? ____

## Self Quiz

____ 1. Which of the following does not occur in prophase I of meiosis?
    a. a cytoplasmic division
    b. a cluster of four chromatids
    c. homologues pairing tightly
    d. crossing over

____ 2. Crossing over is one of the most important events in meiosis because _____ .
    a. it produces new combinations of alleles on chromosomes
    b. homologous chromosomes must be separated into different daughter cells
    c. the number of chromosomes allotted to each daughter cell must be halved
    d. homologous chromatids must be separated into different daughter cells

____ 3. Crossing over _____ .
    a. generally results in pairing of homologues and binary fission
    b. is accompanied by gene-copying events
    c. involves breakages and exchanges between sister chromatids
    d. alters the composition of chromosomes and results in new combinations of alleles being channeled into the daughter cells

____ 4. The appearance of chromosome ends lapped over each other in meiotic prophase I provides evidence of _____ .
    a. meiosis
    b. crossing over
    c. chromosomal aberration
    d. fertilization
    e. spindle fiber formation

____ 5. Which of the following does not increase genetic variation?
    a. Crossing over
    b. Random fertilization
    c. Prophase of mitosis
    d. Random homologue alignments at metaphase I

____ 6. Which of the following is the most correct sequence of events in animal life cycles?
    a. meiosis → fertilization → gametes → diploid organism
    b. diploid organism → meiosis → gametes → fertilization
    c. fertilization → gametes → diploid organism → meiosis
    d. diploid organism → fertilization → meiosis → gametes

____ 7. In sexually reproducing organisms, the zygote is _____ .
    a. an exact genetic copy of the female parent
    b. an exact genetic copy of the male parent
    c. unlike either parent genetically
    d. a genetic mixture of male parent and female parent

____ 8. Which of the following is the most correct sequence of events in plant life cycles?
    a. fertilization → zygote → sporophyte → meiosis → spores → gametophytes → gametes
    b. fertilization → sporophyte → zygote → meiosis → spores → gametophytes → gametes
    c. fertilization → zygote → sporophyte → meiosis → gametes → gametophyte → spores
    d. fertilization → zygote → gametophyte → meiosis → gametes → sporophyte → spores

___ 9. The cell in the diagram is a diploid that has three pairs of chromosomes. From the number and pattern of chromosomes, the cell _____ .
   a. could be in the first division of meiosis
   b. could be in the second division of meiosis
   c. could be in mitosis
   d. could be in neither mitosis nor meiosis, because this stage is not possible in a cell with three pairs of chromosomes

___10. You are looking at a cell from the same organism as in the previous question. Now the cell _____ .
   a. could be in the first division of meiosis
   b. could be in the second division of meiosis
   c. could be in mitosis
   d. could be in neither mitosis nor meiosis, because this stage is not possible in this organism

## Chapter Objectives/Review Questions

| Page | | Objectives/Questions |
|---|---|---|
| (161) | 1. | Distinguish between germ cells and gametes. |
| (162) | 2. | "One parent always passes on a duplicate of all its genes to its offspring" describes _____ reproduction. |
| (162) | 3. | _____ reproduction puts together new combinations of _____ in offspring. |
| (162) | 4. | _____ divides chromosomes into separate parcels not once but twice prior to cell division. |
| (162) | 5. | Describe the relationship between the following terms: homologous chromosomes, diploid number, and haploid number. |
| (162) | 6. | If the diploid chromosome number for a particular plant species is 18, the haploid gamete number is _____ . |
| (162–163) | 7. | During interphase a germ cell duplicates its DNA; a duplicated chromosome consists of two DNA molecules that remain attached to a constriction called the _____ . |
| (163) | 8. | As long as the two DNA molecules remain attached they are referred to as _____ _____ of the chromosome. |
| (163) | 9. | During meiosis I, homologous chromosomes pair; each homologue consists of _____ chromatids. |
| (163) | 10. | During meiosis II, the two sister _____ of each _____ are separated from each other. |
| (166) | 11. | _____ _____ breaks up old combinations of alleles and puts new ones together in pairs of homologous chromosomes. |
| (167) | 12. | The _____ attachment and subsequent positioning of each pair of maternal and paternal chromosomes at metaphase I leads to different _____ of maternal and paternal traits in offspring. |
| (168–169) | 13. | Using the special terms for the cells at the various stages, describe spermatogenesis in male animals and oogenesis in female animals. |
| (168–169) | 14. | Meiosis in the animal life cycle results in haploid _____ ; meiosis in the plant life cycle results in haploid _____ . |
| (169) | 15. | Crossing over, the distribution of random mixes of homologous chromosomes into gametes, and fertilization contribute to _____ in the traits of offspring. |
| (170) | 16. | Mitotic cell division produces _____ ; meiosis and subsequent fertilization promotes _____ in traits among offspring. |
| (170) | 17. | Be able to list three ways that meiosis promotes variation in offspring. |

## Integrating and Applying Key Concepts

A few years ago, it was claimed that the actual cloning of a human being had been accomplished. Later, this claim was admitted to be fraudulent. Recently, in 1997, a mammal was cloned. A sheep named "Dolly" resulted from a cloning experiment in Scotland. If sometime in the future cloning of humans becomes possible, speculate about the effects of reproduction without sex on human populations.

# 11

# OBSERVABLE PATTERNS OF INHERITANCE

## Interactive Exercises

*Selected Words:* *observable* evidence, <u>Pisum</u> <u>sativum</u>, *homozygous* condition, *heterozygous* condition, *dominant* allele, *recessive* allele, $P$, $F_1$, $F_2$

## Boldfaced, Page-Referenced Terms

(177) genes _____

_____

(177) alleles _____

_____

(177) true-breeding lineage _____

_____

(177) hybrid offspring _____

_____

(177) homozygous dominant _____

_____

(177) homozygous recessive _____

_____

(177) heterozygous _____

_____

(177) genotype _____

_____

(177) phenotype _____

_____

(178) monohybrid crosses _____

_____

(179) probability _____

_____

(179) Punnett-square method _____

_____

(179) testcrosses _____

_____

(180) dihybrid crosses _____

_____

(181) independent assortment _____

_____

## Matching

Choose the most appropriate answer for each.

1. ___ genotype
2. ___ alleles
3. ___ heterozygous
4. ___ dominant allele
5. ___ phenotype
6. ___ genes
7. ___ true-breeding lineage
8. ___ homozygous recessive
9. ___ recessive allele
10. ___ homozygous
11. ___ $P$, $F_1$, $F_2$
12. ___ hybrid offspring
13. ___ diploid cell
14. ___ gene locus
15. ___ homozygous dominant

A. Parental, first-generation, and second-generation offspring
B. All the different molecular forms of a gene that exist
C. Particular location of a gene on a chromosome
D. Describes an individual having a pair of nonidentical alleles
E. An individual with a pair of recessive alleles, such as *aa*
F. Gene whose effect is masked by its partner
G. Offspring of a genetic cross that inherit a pair of nonidentical alleles for a trait
H. Refers to an individual's observable traits
I. Refers to the genes present in an individual organism
J. Gene whose effect "masks" the effect of its partner
K. Describes an individual for which two alleles of a pair are the same
L. An individual with a pair of dominant alleles, such as *AA*
M. Units of information about specific traits; passed from parents to offspring
N. Has a pair of genes for each trait, one on each of two homologous chromosomes
O. When offspring of genetic crosses inherit a pair of identical alleles for a trait, generation after generation

## Fill-in-the-Blanks

Offspring of (16) _____ crosses are heterozygous for the one trait being studied. (17) _____ means that the chance that each outcome of a given event will occur is proportional to the number of ways it can be reached. Because (18) _____ is a chance event, the rules of probability apply to genetics crosses. The separation of *A* and *a* as members of a pair of homologous chromosomes move to different gametes during meiosis is known as Mendel's theory of (19) _____ . Crossing $F_1$ hybrids (possibly of unknown genotype) back to a plant known to be a true-breeding recessive plant is known as a (20) _____ . From this cross, a ratio of (21) _____ is expected. When $F_1$ offspring inherit two gene pairs, each consisting of two nonidentical alleles, the cross is known as a (22) _____ cross. "Gene pairs assorting into gametes independently of other gene pairs located on nonhomologous chromosomes" describes Mendel's theory of (23) _____ _____ .

## Problems

24. In garden pea plants, Tall (*T*), is dominant over dwarf (*t*). In the cross *Tt* × *tt*, the *Tt* parent would produce a gamete carrying *T* (tall) and a gamete carrying *t* (dwarf) through segregation; the *tt* parent could only produce gametes carrying the *t* (dwarf) gene. Use the Punnett square method (refer to Figures 11.4, 11.6, and 11.7 in the text) to determine the genotype and phenotype probabilities of offspring from the above cross, *Tt* × *tt*:

Although the Punnett-square (checkerboard) method is a favored method for solving single-factor genetics problems, there is a quicker way. Only six different outcomes are possible from single factor crosses. Studying the following relationships allows one to obtain the result of any such cross by inspection.

1. *AA* × *AA* = all *AA*
   (Each of the four blocks of the Punnett-square would be *AA*.)
2. *aa* × *aa* = all *aa*
3. *AA* × *aa* = all *Aa*
4. *AA* × *Aa* = 1/2 *AA*; 1/2 *Aa*
   or *Aa* × *AA*
   (Two blocks of the Punnett square are *AA*, and two blocks are *Aa*.)
5. *aa* × *Aa* = 1/2 *aa*; 1/2 *Aa*
   or *Aa* × *aa*
6. *Aa* × *Aa* = 1/4 *AA*; 1/2 *Aa*; 1/4 *aa*
   (One block in the Punnett square is *AA*, two blocks are *Aa*, and one block is *aa*.)

## Complete the Table

25. Using the gene symbols (tall and dwarf pea plants) in exercise 24, apply the six Mendelian ratios listed above to complete the following table of single-factor crosses by inspection. State results as phenotype and genotype ratios.

| Cross | Phenotype Ratio | Genotype Ratio |
|---|---|---|
| a. *Tt* × *tt* | | |
| b. *TT* × *Tt* | | |
| c. *tt* × *tt* | | |
| d. *Tt* × *Tt* | | |
| e. *tt* × *Tt* | | |
| f. *TT* × *tt* | | |
| g. *TT* × *TT* | | |
| h. *Tt* × *TT* | | |

When working genetics problems dealing with two gene pairs, one can visualize the independent assortment of gene pairs located on nonhomologous chromosomes into gametes by use of a fork-line device. Assume that in humans, pigmented eyes (*B*) are dominant (an eye color other than blue) over blue (*b*), and right-handedness (*R*) is dominant over left-handedness (*r*). To learn to solve a problem, cross the parents *BbRr* × *BbRr*. A sixteen-block Punnett square is required with gametes from each parent arrayed on two sides of the Punnett square (refer to Figures 11.8 and 11.9 in the text). The gametes receive genes through independent assortment using a fork-line method:

26. Array the gametes at the right on two sides of the Punnett square; combine these haploid gametes to form diploid zygotes within the squares. In the blank spaces below, enter the probability ratios derived within the Punnett square for the phenotypes listed:

    *B  b  R  r*          ×          *B  b  R  r*

    *BR, Br, bR, br*                 *BR, Br, bR, br*

    a. _____ pigmented eyes, right-handed
    b. _____ pigmented eyes, left-handed
    c. _____ blue-eyed, right-handed
    d. _____ blue-eyed, left-handed

27. Albinos cannot form the pigments that normally produce skin, hair, and eye color, so albinos exhibit white hair, pink eyes and skin (because the blood shows through). To be an albino, one must be homozygous recessive (*aa*) for the pair of genes that code for the key enzyme in pigment production. Suppose a woman of normal pigmentation (*A__* ) with an albino mother marries an albino man. State the possible kinds of pigmentation possible for this couple's children, and specify the ratio of each kind of child the couple is likely to have. Show the genotype(s) and state the phenotype(s). _____

    _____

28. In horses, black coat color is influenced by the dominant allele (*B*), and chestnut coat color is influenced by the recessive allele (*b*). Trotting gait is due to a dominant gene (*T*), pacing gait to the recessive allele (*t*). A homozygous black trotter is crossed to a chestnut pacer.

    a. What will be the appearance of the $F_1$ and $F_2$ generations?

    _____

    b. Which phenotype will be most common? _____
    c. Which genotype will be most common? _____
    d. Which of the potential offspring will be certain to breed true? _____

## 11.4. DOMINANCE RELATIONS (p. 182)

## 11.5. MULTIPLE EFFECTS OF SINGLE GENES (p. 183)

## 11.6. INTERACTIONS BETWEEN GENE PAIRS (pp. 184–185)

*Selected Words:* ABO blood typing, transfusions, sickle-cell anemia, albinism

### Boldfaced, Page-Referenced Terms

(182) incomplete dominance _____

_____

(182) codominance _____

_____

(182) multiple allele system _____

_____

(183) pleiotropy _____

_____

(184) epistasis _____

_____

## Complete the Table

1. Complete the following table by supplying the type of inheritance illustrated by each example. Choose from these gene interactions: pleiotropy, multiple allele system, incomplete dominance, codominance, and epistasis.

| Type of Inheritance | Example |
|---|---|
| a. | Pink-flowered snapdragons produced from red- and white-flowered parents |
| b. | AB type blood from a gene system of three alleles, *A*, *B*, and *O* |
| c. | A gene with three or more alleles such as the *ABO* blood typing alleles |
| d. | Black, brown, or yellow fur of Labrador retrievers and comb shape in poultry |
| e. | The multiple phenotypic effects of the gene causing human sickle-cell anemia |

## Problems

2. Genes that are not always dominant or recessive may blend to produce a phenotype of a different appearance. This is termed *incomplete dominance*. In four o'clock plants, red flower color is determined by gene $R$ and white flower color by $R'$, while the heterozygous condition, $RR'$, is pink. Complete the table below by determining the phenotypes and genotypes of the offspring of the following crosses:

| Cross | Phenotype | Genotype |
|---|---|---|
| a. $RR \times R'R' =$ | | |
| b. $R'R' \times R'R' =$ | | |
| c. $RR \times RR' =$ | | |
| d. $RR \times RR =$ | | |

Sickle-cell anemia is a genetic disease in which children who are homozygous for a defective gene produce defective hemoglobin ($Hb^S/Hb^S$). The genotypes of normal persons is $Hb^A/Hb^A$. If the level of blood oxygen drops below a certain level, in a person with the $Hb^S/Hb^S$ genotype, the hemoglobin chains stiffen and cause the red blood cells to form sickle, or crescent shapes. These cells clog and rupture capillaries, which results in oxygen-deficient tissues where metabolic wastes collect. Several body functions are badly damaged. Severe anemia and other symptoms develop, and death nearly always occurs before adulthood. The sickle-cell gene is considered *pleiotropic*. Persons who are heterozygous ($Hb^A/Hb^S$) are said to possess *sickle-cell trait*. They are able to produce enough normal hemoglobin molecules to appear normal but their red blood cells will sickle if they encounter oxygen tension (such as at high altitudes).

3. A man whose sister died of sickle-cell anemia married a woman whose blood is found to be normal. What advice would you give to this couple about the inheritance of this disease as they plan their family?

_____

_____

_____

4. If a man and a woman, each with sickle-cell trait, planned to marry, what information could you provide for them regarding the genotypes and phenotypes of their future children?

_____

_____

In one example of a *multiple allele system* with *codominance*, the three genes $I^Ai$, $I^Bi$ and $i$ produce proteins found on the surfaces of red blood cells that determine the four blood types in the ABO system, A, B, AB, and O. Genes $I^A$ and $I^B$ are both dominant over $i$ but not over each other. They are codominant. Recognize that blood types A and B may be heterozygous or homozygous ($I^AI^A$, $I^Ai$ or $I^BI^B$, $I^Bi$) while blood type O is homozygous (*ii.*). Indicate the genotypes and phenotypes of the offspring and their probabilities from the parental combinations in exercises 5–9.

5. $I^Ai \times I^AI^B$ = _____
6. $I^Bi \times I^Ai$ = _____
7. $I^AI^A \times ii$ = _____
8. $ii \times ii$ = _____
9. $I^AI^B \times I^AI^B$ = _____

In one type of gene interaction, two alleles of a gene mask the expression of alleles of another gene, and some expected phenotypes never appear. *Epistasis* is the term given such interactions. Work the following problems on scratch paper to understand epistatic interactions.

In sweet peas, genes $C$ and $P$ are necessary for colored flowers. In the absence of either (_ _ *pp* or *cc* _ _), or both (*ccpp*), the flowers are white. What will be the color of the offspring of the following crosses and in what proportions will they appear?

10. $CcPp \times ccpp$ = _____
11. $CcPP \times Ccpp$ = _____
12. $Ccpp \times ccPp$ = _____

In the inheritance of the coat (fur) color of Labrador retrievers, allele $B$ specifies black that is dominant to brown (chocolate), $b$. Allele $E$ permits full deposition of color pigment but two recessive alleles, *ee*, reduces deposition, and a yellow coat results.

13. Predict the phenotypes of the coat color and their proportions resulting from the following cross: *BbEe* × *Bbee* = _____

_____

_____

_____

In poultry, an epistatic interaction occurs in which two genes produce a phenotype that neither gene can produce alone. The two interacting genes (*R* and *P*) produce comb shape in chickens. The possible genotypes and phenotypes are:

| Genotypes | Phenotypes |
|-----------|------------|
| R__P__ | walnut comb |
| R__pp | rose comb |
| rrP__ | pea comb |
| rrpp | single comb |

14. What are the genotype and phenotype ratios of the offspring of a heterozygous walnut-combed male and a single-combed female?

_____

_____

15. Cross a homozygous rose-combed rooster with a homozygous single-combed hen and list the genotype and phenotype ratios of their offspring.

_____

_____

_____

## 11.7. LESS PREDICTABLE VARIATIONS IN TRAITS (pp. 186–187)

## 11.8. EXAMPLES OF ENVIRONMENTAL EFFECTS ON PHENOTYPE (p. 188)

*Selected Words:* camptodactyly, "bell-shaped" curve, Hydrangea macrophylla

*Boldfaced, Page-Referenced Terms*

(186) continuous variation _____

*Choice*

For questions 1–5, choose from the following primary contributing factors:

a. environment                    b. a number of genes affect a trait

1. ____ Height of human beings

2. ____ Continuous variation in a trait

3. ____ Flower color in *Hydrangea macrophylla*

4. ____ The range of eye colors in the human population

5. ____ Heat-sensitive version of one of the enzymes required for melanin production in Himalayan rabbits

# Self-Quiz

_____ 1. The best statement of Mendel's principle of independent assortment is that _____ .
    a. one allele is always dominant to another
    b. hereditary units from the male and female parents are blended in the offspring
    c. the two hereditary units that influence a certain trait separate during gamete formation
    d. each hereditary unit is inherited separately from other hereditary units

_____ 2. One of two or more alternative forms of a gene for a single trait is a(n) _____ .
    a. chiasma
    b. allele
    c. autosome
    d. locus

_____ 3. In the $F_2$ generation of a monohybrid cross involving complete dominance, the expected phenotypic ratio is _____ .
    a. 3:1
    b. 1:1:1:1
    c. 1:2:1
    d. 1:1

_____ 4. In the $F_2$ generation of a cross between a red-flowered four o'clock (homozygous) and a white-flowered four o'clock, the expected phenotypic ratio of the offspring is _____ .
    a. 3/4 red, 1/4 white
    b. 100 percent red
    c. 1/4 red, 1/2 pink, 1/4 white
    d. 100 percent pink

_____ 5. In a testcross, $F_1$ hybrids are crossed to an individual known to be _____ for the trait.
    a. heterozygous
    b. homozygous dominant
    c. homozygous
    d. homozygous recessive

_____ 6. The tendency for dogs to bark while trailing is determined by a dominant gene, _S_, while silent trailing is due to the recessive gene, _s_. In addition, erect ears, _D_, is dominant over drooping ears, _d_. What combination of offspring would be expected from a cross between two erect-eared barkers who are heterozygous for both genes?

    a. 1/4 erect barkers, 1/4 drooping barkers, 1/4 erect silent, 1/4 drooping silent
    b. 9/16 erect barkers, 3/16 drooping barkers, 3/16 erect silent, 1/16 drooping silent
    c. 1/2 erect barkers, 1/2 drooping barkers
    d. 9/16 drooping barkers, 3/16 erect barkers, 3/16 drooping silent, 1/16 erect silent

_____ 7. A man with type A blood could be the father of _____ .
    a. a child with type A blood
    b. a child with type B blood
    c. a child with type O blood
    d. a child with type AB blood
    e. all of the above

_____ 8. A single gene that affects several seemingly unrelated aspects of an individual's phenotype is said to be _____ .
    a. pleiotropic
    b. epistatic
    c. mosaic
    d. continuous

_____ 9. Suppose two individuals, each heterozygous for the same characteristic, are crossed. The characteristic involves complete dominance. The expected genotypic ratio of their progeny is _____ .
    a. 1:2:1
    b. 1:1
    c. 100 percent of one genotype
    d. 3:1

_____10. If the two homozygous classes in the $F_1$ generation of the cross in exercise 9 are allowed to mate, the observed genotypic ratio of the offspring will be _____ .
    a. 1:1
    b. 1:2:1
    c. 100 percent of one genotype
    d. 3:1

_____11. Applying the types of inheritance learned in this chapter in the text, the skin color trait in humans exhibits _____ .
    a. pleiotropy
    b. epistasis
    c. environmental effects
    d. continuous variation

# Chapter Objectives/Review Questions

*Page    Objectives/Questions*

(176)  1.  What was the prevailing method of explaining the inheritance of traits before Mendel's work with pea plants?

(177)  2.  Garden pea plants are naturally _____-fertilizing, but Mendel took steps to _____-fertilize them for his experiments.

(177)  3.  _____ are units of information about specific traits; they are passed from parents to offspring.

(177)  4.  What is the general term applied to the location of a gene on a chromosome?

(177)  5.  Define *allele*; how many types of alleles are present in the genotypes *Tt? tt? TT?*

(177)  6.  Explain the meaning of a true-breeding lineage.

(177)  7.  When two alleles of a pair are identical, it is a _____ condition; if the two alleles are different, then it is a _____ condition.

(177)  8.  Distinguish a dominant allele from a recessive allele.

(177)  9.  _____ refers to the genes present in an individual; _____ refers to an individual's observable traits.

(178) 10.  Offspring of _____ crosses are heterozygous for the one trait being studied.

(179) 11.  Explain why probability is useful to genetics.

(179) 12.  Be able to use the Punnett-square method of solving genetics problems.

(179) 13.  Define the *testcross* and cite an example.

(179) 14.  Mendel's theory of _____ states that during meiosis, the two genes of each pair separate from each other and end up in different gametes.

(179) 15.  Describe the testcross.

(180) 16.  Be able to solve dihybrid genetic crosses.

(181) 17.  Mendel's theory of _____ _____ states that gene pairs on homologous chromosomes tend to be sorted into one gamete or another independently of how gene pairs on other chromosomes are sorted out.

(182) 18.  Distinguish between complete dominance, incomplete dominance, and codominance.

(182) 19.  Define *multiple allele system* and cite an example.

(183) 20.  Explain why sickle-cell anemia is a good example of pleiotropy.

(184) 21.  Gene interaction involving two alleles of a gene that mask alleles of another gene is called _____ .

(186) 22.  List possible explanations for less predictable trait variations that are observed.

(186) 23.  List two human traits that are explained by continuous variation.

(188) 24.  Himalayan rabbits and garden hydrangeas are good examples of environmental effects on _____ expression.

# Integrating and Applying Key Concepts

Solve the following genetics problem:

In garden peas, one pair of alleles controls the height of the plant and a second pair of alleles controls flower color. The allele for tall (*D*) is dominant to the allele for dwarf (*d*), and the allele for purple (*P*) is dominant to the allele for white (*p*). A tall plant with purple flowers crossed with a tall plant with white flowers produces 3/8 tall purple, 3/8 tall white, 1/8 dwarf purple, and 1/8 dwarf white. What are the genotypes of the parents?

# 12

# CHROMOSOMES AND HUMAN GENETICS

## Interactive Exercises

**Selected Words:** *Hutchinson-Gilford progeria syndrome, wild type* allele, Colchicum autumnale

**Boldfaced, Page-Referenced Terms**

(192) syndrome _____

_____

(194) genes _____

_____

(194) homologous chromosomes _____

_____

(194) alleles _____

_____

(194) crossing over _____

_____

(194) genetic recombination _____

_____

(194) X chromosome _____

_____

(194) Y chromosome _____

_____

(194) sex chromosomes _____

_____

(194) autosomes _____

_____

(194) karyotype _____

_____

## Fill-in-the-Blanks

Heritable traits arise from units of molecular information (located on chromosomes) that are known as (1) _____ . A pair of chromosomes that have the same length, shape, gene sequence, and interact during meiosis, are known as (2) _____ chromosomes. (3) _____ are slightly different forms of the same gene that arose through mutation. The most common form of an allele is called a (4) _____ type allele. An event known as (5) _____ _____ occurring early in the meiotic process functions to exchange gene segments between the members of a pair and thus provide genetic (6) _____ . Human X and Y chromosomes that determine gender are examples of (7) _____ chromosomes. Chromosomes that are identical in both sexes of a species are called (8) _____ . The (9) _____ of an individual (or species) is a preparation of sorted chromosomes. To prepare a karyotype, technicians culture cells (10) _____ _____ (to culture in glass). During culture, a chemical compound named (11) _____ is added to the culture medium. The purpose of this chemical is to block spindle formation and arrest nuclear divisions at the (12) _____ stage of mitosis when the chromosomes have condensed to their maximum for viewing. Cells are then moved to the bottoms of culture tubes by a spinning force called (13) _____. Following saline treatment to separate the chromosomes, cells are fixed, stained, and photographed for study.

## 12.3. SEX DETERMINATION IN HUMANS (pp. 196–197)
## 12.4. EARLY QUESTIONS ABOUT GENE LOCATIONS (pp. 198–199)

**Selected Words:** *nonsexual traits,* SRY gene, <u>Drosophila</u> <u>melanogaster,</u> <u>Zea</u> <u>mays</u>

### Boldfaced, Page-Referenced Terms

(198)  X-linked genes _____

_____

(198)  Y-linked genes _____

_____

(198)  linkage group _____

_____

(199)  cytological markers _____

_____

### Matching

Choose the most appropriate letter.

1. ___X-linked genes
2. ___linkage group
3. ___SRY
4. ___Y-linked genes
5. ___reciprocal crosses
6. ___cytological markers

A. Male-determining gene found on the Y chromosome; expression leads to testes formation
B. In the first cross one parent displays the trait of interest whereas in the second cross, the other parent displays it
C. The block of genes located on each type of chromosome; these genes tend to travel together in inheritance
D. Distinguishing physical differences seen when comparing chromosomes with a microscope
E. Found only on the Y chromosome
F. Found only on the X chromosome

### Complete the Table

7. Complete the Punnett square table at the right which brings Y-bearing and X-bearing sperm together randomly in fertilization.

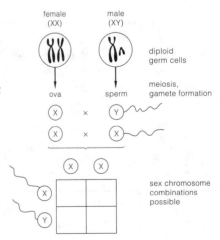

### Dichotomous Choice

Answer the following questions related to the Punnett Square table just completed.

8. Male humans transmit their Y chromosome only to their (sons/ daughters).
9. Male humans receive their X chromosome only from their (mothers/fathers).
10. Human mothers and fathers each provide an X chromosome for their (sons/daughters).

## Problems

11. In Morgan's experiments with Drosophila, white eyes are determined by a recessive X-linked gene and the wild-type or normal brick-red eyes are due to its dominant allele. Use symbols of the following types: $X^wY$ = a white-eyed male; $X^WX^W$ = a homozygous normal red female.
    a. What offspring can be expected from a cross of a white-eyed male and a homozygous normal female?
    b. In addition, show the genotypes and phenotypes of the $F_2$ offspring.

12. Which of the following represents a chromosome that has undergone crossover and recombination? It is assumed that the organism involved is heterozygous with the genotype: $\begin{array}{c|c} A & a \\ B & b \end{array}$ ──────

    a. $\begin{array}{c} A \\ B \end{array}$        b. $\begin{array}{c} A \\ B \end{array}$        c. $\begin{array}{c} A \\ B \end{array}$        d. $\begin{array}{c} A \\ B \end{array}$

## 12.5. RECOMBINATION PATTERNS AND CHROMOSOME MAPPING (pp. 200–201)

## 12.6. HUMAN GENETIC ANALYSIS (pp. 202–203)

*Selected Words:* crossover probability, map distance, physical distance, color blindness, hemophilia

### Boldfaced, Page-Referenced Terms

(201) linkage mapping _____

_____

(202) pedigree _____

_____

(202) genetic abnormality _____

_____

(203) genetic disorder _____

_____

(203) genetic disease _____

_____

Early investigators realized that the frequency of crossing over (data from actual genetic crosses) could be used as a tool to map genes on chromosomes. Genetic maps are graphic representations of the relative positions and distances between genes on each chromosome (linkage group). Geneticists arbitrarily equate l percent of crossover (recombination) with l genetic map unit. In the following questions, a line is used to represent a chromosome with its linked genes.

## Problems

1. If genes A and B are twice as far apart on a chromosome as genes C and D, how often would it be expected that crossing over occurs between genes A and B as between genes C and D? _____

_____

2. Following breeding experiments and a study of crossovers, the following map distances were recorded: gene C to gene B = 35 genetic map units; B to A = 10 map units, and A to C = 45 map units. Which of the following maps is correct?

       A  B  C              B  A  C              B  C  A              B  A  C
    a. └──┴──┘          b. └──┴──┘          c. └──┴──┘          d. └──┴──┘

Individuals with dominant gene N have the nail-patella syndrome and exhibit abnormally developed fingernails and absence of kneecaps; the recessive gene n is normal. Geneticists have learned that the genes determining the ABO blood groups and the nail-patella gene are linked, both found on chromosome 9. A woman with blood type A and nail-patella syndrome marries a man who has blood type O and has normal fingernails and kneecaps. This couple has three children. One daughter has blood type A like her mother but has normal fingernails and kneecaps. A second daughter has blood type O with nail-patella syndrome. A son has blood type O and normal fingernails and kneecaps. The chromosomes of the parents are diagrammed below.

3. Complete the exercise by showing the linked genes of each of the children below.

♀                          ♂

A        n          O        n
_____              _____
           X                     (or An/ON X On/On)
_____              _____
O        N          O        n

Daughter 1    Daughter 2    Son
_____        _____        _____

_____        _____        _____

4. Explain how the son has the genotype On/On when only one parent has a chromosome with genes O and n linked.

## Matching

The following standardized symbols are used in the construction of pedigree charts. Choose the appropriate description for each.

5. __ ■ ●          A. Individual showing the trait being tracked
                   B. Male
6. __ ○            C. Successive generations
7. __ ◇            D. Offspring with birth order left to right
                   E. Female
8. __ □            F. Sex unspecified; numerals present indicates number of children
9. __ I, II, III, IV...    G. Marriage/mating

10. __ □—○

11. __ ○■■○

## Short Answer

12. Distinguish between these terms: genetic abnormality, genetic disorder, and genetic disease.

_____

_____

_____

_____

_____

## Complete the Table

13. Complete the table below by indicating whether the genetic disorder listed is due to inheritance that is autosomal recessive, autosomal dominant, X-linked recessive, X-linked dominant, or due to changes in chromosome number or changes in chromosome structure. Refer to Table 12.1 in the text.

| Genetic Disorder | Inheritance Pattern |
|---|---|
| a. Galactosemia | |
| b. Achondroplasia | |
| c. Hemophilia | |
| d. Anhidrotic ectodermal dysplasia | |
| e. Huntington disorder | |
| f. Duchenne muscular dystrophy | |
| g. Turner and Down syndrome | |
| h. Faulty enamel trait | |
| i. Cri-du-chat syndrome | |
| j. XYY condition | |
| k. Color blindness | |
| l. Fragile X syndrome | |
| m. Progeria | |
| n. Klinefelter syndrome | |
| o. Tay-Sachs disorder | |
| p. Phenylketonuria | |

# 12.7. PATTERNS OF AUTOSOMAL INHERITANCE (pp. 204–205)
# 12.8. PATTERNS OF X-LINKED INHERITANCE (pp. 206–207)

**Selected Words:** *galactosemia, Tay-Sachs disorder,* Huntington disorder, *achondroplasia,* Hemophilia A, *Duchenne muscular dystrophy, faulty enamel trait, amyotrophic lateral disease*

## Choice

For questions 1–18, choose from the following patterns of inheritance; some items may require more than one letter.

a. autosomal recessive      b. autosomal dominant      c. X-linked recessive      d. X-linked dominant

1. ____ The trait is expressed in heterozygous females.

2. ____ Heterozygotes can remain undetected.

3. ____ The trait appears in each generation.

4. ____ The recessive phenotype shows up far more often in males than in females.

5. ____ Both parents may be heterozygous normal.

6. ____ If one parent is heterozygous and the other homozygous recessive, there is a 50 percent chance any child of theirs will be heterozygous.

7. ____ The allele is usually expressed, even in heterozygotes.

8. ____ The trait is expressed in heterozygous females.

9. ____ Heterozygous normal parents can expect that one-fourth of their children will be affected by the disorder.

10. ____ A son cannot inherit the recessive allele from his father, but his daughter can.

11. ____ Females can mask this gene, males cannot.

12. ____ The trait is expressed in heterozygotes of either sex.

13. ____ Heterozygous women transmit the allele to half their offspring, regardless of sex.

14. ____ Individuals displaying this type of disorder will always be homozygous for the trait.

15. ____ The trait is expressed in both the homozygote and the heterozygote.

16. ____ Heterozygous females will transmit the recessive gene to half their sons and half their daughters.

17. ____ If both parents are heterozygous, there is a 50 percent chance that each child will be heterozygous.

18. ____ A son cannot inherit the allele responsible for the trait from an affected father, but all his daughters will.

## Problems

19. The autosomal allele that causes albinism ($a$) is recessive to the allele for normal pigmentation ($A$). A normally pigmented woman whose father is an albino marries an albino man whose parents are normal. They have three children, two normal and one albino. List the genotypes for each person listed.

_____

_____

20. Huntington disorder is a rare form of autosomal dominant inheritance, $H$; the normal gene is $h$. The disease causes progressive degeneration of the nervous system with onset exhibited near middle age. An apparently normal man in his early twenties learns that his father has recently been diagnosed as having Huntington disorder. What are the chances that the son will develop this disorder?

_____

21. A color-blind man and a woman with normal vision whose father was color-blind have a son. Color blindness, in this case, is caused by an X-linked recessive gene. If only the male offspring are considered, what is the probability that their son is color-blind?

_____

22. Hemophilia A is caused by an X-linked recessive gene. A woman who is seemingly normal but whose father was a hemophiliac marries a normal man. What proportion of their sons will have hemophilia? What proportion of their daughters will have hemophilia? What proportion of their daughters will be carriers?

_____

23. Refer to Figure 12.15, p. 206, in the text. The adjacent pedigree shows the pattern of inheritance of color blindness in a family (persons with the trait are indicated by black circles). What is the chance that the third-generation female indicated by the arrow (below) will have a color-blind son if she marries a normal male? A color-blind male?

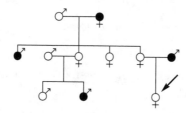

## 12.9. CHANGES IN CHROMOSOME NUMBER (pp. 208–209)
## 12.10. CHANGES IN CHROMOSOME STRUCTURE (pp. 210–211)

**Selected Words:** _tetra_ploid, trisomic, monosomic, _Down syndrome, Turner syndrome, Klinefelter syndrome, cri-du-chat, fragile X syndrome, non_homologous chromosome

### Boldfaced, Page-Referenced Terms

(208) aneuploidy _____

_____

(208) polyploidy _____

_____

(208) nondisjunction _____

_____

(209) double-blind studies _____

_____

(210) deletion _____

_____

(210) duplication _____

_____

(210) inversion _____

_____

(210) translocation _____

_____

## Complete the Table

1. Complete the table below to summarize the major categories and mechanisms of chromosome number change in organisms.

| Category of Change | Description |
|---|---|
| a. Aneuploidy | |
| b. Polyploidy | |
| c. Nondisjunction | |

## Short Answer

2. If a nondisjunction occurs at anaphase I of the first meiotic division, what will be the proportion of abnormal gametes (for the chromosomes involved in the nondisjunction)? (p. 208) _____

_____

3. If a nondisjunction occurs at anaphase II of the second meiotic division, what will be the proportion of abnormal gametes (for the chromosomes involved in the nondisjunction)? (p. 208) _____

_____

4. Contrast the effects of polyploidy in plants and humans. (p. 208) _____

_____

5. Define the following terms: *tetraploid, trisomic,* and *monosomic.* (p. 208) _____

_____

## Choice

For questions 6–15, choose from the following:

a. Down syndrome      b. Turner syndrome      c. Klinefelter syndrome      d. XYY condition

6. ____ XXY male

7. ____ Ovaries nonfunctional and secondary sexual traits fail to develop at puberty

8. ____ Testes smaller than normal, sparse body hair, and some breast enlargement

9. ____ Could only be caused by a nondisjunction in males

10. ____ Older children smaller than normal with distinctive facial features; small skin fold over the inner corner of the eyelid

11. ____ X0 female; often abort early; distorted female phenotype

12. ____ Males that tend to be taller than average; some mildly retarded but most are phenotypically normal

13. ____ Injections of testosterone reverse feminized traits but not the mental retardation

14. ____ Trisomy 21; skeleton develops more slowly than normal with slack muscles

15. ____ At one time these males were thought to be genetically predisposed to become criminals

## Label-Match

On rare occasions, chromosome structure becomes abnormally rearranged. Such changes may have profound effects on the phenotype of an organism. Label the following diagrams of abnormal chromosome structure as a deletion, a duplication, an inversion, or a translocation. Complete the exercise by matching and entering the letter of the proper description in the parentheses following each label.

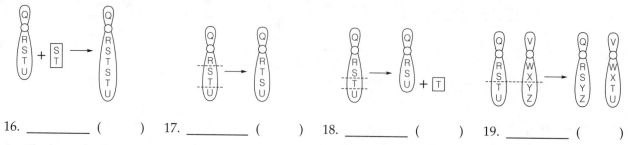

16. _____ (    )    17. _____ (    )    18. _____ (    )    19. _____ (    )

A. The loss of a chromosome segment; an example is cri-du-chat disorder
B. A gene sequence in excess of its normal amount in a chromosome; an example is the fragile X syndrome
C. A chromosome segment that separated from the chromosome and then was inserted at the same place, but in reverse; this alters the position and order of the chromosome's genes; possibly promoted human evolution
D. The transfer of part of one chromosome to a nonhomologous chromosome; an example is when chromosome 14 ends up with a segment of chromosome 8; translocation Down syndrome is an example

## 12.11. Focus on Science: PROSPECTS IN HUMAN GENETICS (pp. 212–213)

*Selected Words:* PKU, *cleft lip, genetic counseling, prenatal diagnosis, preimplantation diagnosis,* pre-pregnancy stage, *test-tube babies*

### Boldfaced, Page-Referenced Terms

(213) amniocentesis _____

_____

(213) chorionic villi sampling (CVS) _____

_____

(213) abortion _____

_____

(213) in-vitro fertilization _____

_____

## Complete the Table

1. Complete the table below, which summarizes methods of dealing with the problems of human genetics. Choose from phenotypic treatments, genetic screening, genetic counseling, and prenatal diagnosis.

| Method | Description |
| --- | --- |
| a. | Detects genetic disorders before birth; may use karyotypes, biochemical tests, amniocentesis, CVS, in-vitro fertilization, and possibly abortion; an example is a pregnancy at risk in a mother forty-five years old |
| b. | Parents at risk requesting emotional support, advice, and risk predictions from clinical psychologists, geneticists, and social workers |
| c. | Suppressing or minimizing symptoms of genetic disorders by surgical intervention, controlling diet or environment, or chemically modifying genes; PKU and cleft lip are examples |
| d. | Large-scale programs to detect affected persons or carriers in a population; early detection may allow introduction of preventive measures before symptoms develop; PKU screening for newborns is an example |

# Self-Quiz

___ 1. All the genes located on a given chromosome compose a _____ .
    a. karyotype
    b. bridging cross
    c. wild-type allele
    d. linkage group

___ 2. Chromosomes other than those involved in sex determination are known as _____.
    a. nucleosomes
    b. heterosomes
    c. alleles
    d. autosomes

___ 3. The farther apart two genes are on a chromosome, _____ .
    a. the less likely that crossing over and recombination will occur between them
    b. the greater will be the frequency of crossing over and recombination between them
    c. the more likely they are to be in two different linkage groups
    d. the more likely they are to be segregated into different gametes when meiosis occurs

___ 4. Karyotype analysis is _____ .
    a. a means of detecting and reducing mutagenic agents
    b. is a surgical technique that separates chromosomes that have failed to segregate properly during meiosis II
    c. used in prenatal diagnosis to detect chromosomal mutations and metabolic disorders in embryos
    d. a process that substitutes defective alleles with normal ones

___ 5. Which of the following did Morgan and his research group not do?
    a. They isolated and kept under culture fruit flies with the sex-linked recessive white-eyed trait.
    b. They developed the technique of amniocentesis.
    c. They discovered X-linked genes.
    d. Their work reinforced the concept that each gene is located on a specific chromosome.

6. Red-green color blindness is a sex-linked recessive trait in humans. A color-blind woman and a man with normal vision have a son. What are the chances that the son is color-blind? If the parents ever have a daughter, what is the chance for each birth that the daughter will be color-blind? (Consider only the female offspring).
   a. 100 percent, 0 percent
   b. 50 percent, 0 percent
   c. 100 percent, 100 percent
   d. 50 percent, 100 percent
   e. none of the above

7. Suppose that a hemophilic male (X-linked recessive allele) and a female carrier for the hemophilic trait have a nonhemophilic daughter with Turner syndrome. Nondisjunction could have occurred in _____.
   a. both parents
   b. neither parent
   c. the father only
   d. the mother only

8. Nondisjunction involving the X chromosome occurs during oogenesis and produces two kinds of eggs, XX and O (no X chromosome). If normal Y sperm fertilize the two types, which genotypes are possible?
   a. XX and XY
   b. XXY and YO
   c. XYY and XO
   d. XYY and YO

9. Of all phenotypically normal males in prisons, the type once thought to be genetically predisposed to becoming criminals was the group with _____ .
   a. XXY disorder
   b. XYY disorder
   c. Turner syndrome
   d. Down syndrome

10. Amniocentesis is _____ .
   a. a surgical means of repairing deformities
   b. a form of chemotherapy that modifies or inhibits gene expression or the function of gene products
   c. used in prenatal diagnosis to detect chromosomal mutations and metabolic disorders in embryos
   d. a form of gene-replacement therapy

## Chapter Objectives/Review Questions

*Page*      *Objectives/Questions*

(192)      1. A _____ is a set of symptoms that characterize a disorder.
(194)      2. The units of information about heritable traits are known as _____ .
(194)      3. Diploid (2n) cells have pairs of _____ chromosomes.
(194)      4. _____ are different molecular forms of the same gene that arise through mutation; a _____ type allele is the most common form of a gene.
(194)      5. State the circumstances required for crossing over and describe the results.
(194)      6. Name and describe the sex chromosomes in human males and females.
(194)      7. Human X and Y chromosomes fall in the general category of _____ chromosomes; all other chromosomes in an individual's cells are the same in both sexes and are called _____ .
(194–195)  8. Define karyotype; briefly describe its preparation and value.
(196)      9. Explain meiotic segregation of sex chromosomes to gametes and the subsequent random fertilization that determines sex in many organisms.
(196)      10. A newly identified region of the Y chromosome called _____ appears to be the master gene for sex determination.
(198)      11. All the genes on a specific chromosome are called a _____ group.
(198)      12. In whose laboratory was sex linkage in fruit flies discovered? When?
(199)      13. Define and give examples of cytological markers.
(200–201) 14. State the relationship between crossover frequency and the location of genes on a chromosome.
(202)      15. A _____ chart or diagram is used to study genetic connections between individuals.

(202–203) 16. A genetic _____ is a rare, uncommon version of a trait whereas a genetic _____ causes mild to severe medical problems.

(203) 17. Describe what is meant by a genetic disease.

(204–207) 18. As viewed on a pedigree chart, describe the characteristics of autosomal recessive inheritance, autosomal dominant inheritance, as well as recessive and dominant X-linked inheritance; cite one example of each.

(208) 19. When gametes or cells of an affected individual end up with one extra or one less than the parental number of chromosomes, it is known as _____ ; relate this concept to monosomy and trisomy.

(208) 20. Having three or more complete sets of chromosomes is called _____ .

(208) 21. _____ is the failure of the chromosomes to separate at either meiosis I or meiosis II.

(208–209) 22. Trisomy 21 is known as _____ syndrome; Turner syndrome has the chromosome constitution, _____ ; XXY chromosome constitution is _____ syndrome; taller than average males with sometimes slightly depressed IQ's have the _____ condition.

(209) 23. Explain what is meant by double-blind studies.

(210–211) 24. A(n) _____ is a loss of a chromosome segment; a(n) _____ is a gene sequence separated from a chromosome but which then was inserted at the same place, but in reverse; a(n) _____ is a repeat of several gene sequences on the same chromosome; a(n) _____ is the transfer of part of one chromosome to a nonhomologous chromosome.

(212) 25. List some benefits of genetic screening and genetic counseling to society.

(212) 26. Define phenotypic treatment and describe one example.

(213) 27. Explain the procedures used in two types of prenatal diagnosis, amniocentesis and chorionic villi analysis; compare the risks.

(213) 28. Discuss some of the ethical considerations that might be associated with a decision of induced abortion.

(213) 29. _____ _____ fertilization is the fertilizing of eggs in a petri dish.

---

# Integrating and Applying Key Concepts

1. The parents of a young boy bring him to their doctor. They explain that the boy does not seem to be going through the same vocal developmental stages as his older brother. The doctor orders a common cytogenetics test to be done, and it reveals that the young boy's cells contain two X chromosomes and one Y chromosome. Describe the test that the doctor ordered and explain how and when such a genetic result, XXY, most logically occurred.

2. Solve the following genetics problem. Show rationale, genotypes, and phenotypes. A husband sues his wife for divorce, arguing that she has been unfaithful. His wife gave birth to a girl with a fissure in the iris of her eye, an X-linked recessive trait. Both parents have normal eye structure. Can the genetic facts be used to argue for the husband's suit? Explain your answer.

# 13

# DNA STRUCTURE AND FUNCTION

*Cardboard Atoms and Bent-Wire Bonds*

**DISCOVERY OF DNA FUNCTION**
    Early and Puzzling Clues
    Confirmation of DNA Function

**DNA STRUCTURE**
    Components of DNA
    Patterns of Base Pairing

**DNA REPLICATION AND REPAIR**
    How a DNA Molecule Gets Duplicated
    Monitoring and Fixing the DNA
    Regarding the Chromosomal Proteins

*Focus on Science:* WHEN DNA CAN'T BE FIXED

## Interactive Exercises

*Cardboard Atoms and Bent-Wire Bonds* (pp. 216–217)

## 13.1. DISCOVERY OF DNA FUNCTION (pp. 218–219)

*Selected Words:* Streptococcus pneumoniae; *pathogenic,* Escherichia coli.

*Boldfaced, Page-Referenced Terms*

(216) DNA, deoxyribonucleic acid _____

_____

(218) bacteriophages _____

_____

1.  Complete the table below which traces the discovery of DNA function.

| Investigators | Year(s) | Contribution |
|---|---|---|
| Miescher | 1868 | a. |
| b. | 1928 | discovered the transforming principle in *Streptococcus pneumoniae;* live, harmless R cells were mixed with dead S cells, R cells became S cells |
| Avery (also MacLeod and McCarty) | 1944 | c. |
| d. | mid-1940s | studied the infectious cycle of bacteriophages |
| Hershey and Chase | 1952 | e. |

## 13.2. DNA STRUCTURE (pp. 220–221)

### Boldfaced, Page-Referenced Terms

(220) nucleotide _____

_____

(220) adenine, A _____

_____

(220) guanine, G _____

_____

(220) thymine, T _____

_____

(220) cytosine, C _____

_____

(220) X-ray diffraction images _____

_____

1. List the three parts of a nucleotide. _____

_____

### Labeling

Four nucleotides are illustrated below. In the blank, label each nitrogen-containing base correctly as guanine, thymine, cytosine, or adenine. In the parentheses following each blank, indicate whether that nucleotide base is a purine (pu) or a pyrimidine (py). See p. 220 in the main text for definitions.

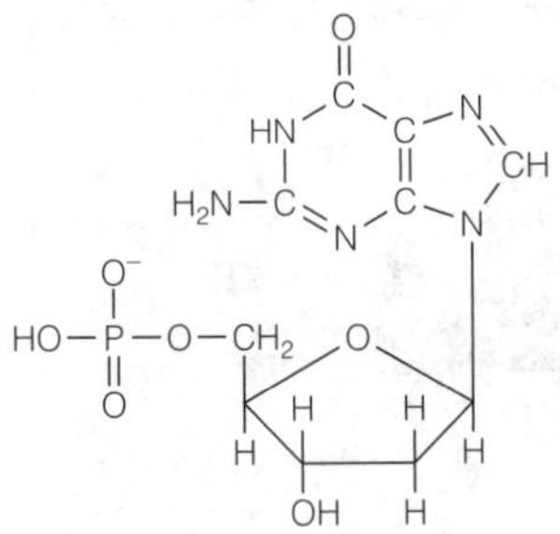

2. 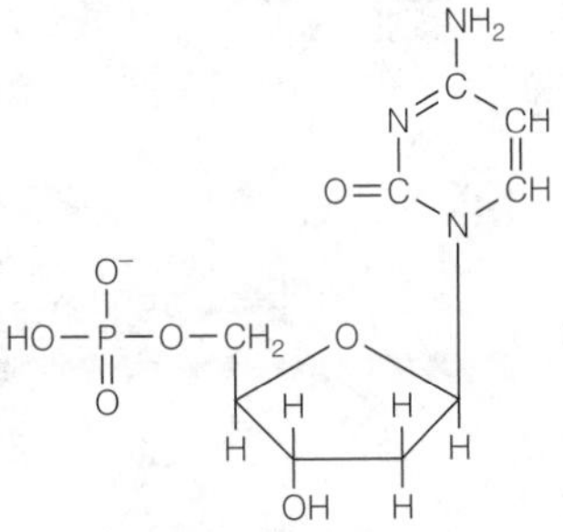 _____ (    )    3. 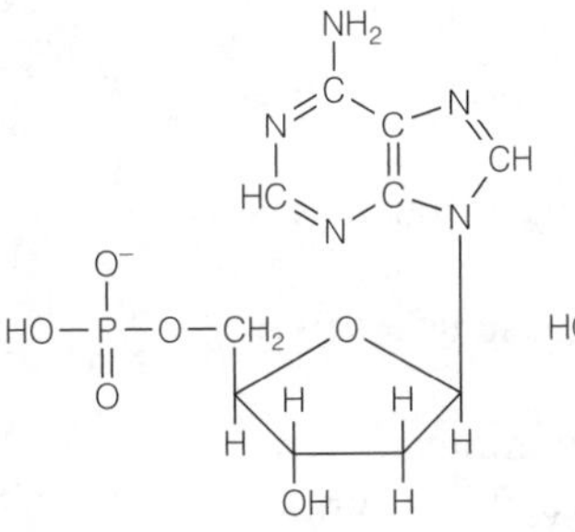 _____ (    )    4. 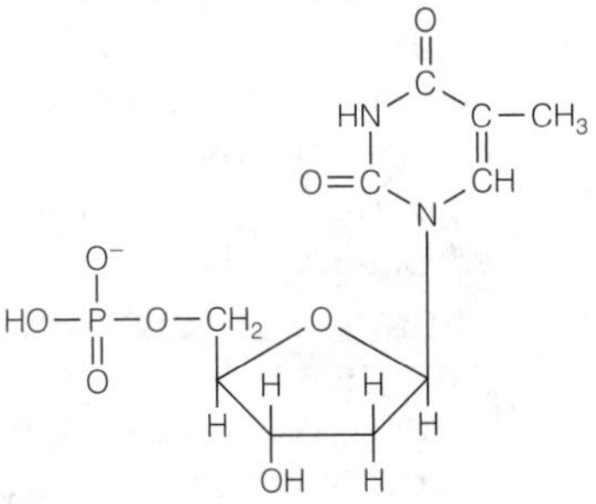 _____ (    )    5. _____ (    )

## Label-Match

Identify each indicated part of the righthand DNA illustration. Choose from these answers: phosphate group, purine, pyrimidine, nucleotide, and deoxyribose. Complete the exercise by matching and entering the letter of the proper structure description in the parentheses following each label.

The following DNA memory devices may be helpful: Use *pyr*CUT to remember that the single-ring **pyr**imidines are **c**ytosine, **u**racil, and **t**hymine; use *pur*AG to remember that the double-ring **pur**ines are **a**denine and **g**uanine; pyrimidine is a *long* name for a *narrow* molecule; purine is a *short* name for a *wide* molecule; to recall the number of hydrogen bonds between DNA bases, remember that AT = 2 and CG = 3.

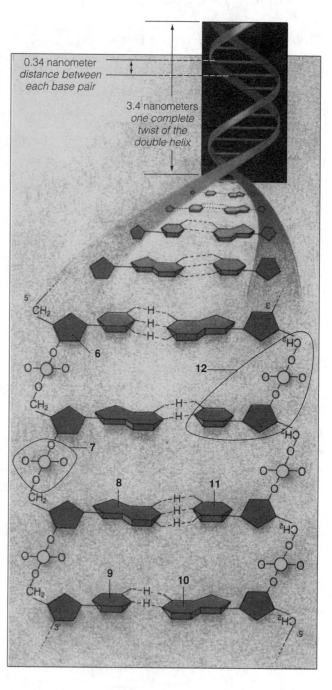

6. _____ ( )

7. _____

   _____ ( )

8. _____ ( )

9. _____ ( )

10. _____ ( )

11. _____ ( )

12. _____ ( )

A. The pyrimidine is thymine because it has two hydrogen bonds.
B. A five-carbon sugar joined to two phosphate groups in the upright portion of the DNA ladder.
C. The purine is guanine because it has three hydrogen bonds.
D. The pyrimidine is cytosine because it has three hydrogen bonds.
E. The purine is adenine because it has two hydrogen bonds.
F. Composed of three smaller molecules: a phosphate group, five-carbon deoxyribose sugar, and a nitrogenous base (in this case, a pyrimidine).
G. A chemical group that joins two sugars in the upright portion of the DNA ladder.

## True-False

If the statement is true, write a T in the blank. If false, correct it by changing the underlined word(s) and writing the correct word(s) in the blank.

_____13. DNA is composed of four different types of <u>nucleotides</u>.

_____14. In the <u>DNA</u> of every species, the amount of adenine present always equals the amount of thymine, and the amount of cytosine always equals the amount of guanine (A = T and C = G).

_____15. In a nucleotide, the phosphate group is attached to the <u>nitrogen-containing base</u>, which is attached to the five-carbon sugar.

_____16. Watson and Crick built their model of DNA in the early <u>1950s</u>.

_____17. Guanine pairs with cytosine and adenine pairs with <u>thymine</u> by forming hydrogen bonds between them.

## Fill-in-the-Blanks

Base (18) _____ between the two nucleotide strands in DNA is (19) _____ for all species (A-T; G-C).

The base (20) _____ (determining which base follows the next in a nucleotide strand) is (21) _____

from species to species.

## Short Answer

22. Explain why understanding the structure of DNA helps scientists understand how living organisms can have so much in common at the molecular level and yet be so diverse at the whole organism level (p. 221).

_____

_____

## 13.3. DNA REPLICATION AND REPAIR (pp. 222–223)

## 13.4. *Focus on Health:* WHEN DNA CAN'T BE FIXED (p. 224)

**Selected Words:** *semiconservative, origins, replication forks, "start" tags, continuous assembly, discontinuous assembly of bases, malignant melanoma, xeroderma pigmentosum, basal cell carcinoma, squamous cell carcinoma.*

## Boldfaced, Page-Referenced Terms

(222) DNA replication _____

_____

(222) DNA polymerases _____

_____

(222) DNA ligases _____

_____

(222) DNA repair _____

_____

## Label

1. The term semiconservative replication refers to the fact that each new DNA molecule resulting from the replication process is "half-old, half-new." In the illustration below, complete the replication required in the middle of the molecule by adding the required letters representing the missing nucleotide bases. Recall that ATP energy and the appropriate enzymes are actually required in order to complete this process.

| T- | ___ | | ___ | -A |
| G- | ___ | | ___ | -C |
| A- | ___ | | ___ | -T |
| C- | ___ | | ___ | -G |
| C- | ___ | | ___ | -G |
| C- | ___ | | ___ | -G |
| old | new | | new | old |

## True-False

If false, explain why.

_____ 2. The hydrogen bonding of adenine to guanine is an example of complementary base pairing.

_____ 3. The replication of DNA is considered a conserving process because the same four nucleotides are used again and again during replication.

_____ 4. Each parent strand remains intact during replication, and a new companion strand is assembled on each of those parent strands.

_____ 5. Some of the enzymes associated with DNA assembly repair errors during the replication process.

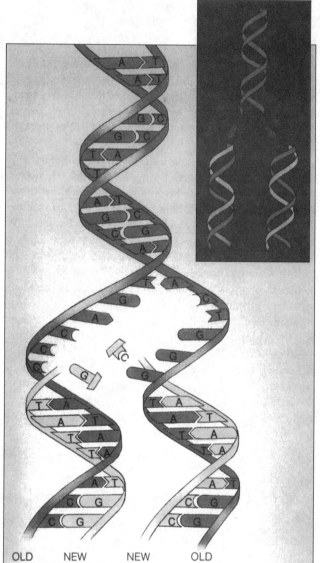

## Matching

Match the numbers on the illustration below to the letters representing structures and functions.

6. ___     A. The enzyme (ligase) links short DNA segments together.
           B. One parent DNA strand.
7. ___     C. Enzymes unwind DNA double helix.
8. ___     D. Enzymes add short "primer" segments to begin chain assembly.
           E. Enzymes (DNA polymerase) assemble new DNA strands.
9. ___     F. Newly-forming DNA strand.

10. ___

11. ___

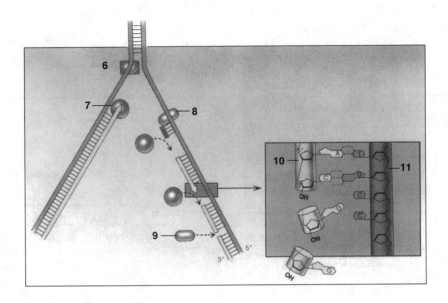

---

# Self-Quiz

___ 1. Each DNA strand has a backbone that consists of alternating _____ .
   a. purines and pyrimidines
   b. nitrogen-containing bases
   c. hydrogen bonds
   d. sugar and phosphate molecules

___ 2. In DNA, complementary base pairing occurs between _____ .
   a. cytosine and uracil
   b. adenine and guanine
   c. adenine and uracil
   d. adenine and thymine

___ 3. Adenine and guanine are _____ .
   a. double-ringed purines
   b. single-ringed purines
   c. double-ringed pyrimidines
   d. single-ringed pyrimidines

___ 4. Franklin used the technique known as _____ to determine many of the physical characteristics of DNA.
   a. transformation
   b. transmission electron microscopy
   c. density-gradient centrifugation
   d. X-ray diffraction

___ 5. The significance of Griffith's experiment that used two strains of pneumonia-causing bacteria is that _____ .
a. the conserving nature of DNA replication was finally demonstrated
b. it demonstrated that harmless cells had become permanently transformed through a change in the bacterial hereditary system
c. it established that pure DNA extracted from disease-causing bacteria transformed harmless strains into "pathogenic strains"
d. it demonstrated that radioactively labeled bacteriophages transfer their DNA but not their protein coats to their host bacteria

___ 6. The significance of the experiments in which $^{32}P$ and $^{35}S$ were used is that _____ .
a. the semiconservative nature of DNA replication was finally demonstrated
b. it demonstrated that harmless cells had become permanently transformed through a change in the bacterial hereditary system
c. it established that pure DNA extracted from disease-causing bacteria transformed harmless strains into "killer strains"
d. it demonstrated that radioactively labeled bacteriophages transfer their DNA but not their protein coats to their host bacteria

___ 7. Franklin's research contribution was essential in _____ .
a. establishing the double-stranded nature of DNA
b. establishing the principle of base pairing
c. establishing most of the principal structural features of DNA
d. all of the above

___ 8. When Griffith injected mice with a mixture of dead pathogenic cells—encapsulated S cells and living, unencapsulated R cells of pneumonia bacteria—he discovered that _____ .
a. the previously harmless strain had permanently inherited the capacity to build protective capsules
b. the dead mice teemed with living pathogenic (R) cells
c. the killer strain R was encased in a protective capsule
d. all of the above

___ 9. A single strand of DNA with the base-pairing sequence C-G-A-T-T-G is compatible only with the sequence _____ .
a. C-G-A-T-T-G
b. G-C-T-A-A-G
c. T-A-G-C-C-T
d. G-C-T-A-A-C

___10. Which of the following statements is not true?
a. The DNA molecules of each species show unique differences in the sequence of base pairs along their lengths.
b. One strand of a DNA molecule runs in the $5' \rightarrow 3'$ direction and the other runs in the $3' \rightarrow 5'$ direction.
c. During cell division, enzymes unwind and separate DNA's two strands. Each strand remains intact throughout the process and enzymes assemble a new, complementary strand on each parent strand.
d. Enzymes involved in replication also repair base-pairing errors which have appeared in the nucleotide sequence of DNA.

## Crossword Puzzle: DNA Structure and Function

## ACROSS

3. Forms two hydrogen bonds with its mate
5. Five carbon sugar + phosphate group + nitrogenous base
8. Developed the best X-ray diffraction images of DNA and deduced its basic dimensions
9. Purine or pyrimidine
10. DNA polymerases and DNA ligases _____ DNA
11. _____ diffraction was used to deduce the width and periodicity of DNA.
12. A form of cancer
14. Watson and _____
15. DNA _____ assembles DNA nucleotides into a DNA strand.
16. Forms three hydrogen bonds with its mate
20. Duplication of hereditary material during the S-phase of the cell cycle
21. A new DNA strand is assembled on each of the two parent strands of DNA.
22. Single-ringed nitrogenous base
23. Forms two hydrogen bonds with its mate

## DOWN

1. X-ray _____ is a technique that can yield clues about molecular structure.
2. Hershey and _____ demonstrated that DNA contained the hereditary instructions for producing new bacteriophages.
4. Coiled structure, like a circular stairway
6. In most organisms, the molecule that contains the genetic code
7. A "start" tag
9. A class of viruses that infect bacteria
10. Transcription produces this code molecule
13. Forms three hydrogen bonds with its mate
17. Built a model of DNA structure; won Nobel Prize
18. DNA _____ is an enzyme that joins new short stretches of nucleotides into a continuous strand.
19. Double-ringed nitrogenous base

# Chapter Objectives/Review Questions

This section lists general and detailed chapter objectives that can be used as review questions. You can make maximum use of these items by writing answers on a separate sheet of paper. Fill in answers where blanks are provided. To check for accuracy, compare your answers with information given in the chapter or glossary.

| Page | Objectives/Questions |
|---|---|
| (216) | 1. Before 1952, _____ molecules and _____ molecules were suspected of housing the genetic code. |
| (216–219) | 2. Summarize the research carried out by Miescher, Griffith, Avery and colleagues, and Hershey and Chase; state the specific advances made by each in the understanding of genetics. |
| (218) | 3. Viruses called _____ were used in early research efforts to discover the genetic material. |
| (218–219) | 4. Summarize the specific research that demonstrated that DNA, not protein, governed inheritance. |
| (220) | 5. DNA is composed of double-ring nucleotides known as _____ and single-ring nucleotides known as _____ ; the two purines are _____ and _____ , while the two pyrimidines are _____ and _____ . |
| (220) | 6. Draw the basic shape of a deoxyribose molecule and show how a phosphate group is joined to it when forming a nucleotide. |
| (220) | 7. Show how each nucleotide base would be joined to the sugar-phosphate combination drawn in objective 6. |
| (221) | 8. List the pieces of information about DNA structure that Rosalind Franklin discovered through her X-ray diffraction research. |
| (216, 221) | 9. The two scientists who assembled the clues to DNA structure and produced the first model were _____ and _____ . |
| (221–222) | 10. Explain what is meant by the pairing of nitrogen-containing bases (base pairing), and explain the mechanism that causes bases of one DNA strand to join with bases of the other strand. |
| (221) | 11. Describe how the Watson-Crick model reflects the constancy in DNA observed from species to species yet allows for variations from species to species. |
| (221) | 12. Assume that the two parent strands of DNA have been separated and that the base sequence on one parent strand is A-T-T-C-G-C; the base sequence that will complement that parent strand is _____ . |
| (222–223) | 13. Describe how double-stranded DNA replicates from stockpiles of nucleotides. |
| (222) | 14. Explain what is meant by "each parent strand is conserved in each new DNA molecule." |
| (222) | 15. DNA replication is specifically referred to as _____ replication. |
| (222–223) | 16. List four functions that enzymes perform during the process of replication. |
| (222) | 17. During DNA replication, enzymes called DNA _____ assemble new DNA strands. |
| (222–223) | 18. Describe how DNA's structure is monitored and repaired if errors occur. |

# Integrating and Applying Key Concepts

Review the stages of mitosis and meiosis, as well as the process of fertilization. Include what has now been learned about DNA replication and the relationship of DNA to a chromosome. As you cover the stages, be sure each cell receives the proper number of DNA threads.

# 14

# FROM DNA TO PROTEINS

## Interactive Exercises

*Beyond Byssus* (pp. 226–227)

### 14.1. *Focus on Science:* DISCOVERING THE CONNECTION BETWEEN GENES AND PROTEINS (pp. 228–229)

### 14.2. TRANSCRIPTION OF DNA INTO RNA (pp. 230–231)

*Selected Words:* *"code words", "one gene, one-enzyme" hypothesis, sickle-cell anemia, "poly-A tail", pre-mRNA cap.*

*Boldfaced, Page-Referenced Terms*

(227) base sequence _____

_____

(227) transcription _____

_____

(227) translation _____

_____

(227) ribonucleic acid (RNA) _____

_____

(228) gel electrophoresis _____

_____

(230) messenger RNA (mRNA) _____

_____

(230) ribosomal RNA (rRNA) _____

_____

(230) transfer RNA (tRNA) _____

_____

(230) uracil _____

_____

(230) RNA polymerases _____

_____

(230) promoter _____

_____

(231) introns _____

_____

(231) exons _____

_____

## Fill-in-the-Blanks

Which base follows the next in a strand of DNA is referred to as the base (1) _____ . The region of
DNA that calls for the assembly of specific amino acids into a polypeptide chain is a (2) _____ . The
two steps from genes to proteins are called (3) _____ and (4) _____ . In eukaryotes, during
(5) _____ , single-stranded molecules of RNA are assembled on DNA templates in the nucleus. During
(6) _____ , the RNA molecules are shipped from the nucleus into the cytoplasm, where they are used as
templates for assembling (7) _____ chains. Following translation, one or more chains become
(8) _____ into the three-dimensional shape of protein molecules. Proteins have (9) _____ and
(10) _____ roles in cells, including control of DNA.

## Complete the Table

11. Three types of RNA are transcribed from DNA in the nucleus (from genes that code only for RNA). Complete the following table, which summarizes information about these molecules.

| RNA Molecule | Abbreviation | Description/Function |
|---|---|---|
| a. Ribosomal RNA | | |
| b. Messenger RNA | | |
| c. Transfer RNA | | |

## Short Answer

12. List three ways in which a molecule of RNA is structurally different from a molecule of DNA.

_____

_____

_____

13. Cite two similarities in DNA replication and transcription. _____

_____

_____

14. What are the three key ways in which transcription differs from DNA replication? _____

_____

_____

## Sequence

Arrange the steps of transcription in correct chronological sequence. Write the letter of the first step next to 15, the letter of the second step next to 16, and so on.

15. ___     A. The RNA strand grows along exposed bases until RNA polymerase meets a DNA base sequence that signals "stop."

16. ___     B. RNA polymerase binds with the DNA promoter region to open up a local region of the DNA double helix.

17. ___     C. An RNA polymerase enzyme locates the DNA bases of the promoter region of one DNA strand by recognizing DNA-associated proteins near a promoter.

18. ___     D. RNA is released from the DNA template as a free, single-stranded transcript.

19. ___     E. RNA polymerase moves stepwise along exposed nucleotides of one DNA strand; as it moves, the DNA double helix keeps unwinding.

## Completion

20. Suppose the line below represents the DNA strand that will act as a template for the production of mRNA through the process of transcription. Fill in the blanks below the DNA strand with the sequence of complementary bases that will represent the message carried from DNA to the ribosome in the cytoplasm.

T A C A A G A T A A C A T T A T T T C C T A C C G T C A T C

(transcribed single-strand of mRNA)

## Label-Match

Newly transcribed mRNA contains more genetic information than is necessary to code for a chain of amino acids. Before the mRNA leaves the nucleus for its ribosome destination, an editing process occurs as certain portions of nonessential information are snipped out. Identify each indicated part of the illustration below; use abbreviations for the nucleic acids. Complete the exercise by matching and entering the letter of the description in the parentheses following each label.

21. _____ (  )

22. _____ (  )

23. _____ (  )

24. _____ (  )

25. _____ (  )

26. _____ _____

_____ (  )

A. The actual coding portions of mRNA
B. Noncoding portions of the newly transcribed mRNA
C. Presence of cap and tail, introns snipped out and exons spliced together
D. Acquiring of a poly-A tail by the modified mRNA transcript
E. The region of the DNA template strand to be copied
F. Reception of a nucleotide cap by the 5′ end of mRNA (the first synthesized)

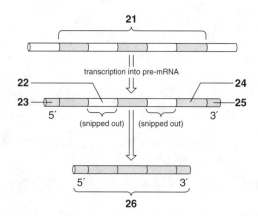

## 14.3. DECIPHERING THE mRNA TRANSCRIPTS (pp. 232–233)

## 14.4. STAGES OF TRANSLATION (pp. 234–235)

*Selected Words:* *"three-bases-at-a-time", "start" signal, tRNA "hook", "wobble effect", initiation, elongation, termination.*

### Boldfaced, Page-Referenced Terms

(232) codons _____

_____

(232) genetic code _____

_____

(232) anticodon _____

_____

(235) polysomes _____

_____

## Matching

1. ___codon
2. ___three at a time
3. ___sixty-one
4. ___the genetic code
5. ___release factors
6. ___ribosome
7. ___anticodon
8. ___the "stop" codons

A. Composed of two subunits, the small subunit with P and A amino acid binding sites as well as a binding site for mRNA
B. Reading frame of the nucleotide bases in mRNA
C. Detach protein and mRNA from the ribosome
D. UAA, UAG, UGA
E. A sequence of three nucleotide bases that can pair with a specific mRNA codon
F. Name for each base triplet in mRNA
G. The number of codons that actually specify amino acids
H. Term for how the nucleotide sequences of DNA and then mRNA correspond to the amino acid sequence of a polypeptide chain

## Complete the Table

9. Complete the following table, which distinguishes the stages of translation.

| Translation Stage | Description |
|---|---|
| a. | Special initiator tRNA loads onto small ribosomal subunit and recognizes AUG; small subunit binds with mRNA, and large ribosomal subunit joins small one. |
| b. | Amino acids are strung together in sequence dictated by mRNA codons as the mRNA strand passes through the two ribosomal subunits; two tRNAs interact at P and A sites. |
| c. | mRNA "stop" codon signals the end of the polypeptide chain; release factors detach the ribosome and polypeptide chain from the mRNA. |

## Completion

10. Given the following DNA sequence, deduce the composition of the mRNA transcript:

TAC   AAG   ATA   ACA   TTA   TTT   CCT   ACC   GTC   ATC

___   ___   ___   ___   ___   ___   ___   ___   ___   ___
(mRNA transcript)

11. Deduce the composition of the tRNA anticodons that would pair with the above specific mRNA codons as these tRNAs deliver the amino acids (identified below) to the P and A binding sites of the small ribosomal subunit.

___   ___   ___   ___   ___   ___   ___   ___   ___   ___
(tRNA anticodons)

12. From the mRNA transcript in exercise 10, use Figure 14.11 of the text to deduce the composition of the amino acids of the polypeptide sequence.

___   ___   ___   ___   ___   ___   ___   ___   ___   ___
(amino acids)

## Fill-in-the-Blanks

The order of (13) _____ _____ in a protein is specified by a sequence of nucleotide bases. The genetic code is read in units of (14) _____ nucleotides; each unit of three codes for (15) _____ amino acid(s). In the table that showed which triplet specified a particular amino acid, the triplet code was incorporated in (16) _____ molecules. Each of these triplets is referred to as a(n) (17) _____ .
(18) _____ alone carries the instructions for assembling a particular sequence of amino acids from the DNA to the ribosomes in the cytoplasm, where (19) _____ of the polypeptide occurs. (20) _____ RNA acts as a shuttle molecule as each type brings its particular (21) _____ _____ to the ribosome where it is to be incorporated into the growing (22) _____ . A(n) (23) _____ is a triplet on mRNA that forms hydrogen bonds with a(n) (24) _____ , which is a triplet on tRNA.

## 14.5. HOW MUTATIONS AFFECT PROTEIN SYNTHESIS (pp. 236–237)
## SUMMARY OF PROTEIN SYNTHESIS (p. 238)

*Selected Words:* *"fix" the wrong base, "frameshift mutation".*

### Boldfaced, Page-Referenced Terms

(236) gene mutations _____

_____

(236) base-pair substitutions _____

_____

(236) insertions _____

_____

(236) deletions _____

_____

(236) transposable elements _____

_____

(237) mutation rate _____

_____

(237) ionizing radiation _____

_____

(237) alkylating agents _____

_____

(237) carcinogens _____

_____

*Short Answer*

1. Cite several examples of mutagens. _____

_____

*Fill-in-the-Blanks*

In addition to changes in chromosomes (crossing over, recombination, deletion, addition, translocation, and inversion), changes can also occur in the structure of DNA; these modifications are referred to as gene mutations. Complete the following exercise on types of spontaneous gene mutations.

Viruses, ultraviolet radiation, and certain chemicals are examples of environmental agents called

(2) _____ that may enter cells and damage strands of DNA. If A becomes paired with C instead of T

during DNA replication, this spontaneous mutation is a base-pair (3) _____ . Sickle-cell anemia is a

genetic disease whose cause has been traced to a single DNA base pair; the result is that one (4) _____

_____ is substituted for another in the beta chain of (5) _____ .

A(n) (6) _____ mutation is defined as the insertion or deletion of one to several DNA base pairs; this

puts the nucleotide sequence out of phase, and abnormal proteins are produced. Some DNA regions "jump"

to new DNA locations and often inactivate the genes in their new environment; such (7) _____

elements may give rise to observable changes in the phenotype of an organism.

(8) _____ is the source of the unity of life; (9) _____ _____ are the original source of life's

diversity; they are heritable, small-scale alterations in the (10) _____ sequence of DNA.

Each gene has a characteristic (11) _____ _____ , which is the probability it will mutate

spontaneously during a given interval of time. But not all mutations are spontaneous. (12) _____

radiation causes base substitutions or breaks in one or both strands.

## Label-Match

A summary of the flow of genetic information in protein synthesis is useful as an overview. Identify the indicated parts of the illustration on the next page by filling in the blanks with the names of the appropriate structures or functions. Choose from the following: DNA, mRNA, tRNA, polypeptide, rRNA subunits, intron, exon, mature mRNA transcript, new mRNA transcript, anticodon, amino acids, ribosome-mRNA complex. Complete the exercise by matching and entering the letter of the description in the parentheses following each label.

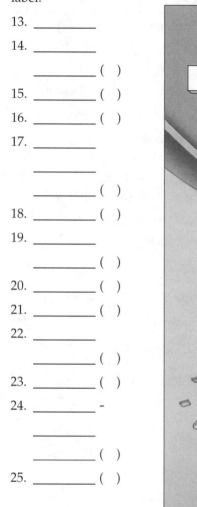

13. _____

14. _____

_____ (  )

15. _____ (  )

16. _____ (  )

17. _____

_____

_____ (  )

18. _____ (  )

19. _____

_____ (  )

20. _____ (  )

21. _____ (  )

22. _____

_____ (  )

23. _____ (  )

24. _____ -

_____

_____ (  )

25. _____ (  )

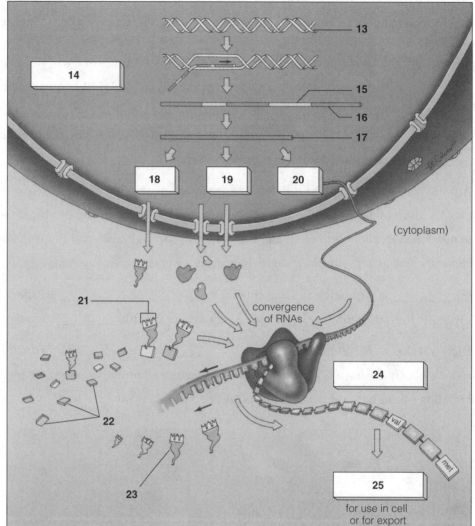

A. Coding portion of mRNA that will translate into proteins
B. Carries the genetic code from DNA in the nucleus to the cytoplasm
C. Transports amino acids to the ribosome and mRNA
D. The building blocks of polypeptides
E. Noncoding portions of newly transcribed mRNA
F. tRNA after delivering its amino acid to the ribosome-mRNA complex
G. Join when translation is initiated
H. Holds the genetic code for protein production
I. Place where translation occurs
J. Includes introns and exons
K. A sequence of three bases that can pair with a specific mRNA codon
L. Snipping out of introns, only exons remaining
M. May serve as a functional protein (enzyme) or a structural protein

## Crossword Puzzle

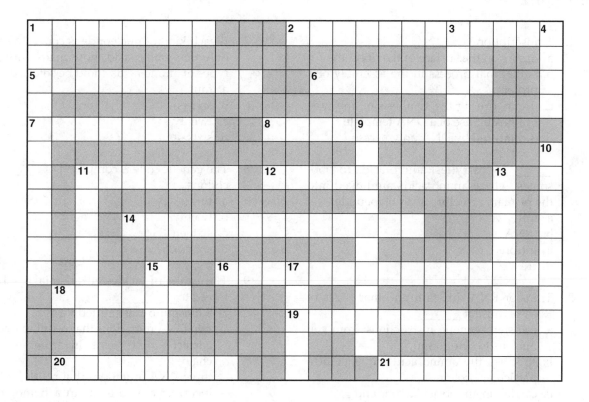

## ACROSS

1. _____ RNA = tRNA.
2. _____ elements may move from one site on a chromosome to another.
5. _____ agents transfer methyl or ethyl groups onto DNA strands and invite mutations to occur.
6. A group of cooperative ribosomes.
7. The base _____ determines the genetic code's recipe to synthesize a protein.
8. The _____ is located on the end of a tRNA.
11. An agent that causes permanent changes in a strand of DNA.
12. One or more bases added into a DNA strand.
14. The DNA code is "rewritten" into an RNA code.
16. A process used to separate components of a mixture that have different weights and electrical charges.
18. An intervening sequence of bases (in a strand of DNA) that separates code sequences of bases.
19. One or more bases is subtracted from a DNA strand.
20. The site where RNA polymerase attaches to DNA and begins transcription.
21. Nitrogen base unique to RNA.

## DOWN

1. The process during which three kinds of RNA cooperate to synthesize a protein product.
*3. _____ contains male animal gametes.
4. Sequences of RNA bases that are used to create proteins.
9. A cancer-causing agent.
10. A _____ mutation changes the "reading frame" in the base sequence downstream.
11. _____ RNA = mRNA.
13. _____ RNA = rRNA.
15. An atom (or group of atoms) in which the positive charges are not equal to the negative charges.
17. A three-base sequence in mRNA that calls for a specific amino acid from its pool.

*Designates a word not included in this chapter's word list.

# Self-Quiz

___ 1. Transcription _____ .
  a. occurs on the surface of the ribosome
  b. is the final process in the assembly of a protein
  c. occurs during the synthesis of any type of RNA by use of a DNA template
  d. is catalyzed by DNA polymerase

___ 2. _____ carry(ies) amino acids to ribosomes, where amino acids are linked into the primary structure of a polypeptide.
  a. mRNA
  b. tRNA
  c. Introns
  d. rRNA

___ 3. Transfer RNA differs from other types of RNA because it _____ .
  a. transfers genetic instructions from cell nucleus to cytoplasm
  b. specifies the amino acid sequence of a particular protein
  c. carries an amino acid at one end
  d. contains codons

___ 4. _____ dominates the process of transcription.
  a. RNA polymerase
  b. DNA polymerase
  c. Phenylketonuria
  d. Transfer RNA

___ 5. _____ and _____ are found in RNA but not in DNA.
  a. Deoxyribose; thymine
  b. Deoxyribose; uracil
  c. Uracil; ribose
  d. Thymine; ribose

___ 6. Each "word" in the mRNA language consists of _____ letters.
  a. three
  b. four
  c. five
  d. more than five

___ 7. If each kind of nucleotide is coded to select only one amino acid, how many different types of amino acids could be selected?
  a. four
  b. sixteen
  c. twenty
  d. sixty-four

___ 8. The genetic code is composed of _____ codons.
  a. three
  b. twenty
  c. sixteen
  d. sixty-four

___ 9. The cause of sickle-cell anemia has been traced to _____ .
  a. a mosquito-transmitted virus
  b. two DNA mutations that result in two incorrect amino acids in a hemoglobin chain
  c. three DNA mutations that result in three incorrect amino acids in a hemoglobin chain
  d. one DNA mutation that results in one incorrect amino acid in a hemoglobin chain

___10. An example of a mutagen is _____ .
  a. a virus
  b. ultraviolet radiation
  c. certain chemicals
  d. all of the above

For the next 5 questions, choose from these possibilities:

  a. Beadle and Tatum
  b. Garrod
  c. Khorana, Nirenberg and others
  d. McClintock
  e. Pauling and Itano

11. In the early 1900s, _____ studied illnesses caused by inheriting one or more defective units of inheritance; concluded that these units function by synthesizing specific enzymes.

12. In the 1940s _____ discovered that transposable elements can move genes from one chromosome to another, or to a different place on the chromosome; finally was awarded the Nobel Prize some fifty years after this news was published.

13. In the late 1940s, _____ developed gel electrophoresis as a way of separating similar proteins from a mixture.

14. In the 1930s, _____ developed the "one gene–one enzyme" hypothesis by studying defective enzymes in the bread mold, *Neurospora crassa*.

15. In the 1950s and 1960s _____ deduced that the mRNA code language formed base triplet "words" and that each "word" put into position on a ribosome selected a particular amino acid type from an amino acid pool.

## Chapter Objectives/Review Questions

This section lists general and detailed chapter objectives that can be used as review questions. You can make maximum use of these items by writing answers on a separate sheet of paper. Fill in answers where blanks are provided. To check for accuracy, compare your answers with information given in the chapter or glossary.

| Page | | Objectives/Questions |
|---|---|---|
| (228–229) | 1. | Cite an example of a change in one DNA base pair that has profound effects on the human phenotype. |
| (230) | 2. | _____ RNA combines with certain proteins to form the ribosome; _____ RNA carries genetic information for protein construction from the nucleus to the cytoplasm; _____ RNA picks up specific amino acids and moves them to the area of mRNA and the ribosome. |
| (230) | 3. | Transcription starts at a _____ , a specific sequence of bases on one of the two DNA strands that signals the start of a gene. |
| (230–231) | 4. | Describe the process of transcription and indicate three ways in which it differs from replication. |
| (231) | 5. | Actual coding portions of a newly transcribed mRNA are called _____ ; _____ are the noncoding portions. |
| (230–231) | 6. | The first end of the mRNA to be synthesized is the ____ end; at the opposite end, the most mature transcripts acquire a _____ tail. |
| (230–231) | 7. | State how RNA differs from DNA in structure and function, and indicate what features RNA has in common with DNA. |
| (230–231) | 8. | What RNA code would be formed from the following DNA code: TAC-CTC-GTT-CCC-GAA? |
| (232) | 9. | Each base triplet in mRNA is called a _____ . |
| (232) | 10. | Scrutinize Figure 14.11 in the text and decide whether the genetic code in this instance applies to DNA, mRNA, or tRNA. |
| (232) | 11. | Explain how the DNA message TAC-CTC-GTT-CCC-GAA would be used to code for a segment of protein, and state what its amino acid sequence would be. |
| (232–233) | 12. | Explain the nature of the genetic code and describe the kinds of nucleotide sequences that code for particular amino acids. |
| (232–233) | 13. | State the relationship between the DNA genetic code and the order of amino acids in a protein chain. |
| (234) | 14. | Describe the process of translation by telling what occurs during each of its three stages. |
| (236) | 15. | Briefly describe the spontaneous DNA mutations known as base-pair substitution, frameshift mutation, and transposable element. |
| (237) | 16. | List some of the environmental agents, or _____ , that can cause mutations. |
| (239) | 17. | Using a diagram, summarize the steps involved in the transformation of genetic messages into proteins (see Figure 14.19 in the text). |

## Integrating and Applying Key Concepts

Genes code for specific polypeptide sequences. Not every substance in living cells is a polypeptide. Explain how genes might be involved in the production of a storage starch (such as glycogen) that is constructed from simple sugars.

# 15

# CONTROLS OVER GENES

## Interactive Exercises

*Here's To Suicidal Cells!* (pp. 242–243)

## 15.1. OVERVIEW OF GENE CONTROLS (pp. 244)

## 15.2. CONTROLS IN BACTERIAL CELLS (pp. 244–245)

*Boldfaced, Page-Referenced Terms*

(242) apoptosis _____

_____

(242) ICE-like proteases _____

_____

(244) regulatory proteins _____

_____

(244) negative control systems _____

_____

(244) positive control systems _____

_____

(244) promoter _____

_____

(244) operator _____

_____

(244) repressor _____

_____

(244) operon _____

_____

(245) activator protein _____

_____

## Complete the Table

1. All the diploid cells in an organism possess the same genes, and every cell utilizes most of the same genes; yet specialized cells must activate only certain genes. Some agents of gene control have been discovered. Transcriptional controls are the most common. Complete the following table to summarize the agents of gene control.

*Agents of Gene Control*        *Method of Gene Control*

| Agents of Gene Control | Method of Gene Control |
|---|---|
| a. Repressor protein | |
| b. | Encourages binding of RNA polymerases to DNA; this is positive control of transcription |
| c. Promoter | |
| d. | Short DNA base sequences between promoter and the start of a gene; a binding site for control agents |

2. When do gene controls come into play? _____

_____

## Fill-in-the-Blanks

A promoter and operator provide (3) _____ _____ . The (4) _____ codes for the formation of

mRNA, which assembles a repressor protein. The affinity of the (5) _____ for RNA polymerase dictates

the rate at which a particular operon will be transcribed. Repressor protein allows (6) _____ _____

over the lactose operon. Repressor binds with operator and overlaps promoter when lactose concentrations

are (7) _____ . This blocks (8) _____ _____ from the genes that will process lactose. This

(9) [choose one] + blocks, + promotes production of lactose-processing enzymes. When lactose is present,

lactose molecules bind with the (10) _____ _____ . Thus, repressor cannot bind to (11) _____ ,

and RNA polymerase has access to the lactose-processing genes. This gene control works well because

lactose-degrading enzymes are not produced unless they are (12) _____ .

## Label-Match

*Escherichia coli*, a bacterial cell living in mammalian digestive tracts, is able to exert a negative type of gene control over lactose metabolism. Use the numbered blanks to identify each part of the illustration below. Use abbreviations for nucleic acids. Choose from the following: lactose, regulator gene, repressor-operator complex, promoter, mRNA transcript, enzyme genes, lactose operon, lactose enzymes, repressor-lactose complex, repressor protein, RNA polymerase, and operator. Complete the exercise by matching and entering the letter of the proper function description in the parentheses following each label.

13. _____ _____ (   )

14. _____ _____ (   )

15. _____ _____ (   )

16. _____ (   )

17. _____ (   )

18. _____ _____ (   )

19. _____ - _____ (   )

20. _____ _____ (   )

21. _____ - _____ _____ (   )

22. _____ (   )

23. _____ _____ (   )

24. _____ _____ (   )

A. Includes promoter, operator, and the lactose-metabolizing enzymes
B. Short DNA base sequence between promoter and the beginning of a gene
C. The nutrient molecule in the lactose operon
D. Major enzyme that catalyzes transcription
E. Capable of preventing RNA polymerases from binding with DNA
F. Prevents repressor from binding to the operator
G. Genes that produce lactose-metabolizing enzymes
H. Catalyze the digestion of lactose
I. Binds to operator and overlaps promoter; this prevents RNA polymerase from binding to DNA and initiating transcription
J. Specific base sequence that signals the beginning of a gene
K. Gene that contains coding for production of repressor protein
L. Carries genetic instructions to ribosomes for production of lactose enzymes

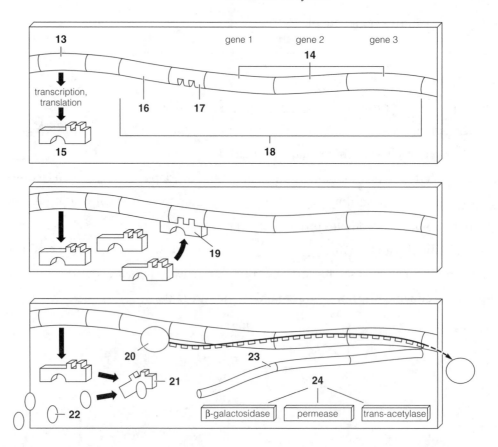

## 15.3. CONTROLS IN EUKARYOTIC CELLS (pp. 246–247)

## 15.4. EVIDENCE OF GENE CONTROL (pp. 248–249)

**Selected Words:** *gene amplification, DNA rearrangements, chemical modification, alternative splicing "masked messengers", "lampbrush" chromosomes, "mosaic" female, "calico" cat, "spotting gene", anhidrotic ectodermal dysplasia.*

### Boldfaced, Page-Referenced Terms

(246) cell differentiation _____

_____

(249) Barr body _____

_____

### Short Answer

1. Although a complex organism such as a human being arises from a single cell, the zygote, differentiation occurs in development. Define *differentiation* and relate it to a definition of *selective gene expression.* _____

_____

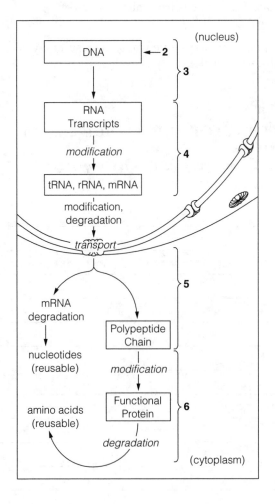

### Label-Match

Identify each numbered part of the illustration to the left showing eukaryotic gene control. Choose from translational control, replicational controls, transcriptional controls, transcript processing controls, and post-translational controls. Complete the exercise by correctly matching and entering the letter of the corresponding gene control description in the parentheses following each label.

2. _____ _____ ( )

3. _____ _____ ( )

4. _____ _____ _____ ( )

5. _____ _____ ( )

6. _____ _____ _____ ( )

A. Govern the rates at which mRNA transcripts that reach the cytoplasm will be translated into polypeptide chains at ribosomes

B. Govern modification of the initial mRNA transcripts in the nucleus

C. Govern how the polypeptide chains become modified into functional proteins

D. Gene amplification can produce many copies of specific genes needed to produce a lot of a specific protein

E. Influence when and to what degree a particular gene will be transcribed (if at all)

*Fill-in-the-Blanks*

All the cells in our bodies contain copies of the same (7) _____ . These genetically identical cells become structurally and functionally distinct from one another through a process called (8) _____ _____ , which arises through (9) _____ gene expression in different cells. Cells depend on (10) _____ , which govern transcription, translation, and enzyme activity. Controls that operate during transcription and transcript processing utilize (11) _____ proteins, especially (12) _____ that are turned on and off by the attachment and detachment of cAMP. (13) _____ is also controlled by the way eukaryotic DNA is packed with proteins in chromosomes. A (14) _____ body is a condensed X chromosome; the process by which an XX person randomly inactivates an X chromosome is called Lyonization. X chromosome inactivation produces adult human females who are (15) _____ for X-linked traits. This effect is shown in human females affected by (16) _____ _____ _____ ; the disorder provides evidence that the process of (17) _____ gene expression causes cells to differentiate.

*True-False*

If false, explain why.

_____ 18. "Lampbrush chromosomes" of meiotic prophase I seen in electron micrographs of amphibian eggs provide evidence of translational control.

_____ 19. When the same primary mRNA transcript is spliced in alternative ways, the result may be different mRNAs, each coding for a slightly different protein.

## 15.5. EXAMPLES OF SIGNALING MECHANISMS (pp. 250–251)
## 15.6. *Focus on Science*: CANCER AND THE CELLS THAT FORGET TO DIE (pp. 252–253)

*Selected Words:* *"polytene chromosome", benign tumors, malignant tumors, myc gene, Burkitt's lymphoma.*

*Boldfaced, Page-Referenced Terms*

(250) hormones _____

_____

(250) enhancers _____

_____

(250) ecdysone _____

_____

(251) phytochrome _____

_____

(252) protein kinases _____

_____

(252) growth factors _____

_____

(252) tumor _____

_____

(252) metastasis _____

_____

(252) cancer _____

_____

(252) oncogene _____

_____

(252) proto-oncogene _____

_____

## Fill-in-the-Blanks

Several kinds of (1) _____ influence gene activity. In both animals and plants, molecules called

(2) _____ stimulate or inhibit gene activity in their target cells. Any cell with (3) _____ for a

specific hormone is a target cell. Sometimes, a hormone molecule must first bind with a(n) (4) _____

(generally, a base sequence on the same DNA molecule that contains the promoter to which RNA

polymerase will attach and begin transcription) before transcription can be either turned on or turned off.

Many insect larvae produce a hormone known as (5) _____ , which is an (4) that binds to a receptor on

salivary gland cells and triggers very rapid transcription of the genes that contain the instructions for

assembling the protein components of saliva. Thus, insect larvae can usually digest leafy vegetation as fast

as they consume it by producing large amounts of salivary proteins. The genes responsible for salivary

protein production have been copied many times by (6) _____ and lie parallel to each other, causing

that portion of the (7) _____ chromosome to appear puffy during transcription.

In humans and other mammals, mammary gland cells do not begin producing milk until the hormone

(8) _____ attaches to receptors on the surfaces of the gland cells and activates the genes responsible for

producing milk proteins. Other cells in mammalian bodies have these genes, but lack the (9) _____ to

which the hormone attaches.

Red wavelengths of sunlight activate (10) _____, a blue-green pigment whose activity influences

transcription of certain genes at certain times of the day and season as enzymes and other proteins play key

roles in germination, growth, and the formation of flowers, fruits and seeds.

Most cells respond to signals that tell them it's time to die, but (11) _____ cells do not. Many (11) cells

form a tissue mass called a (12) _____, that may be either benign or malignant. Malignant cells can

break loose from their original growth area, travel through the body and invade other tissues—a process

called (13) _____ .

## True-False

If the statement is true, write a T in the blank. If false, make it true by changing the underlined word.

_____ 14. <u>Polytene</u> chromosomes of many insect larvae are unusual in that their DNA has been
repeatedly replicated; thus, these chromosomes have multiple copies of the same genes.
When these genes are undergoing transcription, they "puff."

_____15. When cells become cancerous, cell populations <u>decrease</u> to very <u>low</u> densities and <u>stop</u> dividing.

_____16. <u>All</u> abnormal growths and massings of new tissue in any region of the body are called tumors.

_____17. Malignant tumors have cells that <u>migrate and divide</u> in other organs.

_____18. Oncogenes are genes that <u>combat</u> cancerous transformations.

_____19. Proto-oncogenes <u>rarely</u> trigger cancer.

_____20. The normal expression of proto-oncogenes is vital, even though their <u>normal</u> expression may be lethal.

# Self-Quiz

___ 1. _____ refers to the processes by which cells with identical genotypes become structurally and functionally distinct from one another according to the genetically controlled developmental program of the species.
a. Metamorphosis
b. Metastasis
c. Cleavage
d. Differentiation

___ 2. _____ binds to operator whenever lactose concentrations are low.
a. Operon
b. Repressor
c. Promoter
d. Operator

___ 3. Any gene or group of genes together with its promoter and operator sequence is a(n) _____ .
a. repressor
b. operator
c. promoter
d. operon

___ 4. The operon model explains the regulation of _____ in prokaryotes.
a. replication
b. transcription
c. induction
d. Lyonization

___ 5. In multicelled eukaryotes, cell differentiation occurs as a result of _____ .
a. growth
b. selective gene expression
c. repressor molecules
d. the death of certain cells

___ 6. One type of gene control discovered in female mammals is _____ .
a. a conflict in maternal and paternal alleles
b. slow embryo development
c. X chromosome inactivation
d. operon

___ 7. Due to X inactivation of either the paternal or maternal X chromosome, human females with anhidrotic ectodermal dysplasia _____ .
a. completely lack sweat glands
b. develop benign growths
c. have mosaic patches of skin that lack sweat glands
d. develop malignant growths

___ 8. Which of the following characteristics seems to be most uniquely correlated with metastasis?
a. loss of nuclear-cytoplasmic controls governing cell growth and division
b. changes in recognition proteins on membrane surfaces
c. "puffing" in the chromosomes
d. the massive production of benign tumors

___ 9. Genes with the potential to induce cancerous formations are known as ____.
a. proto-oncogenes
b. oncogenes
c. carcinogens
d. malignant genes

___10. ____ controls govern the rates at which mRNA transcripts that reach the cytoplasm will be translated into polypeptide chains at the ribosomes.
a. Transport
b. Transcript processing
c. Translational
d. Transcriptional

# Chapter Objectives/Review Questions

This section lists general and detailed chapter objectives that can be used as review questions. You can make maximum use of these items by writing answers on a separate sheet of paper. Fill in answers where blanks are provided. To check for accuracy, compare your answers with information given in the chapter or glossary.

| Page | Objectives/Questions |
|---|---|
| (244–245) | 1. The negative control of _____ protein prevents the enzymes of transcription from binding to DNA; the positive control of _____ protein enhances the binding of RNA polymerases to DNA. |
| (244) | 2. A _____ is a specific base sequence that signals the beginning of a gene in a DNA strand. |
| (244) | 3. Some control agents bind to _____ , which are short base sequences between promoters and the beginning of genes. |
| (244–245; 250) | 4. Gene expression is controlled through regulatory _____ , _____ , that speed up reaction rates and _____ proteins that bind to enhancer sites on DNA. |
| (244) | 5. A gene control arrangement in which the same promoter-operator sequence services more than one gene is called an _____ . |
| (244–245) | 6. Describe the sequence of events that occurs on the chromosome of *E. coli* after you drink a glass of milk. |
| (244–245) | 7. The cells of *E. coli* manage to produce enzymes to degrade lactose when those molecules are _____ and to stop production of lactose-degrading enzymes when lactose is _____ . |
| (246) | 8. Define *selective gene expression* and explain how this concept relates to cell differentiation in multicelled eukaryotes. |
| (246) | 9. Cell _____ occurs in multicelled eukaryotes as a result of _____ gene expression. |
| (246–247) | 10. Be able to list and define the levels of gene control in eukaryotes. |
| (248) | 11. The "lampbrush" chromosomes seen in amphibian eggs and larvae represent a change in chromosome structure that has been correlated directly with _____ of the genes necessary for growth. |
| (249) | 12. Explain how X chromosome inactivation provides evidence for selective gene expression; use the example of anhidrotic ectodermal dysplasia. |
| (249) | 13. The condensed X chromosome seen on the edge of the nuclei of female mammals is known as the _____ body. |
| (250) | 14. Explain how hormones act as a major agent of gene control. |
| (250) | 15. Describe the gene control mechanism afforded many insect larvae by polytene chromosomes. |
| (252) | 16. Define *tumor* and distinguish between benign and malignant tumors. |
| (252) | 17. _____ is a process in which a cancer cell leaves its proper place and invades other tissues to form new growths. |
| (252) | 18. Describe the relationship of proto-oncogenes, environmental irritants, and oncogenes. |
| (253) | 19. Cigarette smoke, X-rays, gamma rays, and ultraviolet radiation are examples of _____ . |

# Integrating and Applying Key Concepts

Suppose you have been restricting yourself to a completely vegetarian diet for the past six months. Quite unexpectedly, you find yourself in a social situation that requires you to eat a half-pound sirloin steak. Would you expect to digest the steak as easily as you digest soybean burgers? Explain your yes or no answer in terms of transcriptional controls or feedback inhibition.

# 16

# RECOMBINANT DNA AND GENETIC ENGINEERING

## Interactive Exercises

*Mom, Dad, and Clogged Arteries* (pp. 256–257)

## 16.1. RECOMBINATION IN NATURE—AND IN THE LABORATORY (pp. 258–259)

*Selected Words:* HDLs, LDLs, familial cholesterolemia, antibiotics, <u>Escherichia coli</u>, staggered cuts, "recombinant plasmids".

### Boldfaced, Page-Referenced Terms

(257) gene therapy _____

_____

(258) recombinant DNA technology _____

_____

(258) genetic engineering _____

_____

(258) plasmids _____

_____

(259) restriction enzymes _____

(259) DNA ligase _____

(259) DNA library _____

## Short Answer

1. Describe and distinguish between the bacterial chromosome and plasmids present in a bacterial cell.

_____

_____

_____

## True-False

If the statement is true, write T in the blank. If the statement is false, make it correct by changing the under-lined words and writing the correct word(s) in the answer blank.

_____2. Plasmids are <u>organelles on the surfaces of which amino acids are assembled into polypeptides</u>.

_____3. <u>Gene mutations</u> and <u>recombination</u> are common in nature.

## Fill-in-the-Blanks

Genetic experiments have been occurring in nature for billions of years as a result of gene (4) _____ , crossing over and recombination, and other events. Humans now are causing genetic change by using (5) _____ _____ technology in which researchers cut, and splice together gene regions from different (6) _____ , then greatly (7) _____ the number of copies of the genes that interest them. The genes, and in some cases their (8) _____ products, are produced in quantities that are large enough for (9) _____ and for practical applications. (10) _____ _____ involves isolating, modifying, and inserting particular genes back into the same organism or into a different one.

Many bacteria can transfer (11) _____ genes to a bacterial neighbor which may integrate them into the recipient's chromosome, forming a recombinant DNA molecule naturally. In nature, (12) _____ as well as bacteria dabble in gene transfers and recombination of genes, and so do eukaryotic organisms.

## Matching

Match the steps in the formation of a DNA library with the parts of the illustration below.

13. ___
14. ___
15. ___
16. ___
17. ___
18. ___

   A. Joining of chromosomal and plasmid DNA using DNA ligase
   B. Restriction enzyme cuts chromosomal DNA at specific recognition sites
   C. Cut plasmid DNA
   D. Recombinant plasmids containing cloned library
   E. Fragments of chromosomal DNA
   F. Same restriction enzyme is used to cut plasmids

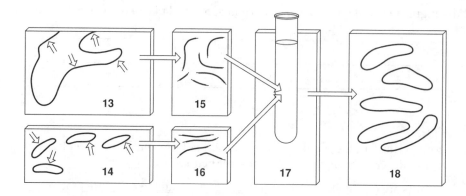

## True-False

A genetic engineer used restriction enzymes to prepare fragments of DNA from two different species that were then mixed. Four of these fragments are illustrated below. Fragments (a) and 9c) are from one species, (b) and (d) from the other species. Answer exercises 19–23. If false, explain why.

a. _____ TACA     b. _____ TTCA     c. _____ CGTA     d. _____ ATGT

_____ 19. Some of the fragments represent sticky ends.

_____ 20. The same restriction enzyme was used to cut fragments (b), (c), and (d).

_____ 21. Different restriction enzymes were used to cut fragments (a) and (d).

_____ 22. Fragment (a) will base-pair with fragment (d) but not with fragment (c).

_____ 23. The same restriction enzyme was used to cut the different locations in the DNA of the two species shown.

## 16.2. WORKING WITH DNA FRAGMENTS (pp. 260–261)

## 16.3. *Focus on Science:* RIFF-LIPS AND DNA FINGERPRINTS (p. 262)

*Selected Words:* "cloned" DNA, Southern blot method.

### Boldfaced, Page-Referenced Terms

(260) DNA amplification _____

_____

(260) polymerase chain reaction (PCR) _____

_____

(260) primers _____

_____

(260) DNA polymerase _____

_____

(260) gel electrophoresis _____

_____

(262) RFLPs, restriction fragment length polymorphisms _____

_____

(262) DNA fingerprint _____

_____

## Complete the Table

1. Complete the table below, which summarizes some of the basic tools and procedures used in recombinant DNA technology.

| Tool/Procedure | Definition and Role in Recombinant DNA Technology |
|---|---|
| a. Restriction enzymes | |
| b. DNA ligase | |
| c. DNA library | |
| d. Gel electrophoresis | |
| e. PCR | |
| f. DNA sequencing | |
| g. RFLP | |
| h. DNA fingerprint | |

## Matching

Match the most appropriate letter with each numbered partner.

2. ___polymerase chain reaction (PCR)

3. ___DNA ligase

4. ___DNA library

5. ___DNA fingerprint

6. ___gel electrophoresis

7. ___recombinant DNA technology

8. ___plasmids

9. ___restriction enzymes

10. ___genome

A. All the DNA in a haploid set of chromosomes
B. Process by which a gene is split into two strands and then copied over and over (most common type of gene amplification) by enzymes
C. Caused by a unique array of RFLPs inherited from each parent
D. Connects DNA fragments
E. Small circular DNA molecules that carry only a few genes
F. Cut DNA molecules
G. A collection of DNA fragments produced by restriction enzymes and incorporated into plasmids
H. Charged molecules move through a viscous medium and are separated according to their different physical and chemical properties
I. Method of genetic engineering

Each person has a genetic (11) _____ : a unique pattern of RFLPs that can be used to map the human genome, apprehend criminals, and resolve cases of disputed paternity and maternity. Identification of the order and identity of nucleotides in DNA is called (12) _____ .

*Fill-in-the-Blanks*

13. From the migration results shown, determine the sequence of nucleotides in the code molecule.

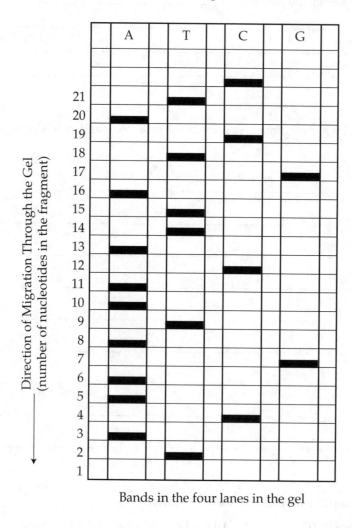

Bands in the four lanes in the gel

First Letter        Last Letter

**16.4. MODIFIED HOST CELLS** (p. 263)

**16.5. BACTERIA, PLANTS, AND THE NEW TECHNOLOGY** (pp. 264–265)

**16.6. GENETIC ENGINEERING OF ANIMALS** (p. 266)

**16.7.** *Focus on Bioethics:* **REGARDING GENE THERAPY** (p. 267)

*Selected Words:* *mature mRNA transcript, reverse transcriptase, diabetics, "fail-safe" genes, hok gene, "ice-minus bacteria", potato blight, "Ti" plasmid,* Agrobacterium tumefaciens, *"Supermouse", "biotech barnyards", Dolly.*

### Boldfaced, Page-Referenced Terms

(263) DNA probes _____

_____

(263) nucleic acid hybridization _____

_____

(263) cDNA _____

_____

(263) reverse transcription _____

_____

(266) human genome project _____

_____

### Complete the Table

1. Complete the table below, which summarizes some of the basic tools and procedures used in recombinant DNA technology.

| Tool/Procedure | Definition and Role in Recombinant DNA Technology |
|---|---|
| a. DNA probe | |
| b. nucleic acid hybridization | |
| c. cDNA | |
| d. reverse transcription | |

## Fill-in-the-Blanks

Host cells that take up modified genes may be identified by procedures involving DNA (2) _____ .
Gene (3) _____ does not automatically follow the successful taking-in of a modified gene by a host cell;
genes must be suitably (4) ____ first, which involves getting rid of the (5) _____ from the mRNA
transcripts, then using the enzyme reverse (6) _____ to produce cDNA.  The process goes like this:

a.

b.

c.

d.

a. An mRNA transcript of a desired gene is used as a (7)
_____ for assembling a DNA strand. An enzyme (6) does the
assembling.

b. An mRNA-(8) _____ hybrid molecule results.

c. (9) _____ action removes the mRNA and assembles a
second strand of (10) _____ on the first strand.

d. The result is double-stranded (11) _____ , "copied" from
an mRNA (12) _____ .

## Short Answer

13. Explain the goal of the human genome project. _____

_____

In exercises 14–19, summarize the results of the given experimentation dealing with genetic modifications of
plants and animals.

14. Cattle may soon be producing human collagen: _____

_____

15. The bacterium, *Agrobacterium tumefaciens:* _____

_____

16. Cotton plants: _____

_____

17. Introduction of the rat and human somatotropin gene into fertilized mouse eggs: _____

_____

18. "Ice-minus" bacteria and strawberry plants: _____

_____

19. Dolly, the cloned lamb: _____

_____

*Fill-in-the-Blanks*

The harmful gene is called the "ice-forming" gene, and the bacteria are known as (20) _____ - _____ bacteria; genetic engineers were able to remove the harmful gene and test the modified bacterium on strawberry plants with no adverse effects. Inserting one or more genes into the (21) _____ _____ of an organism for the purpose of correcting genetic defects is known as (22) _____ _____ . Attempting to modify a human trait by inserting genes into sperm or eggs is called (23) _____ _____ .

# Self-Quiz

___ 1. Small circular molecules of DNA in bacteria are called _____ .
   a. plasmids
   b. desmids
   c. pili
   d. F particles
   e. transferins

___ 2. Base-pairing between nucleotide sequences from different sources is called _____ .
   a. nucleic acid hybridization
   b. reverse replication
   c. heterocloning
   d. plasmid formation

___ 3. Enzymes used to cut genes in recombinant DNA research are _____ .
   a. ligases
   b. restriction enzymes
   c. transcriptases
   d. DNA polymerases
   e. replicases

___ 4. The total DNA in a haploid set of chromosomes of a species is its _____ .
   a. plasmid
   b. enzyme potential
   c. genome
   d. DNA library
   e. none of the above

___ 5. An enzyme that heals random base-pairing of chromosomal fragments and plasmids is _____ .
   a. reverse transcriptase
   b. DNA polymerase
   c. cDNA
   d. DNA ligase

___ 6. A DNA library is _____ .
   a. a collection of DNA fragments produced by restriction enzymes and incorporated into plasmids
   b. cDNA plus the required restriction enzymes
   c. mRNA-cDNA
   d. composed of mature mRNA transcripts

___ 7. Amplification results in _____ .
   a. plasmid integration
   b. bacterial conjugation
   c. cloned DNA
   d. production of DNA ligase

___ 8. Any DNA molecule that is copied from mRNA is known as _____ .
   a. cloned DNA
   b. cDNA
   c. DNA ligase
   d. hybrid DNA

___ 9. The most commonly used method of DNA amplification is _____ .
   a. polymerase chain reaction
   b. gene expression
   c. genome mapping
   d. RFLPs

___10. Restriction fragment length polymorphisms are valuable because_____ .
   a. they reduce the risks of genetic engineering
   b. they provide an easy way to sequence the human genome
   c. they allow fragmenting DNA without enzymes
   d. they provide DNA fragment sizes unique to each person

# Chapter Objectives/Review Questions

This section lists general and detailed chapter objectives that can be used as review questions. You can make maximum use of these items by writing answers on a separate sheet of paper. Fill in answers where blanks are provided. To check for accuracy, compare your answers with information given in the chapter or glossary.

*Page*      *Objectives/Questions*

(258)       1.   List the means by which natural genetic recombination occurs.
(258)       2.   Define *recombinant DNA technology*.
(258)       3.   _____ are small, circular, self-replicating molecules of DNA or RNA within a bacterial cell.
(259)       4.   Some bacteria produce _____ enzymes that cut apart DNA molecules injected into the cell by viruses; such DNA fragments or "_____ ends" often have staggered cuts capable of base-pairing with other DNA molecules cut by the same _____ enzymes.
(259)       5.   Base-pairing between chromosomal fragments and cut plasmids is made permanent by DNA _____ .
(259)       6.   Be able to explain what a DNA library is; review the steps used in creating such a library.
(260)       7.   List and define the two major methods of DNA amplification.
(260)       8.   Polymerase chain reaction is the most commonly used method of DNA _____ .
(260)       9.   Multiple, identical copies of DNA fragments produced by restriction enzymes are known as _____ DNA.
(260–261) 10.   Explain how gel electrophoresis is used to sequence DNA.
(262)      11.   List some practical genetic uses of RFLPs.
(263)      12.   How is a cDNA probe used to identify a desired gene carried by a modified host cell?
(263)      13.   A special viral enzyme, _____ _____ , presides over the process by which mRNA is transcribed into DNA.
(263)      14.   Define *cDNA*.
(263)      15.   Why do researchers prefer to work with cDNA when working with human genes?
(266–267) 16.   Tell about the human genome project and its implications.
(257; 267) 17.   Define *gene therapy* and *eugenic engineering*.

# Integrating and Applying Key Concepts

How could scientists guarantee that *Escherichia coli*, the human intestinal bacterium, will not be transformed into a severely pathogenic form and released into the environment if researchers use the bacterium in recombinant DNA experiments?

# 17

# EMERGENCE OF EVOLUTIONARY THOUGHT

## Interactive Exercises

*Fire, Brimstone, and Human History* (pp. 270–271)

## 17.1. EARLY BELIEFS, CONFOUNDING DISCOVERIES (pp. 272–273)

*Selected Words:* "species"

*Boldfaced, Page-Referenced Terms*

(271) catastrophism _____

_____

(272) biogeography _____

_____

(272) comparative anatomy _____

_____

(273) fossils _____

_____

(273) evolution _____

_____

## Matching

Select the single best answer.

1. ___comparative anatomy

2. ___biogeography

3. ___fossils

4. ___school of Hippocrates

5. ___evolution

6. ___great Chain of Being

7. ___Buffon

8. ___species

9. ___Aristotle

10. ___sedimentary beds

A. Rock layers deposited at different times; may contain fossils
B. Came to view nature as continuum of organization, from lifeless matter through complex forms of plant and animal life
C. Each kind of being; represent links in the great Chain of Being
D. Modification of species over time
E. Extended from the lowest forms of life to humans and on to spiritual beings
F. Suggested that perhaps species originated in more than one place and perhaps had been modified over time
G. Studies of body structure comparisons and patterning
H. Studies of the world distribution of plants and animals
I. Suggested that the gods were not the cause of the sacred disease
J. One type of direct evidence that organisms lived in the past

## 17.2. A FLURRY OF NEW THEORIES (pp. 274–275)

## 17.3. DARWIN'S THEORY TAKES FORM (pp. 276–277)

## 17.4. *Focus on Science:* REGARDING THE FIRST OF THE "MISSING LINKS" (p. 278)

**Selected Words:** *Cuvier, Lamarck, Beagle, Principles of Geology, artificial selection, Darwin, Wallace, natural selection, On the Origin of Species, "missing link", transitional forms,* Archaeopteryx.

### Boldfaced, Page-Referenced Terms

(274) theory of inheritance of acquired characteristics _____

_____

(275) theory of uniformity _____

_____

### Choice

For questions 1–12, choose from the following:

a. Georges Cuvier          b. Jean-Baptiste Lamarck

1.___Stretching directed "fluida" to the necks of giraffes, which lengthened permanently

2.___Acknowledged there were abrupt changes in the fossil record that corresponded to discontinuities between certain layers of sedimentary beds; thought this was evidence of change in populations of ancient organisms

3.___The force for change in organisms is the drive for perfection

4.___There was only one time of creation that populated the world with all species

5.___Catastrophism

6.___Permanently stretched giraffe necks were bestowed on offspring

7.___When a global catastrophe destroyed many organisms, a few survivors repopulated the world

8.____Theory of inheritance of acquired characteristics

9.____The survivors of a global catastrophe were not new species (many differed from related fossils) but naturalists had not discovered them yet

10.____During a lifetime, environmental pressures and internal "desires" bring about permanent changes

11.____The force for change is a drive for perfection, up the Chain of Being

12.____The drive to change is centered in nerves that direct an unknown "fluida" to body parts in need of change

## Complete the Table

13. Several key players and events in the life of Charles Darwin led him to his conclusions about natural selection and evolution. Summarize these influences by completing the table below.

| Event/Person | Importance to Synthesis of Evolutionary Theory |
|---|---|
| a. | Botanist at Cambridge University who perceived Darwin's real interests and arranged for Darwin to become a ship's naturalist |
| b. | Where Darwin earned a degree in theology but also developed his love for natural history |
| c. | British ship that carried Darwin (as a naturalist) on a five-year voyage around the world |
| d. | Wrote *Principles of Geology*; advanced the theory of uniformity; suggested that Earth was much older than 6,000 years |
| e. | Wrote an influential essay (read by Darwin) on human populations asserting that people tend to produce children faster than food supplies, living space, and other resources can be sustained |
| f. | Volcanic islands 900 kilometers from the South American coast where Darwin correlated differences in various species of finches with their environmental challenges |
| g. | The key point in Darwin's theory of evolution; involves reproductive capacity, heritable variations, and adaptive traits |
| h. | English naturalist contemporary with Darwin; independently developed Darwin's theory of evolution before Darwin published |
| i. | Unearthed in 1861; the first transitional fossil (between reptiles and birds); provided evidence for Darwin's theory |

## Sequence

Read Ideas A–G through first, then put them in an order that logically develops the Darwin-Wallace theory of evolution in correct sequence. Number 14 is the first, most fundamental idea. Idea number 15 set the stage for and underlies the remaining ideas.

14. ___The first idea in the series

15. ___The second idea in the series

16. ___The third idea in the series

17. ___The fourth idea in the series

18. ___The fifth idea in the series

19. ___The sixth idea in the series

20. ___The concluding idea in the series

A. Nature "selects" those individuals with traits that allow them to obtain the resources they need. They live longer and produce more offspring than others in the population that cannot get the resources they need to live and reproduce.

B. As population size increases, available resources dwindle.

C. A population is evolving when the forms of its heritable traits are changing over successive generations.

D. Animal populations tend to reproduce faster than food supplies, living space and other resources can sustain the populations.

E. The struggle for existence intensifies.

F. There is genetic variation in all sexually-reproducing populations; variations in traits might affect individuals' abilities to get resources, and therefore to survive and reproduce in particular environments.

G. Over time the more successful phenotypes will dominate the population that exists in that particular environment.

---

# Self-Quiz

___ 1. An acceptable definition of evolution is _____ .

    a. changes in organisms that are extinct
    b. changes in organisms since the flood
    c. changes in organisms over time
    d. changes in organisms in only one place

___ 2. The two scientists most closely associated with the concept of evolution are _____ .

    a. Lyell and Malthus
    b. Henslow and Cuvier
    c. Henslow and Malthus
    d. Darwin and Wallace

___ 3. Studies of the distribution of organisms on the earth is known as _____ .

    a. a great Chain of Being
    b. stratification
    c. geological evolution
    d. biogeography

___ 4. Buffon suggested that _____ .

    a. tail bones in a human have no place in a perfectly designed body
    b. perhaps species originated in more than one place and may have been modified over time
    c. the force for change in organisms was a built-in drive for perfection, up the Chain of Being
    d. gradual processes now molding Earth's surface had also been at work in the past

___ 5. In Lyell's book, *Principles of Geology*, he suggested that _____ .

    a. tail bones in a human have no place in a perfectly designed body
    b. perhaps species originated in more than one place and may have been modified over time
    c. the force for change in organisms was a built-in drive for perfection, up the Chain of Being
    d. gradual processes now molding the earth's surface had also been at work in the past

___ 6. One of the central ideas of Lamarck's theory of desired evolution was that _____ .
   a. tail bones in a human have no place in a perfectly designed body
   b. perhaps species originated in more than one place and may have been modified over time
   c. the force for change in organisms was a built-in drive for perfection, up the Chain of Being
   d. gradual processes now molding the earth's surface had also been at work in the past

___ 7. The theory of catastrophism is associated with _____ .
   a. Darwin
   b. Cuvier
   c. Buffon
   d. Lamarck

___ 8. The idea that any population tends to outgrow its resources, and its members must compete for what is available, belonged to _____ .
   a. Malthus
   b. Darwin
   c. Lyell
   d. Henslow

___ 9. The ideas that all natural populations have the reproductive capacity to exceed the resources required to sustain them and members of a natural population show great variation in their traits are associated with _____ .
   a. catastrophism
   b. desired evolution
   c. natural selection
   d. Malthus's idea of survival

___10. *Archaeopteryx* represents evidence for _____ .
   a. birds descending from reptiles
   b. evolution
   c. an evolutionary "link" between two major groups of organisms
   d. the existence of fossils
   e. all of the above

# Chapter Objectives/Review Questions

This section lists general and detailed chapter objectives that can be used as review questions. You can make maximum use of these items by writing answers on a separate sheet of paper. Fill in answers where blanks are provided. To check for accuracy, compare your answers with information given in the chapter or glossary.

| Page | | Objectives/Questions |
|---|---|---|
| (272) | 1. | Relate Aristotle's view of nature and describe what is meant by the great "Chain of Being." |
| (272) | 2. | It was the study of _____ that raised this question: How did so many species get from the center of creation to islands and other isolated places? |
| (272–273) | 3. | Scientists working in a field known as _____ _____ wondered why animals as different as humans, whales, and bats had so many similarities. |
| (273) | 4. | Give two examples of the type of evidence found by comparative anatomists that suggested living things may have changed with time. |
| (273) | 5. | _____ _____ consist of distinct, multiple layers of sedimentary rock that were originally deposited at different time periods. |
| (273) | 6. | A _____ is defined as the recognizable remains or body impressions of organisms that lived in the past. |
| (273) | 7. | The modification of species over time is called _____ . |
| (274) | 8. | Be able to discuss Cuvier's theory of catastrophism. |
| (274) | 9. | What was the main force for change in organisms as described in Lamarck's theory of desired evolution? |
| (274–278) | 10. | Be able to state the significance of the following to the theory of evolution: Henslow, H.M.S. *Beagle*, Lyell, Darwin, Malthus, Wallace, and *Archaeopteryx*. |
| (276) | 11. | To what conclusion was Darwin led when he considered the various species of finches living on the separate islands of the Galapagos? |
| (275–277) | 12. | Outline the key observations and inferences of Darwin's theory of evolution by natural selection. |

# Integrating and Applying Key Concepts

An Austrian biologist named Paul Kammerer (1880–1926) once reported that he had found solid evidence that the nuptial pads on the fore-limbs of the male midwife toad (used to grip the female during mating) were *acquired* by the "energy and diligence" of the parent toad and were then transmitted to offspring for the benefit of future generations. When he presented his evidence, several observing scientists suspected his specimens were forgeries. Scientists attempted to repeat Kammerer's experiments but they were unrepeatable. Kammerer incurred the wrath of the international scientific community and, in disgrace, he eventually ended his life by his own hand. Considering what you have learned in this chapter, what is your assessment of Kammerer's claim?

# 18

# MICROEVOLUTION

## Interactive Exercises

*Designer Dogs* (pp. 280–281)

## 18.1. INDIVIDUALS DON'T EVOLVE—POPULATIONS DO (pp. 282–283)

## 18.2. *Focus On Science*: WHEN IS A POPULATION <u>NOT</u> EVOLVING? (pp. 284–285)

*Selected Words:* *"agent", morphological traits; physiological traits; behavioral traits, morph, continuous variation, natural selection, gene flow, genetic drift.*

### Boldfaced, Page-Referenced Terms

(282) population _____

_____

(282) polymorphism _____

_____

(282) gene pool _____

_____

(282) alleles _____

_____

(283) allele frequencies _____

_____

(283) genetic equilibrium _____

_____

(283) microevolution _____

_____

(283) mutation rate _____

_____

(283) lethal mutation _____

_____

(283) neutral mutation _____

_____

(284) Hardy-Weinberg rule _____

_____

## Labeling

The traits of individuals in a population are often classified as being morphological, physiological, or behavioral traits. In the list of traits below, enter "M" if the trait seems to be morphological, "P" if the trait is physiological, and "B" if the trait is behavioral.

1. ____ Frogs have a three-chambered heart.

2. ____ The active transport of $Na^+$ and $K^+$ ions is unequal.

3. ____ Humans and orangutans possess an opposable thumb.

4. ____ An organism periodically seeking food.

5. ____ Some animals have a body temperature that fluctuates with the environmental temperature.

6. ____ The platypus is a strange mammal. It has a bill like a duck and lays eggs.

7. ____ Some vertebrates exhibit greater parental protection for offspring than others.

8. ____ During short photoperiods, the pituitary gland releases small quantities of gonadotropins.

9. ____ Red grouse defend large multipurpose territories where they forage, mate, nest, and rear young.

10. ____ Lampreys have an elongated cylindrical body without scales.

## Matching

Select the one most appropriate answer to match the sources of genetic variation.

11. ___ fertilization

12. ___ changes in chromosome structure or number

13. ___ crossing over at meiosis

14. ___ gene mutation

15. ___ independent assortment at meiosis

A. Leads to mixes of paternal and maternal chromosomes in gametes
B. Produces new alleles
C. Leads to the loss, duplication, or alteration of alleles
D. Brings together combinations of alleles from two parents
E. Leads to new combinations of alleles in chromosomes

## Short Answer

16. Briefly discuss the role of the environment and genetics in the production of an organism's phenotype.

_____

_____

17. List the conditions (in any order) that must be met before genetic equilibrium (or nonevolution) will occur.

_____

18. For the following situation, assume that the conditions listed in question 17 do exist; therefore, there should be no change in gene frequency, generation after generation. Consider a population of hamsters in which dominant gene $B$ produces black coat color and recessive gene $b$ produces gray coat color (two alleles are responsible for color). The dominant gene has a frequency of 80 percent (or .80). It would follow that the frequency of the recessive gene is 20 percent (or .20). From this, the assumption is made that 80 percent of all sperm and eggs have gene B. Also, 20 percent of all sperm and eggs carry gene $b$. (See page 284 in the text.)

a. Calculate the probabilities of all possible matings in the Punnett square.
b. Summarize the genotype and phenotype frequencies of the $F_1$ generation.

| Genotypes | Phenotypes |
|-----------|------------|
| ____ BB   |            |
| ____ Bb   | ____ % black |
| ____ bb   | ____ % gray |

Sperm

|        |        | 0.80 B | 0.20 b |
|--------|--------|--------|--------|
| Eggs   | 0.80 B | BB     | Bb     |
|        | 0.20 b | Bb     | bb     |

c. Further assume that the individuals of the $F_1$ generation produce another generation and the assumptions of the Hardy-Weinberg rule still hold. What are the frequencies of the sperm produced?

| Parents ($F_1$) | B sperm | b sperm |
|-----------------|---------|---------|
| ___ BB          | ___     | ___     |
| ___ Bb          | ___     | ___     |
| ___ bb          | ___     | ___     |
| Totals =        | ___     | ___     |

The egg frequencies may be similarly calculated. Note that the gamete frequencies of the $F_2$ are the same as the gamete frequencies of the last generation. Phenotype percentage also remains the same. Thus, the gene frequencies did not change between the $F_1$ and the $F_2$ generation. Again, given the assumptions of the Hardy-Weinberg equilibrium, gene frequencies do not change generation after generation.

19. In a population, 81 percent of the organisms are homozygous dominant, and 1 percent are homozygous recessive. Find the following
    a. the percentage of heterozygotes
    b. the frequency of the dominant allele
    c. the frequency of the recessive allele

20. In a population of 200 individuals, determine the following for a particular locus if $p = 0.80$.
    a. the number of homozygous dominant individuals
    b. the number of homozygous recessive individuals
    c. the number of heterozygous individuals

21. If the percentage of gene $D$ is 70 percent in a gene pool, find the percentage of gene $d$.

_____

_____

22. If the frequency of gene $R$ in a population is 0.60, what percentage of the individuals are heterozygous $Rr$?

_____

_____

## Matching

Select the most appropriate answer for each.

23. ___gene flow
24. ___allele frequencies
25. ___neutral mutations
26. ___microevolution
27. ___lethal mutation
28. ___alleles
29. ___mutation
30. ___genetic drift
31. ___genetic equilibrium
32. ___natural selection

A. A heritable change in DNA
B. Zero evolution
C. Different molecular forms of a gene
D. Change in allele frequencies as individuals leave or enter a population
E. Change or stabilization of allele frequencies due to differences in survival and reproduction among variant members of a population
F. Random fluctuation in allele frequencies over time due to chance
G. Change in which expression of mutated gene always leads to the death of the individual
H. The abundance of each kind of allele in the entire population
I. Neither harmful nor helpful to the individual
J. Changes in allele frequencies brought about by mutation, genetic drift, gene flow, and natural selection

## Fill-in-the-Blanks

A(n) (33) _____ is a group of individuals of the same species that occupy a given area at a specific time. The (34) _____ - _____ principle allows researchers to establish a theoretical reference point (baseline) against which changes in allele frequency can be measured. Variation can be expressed in terms of (35) _____ _____ , which means the relative abundance of different alleles carried by the individuals in that population. The stability of allele ratios that would occur if all individuals had equal probability of surviving and reproducing is called (36) _____ _____ . Over time, allele frequencies tend to change through infrequent but inevitable (37) _____ , which are the original source of genetic variation. Random fluctuation in allele frequencies over time due to chance occurrence alone is called (38) _____ _____ ; it is more pronounced in small populations than in large ones. (39) _____ flow associated with immigration and/or emigration also changes allele frequencies. (40) _____ is the differential survival and reproduction of individuals of a population that differ in one or more traits. (41) _____ _____ is the most important microevolutionary process.

## 18.3. NATURAL SELECTION REVISITED (p. 285)
## 18.4. DIRECTIONAL CHANGE IN THE RANGE OF VARIATION (p. 286–287)
## 18.5. SELECTION AGAINST OR IN FAVOR OF EXTREME PHENOTYPES (p. 288–289)
## 18.6. SPECIAL TYPES OF SELECTION (p. 290–291)

*Selected Words:* Biston betularia, *mark-release-recapture, biological control, sickle-cell anemia, malaria,* Plasmodium

## Boldfaced, Page-Referenced Terms

(285) fitness _____

_____

(278) natural selection _____

_____

(286) directional selection _____

_____

(287) pest resurgence _____

_____

(287) antibiotics _____

_____

(288) stabilizing selection _____

_____

(289) disruptive selection _____

_____

(290) sexual dimorphism _____

_____

(290) sexual selection _____

_____

(290) balancing selection _____

_____

(290) balanced polymorphism _____

_____

## Complete the Table

1. Complete the following table, which defines three types of natural selection effects.

| Effect | Characteristics | Example |
|---|---|---|
| a. Stabilizing selection | | |
| b. Directional selection | | |
| c. Disruptive selection | | |

## Labeling

2. Identify the three curves below as stabilizing selection, directional selection, or disruptive selection.

a. 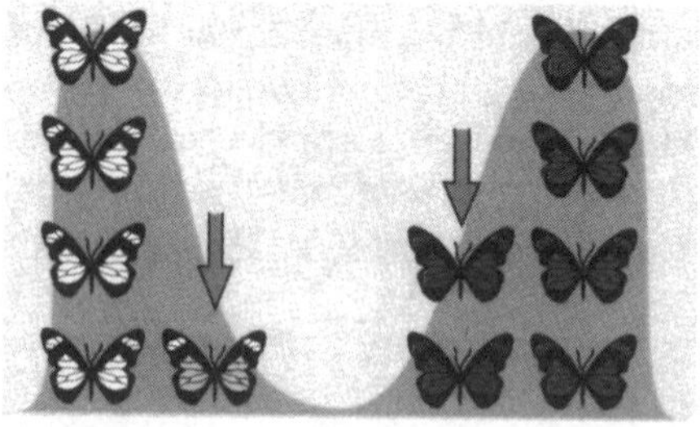 _____    b.  _____    c. _____

## Fill-in-the-Blanks

When individuals of different phenotypes in a population differ in their ability to survive and reproduce, their alleles are subject to (3) _____ selection. (4) _____ selection favors the most common phenotypes in the population. (5) _____ selection occurs when a specific change in the environment causes a heritable trait to occur with increasing frequency and the whole population tends to shift in a parallel direction. (6) _____ selection favors the development of two or more distinct polymorphic varieties such that they become increasingly represented in a population and the population splits into different phenotypic variations. (7) _____ provide an excellent example of stabilizing selection, because extreme phenotypes of the gallfly are eliminated by birds and wasps. (8) _____ - _____

_____ , a genetic disorder that produces an abnormal form of hemoglobin, is an example of

(9) _____ selection that maintains a high frequency of the harmful (10) _____ along with the

normal gene. Such a genetic juggling act is called a (11) _____ _____ . Differences in appearance

between males and females of a species are known as (12) _____ _____ . Among birds and

mammals, females act as agents of (13) _____ when they choose their mates. (14) _____ selection

is based on any trait that gives the individual a competitive edge in mating and producing offspring. The

more colorful, showier, and larger appearance of male pheasants when compared with females is the result

of (15) _____ selection. Any alleles in an individual that help it obtain resources, survive and

reproduce increase its (16) _____ and tend to be in phenotypes that become more numerous in the

population as it adapts to its environment.

## Choice

For questions 17–30, choose from the following categories of natural selection; in some cases, two letters may
be correct.

a. directional selection    b. stabilizing selection    c. disruptive selection
d. balanced polymorphism        e. sexual selection

17.____May result even when heterozygotes are selected against; unstable *Rh* markers in the human
population provide an example

18.____Sexual dimorphism

19.____Phenotypic forms at both ends of the variation range are favored and intermediate forms are selected
against

20.____May account for the persistence of certain phenotypes over time; horsetail plants still bear strong
resemblance to their ancient relatives

21.____Selection maintains two or more alleles for the same trait in steady fashion, generation after
generation

22.____The most common forms of a trait in a population are favored; over time, alleles for uncommon
forms are eliminated

23.____Allele frequencies shift in a steady, consistent direction; this may be due to a change in the
environment

24.____Often results when environmental conditions favor homozygotes over heterozygotes

25.____Counters the effects of mutation, genetic drift, and gene flow

26.____Human newborns weighing an average of 7 pounds are favored

27.____In West Africa, finches known as black-bellied seedcrackers have either large or small bills

28.____When a trait gives an individual an advantage in reproductive success

29.____The most frequent wing color of peppered moths shifted from a light form to a dark form as tree
trunks became soot-darkened due to coal used for fuel during the industrial revolution

30.____Females of a species choosing mates and directly affecting reproductive success

## 18.7. GENE FLOW (p. 291)
## 18.8. GENETIC DRIFT (pp. 292–293)

*Selected Words:* "*immigrant acorns*", *Ellis-van Creveld syndrome, feline infectious peritonitis*

### Boldfaced, Page-Referenced Terms

(291) gene flow _____

_____

(292) genetic drift _____

_____

(292) sampling error _____

_____

(292) fixation _____

_____

(292) bottleneck _____

_____

(293) founder effect _____

_____

(293) inbreeding _____

_____

(293) endangered species _____

_____

### Choice

For questions 1–10, choose from the following microevolutionary forces that can change gene frequency:

a. genetic drift       b. gene flow

1. ____ A random change in allele frequencies over the generations, brought about by chance alone

2. ____ Emigration

3. ____ Prior to the turn of the century hunters killed all but twenty of a large population of northern elephant seals

4. ____ When allele frequencies change due to individuals leaving a population or new individuals enter it

5. ____ Long ago, seabirds, winds, or ocean currents carried a few seeds from the Pacific Northwest to the Hawaiian Islands to establish plant populations

6. ____ Sometimes results in producing endangered species

7. ____ Founder effects and bottlenecks are two extreme cases

8. ____ Immigration

9. ____ Only 20,000 cheetahs have survived to the present

10. ____ Scrubjays make hundreds of round trips carrying acorns from oak trees as much as a mile away to soil in their home territories for winter storage; this introduces new alleles to oak forests

*Short Answer*

11. Distinguish between the two extreme cases of genetic drift: the founder effect and bottlenecks.

_____

_____

_____

# Self-Quiz

For questions 1–3, choose from these answers:
   a. gene pool
   b. genetic variation
   c. gene flow
   d. allele frequency

___ 1. Differences in the combinations of alleles carried by the individuals of a population would be its _____ .

___ 2. For sexually reproducing species, the _____ is a source of potentially enormous variation in traits.

___ 3. The relative abundance of each type of allele in a population is the _____ .

___ 4. $p$ plus $q = 1$ expresses the _____ of a population; $p^2$ plus $2pq$ plus $q^2$ expresses the _____ of a population.
   a. genotype frequency; allele frequency
   b. genotype frequency; genotype frequency
   c. allele frequency; genotype frequency
   d. allele frequency; allele frequency

___ 5. The unit used in studying evolution is the _____ .
   a. population
   b. individual
   c. fossil
   d. missing link

___ 6. Changes in allele frequencies brought about by mutation, genetic drift, gene flow, and natural selection are called _____ processes.
   a. genetic equilibrium
   b. microevolution
   c. founder effects
   d. independent assortment

For questions 7–9, choose from these answers:
   a. disruptive selection
   b. genetic drift
   c. directional selection
   d. stabilizing selection

___ 7. Cockroaches becoming increasingly resistant to pesticides is an example of _____ .

___ 8. The founder effect is a special case of _____ .

___ 9. Different alleles that code for different forms of hemoglobin have led to _____ in the midst of the malarial belt that extends through West and Central Africa.

For questions 10–12, choose from these answers:
   a. balanced polymorphism
   b. sexual dimorphism
   c. genetic equilibrium
   d. sexual selection

___10. Hardy and Weinberg invented an ideal population that was in _____ and used it as a baseline against which to measure evolution in real populations.

___11. The term _____ is applied when a population has nonidentical alleles for a trait, and both alleles are maintained at frequencies greater than 1 percent.

___12. Phenotypic differences between the males and females of a species are known as _____ .

# Chapter Objectives/Review Questions

This section lists general and detailed chapter objectives that can be used as review questions. You can make maximum use of these items by writing answers on a separate sheet of paper. Fill in answers where blanks are provided. To check for accuracy, compare your answers with information given in the chapter or glossary.

| Page | | Objectives/Questions |
|---|---|---|
| (282) | 1. | A _____ is a group of individuals occupying a given area and belonging to the same species. |
| (282) | 2. | Distinguish between morphological, physiological, and behavioral traits. |
| (282) | 3. | All of the genes of an entire population belong to a _____ . |
| (282) | 4. | Each kind of gene usually exists in one or more molecular forms, called _____ . |
| (282) | 5. | Review the five categories through which genetic variation occurs among individuals. |
| (283) | 6. | The abundance of each kind of allele in the whole population is referred to as allele _____ . |
| (283) | 7. | A point at which allele frequencies for a trait remain stable through the generations is called genetic _____ . |
| (283) | 8. | Be able to list the five conditions that must be met before conditions for the Hardy-Weinberg Rule are met. |
| (283) | 9. | Changes in allele frequencies brought about by mutation, genetic drift, gene flow, and natural selection are called _____ . |
| (283) | 10. | A _____ is a random, heritable change in DNA. |
| (283) | 11. | Distinguish a lethal mutation from a neutral mutation. |
| (284–285) | 12. | Be able to calculate allele and genotype frequencies when provided with the genotype frequency of the recessive allele. |
| (285) | 13. | _____ _____ is defined as the change or stabilization of allele frequencies due to differences in survival and reproduction among variant members of a population. |
| (286–289) | 14. | Be able to define and provide an example of directional selection, stabilizing selection, and disruptive selection. |
| (290) | 15. | Define balanced polymorphism. |
| (290) | 16. | The occurrence of phenotypic differences between males and females of a species is called _____ _____ . |
| (290) | 17. | _____ selection is based on any trait that gives an individual a competitive edge in mating and producing offspring. |
| (291) | 18. | Allele frequencies change as individuals leave or enter a population; this is gene _____ . |
| (292) | 19. | Random fluctuations in allele frequencies over time, due to chance, is called _____ . |
| (292–293) | 20. | Distinguish the founder effect from a bottleneck. |
| (291) | 21. | Relate bottlenecks to endangered species. |

# Integrating and Applying Key Concepts

Can you imagine any way in which directional selection may have occurred or may be occurring in humans? Which factors do you suppose are the driving forces that sustain the trend? Do you think the trend could be reversed? If so, by what factor(s)?

# 19

# SPECIATION

## Interactive Exercises

*The Case of the Road-Killed Snails* (pp. 296–297)

### 19.1. ON THE ROAD TO SPECIATION (pp. 298–299)

### 19.2. REPRODUCTIVE ISOLATING MECHANISMS (pp. 300–301)

***Selected Words:*** <u>Helix aspersa</u>, *prezygotic isolation, sibling species, postzygotic isolation,* <u>Magicicada septendecim</u>, *zebroid.*

### *Boldfaced, Page-Referenced Terms*

(297) speciation _____

_____

(298) species _____

_____

(298) biological species concept _____

_____

(299) gene flow _____

_____

(299) genetic divergence _____

_____

(300) reproductive isolating mechanisms _____

_____

## Choice

For questions 1–10, choose from the following:

a. Mayr's biological species       b. genetic divergence       c. isolating mechanisms

1. ____Heritable aspect of body form, physiology, or behavior that prevents gene flow between genetically divergent populations

2. ____Groups of interbreeding natural populations

3. ____May occur as a result of stopping gene flow

4. ____Phenotypically very different but can interbreed and produce fertile offspring

5. ____Defined as the buildup of differences in the separated pools of alleles of populations

6. ____A "kind" of living organism

7. ____Factors that prevent horses and zebras from mating in the wild

8. ____Prezygotic

9. ____Does not work well with organisms that reproduce solely by asexual means or those known only from the fossil record

10. ____Postzygotic

## Matching

Select the most appropriate answer to match the isolating mechanisms; complete the exercise by entering "pre" in the parentheses if the mechanism is prezygotic and "post" if the mechanism is postzygotic.

11. ___ ecological (    )

12. ___ temporal (    )

13. ___ hybrid inviability (    )

14. ___ mechanical (    )

15. ___ zygote mortality (    )

16. ___ gametic mortality (    )

17. ___ behavioral (    )

18. ___ hybrid offspring (    )

A. Potential mates occupy overlapping ranges but reproduce at different times.

B. The first-generation hybrid forms but shows very low fitness.

C. Potential mates occupy different local habitats within the same area.

D. Potential mates meet but cannot figure out what to do about it.

E. Sperm is transferred but the egg is not fertilized (gametes die or gametes are incompatible).

F. The hybrid is sterile or partially so.

G. Potential mates attempt engagement, but sperm cannot be successfully transferred.

H. The egg is fertilized, but the zygote or embryo dies.

## Choice

For questions 19–25, choose from the following isolating mechanisms:

a. temporal     b. behavioral     c. mechanical     d. gametic mortality     e. ecological     f. postzygotic

19.___Sterile zebroids

20.___Two sage species; each has its flower petals arranged as a "landing platform" for a different pollinator

21.___Two species of cicada; one matures, emerges, and reproduces every thirteen years, the other every seventeen years

22.___Populations of the manzanita shrub *Arctostaphylos patula* and *A. viscida* demonstrate differing tolerances to water stress at varying distances from water sources; speciation may be underway

23.___Pollen grains from one flowering plant species molecularly mismatched with gametes of a different flowering plant species

24.___Prior to copulation, male and female birds engage in complex courtship rituals recognized only by birds of their own species

25.___Sterile mules

## Identification

26. The illustration below depicts the divergence of one species as time passes. Each horizontal line represents a population of that species and each vertical line, a point in time. Answer the questions below the illustration.

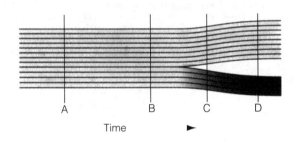

Time ▶

a. How many species are represented at time A?
b. How many species exist at time D?
c. Between what times (letters) does divergence begin?
d. What letter represents the time that complete divergence is reached?
e. Between what two times (letters) is divergence clearly underway but not complete?

## 19.3. SPECIATION IN GEOGRAPHICALLY ISOLATED POPULATIONS (pp. 302–303)

## 19.4. MODELS FOR OTHER SPECIATION ROUTES (pp. 304–305)

***Selected Words:*** *"Pacific enzymes", "Atlantic enzymes", Hawaiian honeycreepers, cichlids, "dosage compensation",* <u>Triticum</u> <u>aestivum</u>, *"secondary contact".*

## *Boldfaced, Page-Referenced Terms*

(302) allopatric speciation _____

(303) archipelago _____

_____

(304) sympatric speciation _____

_____

(304) polyploidy _____

_____

(305) parapatric speciation _____

_____

(305) hybrid zone _____

_____

## Complete the Table

1. Complete the following table, which defines three speciation models.

| Speciation Model | Description |
|---|---|
| a. | Suggests that species can form within the range of an existing species, in the absence of physical or ecological barriers; a controversial model; evidence comes from crater lake fishes |
| b. | Emphasizes disruption of gene flow; perhaps the main speciation route; through divergence and evolution of isolating mechanisms, separated populations become reproductively incompatible and cannot interbreed |
| c. | Occurs where populations share a common border; the borders may be permeable to gene flow; in some, gene exchange is confined to a common border, the hybrid zone |

## Dichotomous Choice

Circle one of two possible answers given between parentheses in each statement.

2. Instant speciation by polyploidy in many flowering plant species represents (parapatric / sympatric) speciation.
3. In the past, hybrid orioles existed in a hybrid zone between eastern Baltimore orioles and western Bullock's orioles; today, hybrid orioles are becoming less frequent in this example of (allopatric / parapatric) speciation.
4. In the absence of gene flow between geographically separate populations, genetic divergence may become sufficient to bring about (allopatric / parapatric) speciation.
5. The uplifted ridge of land we now call the Isthmus of Panama divides the Atlantic Ocean from the Pacific Ocean and formed a natural laboratory for biologists to study (allopatric / sympatric) speciation.
6. In the 1800s, a major earthquake changed the course of the Mississippi River; this isolated some populations of insects that could not swim or fly, and set the stage for (parapatric / allopatric) speciation.

## Short Answer

7. Study the illustration below, which shows the possible course of wheat evolution; similar genomes (a genome is a complete set of chromosomes) are designated by the same letter. Answer the questions below the diagram.

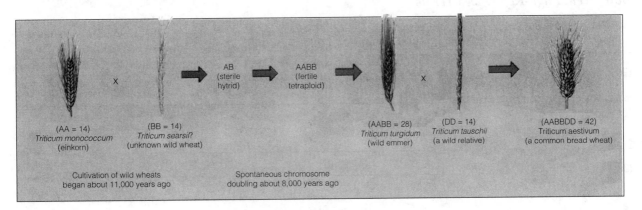

a. How many pairs of chromosomes are found in *Triticum monococcum*?

_____

b. How are the *T. monococcum* chromosomes designated?

_____

c. How many pairs of chromosomes are found in *T. searsii*?

_____

d. How are the *T. searsii* chromosomes designated?

_____

e. Why is the hybrid designated AB sterile?

_____

f. What cellular event must have occurred to create the *T. turgidum* genome?

_____

g. Describe the genome of the plant arising from the cross of *T. turgidum* and *T. tauschii* (not shown on the diagram)?

_____

h. What cellular event must have occurred to create the *T. aestivum* genome?

_____

i. What is the source of the A genome in *T. aestivum*? The source of the B genome in *T. aestivum*? The source of the D genome in *T. aestivum*?

_____

## 19.5. PATTERNS OF SPECIATION (pp. 306–307)

*Selected Words:* pattern of speciation, branch, branch point, physical access; evolutionary access; ecological access

### Boldfaced, Page-Referenced Terms

(306) cladogenesis _____

_____

(306) anagenesis _____

_____

(306) evolutionary tree _____

_____

(306) gradual model of speciation _____

_____

(306) punctuation model of speciation _____

_____

(307) adaptive radiation _____

_____

(307) adaptive zone _____

_____

(307) key innovation _____

_____

(307) background extinction _____

_____

(307) mass extinction _____

_____

### Complete the Table

1. Complete the table below to summarize information to interpret evolutionary tree diagrams. Refer to the text, pp. 306–307.

| Branch Form | Interpretation |
|---|---|
| a. | Traits changed rapidly around the time of speciation |
| b. | An adaptive radiation occurred |
| c. | Extinction |
| d. | Speciation occurred through gradual changes in traits over geologic time |
| e. | Traits of the new species did not change much thereafter |
| f. | Evidence of this presumed evolutionary relationship is only sketchy |

## Matching

Select the most appropriate answer for each.

2. ___cladogenesis

3. ___evolutionary tree diagrams

4. ___gradual model of speciation

5. ___punctuation model of speciation

6. ___adaptive radiation

7. ___adaptive zones

8. ___background extinction

9. ___mass extinction

10. ___physical, evolutionary, and ecological access

11. ___anagenesis

A. Expected rate of species disappearance as local conditions change
B. Conditions allowing a lineage to radiate into an adaptive zone
C. Speciation in which most morphological changes are compressed into a brief period of hundreds or thousands of years when populations first start to diverge; evolutionary tree branches make abrupt ninety-degree turns
D. Ways of life such as "burrowing in the seafloor"
E. Speciation with slight angles on the evolutionary tree, which conveys many small changes in form over long time spans
F. An abrupt rise in the rates of species disappearance above background level
G. Speciation occurring within a single, unbranched line of descent
H. A method of summarizing information about the continuities of relationships among species
I. A branching speciation pattern; isolated populations diverge
J. A burst of microevolutionary activity within a lineage; results in the formation of new species in a wide range of habitats

## Fill-in-the-Blanks

(12) _____ is the process whereby species are formed. A (13) _____ is composed of one or more populations of individuals who can interbreed under natural conditions and produce fertile, reproductively isolated offspring. (14) _____ can be described as a process in which differences in alleles accumulate between populations. Any aspect of structure, function, or behavior that prevents interbreeding is a (15) _____ _____ mechanism. Physical barriers that prevent gene flow as in the case of large rivers changing course or forests giving way to grasslands are (16) _____ barriers. Wheat is an example of a species formed by (17) _____ and (18) _____.

---

# Self-Quiz

___ 1. Of the following, _____ is (are) not a major consideration of the biological species concept.
a. groups of populations
b. reproductive isolation
c. asexual reproduction
d. interbreeding natural populations
e. production of fertile offspring

___ 2. When something prevents gene flow between two populations or subpopulations, _____ may occur.
a. genetic drift, natural selection, and mutation
b. genetic divergence
c. a buildup of differences in the separated gene pools of alleles
d. all of the above

___ 3. When potential mates occupy different local habitats within the same area, it is termed _____ isolation.
   a. temporal
   b. behavioral
   c. mechanical
   d. ecological

___ 4. "Potential mates occupy overlapping ranges but reproduce at different times" is a description of _____ isolation.
   a. temporal
   b. behavioral
   c. mechanical
   d. ecological

___ 5. Of the following, _____ is a postzygotic isolating mechanism.
   a. behavioral isolation
   b. gametic mortality
   c. hybrid inviability
   d. mechanical isolation

For questions 6–8, choose from the following answers:
   a. parapatric speciation
   b. sympatric speciation
   c. allopatric speciation

___ 6. When the Isthmus of Panama was formed, it provided contemporary scientists with an ideal natural laboratory to study _____ .

___ 7. Polyploidy and cichlid fishes of crater lakes provide evidence for _____ .

___ 8. The identification of a hybrid zone between the ranges of eastern Baltimore orioles and western Bullock orioles is an example of _____ .

___ 9. The branching speciation pattern revealed by the fossil record is called _____ .
   a. anagenesis
   b. the gradual model
   c. cladogenesis
   d. the punctuation model

___ 10. Species formation by many small changes over long time spans fit the _____ model of speciation; rapid speciation with morphological changes compressed into a brief period of population divergence describes the _____ model of speciation.
   a. anagenesis; punctuation
   b. punctuation; gradual
   c. gradual; cladogenesis
   d. gradual; punctuation

___ 11. Adaptive radiation is _____ .
   a. a burst of microevolutionary activity within a lineage
   b. the formation of new species in a wide range of habitats
   c. the spreading of lineages into unfilled adaptive zones
   d. a lineage radiating into an adaptive zone when it has physical, ecological, or evolutionary access to it
   e. all of the above

___ 12. The expected rate of species disappearance over time is called _____ .
   a. mass extinction
   b. reverse background radiation
   c. background extinction
   d. speciation extinction

For questions 13–14, choose from the following answers:
   a. stabilizing selection
   b. divergence
   c. a reproductive isolating mechanism
   d. polyploidy

___ 13. Two species of sage plants with differently shaped floral parts that prevent pollination by the same pollinator serve as an example of _____ .

___ 14. _____ occurs when isolated populations accumulate allele frequency differences between them over time.

# Chapter Objectives/Review Questions

This section lists general and detailed chapter objectives that can be used as review questions. You can make maximum use of these items by writing answers on a separate sheet of paper. Fill in answers where blanks are provided. To check for accuracy, compare your answers with information given in the chapter or glossary.

| Page | | Objectives/Questions |
|---|---|---|
| (297) | 1. | The process by which species are formed is known as _____ . |
| (298) | 2. | Be able to give the biological species concept as stated by Ernst Mayr. |
| (299) | 3. | _____ _____ is a buildup of differences in separated pools of alleles. |
| (300–301) | 4. | The evolution of reproductive _____ _____ paves the way for genetic divergence and speciation. |
| (302) | 5. | _____ speciation occurs when daughter species form gradually by divergence in the absence of gene flow between geographically separate populations. |
| (304) | 6. | In _____ speciation, daughter species arise, sometimes rapidly, from a small proportion of individuals within an existing population. |
| (304–305) | 7. | Explain why sympatric speciation by polyploidy is a rapid method of speciation. |
| (305) | 8. | When daughter species form from a small proportion of individuals along a common border between two populations, it is called _____ speciation. |
| (305) | 9. | In some cases of parapatric speciation, gene exchange between two species is confined to a _____ zone. |
| (306) | 10. | Distinguish cladogenesis from anagenesis. |
| (306) | 11. | Be able to explain the use of evolutionary tree diagrams and the symbolism used. |
| (306) | 12. | Explain why branches on an evolutionary tree that have slight angles indicate the _____ model of speciation. |
| (306) | 13. | Branches on an evolutionary tree that turn abruptly with 90-degree turns are consistent with the _____ model of speciation. |
| (307) | 14. | An adaptive radiation is a burst of _____ activity within a lineage that results in the formation of new species in a variety of habitats. |
| (307) | 15. | Cite examples of adaptive zones. |
| (307) | 16. | How does background extinction differ from mass extinction? |
| (308) | 17. | Be able to briefly define the major categories of prezygotic and postzygotic isolating mechanisms (see text, Table 19.2). |

# Integrating and Applying Key Concepts

*Systematics* is defined as the practice of describing, naming, and classifying living things; this includes the comparative study of organisms and all relationships among them. Plant systematists who work with flowering plants readily accept Ernst Mayr's definition of a "biological species" as described in this chapter. However, in real practice, systematists often must also work with a concept known as the "morphological species." For example, the statement is sometimes made that "two plant specimens belong to the same morphological species but not to the same biological species." Explain the meaning of that statement. What type of experimental evidence would be necessary as a basis for this statement? From your study of this chapter, can you suggest a reason that application of the term "morphological species" is sometimes necessary? What difficulties and inaccuracies, in terms of identifying a species, might the use of both species concepts present?

# 20

# THE MACROEVOLUTIONARY PUZZLE

*Of Floods and Fossils*

**FOSSILS—EVIDENCE OF ANCIENT LIFE**
Fossilization
Interpreting the Geologic Tombs
Interpreting the Fossil Record

**EVIDENCE OF A CHANGING EARTH**
An Outrageous Hypothesis
Drifting Continents and Changing Seas

**EVIDENCE FROM COMPARATIVE EMBRYOLOGY**
Developmental Program of Larkspurs
Developmental Program of Vertebrates

**EVIDENCE OF MORPHOLOGICAL DIVERGENCE**
Homologous Structures
Potential Confusion from Analogous Structures

**EVIDENCE FROM COMPARATIVE BIOCHEMISTRY**

Molecular Clocks
Protein Comparisons
Nucleic Acid Comparisons

**IDENTIFYING SPECIES, PAST AND PRESENT**
Assigning Names to Species
Groupings of Species – The Higher Taxa
Putting Classification Schemes to Work

**FINDING EVOLUTIONARY RELATIONSHIPS AMONG SPECIES**
The Systematics Approach
Actually, There is No Single Systematics Approach

*Focus on Science:* CONSTRUCTING A CLADOGRAM

**HOW MANY KINGDOMS?**
Whittaker's Five-Kingdom Scheme
Along Came the Archaebacteria

---

## Interactive Exercises

---

*Of Floods and Fossils* (pp. 310–311)

## 20.1. FOSSILS—EVIDENCE OF ANCIENT LIFE (pp. 312–313)

## 20.2. EVIDENCE OF A CHANGING EARTH (pp. 314–315)

*Selected Words:* species, trace fossils, <u>Cooksonia,</u> Cenozoic, Mesozoic, Paleozoic, Proterozoic, Archean, Precambrian, Carboniferous, Permian, Jurassic, Cretaceous, Wegener, <u>Glossopteris</u>, superplumes, "hot spots".

### Boldfaced, Page-Referenced Terms

(311) lineages _____

_____

(311) macroevolution _____

_____

(312) fossils _____

_____

(312) fossilization _____

_____

(313) stratification _____

_____

(313) geologic time scale _____

_____

(314) theory of uniformity _____

_____

(314) continental drift _____

_____

(314) Pangea _____

_____

(314) sea-floor spreading _____

_____

(314) plate tectonics _____

_____

(315) Gondwana _____

_____

## Matching

Select the most appropriate answer

1. ___fossilization
2. ___stratification
3. ___trace fossils
4. ___continuity of relationship
5. ___macroevolution
6. ___fossils
7. ___biased fossil record
8. ___formation of sedimentary deposits
9. ___oldest fossils
10. ___rupture or tilt of sedimentary layers

A. Evolutionary connectedness between organisms, past to present
B. Found in the deepest rock layers
C. Affected by movements of Earth's crust, hardness of organisms, population sizes, and certain environments
D. Produced by gradual deposition of silt, volcanic ash, and other materials
E. Refers to large-scale patterns, trends, and rates of change among groups of species
F. A process favored by rapid burial in the absence of oxygen
G. Evidence of later geologic disturbance
H. Recognizable, physical evidence of ancient life; the most common are bones, teeth, shells, spore capsules, seeds, and other hard parts
I. Includes the indirect evidence of coprolites and imprints of leaves, stems, tracks, trails, and burrows
J. A layering of sedimentary deposits

## Sequence

Earth history has been divided into four great eras that are based on four abrupt transitions in the fossil record. The oldest era has been subdivided. Arrange the eras in correct chronological sequence from the oldest to the youngest.

11. ___     A. Mesozoic

12. ___     B. Cenozoic

13. ___     C. Proterozoic

14. ___     D. Archean

15. ___     E. Paleozoic

## Matching

Choose the one most appropriate answer for each.

16. ___ Archean

17. ___ Mesozoic

18. ___ Pangea

19. ___ Proterozoic

20. ___ plate tectonic theory

21. ___ Paleozoic

22. ___ Gondwana

23. ___ Cenozoic

24. ___ influenced the evolution of life

25. ___ geologic time scale

A. The "modern era" of geologic time
B. Source of the most ancient fossils prior to subdivision of this era
C. An ancient continent that drifted southward from the tropics, across the south polar region, then northward
D. The boundaries mark the times of mass extinctions
E. The first era of the geologic time scale; a recent subdivision of the Proterozoic
F. An era that follows the Paleozoic
G. Plumes of molten rock from Earth's mantle spread out beneath the crust or break through it
H. An era whose fossils followed those of the Proterozoic
I. Formed from fusion with Gondwana and other land masses; a single world continent that extended from pole to pole
J. Changes in land masses, the oceans, and the atmosphere

*Complete the Table*

In the geologic time scale table below, fill in the missing information.

| | Period | Epoch | Millions of Years Ago (mya) |
|---|---|---|---|
| **(26)** ERA | QUATERNARY | Recent | 0.01– |
| | | Pleistocene | 1.55 |
| | TERTIARY | Pliocene | 5 |
| | | Miocene | 25 |
| | | Oligocene | 38 |
| | | Eocene | 54 |
| | | Paleocene | |
| | | | **(35)** |
| **(27)** ERA | **(31)** | Late | 100 |
| | | Early | 138 |
| | **(32)** | | 205 |
| | TRIASSIC | | |
| | | | **(36)** |
| **(28)** ERA | **(33)** | | 290 |
| | CARBONIFEROUS | | 360 |
| | DEVONIAN | | 410 |
| | SILURIAN | | 435 |
| | ORDOVICIAN | | 505 |
| | **(34)** | | |
| | | | **(37)** |
| **(29)** EON | | | |
| | | | **(38)** |
| **(30)** EON | | | |
| | | | **(39)** |

## 20.3. EVIDENCE FROM COMPARATIVE EMBRYOLOGY (pp. 316–317)

## 20.4. EVIDENCE OF MORPHOLOGICAL DIVERGENCE (pp. 318–319)

## 20.5. EVIDENCE FROM COMPARATIVE BIOCHEMISTRY (pp. 320–321)

*Selected Words:* <u>Delphinium</u> *(larkspur), vertebrate, cytochrome <u>c</u> molecules, giant panda.*

*Boldfaced, Page-Referenced Terms*

(316) comparative morphology _____

_____

(318) homologous structures _____

_____

(318) morphological divergence _____

_____

(318) homology _____

_____

(318) analogous structures _____

_____

(319) morphological convergence _____

_____

(319) analogy _____

_____

(320) neutral mutations _____

_____

(320) molecular clock _____

_____

(321) DNA-DNA hybridization _____

_____

## True-False

If the statement is true, write a T in the blank. If the statement is false, correct it by changing the underlined word(s) and writing the correct word(s) in the answer blank.

_____1. The extent to which a strand of DNA or RNA isolated from individuals of one species will base-pair with a comparable strand from individuals of a different species is a rough measure of the evolutionary <u>distance</u> between them.

_____2. The amino acid sequence of cytochrome *c* is very similar in such distantly related organisms as (1) yeast, (2) wheat and (3) humans; the probability that this striking resemblance resulted from chance alone is very <u>high</u>.

_____3. Analogous structures in two different lineages indicate strong evidence of morphological <u>divergence</u>.

_____4. Body parts that have similar form and function in different lineages yet develop from similar tissues in their embryos are homologous structures that provide very strong evidence of morphological <u>divergence</u>.

_____5. Biochemical <u>similarities</u> are fewest among the most closely related species and most numerous among the most distantly related.

## Choice

a. analogy   b. homology   c. an example of a lineage   d. a specific example of morphological convergence
                    e. a specific example of morphological divergence

6. ____Sharks, penguins and porpoises.

7. ____Ancestral ape-human, *Homo erectus*, and modern humans (*Homo sapiens*).

8. ____Body parts that were once quite different among distantly related lineages but are now quite similar in function and in apparent structure.

9. ____The human arm and hand, the porpoise's front flipper, and the bird's wing.

10. ____Body parts of different organisms that develop from similar embryonic parts yet might result in quite differently appearing adult structures that would probably not have similar functions.

## Labeling

Evidence for macroevolution comes from comparative morphology and comparative biochemistry. For each of the following listed items, place an "M" in the blank if the evidence comes from comparative morphology or a "B" if the evidence comes from comparative biochemistry.

11.____Early in the vertebrate developmental program, embryos of different lineages proceed through strikingly similar stages.

12.____Comparison of the developmental rates of changes in a chimpanzee skull and a human skull.

13.____The evolution of different rates of flower development in two larkspur species also led to the coevolution of two different pollinators, honeybees and hummingbirds.

14.____Using accumulated neutral mutations to date the divergence of two species from a common ancestor.

15.____The amino acid sequence in the cytochrome $c$ of humans precisely matches the sequence in chimpanzees.

16.____Establishing a rough measure of evolutionary distance between two organisms by use of DNA-DNA hybridization studies that demonstrate the extent to which the DNA from one species base-pairs with another.

17.____Regulatory genes control the rate of growth of different body parts and can produce large differences in the development of two very similar embryos.

18.____Results from DNA-DNA hybridization studies supported an earlier view that giant pandas, bears, and red pandas are related by common descent.

19.____Distantly related vertebrates, such as sharks, penguins, and porpoises, show a similarity to one another in their proportion, position, and function of body parts.

20.____Finding similarities in vertebrate forelimbs when comparing the wings of pterosaurs, birds, bats, and porpoise flippers.

## Short Answer

21. List three methods employed by systematics to elucidate the patterns of diversity in an evolutionary context. _____

_____

_____

_____

_____

_____

## 20.6. IDENTIFYING SPECIES, PAST AND PRESENT (pp. 322–323)

## 20.7. FINDING EVOLUTIONARY RELATIONSHIPS AMONG SPECIES (pp. 324–325)

## 20.8. *Focus on Science:* CONSTRUCTING A CLADOGRAM (pp. 326–327)

## 20.9. HOW MANY KINGDOMS? (pp. 328–329)

**Selected Words:** *Linnaeus, classical taxonomy, cladistic taxonomy, in-group, out-group, monophyletic group, ancestor, Whittaker's fire-kingdom classification scheme, Woese, Bult,* Methanococcus jannaschii, *three-domain scheme, species.*

## Boldfaced, Page-Referenced Terms

(322) taxonomy _____

_____

(322) binomial system _____

_____

(322) genus, genera (pl.) _____

_____

(322) specific name _____

_____

(323) classification schemes _____

_____

(323) higher taxa _____

_____

(323) phylogeny _____

_____

(324) evolutionary systematics _____

_____

(324) derived trait _____

_____

(325) cladograms _____

_____

(328) Monera _____

_____

(328) Protista _____

_____

(328) Fungi _____

_____

(328) Plantae _____

_____

(328) Animalia _____

_____

(328) Archaebacteria _____

_____

(329) six-kingdom classification scheme _____

_____

(329) Eubacteria _____

_____

## Matching

Choose the single most appropriate letter.

1. ___black bear
2. ___phylogeny
3. ___binomial system
4. ___*Pinus strobus* and *Pinus banksiana*
5. ___species
6. ___genus
7. ___family
8. ___taxa
9. ___*Ursus americanus* (black bear) and *Homarus americanus* (Atlantic lobster)
10. ___Linnaeus

A. A group of similar species
B. The taxon most often studied
C. Evolutionary relationships among species, reflected in classification schemes
D. The organizing units of classification schemes
E. Classification level that includes similar genera
F. Originated the modern practice of providing two scientific names for organisms
G. An example of a common name
H. Two species belonging to one genus
I. An example of two very different organisms having the same specific epithet (species name)
J. System of assigning a two-part Latin name to a species

## Short Answer

11. Arrange the following jumbled taxa in proper order, with the most inclusive first: family, class, species, kingdom, genus, phylum (or division), and order. _____

_____

_____

_____

12. Arrange the following jumbled taxa and their categories in proper order, the most inclusive first: genus: *Archibaccharis*; kingdom: Plantae; order: Asterales; species: *lineariloba*; class: Dicotyledonae; division: Anthophyta; family: Asteraceae. _____

_____

_____

_____

13. List three methods employed by systematics to elucidate the patterns of diversity in an evolutionary context. _____

_____

_____

## Complete the Table.

14. Complete the table below to define two different approaches to the problem of reconstructing the history of life.

| School of Systematics | Definition |
|---|---|
| a. Evolutionary systematics | |
| b. Cladistics | |

## Short Answer

15. What are monophyletic lineages? _____

_____

16. Describe the construction of a cladogram. _____

_____

17. What is an outgroup? _____

_____

18. What is meant by a derived trait? _____

_____

19. What is the essential information portrayed by a cladogram? _____

_____

## Cladogram Interpretation

20. Study the cladogram of seven taxa below. The vertical bars on the stem of the cladogram represent shared derived traits. Various taxa are indicated by numbers. Answer the questions following the cladogram.

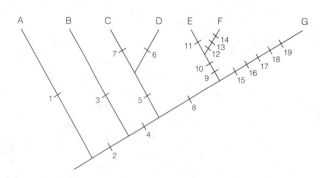

   a. On the cladogram above, what types of information could be represented by the numbered traits?

_____

_____

b. Which derived trait is shared by taxa EFG?

c. Which derived trait is shared by taxa CDEFG?

d. Which is the unique derived trait shared by taxa C and D?

_____

e. What does it mean if some taxa are closer together on the cladogram than others?

_____

f. Which taxon on the cladogram represents the outgroup condition?

_____

g. What is a monophyletic taxon?

_____

h. For this cladogram, list the monophyletic taxa shown as well as shared derived traits.

_____

## Matching

Choose the one best answer for each.

21. ___ Monera

22. ___ Protista

23. ___ Fungi

24. ___ Plantae

25. ___ Animalia

A. Multicelled heterotrophs that feed by extracellular digestion and absorption
B. Single-celled prokaryotes; some are autotrophs, others heterotrophs
C. Diverse multicelled heterotrophs, including predators and parasites
D. Multicelled photosynthetic autotrophs
E. Diverse single-celled eukaryotes; some are photosynthetic autotrophs, many heterotrophs

# Self-Quiz

___ 1. Phylogeny is _____ .
   a. the identification of organisms and assigning names to them
   b. producing a "retrieval system" of several levels
   c. an evolutionary history of organism(s), both living and extinct
   d. a study of the adaptive responses of various organisms

___ 2. The significant contribution by Linnaeus was to develop _____ .
   a. the idea that organisms should be placed in two kingdoms
   b. the binomial system
   c. the theory of evolution
   d. the idea that a species could have several common names

___ 3. The process of grouping organisms according to similarities derived from a common ancestor is known as _____ .
   a. taxonomy
   b. classification
   c. a binomial system
   d. macroevolution

___ 4. Lineages sharing a common evolutionary heritage are termed _____ .
   a. phenetic
   b. polyphyletic
   c. monophyletic
   d. homologous

___ 5. All living organisms have eukaryotic cell structure except the _____ .
   a. Animalia
   b. Plantae
   c. Fungi
   d. Monera
   e. Protista

For questions 6–7, choose from the following answers.
   a. convergence
   b. divergence

___ 6. The wings of pterosaurs, birds, and bats serve as examples of _____ .

___ 7. Penguins and porpoises serve as examples of _____ .

For questions 8–12, choose from the following answers.
   a. DNA-DNA hybridization
   b. homologous
   c. macroevolution
   d. analogous
   e. regulatory genes

___ 8. _____ structures resemble one another due to common descent.

___ 9. A rough measure of the evolutionary distance between two species is provided by _____ .

___10. _____ control the rate of growth in different body parts.

___11. Examples of _____ structures are the shark fin and the penguin wing.

___12. _____ refers to changes in groups of species.

___13. Early embryos of vertebrates strongly resemble one another *because* _____ .
   a. the genes that guide early embryonic development are the same (or similar) in all vertebrates
   b. the analogous structures they each contain are evidence of morphological convergence
   c. the homologous structures they each contain provide very strong evidence of morphological divergence
   d. DNA-DNA hybridization techniques reveal many structural alterations that have resulted from gene mutations in early embryos

# Chapter Objectives/Review Questions

This section lists general and detailed chapter objectives that can be used as review questions. You can make maximum use of these items by writing answers on a separate sheet of paper. Fill in answers where blanks are provided. To check for accuracy, compare your answers with information given in the chapter or glossary.

*Page*      *Objectives/Questions*

(311)     1.  _____ refers to large-scale patterns, trends, and rates of change among groups of species.
(312)     2.  Give a generalized definition of a "fossil"; cite examples.
(312)     3.  _____ begins with rapid burial in sediments or volcanic ash.
(313)     4.  What is the name given to the layering of sedimentary deposits?
(313)     5.  Are the oldest fossils found in the deepest layers of sedimentary rocks or nearer the surface?
(314)     6.  Present the evidence that has caused geologists to amend the theory of uniformity and accept the ideas of plate tectonics as they attempt to decipher Earth's evolutionary history.
(317)     7.  Early in the vertebrate developmental program, _____ of different lineages proceed through strikingly similar stages.
(317)     8.  Mutation of _____ genes explains the differences in the timing of growth sequences of chimps and humans (even though they have nearly identical genes).
(316)     9.  Explain why two species of larkspurs have flowers with different shapes that attract different pollinators.
(318)    10.  The macroevolutionary pattern of change from a common ancestor is known as morphological _____ .
(318–319) 11.  Distinguish homologies from analogies as applied to organisms.

(319)   12. A pattern of long-term change in similar directions among remotely related lineages is called morphological _____.

(320)   13. Define *neutral mutations* and discuss their use as molecular clocks.

(321)   14. Some nucleic acid comparisons are based on _____ hybridization.

(322)   15. Describe the binomial system as developed by Linnaeus; why is Latin used as the language of binomials?

(322)   16. Be able to arrange the classification groupings correctly, from most to least inclusive.

(323)   17. Modern classification schemes reflect _____, the evolutionary relationships among species, starting with the most ancestral and including all the branches leading to their descendants.

(322)   18. Explain the difficulties taxonomists may encounter identifying species.

(324)   19. _____ is a branch of biology that deals with patterns of diversity in an evolutionary context.

(324–327) 20. Be able to describe the construction, appearance, and use of cladograms.

(328)   21. List the five kingdoms of life in the Whittaker system and cite examples of the organisms in each.

---

## Integrating and Applying Key Concepts

Water crowfoot plants (family Ranunculaceae) often have two or more distinct leaf types within the same species and on the same plant. One type is capillary (hair-like), found submerged in the water or growing well above the water. The other type is laminate (flat and expanded), and is found floating on the water or submerged. In some *Ranunculus* species three different leaf shapes form on a plant in a sequence. Suppose two taxonomists describe the same crowfoot plant as being two different and distinct species due to these two different leaf shapes. Can you think of difficulties that might arise when other researchers are required to refer to these "two species" in the course of their work?

# 21

# THE ORIGIN AND EVOLUTION OF LIFE

## Interactive Exercises

*In the Beginning . . .* (pp. 332–333)

### 21.1. CONDITIONS ON THE EARLY EARTH (pp. 334–335)

### 21.2. EMERGENCE OF THE FIRST LIVING CELLS (pp. 336–337)

*Selected Words: core, "proteinoids", porphyrin ring.*

*Boldfaced, Page-Referenced Terms*

(333) Big Bang _____

(334) crust _____

(334) mantle _____

(337) RNA world _____

(337) proto-cells _____

## Choice

Answer questions 1–20 by choosing from the following probable stages of the physical and chemical evolution of life.

a. Formation of Earth
b. Early Earth and the first atmosphere
c. The synthesis of organic compounds

d. Origin of agents of metabolism
e. Origin of self-replicating systems
f. Origin of the first plasma membranes

1. ____ Most likely, the proto-cells were little more than membrane sacs protecting information-storing templates and various metabolic agents from the environment.

2. ____ Simple systems (of this type) of RNA, enzymes, and coenzymes have been created in the laboratory.

3. ____ Four billion years ago, Earth was a thin-crusted inferno.

4. ____ Sunlight, lightning, or heat escaping from Earth's crust could have supplied the energy to drive their condensation into complex organic molecules.

5. ____ During the first 600 million years of Earth history, enzymes, ATP, and other molecules could have assembled spontaneously at the same locations.

6. ____ Was gaseous oxygen ($O_2$) present? Probably not.

7. ____ Sidney Fox heated amino acids under dry conditions to form protein chains, which he placed in hot water. The cooled chains self-assembled into small, stable spheres. The spheres were selectively permeable.

8. ____ Stanley Miller mixed hydrogen, methane, ammonia, and water in a reaction chamber that recirculated the mixture and bombarded it with a spark discharge. Within a week, amino acids and other small organic compounds had been formed.

9. ____ Without an oxygen-free atmosphere, the organic compounds that started the story of life never would have formed on their own. Free oxygen would have attacked them.

10. ____ Much of Earth's inner rocky material melted.

11. ____ Imagine an ancient sunlit estuary, rich in clay deposits. Countless aggregations of organic molecules stick to the clay.

12. ____ In other experiments, fatty acids and glycerol combined to form long-tail lipids under conditions that simulated evaporating tidepools. The lipids self-assembled into small, water-filled sacs, which in many were like cell membranes.

13. ____ We still don't know how DNA entered the picture.

14. ____ Rocks collected from Mars, meteorites, and the moon contain chemical precursors that must have been present on early Earth.

15. ____ This differentiation resulted in the formation of a crust of basalt, granite, and other types of low-density rock, a rocky region of intermediate density (the mantle), and a high-density, partially molten core of nickel and iron.

16. ____ When the crust finally cooled and solidified, water condensed into clouds and the rains began. For millions of years, runoff from rains stripped mineral salts and other compounds from Earth's parched rocks.

17. ____ The close association of enzymes, ATP, and other molecules would have promoted chemical interactions. Sunlight energy alone could have driven the spontaneous formation of RNA molecules.

18. ____ Even if amino acids did form in the early seas, they wouldn't have lasted long. In water, the favored direction of most spontaneous reactions is toward hydrolysis, not condensation.

19.___Between 4.6 and 4.5 billion years ago, the outer regions of the cloud cooled.

20.___At first, it probably consisted of gaseous hydrogen ($H_2$), nitrogen ($N_2$), carbon monoxide (CO), and carbon dioxide ($CO_2$).

## 21.3. ORIGIN OF PROKARYOTIC AND EUKARYOTIC CELLS (pp. 338–339)

## 21.4. *Focus on Science:* WHERE DID ORGANELLES COME FROM? (pp. 340–341)

*Selected Words:* *"Precambrian", Laurentia,* Euglena, *Lynn Margulis, "mitochondrial code",* Cyanophora paradoxa, *Pangea, Tethys Sea.*

### *Boldfaced, Page-Referenced Terms*

(338) Archean _____

_____

(338) prokaryotic cells _____

_____

(338) eubacteria _____

_____

(338) archaebacteria _____

_____

(338) eukaryotic cells _____

_____

(338) stromatolites _____

_____

(339) Proterozoic _____

_____

(341) endosymbiosis _____

_____

*Choice*

For questions 1–15, choose from the following answers:

a. Archean     b. Proterozoic

1. ____ By 2.5 billion years ago, the noncyclic pathway of photosynthesis had evolved in some eubacterial species.

2. ____ By 1.2 billion years ago, eukaryotes had originated.

3. ____ The first prokaryotic cells emerged.

4. ____ Oxygen, one of the by-products of photosynthesis, began to accumulate.

5. ____ There was an absence of free oxygen.

6. ____ There was a divergence of the original prokaryotic lineage into three evolutionary directions.

7. ____ About 800 million years ago, stromatolites began a dramatic decline.

8. ____ Between 3.5 and 3.2 billion years ago, the cyclic pathway of photosynthesis evolved in some species of eubacteria.

9. ____ Fermentation pathways were the most likely sources of energy.

10. ____ An oxygen-rich atmosphere stopped the further chemical origin of living cells.

11. ____ Food was available, predators were absent, and biological molecules were free from oxygen attacks.

12. ____ Aerobic respiration became the dominant energy-releasing pathway.

13. ____ Archaebacteria and eubacteria arose.

14. ____ The first cells may have originated in tidal flats or on the seafloor, in muddy sediments warmed by heat from volcanic vents.

15. ____ Near the shores of the supercontinent Laurentia, small, soft-bodied animals were leaving tracks and burrows on the seafloor.

## Fill-in-the Blanks

(See *Focus on Science:* Where Did Organelles Come From? [pp. 340–341])

The most important feature of eukaryotic cells is the abundance of membrane-bound (16) _____ in the cytoplasm. The origin of these structures remains a mystery. Some probably evolved gradually, through (17) _____ mutations and natural selection. Some extant prokaryotic cells do possess infoldings of the (18) _____ membrane, a site where enzymes and other metabolic agents are embedded. In prokaryotic cells that were ancestral to eukaryotic cells, similar membranous infoldings may have served as a (19) _____ from the surface to deep within the cell. These membranes may have evolved into (20) _____ channels and into an envelope around the DNA. The advantage of such membranous enclosures may have been to protect (21) _____ and their products from foreign invader cells. Evidence for this is that yeasts, simple nucleated eukaryotic cells, have an abundance of (22) _____ which they transfer from cell to cell much as prokaryotic cells do.

As the evolution of eukaryotes proceeded, accidental partnerships between different (23) _____ species must have formed countless times. Some partnerships resulted in the origin of mitochondria, chloroplasts, and other organelles. This is the theory of (24) _____ , developed in greatest detail by Lynn Margulis. According to this theory, (25) _____ arose after the noncyclic pathway of photosynthesis emerged and oxygen had accumulated to significant levels. (26) _____ transport systems had been expanded in some bacteria to include extra cytochromes to donate electrons to (27) _____ in a system of aerobic respiration. By 1.2 billion years ago or earlier, amoebalike ancestors of eukaryotes were engulfing aerobic (28) _____ and perhaps forming endocytic vesicles around the food for delivery to the cytoplasm for digestion. Some aerobic bacteria resisted digestion and thrived in a new, protected, nutrient-rich environment. In time they released extra (29) _____—which the hosts came to depend on for growth, greater activity, and assembly of other structures. The guest aerobic bacteria came to depend on the metabolic functions of their (30) _____ cell. The aerobic and anaerobic cells were now (31) _____ of independent existence. The guests had become (32) _____ , supreme suppliers of (33) _____ . Mitochondria are bacteria-size and replicate their own (34) _____ dividing independently of the host cell's division process. In addition, (35) _____ may be descended from aerobic eubacteria that engaged in oxygen-producing photosynthesis. Such cells may have been engulfed, resisted digestion, and provided their respiring (36) _____ cells with needed oxygen. Chloroplasts resemble some (37) _____ in metabolism and overall nucleic acid sequence. Chloroplasts have self-replicating DNA and they divide independently of the host cell's division.

New cells appeared on the evolutionary stage that were equipped with a nucleus, ER membranes, and mitochondria or chloroplasts (or both); they were the first eukaryotic cells, the first (38) ____.

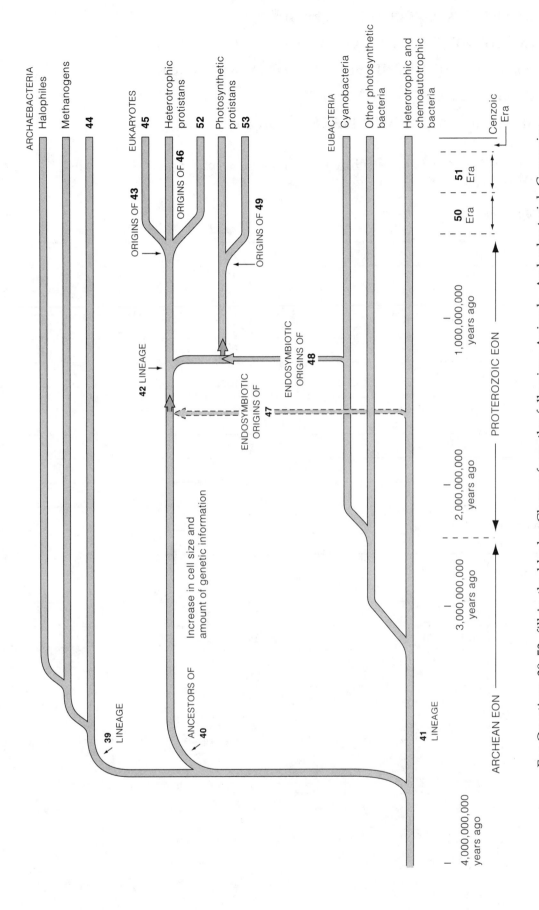

For Questions 39–53, fill in the blanks. Choose from the following: Animals, Archaebacterial, Cenozoic, Chloroplasts, Eubacterial, Eukaryote, Eukaryotes, Fungi, Mesozoic, Mitochondria, Paleozoic, Plants

**21.5. LIFE IN THE PALEOZOIC ERA** (pp. 342–343)

**21.6. LIFE IN THE MESOZOIC ERA** (pp. 344–345)

**21.7.** *Focus on Science:* **HORRENDOUS END TO DOMINANCE** (p. 346)

**21.8. LIFE IN THE CENOZOIC ERA** (p. 347)

**21.9. SUMMARY OF EARTH AND LIFE HISTORY** (pp. 348–349)

*Selected Words:* *trilobites, nautiloids,* Psilophyton, Dimetrodon, *Pangea, Tethys Sea,* Triceratops, Velociraptor, *Melosh,* Smilodon.

*Boldfaced Page-Referenced Terms*

(342) Paleozoic _____

_____

(342) Ediacarans _____

_____

(344) Mesozoic _____

_____

(344) flowering plants _____

_____

(344) dinosaurs _____

_____

(346) asteroids _____

_____

(346) K - T asteroid impact theory _____

_____

(346) global broiling theory _____

_____

(347) Cenozoic _____

_____

## Sequence

Refer to Figure 21.20 in the text. Study of the geologic record reveals that as the major events in the evolution of Earth and its organisms occurred, there were periodic major *extinctions* of organisms followed by major *radiations* of organisms. Using the list below, arrange the letters of the extinctions and radiations in the approximate order in which they occurred, from most recent to most ancient.

1. ___     A. Pangea, worldwide ocean forms; shallow seas squeezed out. Major radiations of reptiles, gymnosperms.

2. ___     B. Land masses dispersed near equator. Simple marine communities only. Origin of animals with hard parts.

3. ___     C. Mass extinction of many marine invertebrates, most fishes.

4. ___     D. Pangea breakup begins. Rich marine communities. Major radiations of dinosaurs.

5. ___     E. Asteroid impact? Mass extinction of all dinosaurs and many marine organisms.

6. ___     F. Recovery, radiations of marine invertebrates, fishes, dinosaurs. Gymnosperms the dominant land plants. Origin of mammals.

7. ___     G. Major glaciations. Modern humans emerge and begin what may be the greatest mass extinction of all time on land, starting with Ice Age hunters.

8. ___     H. Unprecedented mountain building as continents rupture, drift, collide. Major climatic shifts; vast grasslands emerge. Major radiations of flowering plants, insects, birds, mammals. Origin of earliest human forms.

9. ___     I. Oxygen accumulating in atmosphere. Origin of aerobic metabolism.

10. ___    J. Laurasia forms. Gondwana crosses South Pole. Mass extinction of many marine species. Vast swamplands, early vascular plants. Radiation of fishes continues. Origin of amphibians.

11. ___    K. Pangea breakup continues, broad inland seas form. Major radiations of marine invertebrates, fishes, insects, dinosaurs, origin of angiosperms (flowering plants).

12. ___    L. Asteroid impact? Mass extinction of many organisms in seas, some on land; dinosaurs, mammals survive.

13. ___    M. Mass extinction. Nearly all species in seas and on land perish.

14. ___    N. Tethys sea forms. Recurring glaciations. Major radiations of insects, amphibians. Spore-bearing plants dominant; gymnosperms present; origin of reptiles.

## Chronology of Events – Geologic Time

Refer to Figure 21.20 in the text. From the list of evolutionary events above (A - N), select the letters occurring in a particular era of time by circling the appropriate letters.

15. Cenozoic:                          A - B - C - D - E - F - G - H - I - J - K - L - M - N
16. Border of Cenozoic-Mesozoic:       A - B - C - D - E - F - G - H - I - J - K - L - M - N
17. Mesozoic:                          A - B - C - D - E - F - G - H - I - J - K - L - M - N
18. Border of Mesozoic-Paleozoic:      A - B - C - D - E - F - G - H - I - J - K - L - M - N
19. Paleozoic:                         A - B - C - D - E - F - G - H - I - J - K - L - M - N

## Complete the Table

20. Refer to Figure 21.20 in the text. To review some of the events that occurred in the geologic past, complete the table below by entering the geologic era (Archean, Proterozoic, Paleozoic, Mesozoic, and Cenozoic) and the *approximate* time in millions of years since the time of the events.

| Era/Eon | Time | Events |
|---|---|---|
| a. | | A few reptile lineages give rise to mammals and the dinosaurs. |
| b. | | Formation of Earth's crust, early atmosphere, oceans; chemical evolution leading to the origin of life. |
| c. | | Origin of amphibians. |
| d. | | Origin of animals with hard parts. |
| e. | | Rocks 3.5 billion years old contain fossils of well-developed prokaryotic cells that probably lived in tidal mud flats. |
| f. | | Flowering plants emerge, gymnosperms begin their decline. |
| g. | | In the Carboniferous, there were major radiations of insects and amphibians; gymnosperms present, origin of reptiles. |
| h. | | Oxygen accumulated in the atmosphere. |
| i. | | Before the close of this era, the first photosynthetic bacteria had evolved. |
| j. | | Insects, amphibians, and early reptiles flourished in the swamp forests of the Permian. |
| k. | | Most of the major animal phyla evolved in rather short order. |
| l. | | Dinosaurs ruled. |
| m. | | Origin of aerobic metabolism; origin of protistans algae, fungi, animals. |
| n. | | Grasslands emerge and serve as new adaptive zones for plant-eating mammals and their predators. |
| o. | | Humans destroy habitats and many species. |
| p. | | The first ice age initiated the first global mass extinction; reef life everywhere collapsed. |
| q. | | The invasion of land begins; small stalked plants establish themselves along muddy margins, and the lobe-finned fishes ancestral to amphibians move onto land. |

*Crossword*

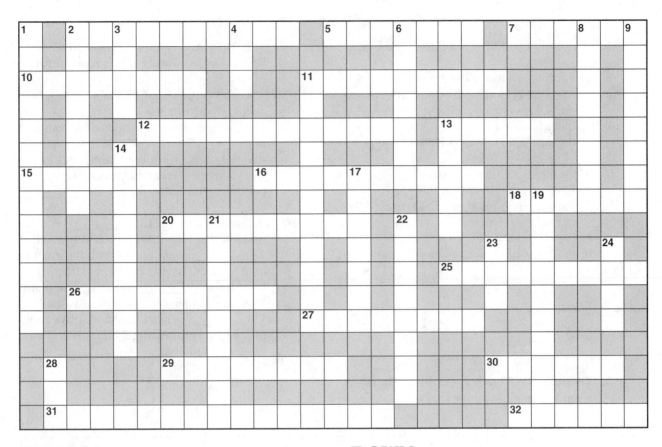

## ACROSS

2. A conifer, cycad or <u>Ginkgo</u>, for example
5. The most recent period of the Paleozoic, at the end of which nearly all species on land and in the sea perish
7. Concerning all of Earth
10. Mesozoic reptile with a "three-toed" foot print
11. Ancestor of a cell
12. Mat of calcium-hardened photosynthetic eubacteria
13. A gymnosperm that resembles a squat palm tree.
15. A layer lying between the mantle and core of Earth
16. The most recent period of the Mesozoic era
18. A massive continent that included the "entire Earth"
20. An era during which fish, amphibians, and reptiles first appeared.
25. Single-celled eukaryote
26. A fossil of an organism that lived 610 → 550 million years ago
27. The most recent era
29. A rocky, metallic body hurtling through space
30. A statement about how events occur in the natural world
31. A halophile, methanogen or thermophile
32. hit; affect

## DOWN

1. Theory of _____ , best developed by Lynn Margulis
2. Southern supercontinent of the Paleozoic era
3. Small leafy nonflowering plant
4. A long period of time
6. An epoch of the Tertiary Period of the Cenozoic Era
8. Cooking on a very hot surface
9. A supercontinent that became part of Pangea and later formed the continents of Earth's northern hemisphere
11. $O_2$ accumulated in Earth's atmosphere during the eon.
13. Hardened, crunchy surface
14. _____ plants radiated dramatically during the Cenozoic
17. Photosynthetic bacteria orginated during this eon
19. Flowering plant
21. Supercontinent of Proterozoic
22. Dinosaurs and gymnosperms dominated this era
23. A longer period of time than 4 down
24. The Big _____
28. The _____ world was based on ribonucleic acid template use.

## Self-Quiz

_____ 1. Between 4.6 and 4.5 billion years ago, _____ .
  a. Earth was a thin-crusted inferno
  b. the outer regions of the cloud from which our solar system formed was cooling
  c. life originated
  d. the atmosphere was laden with oxygen

_____ 2. More than 3.8 billion years ago, _____ .
  a. Earth was a thin-crusted inferno
  b. the outer regions of the cloud from which our solar system formed was cooling
  c. life originated
  d. the atmosphere was laden with oxygen

For questions 3–10, choose from these answers:
  a. Archean
  b. Cenozoic
  c. Mesozoic
  d. Paleozoic
  e. Proterozoic

_____ 3. Dinosaurs and gymnosperms were the dominant forms of life during the _____ era.

_____ 4. The Alps, Andes, Himalayas, and Cascade Range formed during major reorganization of land masses early in the _____ era.

_____ 5. The composition of Earth's atmosphere changed during the _____ era from one that was anaerobic to one that was aerobic.

_____ 6. Invertebrates, primitive plants, and primitive vertebrates were the principal groups of organisms on Earth during the _____ era.

_____ 7. Before the close of the _____ era, the first photosynthetic bacteria had evolved.

_____ 8. The _____ era ended with the greatest of all extinctions, the Permian extinction.

_____ 9. Late in the _____ era, flowering plants arose and underwent a major radiation.

_____10. The _____ era included adaptive zones into which plant-eating mammals and their predators radiated.

## Chapter Objectives/Review Questions

This section lists general and detailed chapter objectives that can be used as review questions. You can make maximum use of these items by writing answers on a separate sheet of paper. Fill in answers where blanks are provided. To check for accuracy, compare your answers with information given in the chapter or glossary.

| Page | | Objectives/Questions |
|---|---|---|
| (334) | 1. | Cloudlike remnants of stars are mostly _____ , but they also contain water, iron, silicates, hydrogen cyanide, methane, ammonia, formaldehyde, and many other simple inorganic and organic substances. |
| (334) | 2. | Describe the formation of early Earth prior to the formation of the first atmosphere. |
| (334) | 3. | Be able to list the probable chemical constituents of Earth's first atmosphere. |
| (334) | 4. | _____ oxygen was probably not present in early Earth's atmosphere. |
| (334–335) | 5. | Describe experimental evidence provided by Stanley Miller (and others) that the formation of biological molecules from simple precursor molecules might have occurred on early Earth. |
| (335) | 6. | _____ might have been the medium on which condensation reactions yielding complex organic compounds occurred. |
| (336) | 7. | During the first _____ years of Earth's history, enzymes, ATP, and other molecules could have assembled at the same location; their close association would have promoted chemical interactions—and the beginning of _____ pathways. |
| (337) | 8. | Although simple, self-replicating systems of RNA, enzymes, and coenzymes have been created in the laboratory, the chemical ancestors of _____ and DNA remain unknown. |
| (337) | 9. | Describe the experiments performed by Sidney Fox that aided understanding of the origin of plasma membranes. |
| (338) | 10. | The first eon of the geologic time scale is now called the _____ , "the beginning." |

(338)     11.  List the three evolutionary directions of the original prokaryotic lineage during the early Archean era.

(338)     12.  Populations of some early anaerobic eubacteria utilized the cyclic photosynthetic pathway; their populations formed huge mats called _____ .

(340–341) 13.  Describe how the endosymbiosis theory may help to explain the origin of eukaryotic cells.

(342–349) 14.  Be able to generally discuss the important geological and biological events occurring throughout the Archean, Proterozoic, Paleozoic, Mesozoic, and Cenozoic eras.

(348)     15.  Be able to list the three geologic eras and two geologic eons in proper order, from oldest to youngest.

---

## Integrating and Applying Key Concepts

As Earth becomes increasingly loaded with carbon dioxide and various industrial waste products, how do you think living forms on Earth will evolve to cope with these changes?

# 22

# BACTERIA AND VIRUSES

## Interactive Exercises

*The Unseen Multitudes* (pp. 352–353)

## 22.1. CHARACTERISTICS OF BACTERIA (pp. 354–355)
## 22.2. BACTERIAL GROWTH AND REPRODUCTION (pp. 356–357)

*Selected Words: archaebacteria, eubacteria,* <u>Escherichia</u> <u>coli</u>, *photoautotrophic, chemoautotrophic, photoheterotrophic, chemoheterotrophic, parasitic, saprobic, peptidoglycan, Gram-positive, Gram-negative, staphylococci, binary fission,* <u>Streptococcus</u>.

### Boldfaced, Page-Referenced Terms

(352) microorganisms _____

_____

(352) pathogens _____

_____

(354) coccus, cocci _____

_____

(354) bacillus, bacilli _____

_____

(354) spirillum, spirilla _____

_____

(355) prokaryotic cells _____

_____

(355) cell wall _____

_____

(355) Gram stain _____

_____

(355) glycocalyx _____

_____

(355) bacterial flagella (sing., flagellum) _____

_____

(355) pilus, pili _____

_____

(356) bacterial chromosome _____

_____

(356) prokaryotic fission _____

_____

(357) plasmid _____

_____

(357) bacterial conjugation _____

_____

## Choice

a. archaebacteria          b. chemoautotrophic eubacteria          c. chemoheterotrophic eubacteria
          d. photoautotrophic eubacteria          e. photoheterotrophic eubacteria

1____ Use $CO_2$ from the environment as a source of carbon atoms and use electrons, hydrogen, and energy released from reactions between various inorganic substances to assemble chains of carbon (food storage)

2____ Use $CO_2$ and $H_2O$ from the environment as sources of carbon, hydrogen and oxygen atoms and use sunlight to power the assembly of food storage molecules

3____ Cannot use $CO_2$ from the environment to construct own carbon chains; instead obtain nutrients from the products, wastes or remains of other organisms; can break down glucose to pyruvate and follow it with fermentation of some sort or another

4____ Cannot use $CO_2$ from the environment to construct own cellular molecules but can absorb sunlight and transfer some of that energy to the bonds of ATP; must obtain food molecules carbon chains produced by other organisms to construct their own molecules

5____ Are either parasites or saprobes, but are not self-feeders

*Fill-in-the-Blanks*

Bacteria are microscopic, (6) _____ cells having one bacterial chromosome and, often, a number of smaller (7) _____ . The cells of nearly all bacterial species have a(n) (8) _____ around the plasma membrane, and a(n) (9) _____ or slime layer surrounding the cell wall. Typically, the width or length of these cells falls between 1 and 10 (10) (choose one) ☐ millimeters ☐ nanometers ☐ centimeters ☐ micrometers. Most bacteria reproduce by (11) _____ _____ . Spherical bacteria are (12) _____ , rod-shaped bacteria are (13) _____ , and helical bacteria are (14) _____ . (15) Gram-_____ bacteria retain the purple stain when washed with alcohol.

## 22.3. BACTERIAL CLASSIFICATION (p. 357)

## 22.4. EUBACTERIA (pp. 358–359)

## 22.5. ARCHAEBACTERIA (p. 360)

## 22.6. SUMMARY OF MAJOR BACTERIAL GROUPS (p. 361)

*Selected Words:* <u>Anabaena</u>, <u>Lactobacillus</u>, <u>Azospirillum</u>, <u>Rhizobium</u>, <u>Clostridium botulinum</u>, *botulism*, <u>Clostridium tetani</u>, *tetanus*, <u>Borrelia burgdorferi</u>, *Lyme disease*.

*Boldfaced, Page-Referenced Terms*

(357) numerical taxonomy _____

_____

(357) eubacteria _____

_____

(357) archaebacteria _____

_____

(358) heterocysts _____

_____

(358) endospores _____

_____

(359) fruiting bodies _____

_____

(360) methanogens _____

_____

(360) halophiles _____

_____

(360) extreme thermophiles _____

_____

(361) strain _____

_____

## Fill-in-the-Blanks

In many respects, the cell structure, metabolism, and nucleic acid sequences are unique to the

(1) _____ ; for example, none has a cell wall that contains (2) _____ . On the other hand,

(3) _____ are far more common than the three rather unusual types of (1). The most common

photoautotrophic bacteria are the (4) _____ (also called the blue-green algae). *Anabaena* and others of

the (4) group produce oxygen during photosynthesis. Heterocysts are cells in *Anabaena* that carry out

(5) _____ _____ . Many species of chemoautotrophic eubacteria affect the global cycling of

nitrogen, sulfur, (6) _____ and other nutrients. Sugarcane and corn plants benefit from a

(7) _____ -fixing spirochete, *Azospirillum*. Beans and other legumes benefit from the nitrogen-fixing

activities of (8) _____ , which dwells in their roots.

When environmental conditions become adverse, many bacteria form (9) _____ , which resist moisture

loss, irradiation, disinfectants, and even acids. Endospore formation by bacteria can kill humans if they

enter the food supply and are not killed by high temperature and high (10) _____ . Anaerobic bacteria

can live in canned food, reproduce, and produce deadly toxins. Two examples of pathogenic (disease-

causing) bacteria that form endospores harmful to humans are (11) _____ _____ and

(12) _____ _____ . (13) _____ _____ may be the most common tick-borne disease in the

United States by now; tick bites deliver the (14) _____ from one host to another. Bacterial behavior

depends on (15) _____ _____ , which change shape when they absorb or connect with chemical

compounds. Cyanobacteria require (16) _____ as their source of energy that drives their metabolic

activities. Most of the world's bacteria are (17) (choose one) ☐ producers ☐ consumers ☐ decomposers, so

we think of them as "good" heterotrophs. (18) _____ include two of the major producers of antibiotics:

*Actinomyces* and (19) _____ . *Escherichia coli*, which dwell in our gut, synthesize vitamin (20) _____

and substances useful in digesting fats.

## Matching

Match each of the items below with a lowercase letter designating its principal bacterial group and an uppercase letter denoting its best descriptor from the right-hand column.

a. archaebacteria      b. chemoautotrophic eubacteria      c. chemoheterotrophic eubacteria
d. photoautotrophic eubacteria      e. photoheterotrophic eubacteria

21. ___, ___ *Anabaena*

22. ___, ___ *Bacillus, Clostridium*

23. ___, ___ *Escherichia coli*

24. ___, ___ *Halobacterium*

25. ___, ___ *Lactobacillus*

26. ___, ___ *Methanobacterium*

27. ___, ___ *Nitrobacter, Nitrosomonas*

28. ___, ___ *Rhizobium, Agrobacterium*

29. ___, ___ *Rhodospirillum*

30. ___, ___ *Salmonella*

31. ___, ___ *Spirochaeta, Treponema*

32. ___, ___ *Staphylococcus, Streptococcus*

33. ___, ___ *Streptomyces, Actinomyces*

34. ___, ___ *Thermoplasma, Sulfolobus*

A. Lives in anaerobic sediments of lakes and in animal gut; chemosynthetic; used in sewage treatment facilities

B. Purple; generally in anaerobic sediments of lakes or ponds; do not produce oxygen, do not use water as a source of electrons

C. Endospore—forming rods and cocci that live in the soil and in the animal gut; some major pathogens

D. Gram-positive cocci that live in the soil and in the skin and mucous membranes of animals; some major pathogens

E. Gram-positive nonsporulating rods that ferment plant and animal material; some are important in dairy industry; others contaminate milk, cheese

F. In acidic soil, hot springs, hydrothermal vents on seafloor; may use sulfur as a source of electrons for ATP formation

G. Live in extremely salty water; have a unique form of photosynthesis

H. Gram-negative aerobic rods and cocci that live in soil or aquatic habitats or are parasites of animals and/or plants; some fix nitrogen

I. Nitrifying bacteria that live in the soil, fresh water, and marine habitats; play a major role in the nitrogen cycle

J. Gram-negative anaerobic rod that inhabits the human colon where it produces vitamin K

K. Major gram-negative pathogens of the human gut that cause specific types of food poisoning

L. Mostly in lakes and ponds; cyanobacteria; produce $O_2$ from water as an electron donor

M. Major producer of antibiotics; an actinomycete that lives in soil and some aquatic habitats

N. Helically coiled, motile parasites of animals; some are major pathogens

## 22.7. THE VIRUSES (pp. 362–363)

## 22.8. VIRAL MULTIPLICATION CYCLES (pp. 364–365)

## 22.9. *Focus on Health:* THE NATURE OF INFECTIOUS DISEASES (pp. 366–367)

**Selected Words:** *AIDS, helical, polyhedral, enveloped and complex viruses, cold sores,* <u>Herpes</u> <u>simplex</u> *types I and II, kuru, Creutzfeldt-Jakob, scrapie, contagious diseases, sporadic diseases, endemic diseases,* <u>Ebola</u>, <u>Salmonella</u> <u>enteridis</u>, *BSE, mad cow disease.*

## Boldfaced, Page-Referenced Terms

(362) virus _____

_____

(362) bacteriophages _____

_____

(363) prions _____

_____

(368) viroids _____

_____

(364) lytic pathway _____

_____

(364) lysis _____

_____

(364) lysogenic pathway _____

_____

(366) infection _____

_____

(366) disease _____

_____

(366) epidemic _____

_____

(366) pandemic _____

_____

(366) emerging pathogens _____

_____

## Short Answer

1. a. State the principal characteristics of viruses (p. 362). _____

_____

b. Describe the structure of viruses (pp. 362–365). _____

_____

c. Distinguish between the ways viruses replicate themselves (pp. 364–365). _____

_____

2. a. List five specific viruses that cause human illness (pp. 362–363). _____

_____

b. Describe how each virus in (2a) does its dirty work (pp. 366–367). _____

_____

_____

_____

_____

_____

_____

## Fill-in-the-Blanks

A(n) (3) _____ is a noncellular, nonliving infectious agent, each of which consists of a central
(4) _____ _____ core surrounded by a protective (5) _____ _____ . (6) _____
contain the blueprints for making more of themselves but cannot carry on metabolic activities. Chickenpox
and shingles are two infections caused by DNA viruses from the (7) _____ category. Naked strands or
circles of RNA that lack a protein coat are called (8) _____ . (9) _____ are RNA viruses that infect
animal cells, cause AIDS, and follow (10) _____ pathways of replication. (11) _____ are the usual
units of measurement with which to measure viruses, while microbiologists measure bacteria and protistans
in terms of (12) _____ . A bacterium 86 micrometers in length is (13) _____ nanometers long.
During a period of (14) _____ , viral genes remain inactive inside the host cell and any of its
descendants. Pathogenic protein particles are called (15) _____ .

## Identification

Identify the virus that causes the following illnesses by writing the name of the virus in the blank preceding
the disease. Then, tell whether it is a DNA virus or an RNA virus.

_____ _____ 16. Common colds
_____ _____ 17. AIDS, leukemia
_____ _____ 18. Cold sores, chickenpox

## Matching

Match each item below with the correct lettered description.

19. ___antibiotic

20. ___bacteriophage

21. ___endemic

22. ___epidemic

23. ___lysogenic pathway

24. ___lytic pathway

25. ___microorganism

26. ___pathogen

27. ___prion

28. ___sporadic

29. ___viroid

30. ___virus

A. "naked" RNA bits that resemble introns
B. Disease that breaks out irregularly, affects few organisms
C. Chemical substance that interferes with gene expression or other normal functions of bacteria
D. Disease abruptly spreads through large portions of a population
E. Disease that occurs continuously, but is localized to a relatively small portion of the population
F. Small proteins that are altered products of a gene; linked to 8 degenerative diseases of the nervous system
G. Any organism too small to be seen without a microscope
H. A virus that infects a bacterium
I. Damage and destruction to host cells occurs quickly
J. Noncellular infectious agent that must take over a living cell in order to reproduce itself
K. Viral nucleic acid is integrated into the nucleic acid system of the host cell and replicated during this time
L. Any disease-causing organism or agent

# Self-Quiz

## Multiple-Choice

___ 1. Which of the following diseases is *not* caused by a virus?
   a. smallpox
   b. polio
   c. influenza
   d. syphilis

___ 2. Bacteriophages are _____ .
   a. viruses that parasitize bacteria
   b. bacteria that parasitize viruses
   c. bacteria that phagocytize viruses
   d. composed of a protein core surrounded by a nucleic acid coat

## Matching

Match all applicable letters with the appropriate terms. A letter may be used more than once, and a blank may contain more than one letter.

3. _____ *Anabaena, Nostoc*

4. _____ *Clostridium botulinum*

5. _____ *Escherichia coli*

6. _____ *Herpes simplex*

7. _____ HIV

8. _____ *Lactobacillus*

9. _____ *Staphylococcus*

A. Bacteria
B. Virus
C. Cyanobacteria
D. Gram-positive eubacteria
E. Cause cold sores and a type of venereal disease
F. Associated with AIDS, ARC

## Matching

Match the pictures below with their correct names.

10. ___          A. *Bacillus*
11. ___          B. bacteriophage
                 C. *Clostridium tetani*
12. ___          D. Cyanobacterium
                 E. *Herpesvirus*
13. ___          F. HIV
14. ___

15. ___

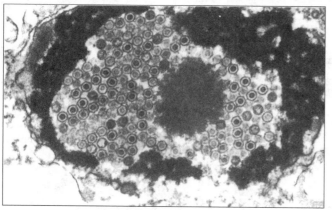

10.

11.

12.

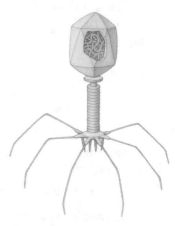

13.

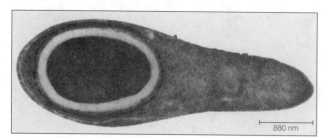

880 nm

14.

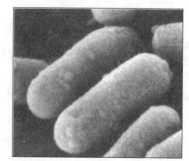

15.

## ACROSS

2. _____ can assemble their own food molecules by using light energy to join together carbon, hydrogen and oxygen atoms.

3. Binary _____ is the process most bacteria use to reproduce.

6. Disease causers

7. Bacterial _____ transfers a plasmid from a donor to a recipient cell.

8. A chemical substance produced by *actinomycetes* and other microorganisms that kills or inhibits the growth of other microorganisms

10. A whip-like organelle of locomotion

12. Boy, young man

14. Group that includes blue-green algae

16. Sticky mesh of polysaccharides, polypeptides, or both; alternate name of either a capsule or slime layer

17. _____ cells existed before the origin of the nucleus and of eukaryotic cells.

19. Cannot synthesize their own food molecules; <u>can</u> use sunlight as an energy source to make ATP, but cannot fix $CO_2$ from their surroundings; must obtain carbon compounds from other organisms

20. Viruses that infect bacterial host cells

23. Any organism that can be seen most clearly by using a microscope

24. A short filamentous protein that projects above a cell wall; helps to tether a bacterium to another bacterium or to a surface.

25. A(n) _____ occurs when, over the entire world, a disease spreads through large portions of many populations for a limited period.

26. _____ are structures that resist heat, drying, boiling and radiation; can give rise to new bacterial cells.

*(continued on next page)*

27. A(n) _____ occurs when a disease abruptly spreads through large portions of a single population for a limited period.
28. Salt-loving archaebacterium
29. Methane-producing archaebacterium
30. Spherical bacterium

# DOWN

1. Rod-shaped bacterium
2. Substance not in archaebacterial cell walls
4. Helical
5. _____-positive bacterial cell walls have an affinity for crystal violet stain.
9. _____ include Earth's major decomposers, nitrogen-fixers and fermenters of milk.

11. The _____ pathway of bacteriophage multiplication assembles new viral particles very soon after infecting the host cell; no integration into host cell's DNA.
13. This group includes the most ancient of Earth's bacteria.
15. Bacteria that can synthesize their own food molecules by using energy released from specific chemical reactions
18. Heat-tolerant bacterial types
21. The _____ pathway of bacteriophage multiplication includes integrating viral DNA into its bacterial host's chromosome.
22. Small loops of DNA in addition to the main bacterial "chromosome"

# Chapter Objectives/Review Questions

This section lists general and detailed chapter objectives that can be used as review questions. You can make maximum use of these items by writing answers on a separate sheet of paper. Fill in answers where blanks are provided. To check for accuracy, compare your answers with information given in the chapter or glossary.

*Page*      *Objectives/Questions*

(354)       1. Distinguish chemoautotrophs from photoautotrophs.
(354–355)   2. Describe the principal body forms of monerans (inside and outside).
(356–357)   3. Explain how, with no nucleus, or few if any, membrane–bound organelles, bacteria reproduce themselves and obtain energy to carry on metabolism.
(358, 360)  4  State the ways in which archaebacteria differ from eubacteria.
(362–363)   5. Describe the shapes of various viral types and explain the ways in which viruses infect their hosts.
(364–365)   6. Distinguish the lytic and lysogenic patterns of viral replication.

# Integrating and Applying Key Concepts

The textbook (Figure 50.15) identifies natural gas as a nonrenewable fuel resource, yet there is a group of archaebacteria that produce methane, the burning of which can serve as a fuel for heating and/or cooking. Recall or imagine how these bacteria could be incorporated into a system that could serve human societies by generating methane in a cycle that is renewable. Why did your text categorize natural gas as a nonrenewable resource? Is methane a constituent of natural gas? Why or why not?

# 23

# PROTISTANS

## Interactive Exercises

*Kingdom at the Crossroads* (pp. 370–371)

### 23.1. PARASITIC AND PREDATORY MOLDS (pp. 372–373)

*Selected Words:* <u>Saprolegnia,</u> <u>Plasmopara</u> <u>viticola</u>, *downy mildew,* <u>Phytophthora</u> <u>infestans</u>, *late blight, cellular slime molds,* <u>Dictyostelium</u> <u>discoideum</u>, *plasmodial slime molds.*

### Boldfaced, Page-Referenced Terms

(370) protistans _____

_____

(372) chytrids _____

_____

(372) water molds _____

_____

(372) slime molds _____

_____

(372) saprobes _____

_____

(372) mycelium _____

_____

## Choice

1. For each structure, indicate with a P if it is a prokaryotic (bacterial) characteristic, and indicate with an E if it is a eukaryotic characteristic.

| | |
|---|---|
| a. double-membraned nucleus | |
| b. mitochondria present | |
| c. reproduce by binary fission | |
| d. engage in mitosis | |
| e. circular chromosome present | |
| f. endoplasmic reticulum present | |
| g. cilia or flagella with 9+2 core | |

2. Fill in each box with letters selected from the grouped items below. Choose from the choices in group I for the boxes in column I, from group II for the boxes in column II, and from group III for the boxes in column III.

| Protistan Group of Fungus-like Organisms | I. Structural Features | II. Behaviors and Typical Habitats | III. Representatives |
|---|---|---|---|
| chytrids (Chytridiomycota) | a. | b. | c. |
| cellular slime molds (Acrasiomycota) | d. | e. | f. |
| plasmodial slime molds (Myxomycota) | g. | h. | i. |
| water molds (Oomycota) | j. | k. | l. |

*Choices*

### I.

A. Mycelium present in most multicelled species
B. Free-living amoeba-like cells present
C. Rhizoids present in many species as adults
D. Single-celled types are globular/spherical
E. Produce flagellated asexual spores

### II.

F. Most are saprobic decomposers
G. Some are parasites on or in living organisms
H. Phagocytic predators
I. Crawl on rotting plant parts during part of their life cycle
J. Some dwell in marine habitats
K. Some dwell in freshwater habitats
L. Most are terrestrial—living in the soil, on rotten logs or other vegetation

### III.

M. <u>Phytophtora</u> <u>infestans</u> (causes late blight in potatoes and tomatoes)
N. <u>Plasmopara</u> <u>viticola</u> (causes downy mildew in grapes)
O. <u>Dictyostelium</u> <u>discoideum</u> (commonly used for laboratory studies of development)
P. <u>Physarum</u>
Q. <u>Saprolegnia</u> (causes "ick" on aquarium fishes)

*Fill-in-the-Blanks*

(3) _____ and (4) _____ _____ are the only fungi that produce motile spores; this is a primitive trait that may resemble ancestral fungi that lived several hundred million years ago in watery habitats.  Water molds are only distantly related to other fungi and are thought to have evolved from (5) _____ algae.  The cells of some (6) _____ _____ differentiate and form (7) _____ _____ : stalked structures bearing spores at their tips; in this manner, they resemble (8) _____ .  Some slime-mold spores resemble the spores of many (9) _____ .  Slime molds also spend part of their life creeping about like (10) _____ and engulfing food.

Many fungi have cells merged lengthwise, forming tubes that have thin transparent walls reinforced with (11) _____ ; so do some (12) _____ .  Some fungal species, such as late blight, are (13) [choose one] □ parasitic, □ saprophytic.  The vegetative body of most true fungi is a (14) _____ , which is a mesh of branched, tubular filaments.  Their metabolic activities enable them to act as (15) _____ in ecosystems.  Fungi secrete (16) _____ into their surroundings, where large organic molecules are broken down into smaller components that the fungal cells then absorb.  While most fungi are (17) _____ (obtaining their nutrients from nonliving organic matter), some are (18) _____ and get their nutrients directly from their living host's tissues.  A common form of asexual reproduction is the growth of a new fungal body from a(n) (19) _____ .

## 23.2. THE ANIMAL-LIKE PROTISTANS (p. 374)

## 23.3. AMOEBOID PROTOZOANS (pp. 374–375)

## 23.4. MAJOR PARASITES—ANIMAL-LIKE FLAGELLATES AND SPOROZOANS
(p. 376)

## 23.5. *Focus on Health:* MALARIA AND THE NIGHT-FEEDING MOSQUITOES (p. 377)

## 23.6. A SAMPLING OF CILIATED PROTOZOANS (pp. 378–379)

**Selected Words:** <u>Entamoeba histolytica</u>, *amoebic dysentery,* <u>Giardia lamblia</u>, <u>Trichomonas vaginalis</u>, <u>Trypanosoma brucei</u>, *African sleeping sickness,* <u>Trypanosoma cruzi</u>, *Chagas disease,* <u>Plasmodium</u>, *malaria,* <u>Cryptosporidium</u>, <u>Toxoplasma</u>, *toxoplasmosis, vaccine,* <u>Paramecium</u>, *hypotrichs, micronucleus, macronucleus.*

## Boldfaced, Page-Referenced Terms

(374) protozoans _____

_____

(374) binary fission _____

_____

(374) multiple fission _____

_____

(374) cysts _____

_____

(374) amoeboid protozoans _____

_____

(374) pseudopods _____

_____

(374) rhizopods _____

_____

(375) actinopods _____

_____

(375) plankton _____

_____

(376) animal-like flagellates _____

_____

(376) sporozoan _____

_____

(378) ciliated protozoans _____

_____

(378) pellicle _____

_____

(379) contractile vacuoles _____

_____

(379) conjugation _____

_____

## Fill-in-the-Blanks

Amoebas move by sending out (1) _____ , which surround food and engulf it. (2) _____ secrete a hard exterior covering of calcareous material that is peppered with tiny holes through which sticky, food-trapping pseudopods extend. Needle-like (3) _____ often support the pseudopods in heliozoans. Accumulated shells of (4) _____ , which generally have a skeleton of silica (glass), and foraminiferans are key components of many oceanic sediments.

Examples of flagellated protozoans that are parasitic include the (5) _____ , two species of which cause African sleeping sickness and Chagas disease.

*Paramecium* is a ciliate that lives in (6) _____ environments and depends on (7) _____ _____ for eliminating the excess water constantly flowing into the cell. *Paramecium* has a (8) _____ , a cavity that opens to the external watery world. Once inside the cavity, food particles become enclosed in (9) _____-_____ _____ , where digestion takes place.

(10) _____ is a famous sporozoan that causes malaria. When a particular (11) _____ draws blood from an infected individual, (12) _____ of the parasite fuse to form zygotes, which eventually develop within the mosquito.

## Matching

Put as many letters in each blank as are applicable.

13. _____ *Amoeba proteus*

14. _____ *Entamoeba histolytica*

15. _____ *foraminiferans*

16. _____ *Paramecium*

17. _____ *Plasmodium*

18. _____ *Trichomonas vaginalis*

19. _____ *Trypanosoma brucei*

A. Ciliophora
B. Mastigophora
C. Sarcodina
D. Apicomplexa
E. Amoeboid protozoans
F. Animal-like flagellates
G. African sleeping sickness
H. Malaria
I. Amoebic dysentery
J. Sporozoans
K. Primary component of many ocean sediments

## Matching

Match the pictures below with the names below.

20. ___
21. ___
22. ___
23. ___
24. ___

A. *Amoeba proteus*
B. Flagellated protozoans
C. Foraminiferans
D. Heliozoans
E. *Paramecium*

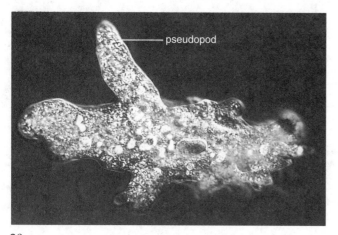

pseudopod

20.

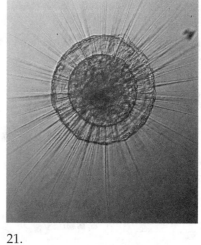

21.

22.

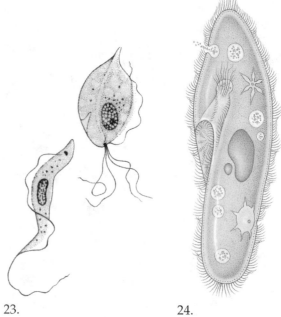

23.                    24.

## 23.7. A SAMPLING OF THE (MOSTLY) SINGLE-CELLED ALGAE (pp. 380–381)
## 23.8. RED ALGAE (p. 382)
## 23.9. BROWN ALGAE (p. 383)
## 23.10. GREEN ALGAE (pp. 384–385)

*Selected Words:* the algae, Euglenophyta, *photosynthetic,* *heterotrophs,* Euglena, *"eyespot",* Chrysophyta.

### Boldfaced, Page-Referenced Terms

(380) phytoplankton _____

_____

(380) euglenoids _____

_____

(380) chrysophytes _____

_____

(380) diatoms _____

_____

(380) golden algae _____

_____

(381) yellow-green algae _____

_____

(381) dinoflagellates _____

_____

(381) red tides _____

_____

(382) red algae _____

_____

(383) brown algae _____

_____

(384) green algae _____

_____

## Fill-in-the-Blanks

Most euglenoids contain (1) _____ , which enable them to carry out photosynthesis. A(An) (2) _____ of carotenoid pigment granules partly shields a light-sensitive receptor and enables *Euglena* to remain where light is optimal for its activities. Some strains of *Euglena* can be converted from photosynthetic, chloroplast-containing forms to strains that are (3) _____ .

The term (4) "_____" no longer has formal classification significance, because organisms once lumped under that term are now assigned to different kingdoms. (5) _____ include 600 species of "yellow-green algae," about 500 species of "golden algae," and more than 5,600 existing species of golden-brown (6) _____ . Except in yellow-green algae, photosynthetic chrysophytes contain xanthophylls and (7) _____ ; those pigments mask the green color of chlorophyll in golden algae and diatoms. Diatom cells have external thin, overlapping "shells" of (8) _____ that fit together like a pill box. 270,000 metric tons of (9) _____ _____ are extracted annually from a quarry near Lompoc, California, and are used to make abrasives, (10) _____ materials, and insulating materials. Dinoflagellates are mostly photosynthetic members of marine (11) _____ and freshwater ecosystems; some forms are also heterotrophic. (12) _____ undergo explosive population growth and color the seas red or brown, causing a red tide that may kill hundreds or thousands of fish and, occasionally, people.

Several species of red algae secrete (13) _____ (used in culture media) as part of their cell walls. Most red algae live in (14) _____ habitats. Some red algae have stonelike cell (15) _____ , participate in coral reef building, and are major producers. The (16) _____ algae live offshore or in intertidal zones and have many representatives with large sporophytes known as kelps; some species produce (17) _____ , a valuable thickening agent. Green algae are thought to be ancestral to more complex plants, because they have the same types and proportions of (18) _____ pigments, have (19) _____ in their cell walls, and store their carbohydrates as (20) _____ .

## Complete the Table

21. Complete the table below.

| Type of alga | Typical pigments | Probably evolved from | Uses by humans | Representatives |
|---|---|---|---|---|
| Red algae (Rhodophyta) | a. | b. | c. | d. |
| Brown algae (Phaeophyta) | e. | f. | g. | h. |
| Green algae (Chlorophyta) | i. | j. | k. | l. |

## Label-Match

Identify each indicated part of the illustration below by entering its name in the appropriate numbered blank. Choose from the following terms: cytoplasmic fusion, asexual reproduction, resistant zygote, fertilization, zygote, meiosis and germination, spore mitosis, gametes meet. Complete the exercise by matching from the list below, entering the correct letter in the parentheses following each label.

22. _____ ( )

23. _____ _____ ( )

24. _____ and _____ ( )

25. _____ _____ ( )

26. _____ _____ ( )

27. _____ _____ ( )

28. _____ _____ ( )

29. _____ ( )

A. Fusion of two gametes of different mating types
B. A device to survive unfavorable environmental conditions
C. More spore copies are produced
D. Fusion of two haploid nuclei
E. Haploid cells form smaller haploid gametes when nitrogen levels are low
F. Formed after fertilization
G. Two haploid gametes coming together
H. Reduction of the chromosome number

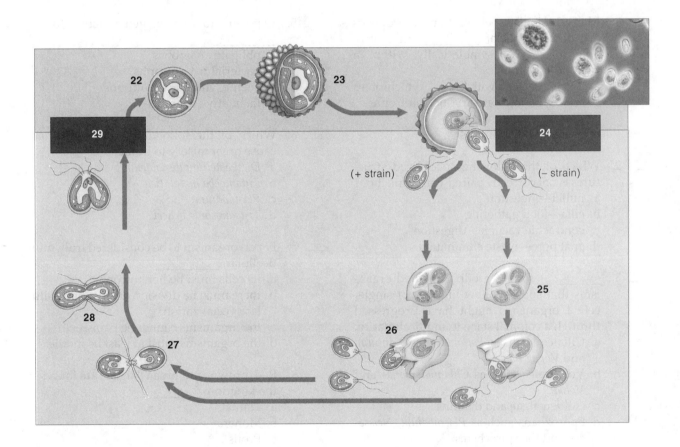

(+ strain)          (− strain)

30. For each group below, indicate with a "+" if it has chloroplasts, and a "−" if members of that group lack chloroplasts and the ability to do photosynthesis.

| | |
|---|---|
| a. brown algae | f. protozoans |
| b. chytrids | g. red algae |
| c. chrysophytes | h. slime molds |
| d. dinoflagellates | i. sporozoans |
| e. green algae | j. water molds |

# Self-Quiz

___ 1. Many biologists believe that chloroplasts are descendants of _____ that were able to survive symbiotically within a predatory host cell.
   a. aerobic bacteria with "extra" cytochromes
   b. endosymbiont prokaryotic autotrophs
   c. dinoflagellates
   d. bacterial heterotrophs

___ 2. Which of the following specialized structures is not correctly paired with a function?
   a. gullet—ingestion
   b. cilia—food gathering
   c. contractile vacuole—digestion
   d. anal pore—waste elimination

___ 3. _____ form a group of related organisms that suggests how lineages of single-celled organisms might have progressed through a colonial stage to multicellularity.
   a. Ciliates such as *Paramecium*, *Didinium*, and *Vorticella*
   b. Volvocales such as *Chlamydomonas* and *Volvox*
   c. Golden algae and diatoms
   d. Sporozoans such as *Plasmodium*, *Neisseria*, and the spirochetes

___ 4. Population "blooms" of _____ cause "red tides" and extensive fish kills.
   a. *Euglena*
   b. specific dinoflagellates
   c. diatoms
   d. *Plasmodium*

___ 5. Exposure to free oxygen is lethal for all _____ .
   a. obligate anaerobes
   b. bacterial heterotrophs
   c. chemosynthetic autotrophs
   d. facultative anaerobes

___ 6. Which of the following protists does *not* cause great misery to humans?
   a. *Dictyostelium discoideum*
   b. *Entamoeba histolytica*
   c. *Plasmodium*
   d. *Trypanosoma brucei*

___ 7. For an organism to be considered truly multicellular, _____ .
   a. its cells must be heterotrophic
   b. there must be division of labor and cellular specialization
   c. the organisms cannot be parasitic
   d. the organisms must at least be motile

___ 8. Red, brown, and green "algae" are found in the kingdom _____ .
   a. Plantae
   b. Monera
   c. Protista
   d. all of the above

___ 9. Red algae _____ .
   a. are primarily marine organisms
   b. are thought to have developed from green algae
   c. contain xanthophyll as their main accessory pigments
   d. all of the above

___10. Stemlike structure, leaflike blades, and gas-filled floats are found in the species of _____ .
   a. red algae
   b. brown algae
   c. bryophytes
   d. green algae

___11. Because of pigmentation, cellulose walls, and starch storage similarities, the _____ algae are thought to be ancestral to more complex plants.
   a. red
   b. brown
   c. blue-green
   d. green

## Matching

Match all applicable letters with the appropriate terms. A letter may be used more than once, and a blank may contain more than one letter.

12. _____ *Amoeba proteus*

13. _____ diatoms

14. _____ *Dictyostelium*

15. _____ foraminifera

16. _____ *Ptychodiscus brevis* (red tide)

17. _____ *Paramecium*

18. _____ *Plasmodium*

19. _____ *Volvox*

A. Protista
B. Slime mold
C. Photosynthetic flagellates
D. Dinoflagellates
E. Obtain food by using pseudopodia
F. Causes malaria
G. A sporozoan
H. A ciliate
I. Live in "glass" houses
J. Live in hardened shells that have thousands of tiny holes, through which pseudopodia protrude

## Matching

20. ___
21. ___
22. ___
23. ___

A. *Amoeba proteus*
B. diatoms
C. foraminiferans
D. *Paramecium*

20.

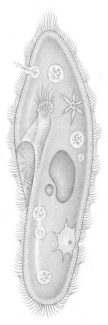

21.

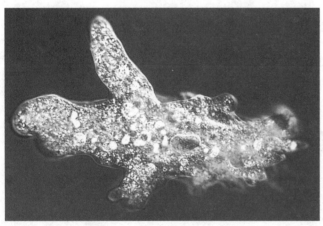

23.

22.

# Chapter Objectives/Review Questions

This section lists general and detailed chapter objectives that can be used as review questions. You can make maximum use of these items by writing answers on a separate sheet of paper. Fill in answers where blanks are provided. To check for accuracy, compare your answers with information given in the chapter or glossary.

| Page | | Objectives/Questions |
|---|---|---|
| (370) | 1. | Explain how heterotrophic protistans could have acquired the capacity for photosynthesis, and state the evidence to support your explanation. |
| (371–385) | 2. | Discuss the contributions that protistans make to Earth's ecosystems and the ways that humans use protistans to make specific products. |
| (374–375) | 3. | State the principal characteristics of the amoebas, radiolarians, and foraminiferans. Indicate how they generally move from one place to another and how they obtain food. |
| (376) | 4. | Two flagellated protozoans that cause human misery are _____ and _____ . |
| (376–377) | 5. | Characterize the sporozoan group, identify the group's most prominent representative, and describe the life cycle of that organism. |
| (378–379) | 6. | List the features common to most ciliated protozoans. |
| (380–381) | 7. | How do golden algae resemble diatoms? |
| (381) | 8. | Explain what causes red tides. |
| (382–385) | 9. | State the outstanding characteristics of organisms of the red, brown, and green algae divisions. |

# Integrating and Applying Key Concepts

Explain why totally submerged aquatic plants that live in deep water never developed heterosporous life cycles.

# 24

# FUNGI

## Interactive Exercises

*Dragon Run* (pp. 388–389)

## 24.1. CHARACTERISTICS OF FUNGI (p. 390)

***Selected Words:*** Gymnophilus, Clavaria, Sarcosoma, Hygrophorus, Boletus, Craterellus, *extracellular digestion and absorption*

### *Boldfaced, Page-Referenced Terms*

(388) fungi _____

_____

(390) zygomycetes _____

_____

(390) sac fungi _____

_____

(390) club fungi _____

_____

(390) saprobes _____

_____

(390) parasites _____

_____

(390) spores _____

_____

(390) mycelium, -a _____

_____

(390) hypha, -e _____

_____

## Matching

Choose the most appropriate answer for each.

1. ___saprobes
2. ___parasites
3. ___mycelium
4. ___fungi
5. ___hypha
6. ___zygomycetes, sac fungi, club fungi
7. ___spores
8. ___extracellular digestion and absorption

A. Nonmotile reproductive cells or multicelled structures; often walled and germinate following dispersal from the parent body
B. Represent major lines of fungal evolution
C. A mesh of branching fungal filaments that grows over and into organic matter, secretes digestive enzymes and functions in food absorption
D. Fungi that obtain nutrients from nonliving organic matter and so cause its decay
E. Fungi that extract nutrients from tissues of a living host
F. External secretion of enzymes prior to nutrient intake
G. Each filament in a mycelium; consists of tube-shaped cells with chitin-reinforced walls
H. A richly diverse group of heterotrophs that are premier decomposers

## 24.2. CONSIDER THE CLUB FUNGI (pp. 390–391)

**Selected Words:** Agaricus brunnescens, Armillaria bulbosa, Amanita, *dikaryotic* mycelium

**Boldfaced, Page-Referenced Terms**

(391) mushrooms _____

_____

(391) basidiospores _____

_____

## Fill-in-the-Blanks

The numbered items on the illustration of a club fungus life cycle below represent missing information; complete the numbered blanks in the narrative below to supply the missing information.

The mature mushroom is actually a short-lived (1) _____ body; each consists of a cap and a stalk. Spore-producing (2) _____-shaped structures develop on the gills. Each bears two haploid ($n + n$) nuclei. (3) _____ fusion occurs within the club-shaped structures, which yields a (4) _____ stage. (5) _____ occurs within the club-shaped structures and four haploid (6) _____ emerge at the tip of each. The spores are released and each may germinate into a haploid ($n$) mycelium. When hyphae of two compatible mating strains meet, (7) _____ fusion occurs. Following this, a (8) "_____" ($n + n$) mycelium gives rise to the spore-bearing mushrooms.

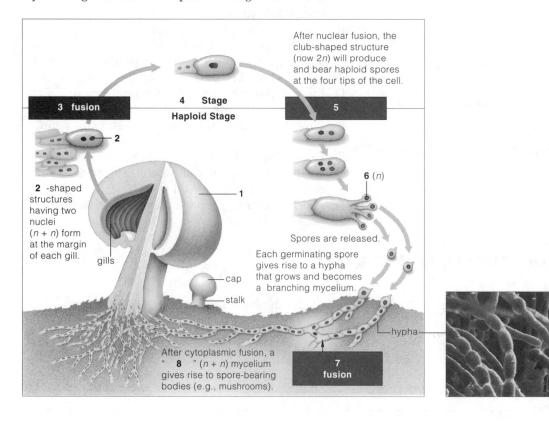

hypha in mycelium

## Matching

Choose the most appropriate answer for each.

9. ___rust and smut fungi

10. ___*Agaricus brunnescens*

11. ___*Armillaria bulbosa*

12. ___*Amanita muscaria* (text, Figure 24.4a)

13. ___*Amanita phalloides*

A. Fly agaric mushroom, causes hallucinations when eaten; ritualistic use
B. Among the oldest and largest of the club fungi
C. Destroys entire fields of wheat, corn, and other major crops
D. Common cultivated mushroom; multimillion dollar business
E. Death cap mushroom; kills humans

## 24.3. SPORES AND MORE SPORES (pp. 392–393)

*Selected Words:* Rhizopus stolonifer, asci, -us, Penicillium, Aspergillus, Neurospora sitophila, N. crassa, Saccharomyces cerevisiae, Candida albicans, Sarcoscypha coccinia, Morchella esculenta, Eupenicillium, Arthrobotrys dactyloides

### Boldfaced, Page-Referenced Terms

(392) zygosporangium _____

_____

(392) ascospores _____

_____

### True/False

If the statement is true, write a T in the blank. If the statement is false, make it correct by changing the underlined word(s) and writing in the correct word(s) in the answer blank.

_____ 1. Most of the sac fungi species are single-celled.

_____ 2. Fermentation carried on by *Saccharomyces cerevisiae* furnishes carbon dioxide for leavening bread and ethanol for alcoholic beverages.

_____ 3. Together, the zygote and its protective wall form a(n) ascospore.

_____ 4. Certain species of *Aspergillus* are used to flavor Camembert and Roquefort cheeses; other species are used to make penicillins, widely used as antibiotics.

_____ 5. *Neurospora sitophila* is a sac fungus important in genetic research.

_____ 6. Trained pigs and dogs are used to snuffle out edible morels.

_____ 7. Multicelled sac fungi species form specialized sexual spores called conidiospores.`

_____ 8. Ascospores are asexual spores formed inside sac-shaped cells.

_____ 9. Spore-producing sacs called asci usually form on the inner surface of reproductive structures shaped like flasks, globes, and shallow cups.

_____10. In the sac fungi, mitosis produces haploid spores that are dispersed from a sac; each spore is capable of germination to form a mycelium that grows through soil, decaying wood, and other substrates.

### Fill-in-the-Blanks

The numbered items on the illustration on the following page (*Rhizopus* life cycle) represent missing information; complete the corresponding numbered blanks in the narrative below to supply missing information about zygomycetes.

The sexual phase begins when haploid (11) _____ of two different mating strains ( + and –) grow into each other and fuse due to a chemical attraction. Two (12) _____ form between two hyphae, and several haploid nuclei are produced inside each. Later, their nuclei fuse, forming a zygote with a thick protective wall called a (13) _____ (the key defining feature of the zygomycetes). This structure may remain dormant for several months. Meiosis proceeds and haploid sexual (14) _____ are produced when this structure germinates. Each gives rise to stalked structures, each with a spore sac on its tip, that can produce many spores, each of which can be the start of an extensive (15) _____ .

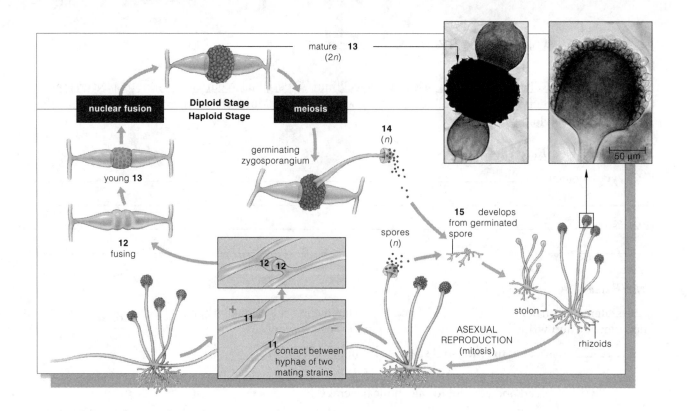

*Short Answer*

16. What criterion is used to determine whether a particular fungal species is an "imperfect" fungus? Cite one example of an "imperfect" fungus. _____

_____

_____

_____

_____

_____

_____

## 24.4. BENEFICIAL ASSOCIATIONS BETWEEN FUNGI AND PLANTS (pp. 394–395)

## 24.5. *Focus on Science:* A LOOK AT THE UNLOVED FEW (p. 396)

*Selected Words:* mycobiont, *myes,* photobiont, Usnea, Cladonia rangiferina, Lobaria, *exo*mycorrhiza, *endo*mycorrhizae, *histoplasmosis, ergotism*

*Boldfaced, Page-Referenced Terms*

(394) symbiosis _____

_____

(394) mutualism _____

_____

(394) lichen _____

_____

(395) mycorrhiza, -e _____

_____

## Fill-in-the-Blanks

(1) _____ refers to species that live in close association. In cases of symbiosis called (2) _____ , both partners benefit. A (3) _____ is commonly called a mutualistic interaction between a fungus and one or more photosynthetic species. The fungal part of a lichen is known as the (4) _____ and the photosynthetic component is the (5) _____ . In almost every instance, the (6) _____ is the largest part of the lichen. Lichens can colonize places that are too (7) _____ for other organisms. The fungus benefits by having a long-term source of nutrients that it absorbs from cells of the (8) _____ . By giving up some nutrients, the (9) _____ suffers a bit in terms of its own growth, although it might benefit some from the (10) _____ sheltering effect. In lichens having more than one fungus present, the added species may be a (11) _____ , or it may be parasitizing the lichen or using the lichen as a (12) _____ . A lichen forms after the tip of a fungal (13) _____ binds with a suitable host cell. Either (14) _____ fusion occurs or the hypha induces the host cell to cup around it. Now the (15) _____ and the (16) _____ grow and multiply together. The structure of the lichen depends on how cells of the (16) _____ become distributed among fungal cells. Sometimes they are distributed more or less uniformly throughout the lichen but, in most cases, the lichen has distinct (17) _____ . The overall (18) _____ pattern may be leaflike, flattened, or pendulous or erect. Lichens absorb (19) _____ _____ from their substrates and (20) _____ from the air. Among the products of their metabolic activities are (21) _____ against bacteria that might decompose the lichen body and (22) _____ that inhibit larval development of invertebrates that might graze on them. Products from lichens contribute to the formation or enrichment of (23) _____ . The metabolic activities of lichens pioneering a habitat can set the stage for (24) _____ by different species. Some ecosystems depend on cyanobacteria-containing lichens as (25) _____ sources. Lichens also serve as early warnings of deteriorating (26) _____ conditions in that their death around cities signals that air pollution is getting bad.

## Matching

Choose the most appropriate answer for each.

27. ___mycorrhizae

28. ___carbohydrates

29. ___minerals

30. ___exomycorrhizae

31. ___endomycorrhizae

32. ___air pollution

A. Fungal hyphae that penetrate plant cells, as they do in lichens
B. Absorption of these benefits the plant
C. A form of mutualism meaning "fungus root"
D. Correlated with a decline in numbers and diversity of fungi
E. Absorption of these benefits the fungus
F. Hyphae form a dense net around living cells in roots but do not penetrate them; can pass phosphorus or mineral ions to the plant when they are scarce

33. After reading the *Focus on Science*, "A Look at the Unloved Few," in the text, p. 396, complete the following table which deals with a few pathogenic and toxic fungi.

| Fungi—Group Name | Description |
|---|---|
| a. | Black stem wheat rust, corn smut, severe mushroom poisoning |
| b. | Plant wilt, various species cause ringworms, including athlete's foot, mucous membrane infections, Histoplasmosis |
| c. | Food spoilage |
| d. | Dutch elm disease, Chestnut blight, Apple scab, Ergot of rye (ergotism), Brown rot of stone fruits |

# Self-Quiz

## Label-Match

In the blank beneath each illustration below (1–8), identify the organism by common name (or scientific name if a common name is unavailable). Then match each organism with the appropriate item (may be used more than once) from the list below the illustrations by entering the letter in the parenthesis.

1. _____ (  )

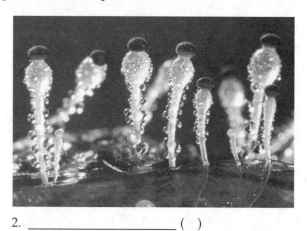

2. _____ (  )

3. _____ (  )

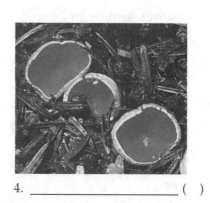

4. _____ (  )

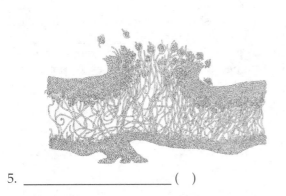

5. _____ (  )

6. _____ (  )

7. _____ (  )

8. _____ (  )

a. sac fungi
b. zygomycetes
c. club fungi
d. imperfect fungi
e. algae and fungi

___ 9. Most true fungi send out cellular filaments called _____ .
a. mycelia
b. hyphae
c. mycorrhizae
d. asci

___10. Heterotrophic species of fungi can be _____ .
a. saprobic
b. parasitic
c. mutualistic
d. all of the above

For questions 11–20, choose from the following.
a. club fungi
b. imperfect fungi
c. sac fungi
d. zygomycetes

___11. The group that includes *Rhizopus stolonifer*, the notorious black bread mold, is _____ .

___12. The group that includes delectable morels and truffles but also includes bakers' and brewers' yeasts is _____ .

___13. The group that includes shelf fungi, which decompose dead and dying trees, and mycorrhizal symbionts that help trees extract mineral ions from the soil is _____ .

___14. The group that includes the commercial mushroom *Agaricus brunnescens*, as well as the death cap mushroom, *Amanita phalloides*, is _____ .

___15. The group that includes *Penicillium*, which has a variety of species that produce penicillin and substances that flavor Camembert and Roquefort cheeses, is _____ .

___16. The group whose spore-producing structures (asci) are shaped like flasks, globes, and cups is _____ .

___17. The group that forms a thick wall around the zygote to produce a zygosporangium is _____ .

___18. The group whose spore-producing structures are club-shaped is _____ .

___19. The groups that are symbiotic with young roots of shrubs and trees in mycorrhizal associations are _____ and _____ .

___20. A group of fungi whose members were assigned to it because their sexual phase was undetected or absent.

## Chapter Objectives/Review Questions

| Page | Objectives/Questions |
|---|---|
| (390) | 1. The common names of the three major lineages of fungi are the _____ , the _____ , and the _____ . |
| (390) | 2. Fungi are heterotrophs; most are _____ and obtain nutrients from nonliving organic matter and so cause its decay. |
| (390) | 3. Other fungi are _____ ; they extract nutrients from tissues of a living host. |
| (390) | 4. Distinguish between the meanings of the following terms: hypha, hyphae, mycelium, and mycelia. |
| (390) | 5. Describe the diverse appearances of the fungi classified as club fungi. |
| (390) | 6. What is the most used reproductive mode in the fungi? |
| (391) | 7. The short-lived reproductive bodies of most club fungi are known as _____ . |
| (391) | 8. The spores produced by members of the club fungi are known as _____ . |
| (385) | 9. A _____ mycelium is one in which the hyphae have undergone cytoplasmic fusion but not nuclear fusion. |
| (391) | 10. Be able to review the generalized life cycle of a club fungus. |
| (392) | 11. The key defining feature of the zygomycetes is the _____ . |
| (392) | 12. Be able to review the life cycle of *Rhizopus*. |
| (392) | 13. Most sac fungi produce sexual spores called _____ . |
| (392) | 14. What are some typical shapes of the reproductive structures enclosing the asci? |
| (393) | 15. Flavoring cheeses, producing citric acid for candies and soft drinks, food spoilage, genetic research, leavening bread, and production of ethanol for wine and beer are all activities associated with fungi known as the _____ . |
| (393) | 16. Explain why a group of fungi known as the imperfect fungi is taxonomically convenient |
| (394) | 17. Define *mutualism* and explain why a lichen fits that definition. |
| (394) | 18. Distinguish the mycobiont from the photobiont. |
| (395) | 19. Describe the fungus–plant root association known as mycorrhizae. |
| (395) | 20. Distinguish exomycorrhizae from endomycorrhizae. |
| (395) | 21. What is the effect of pollution on mycorrhizae? |
| (396) | 22. Give the name of the fungus that causes the disease known as ergotism; list the symptoms of ergotism. |
| (396) | 23. What is the importance of the club fungi in the genus *Amanita*? |

## Integrating and Applying Key Concepts

Suppose humans acquired a few well-placed fungal genes that caused them to reproduce in the manner of a "typical" fungus (Figure 24.3, text). Try to imagine the behavioral changes that humans would likely undergo. Would their food supplies necessarily be different? Table manners? Stages of their life cycle? Courtship patterns? Habitat? Would the natural limits to population increase be the same? Would their body structure change? Would there necessarily have to be separate sexes? Compose a descriptive science-fiction tale about two mutants who find each other and set up "housekeeping" together.

# 25

# PLANTS

## Interactive Exercises

*Pioneers In a New World* (pp. 398–399)

### 25.1. EVOLUTIONARY TRENDS AMONG PLANTS (pp. 400–401)

*Selected Words: non*vascular plants, *seedless* vascular plants, *seed-bearing* vascular plants, *haploid (n)* phase, *diploid (2n)* phase, heterospory, homospory

### Boldfaced, Page-Referenced Terms

(400) vascular plants _____

_____

(400) bryophytes _____

_____

(400) gymnosperms _____

_____

(400) angiosperms _____

_____

(400) root systems _____

_____

(400) shoot systems _____

_____

(400) lignin _____

_____

(400) xylem _____

_____

(400) phloem _____

_____

(400) cuticle _____

_____

(400) stomata _____

_____

(400) gametophytes _____

_____

(400) sporophyte _____

_____

(400) spore _____

_____

(401) pollen grains _____

_____

(401) seed _____

_____

## Matching

Choose the most appropriate answer for each.

1. ___ seedless vascular plants
2. ___ angiosperms
3. ___ vascular plants
4. ___ bryophytes
5. ___ gymnosperms

A. Liverworts, hornworts, and mosses
B. Seed-bearing plants that include cycads, ginkgo, and conifers
C. Whisk ferns, lycophytes, horsetails, and ferns
D. A group of plants producing seeds by means of flowers
E. In general, a large number of plants having internal conducting tissues that conduct water and solutes through the plant body

## Complete the Table

6. As plants evolved, several key evolutionary developments occurred that solved the problems of living in new land environments. Complete the following table to summarize these events. Choose from heterospory, xylem and phloem, seed, shoot systems, spores, lignin, gametophytes, root systems, stomata, pollen grains, and cuticle.

| Evolutionary Trends | Survival Problem Solved |
| --- | --- |
| a. | Provides a large surface area for rapidly taking up soil water and scarce mineral ions; often anchors the plant |
| b. | Consist of stems and leaves that function in the absorption of sunlight energy and $CO_2$ from the air |
| c. | Allows extensive growth of stems and branches; a very hard organic substance that strengthens cell walls, thus allowing erect plant parts to display leaves to sunlight |
| d. | Provides cellular pipelines to distribute water and dissolved ions through plant parts |
| e. | A waxy coat on many plant organs that helps conserve water on hot, dry days |
| f. | Provides the main route for plant absorption of carbon dioxide while restricting evaporative water loss |
| g. | The haploid ($n$) phase that dominates the life cycles of algae |
| h. | The diploid ($2n$) phase that dominates the life cycles of most plants |
| i. | Haploid cells produced by meiosis in sporophyte plants; later divide by mitosis to give rise to gametophytes |
| j. | Condition in some seedless species and seed-bearing plants where two kinds of spores are produced |
| k. | Developed from one type of spore in gymnosperms and angiosperms; in turn develops into mature, sperm-bearing male gametophytes |
| l. | Consists of an embryo sporophyte, nutritive tissues, and a protective coat |

## Crossword Puzzle

7. Complete the crossword puzzle that includes terms and concepts related to plant classification and evolutionary trends. Clues are located below the blank puzzle.

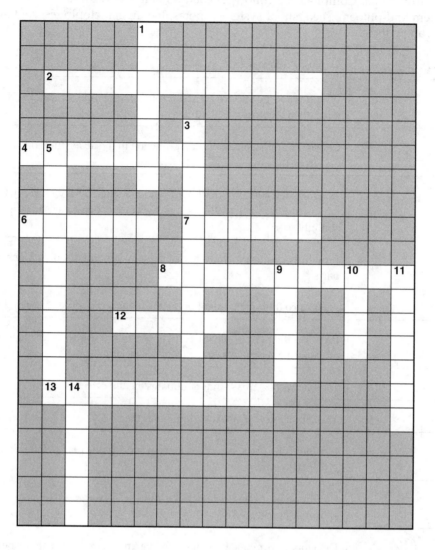

## ACROSS

2. Multicelled, haploid, gamete-producing structures produced from a haploid spore; nourish and protect the forthcoming generation.
4. Plants that possess internal tissues that conduct water and solutes; also have roots, stems, and leaves.
6. An organic compound that strengthens plant cell walls; such reinforced tissues structurally support plant parts that display the leaves to sunlight.
7. A vascular tissue that distributes sugars and other photosynthetic products.
8. Seed-bearing plants that include cycads, ginkgo, gnetophytes, and conifers.
12. A vascular tissue that distributes water and dissolved ions through plant parts.

13. Means "spore-producing body"; forms when a new 2n zygote embarks on a course of mitotic cell divisions, which produces the large multicelled plant body.

## DOWN

1. A waxy surface covering over stems and leaves that reduces water loss on hot, dry days.
3. Includes mosses, liverworts, and hornworts that generally grow in moist habitats; small plants lacking true roots, stems and leaves as well as vascular tissues.
5. Plants that produce seeds by means of flowers; includes two classes: dicots and monocots.
9. Systems that consist of stems and leaves, which function in the absorption of sunlight energy and carbon dioxide from the air.

10. Systems that consist of underground, cylindrical absorptive structures with a large surface area for taking up soil and scarce mineral ions.
11. Numerous tiny passageways in surface tissues of leaves and young stems; the main route for absorbing carbon dioxide and controlling evaporative water loss.

14. Develops from one type of haploid plant spore; become mature, sperm-bearing male gametophytes.

## 25.2. BRYOPHYTES (pp. 402–403)

*Selected Words:* Polytrichum, Sphagnum, Marchantia

*Boldfaced, Page-Referenced Terms*

(402) mosses _____

_____

(402) liverworts _____

_____

(402) hornworts _____

_____

(403) peat bogs _____

_____

## True/False

If the statement is true, write a T in the blank. If the statement is false, make it correct by changing the underlined word(s) and writing the correct word(s) in the answer blank.

_____ 1. Mosses are highly sensitive to <u>water</u> pollution.

_____ 2. Bryophytes <u>have</u> leaflike, stemlike, and rootlike parts although they do not contain xylem or phloem.

_____ 3. Most bryophytes have <u>rhizomes</u>, elongated cells or threads that attach gametophytes to soil and serve as absorptive structures.

_____ 4. Bryophytes are the simplest plants to exhibit a cuticle, cellular jackets around gamete-producing parts, and large gametophytes that retain nutritionally <u>dependent</u> sporophytes.

_____ 5. The true <u>liverworts</u> are the most common bryophytes.

_____ 6. Following fertilization, zygotes give rise to <u>gametophytes</u>.

_____ 7. Each <u>sporophyte</u> consists of a stalk and a jacketed structure in which spores will develop.

_____ 8. <u>Club</u> moss is a bog moss whose large, dead cells in their leaflike parts soak up five times as much water as cotton.

_____ 9. Bryophyte sperm reach eggs by movement through <u>air</u>.

_____10. *Marchantia* is a <u>moss</u> that reproduces asexually by way of gemmae, multicelled vegetative bodies that develop in tiny cups on the plant body.

## Fill-in-the-Blanks

The numbered items on the illustration below represent missing information about a typical moss life cycle; complete the numbered blanks in the narrative below to supply the missing information on the illustration.

The gametophytes are the green leafy "moss plants." Sperms develop in jacketed structures at the shoot tip of the male (11) _____ and eggs develop in jacketed structures at the shoot tip of the female (12) _____ . Raindrops transport (13) _____ to the egg-producing structure. (14) _____ occurs within the egg-producing structure. The (15) _____ grows and develops into a mature (16) _____ (with sporangium and stalk) while attached to the gametophyte. (17) _____ occurs within the sporangium of the sporophyte where haploid (18) _____ form, develop, and are released. The released spores grow and develop into male or female (19) _____ .

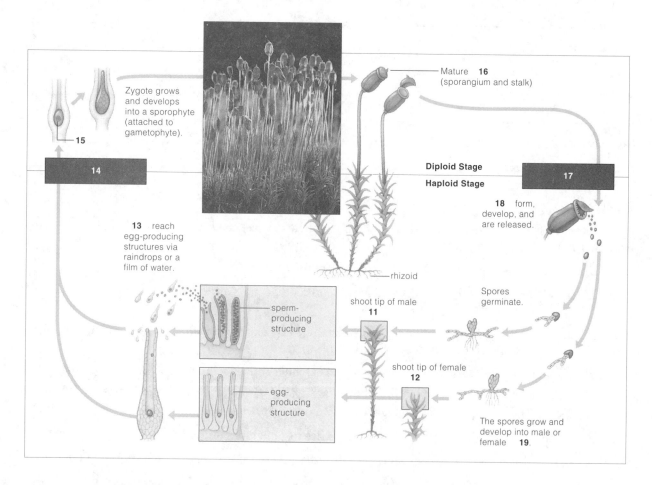

## 25.3. EXISTING SEEDLESS VASCULAR PLANTS (pp. 404–405)

## 25.4. *Focus on Science:* ANCIENT CARBON TREASURES (p. 406)

*Selected Words:*

Cooksonia, Psilotum, Lycopodium, Selaginella, Equisetum, Lepidodendron, Medullosa, *epiphyte*, Calamites

## Boldfaced, Page-Referenced Terms

(404) whisk ferns _____

_____

(404) lycophytes _____

_____

(404) horsetails _____

_____

(404) ferns _____

_____

(404) rhizomes _____

_____

(404) strobilus _____

_____

(406) coal _____

_____

## Choice

For questions 1–20 , choose from the following answers:

a. whisk ferns      b. lycophytes      c. horsetails      d. ferns      e. applies to a, b, c, and d

1. ____ A group in which only the genus *Equisetum* survives
2. ____ Familiar club mosses growing as mats on forest floors
3. ____ Seedless vascular plants
4. ____ *Psilotum*
5. ____ Rust-colored patches, the sori, are on the lower surface of their fronds
6. ____ Some tropical species are the size of trees
7. ____ Ancestral plants living in the Carboniferous; through time, heat, and pressure these plants became peat and coal, the "ancient carbon treasures"
8. ____ Stems were used by pioneers of the American West to scrub cooking pots
9. ____ The sporophytes have no roots or leaves
10. ____ Mature leaves are usually divided into leaflets
11. ____ The sporophyte has vascular tissues
12. ____ When the spore chamber snaps open, spores catapult through the air
13. ____ Grow in mud soil of streambanks and in disturbed habitats, such as roadsides and vacant lots
14. ____ Sporophytes have rhizomes and a hollow photosynthetic aboveground stems with scalelike leaves
15. ____ The sporophyte is the larger, longer-lived phase of the life cycle
16. ____ *Lycopodium*
17. ____ The young leaves are coiled into the shape of a fiddlehead

18.____A germinating spore develops into a small green, heart-shaped gametophyte

19.____*Selaginella* produces two spore types, a heterosporous genus

20.____"Amphibians" of the plant kingdom; life cycles require water

## Fill-in-the-Blanks

The numbered items on the illustration of a generalized fern life cycle below represent missing information; complete the numbered blanks in the narrative below to supply the missing information on the illustration.

Fern leaves (fronds) of the sporophyte are usually divided into leaflets. The underground stem of the sporophyte is termed a (21) _____ . On the undersides of many fern fronds, rust-colored patches, each of which is called a (22) _____ , is composed of spore chambers. (23) _____ of diploid cells within each sporangium produces haploid (24) _____ . The (25) _____ are catapulted into the air when each spore chamber snaps open. A spore may germinate and grow into a (26) _____ that is small, green, and heart-shaped. Jacketed structures develop on the underside of the mature (27) _____ . Each male jacketed structure produces many (28) _____ while each female jacketed structure produces a single (29) _____ . These gametes meet in (30) _____ . The diploid (31) _____ is first formed inside the female jacketed structure; it divides to form the developing (32) _____ , still attached to the gametophyte.

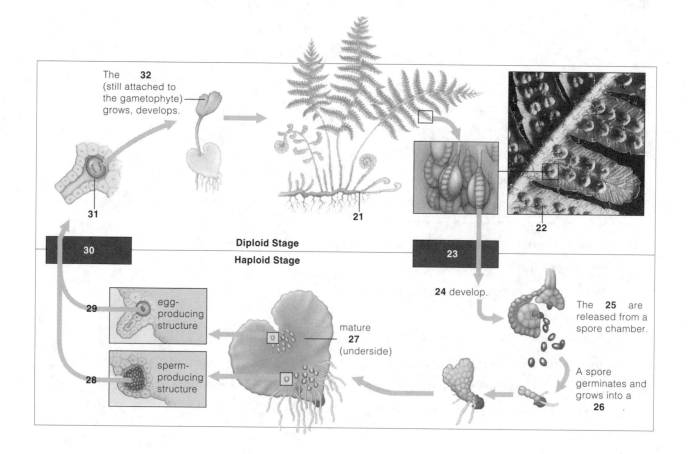

## 25.5. THE RISE OF THE SEED-BEARING PLANTS (p. 407)

*Selected Words:* Pinus, Prunus

*Boldfaced, Page-Referenced Terms*

(407) microspores _____

_____

(407) pollination _____

_____

(407) megaspore _____

_____

(407) ovules _____

_____

(407) seed ferns _____

_____

(407) progymnosperms _____

_____

### Matching

Choose the most appropriate answer for each.

1. ___microspores
2. ___pollination
3. ___megaspore
4. ___ovule
5. ___seed ferns
6. ___progymnosperms

A. May have been the first seed bearing plants; when climates became cooler, they were replaced by the cycads, conifers, and other gymnosperms
B. Spore type found in seed-bearing plants that develop into ovules
C. Female reproductive parts which, at maturity, are seeds
D. Spore type found in seed-bearing plants that gives rise to pollen grains
E. Seed-bearing plants that apparently arose from progymnosperms
F. The name for the arrival of pollen grains on the female reproductive structures

## 25.6. GYMNOSPERMS—PLANTS WITH "NAKED" SEEDS (pp. 408–409)
## 25.7. A CLOSER LOOK AT THE CONIFERS (pp. 410–411)

*Selected Words:* gymnos, sperma, evergreen, deciduous, Zamia, Pinus longaeva, Juniperus, Ginkgo biloba, Ephedra, Welwitschia mirabilis

*Boldfaced, Page-Referenced Terms*

(408) conifers _____

_____

(408) cones _____

_____

(408) cycads _____

_____

(409) ginkgos _____

_____

(409) gnetophytes _____

_____

(411) deforestation _____

_____

## Choice

For questions 1–12, choose from the following answers:

a. cycads    b. ginkgos    c. gnetophytes    d. conifers    e. gymnosperms (includes a, b, c, d)

1.____Fleshy-coated seeds of female trees produce an awful stench

2.____Includes pines, fir, yews, spruces, junipers, larches, cypresses, bald cypress, dawn redwood, and podocarps

3.____Seeds and a flour made from the trunk are edible following removal of poisonous alkaloids

4.____Only a single species survives, the maidenhair tree

5.____Includes *Welwitschia* of hot deserts of south and west Africa, *Gnetum* of humid tropical regions, and *Ephedra* of California deserts and other arid regions

6.____Have pollen-bearing cones and massive, seed-bearing cones that bear ovules; superficially resemble palm trees

7.____Seeds are mature ovules

8.____The favored male trees are now planted in cities; they have attractive, fan-shaped leaves and are resistant to insects, disease, and air pollutants

9.____Their ovules and seeds are not covered; they are borne on surfaces of spore-producing reproductive structures

10.____Some sporophyte plants in this group mainly have a deep-reaching taproot; the exposed part is a woody disk-shaped stem bearing cone-shaped strobili and one or two strap-shaped leaves that split lengthwise repeatedly as the plant ages

11.____Most species are "evergreen" trees and shrubs with needlelike or scalelike leaves

12.____Includes conifers, cycads, ginkgos, and gnetophytes

## Fill-in-the-Blanks

The numbered items on the illustration on the following page represent missing information; complete the numbered blanks in the narrative to supply the missing information on the illustration.

The familiar pine tree, a conifer, represents the mature (13) _____ . Pine trees produce two kinds of spores in two kinds of cones. Pollen grains are produced in male (14) _____ . Ovules are produced in young female (15) _____ . Inside each (16) _____ , (17) _____ occurs to produce haploid megaspores; one develops into a many-celled female (18) _____ that contains haploid (19) _____ . Diploid cells within pollen sacs of male cones undergo (20) _____ to produce haploid microspores. Microspores develop into (21) _____ grains. (22) _____ occurs when spring air currents deposit pollen grains near ovules of female cones. A pollen (23) _____ representing a male gametophyte grows toward the female gametophyte. (24) _____ nuclei form within the pollen tube as it grows toward the egg. (25) _____ follows and the ovule becomes a seed that is composed of an outer seed coat, the (26) _____ diploid sporophyte plant, and nutritive tissue.

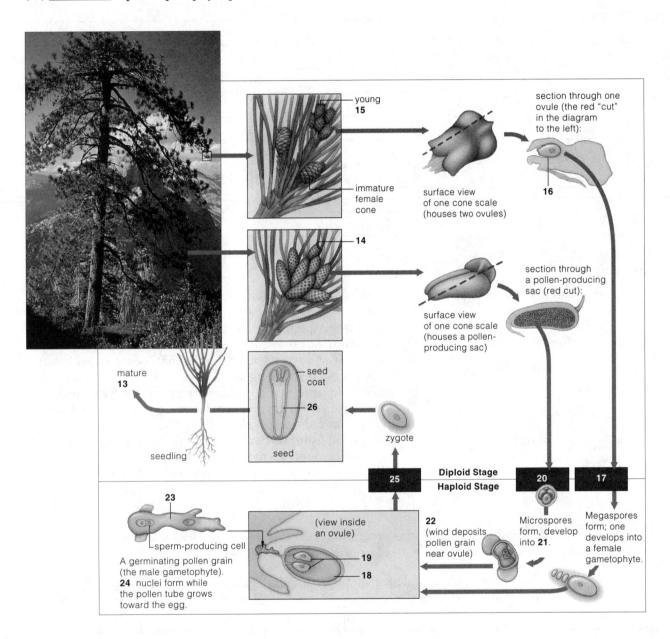

young
**15**

immature
female
cone

**14**

section through one
ovule (the red "cut"
in the diagram
to the left):

**16**

surface view
of one cone scale
(houses two ovules)

section through
a pollen-producing
sac (red cut):

surface view
of one cone scale
(houses a pollen-
producing sac)

mature
**13**

seed
coat

**26**

seed

seedling

zygote

**Diploid Stage**

**25**

**Haploid Stage**

**20**

**17**

Microspores
form, develop
into **21**.

Megaspores
form; one
develops into
a female
gametophyte.

**23**

sperm-producing cell

A germinating pollen grain
(the male gametophyte).
**24** nuclei form while
the pollen tube grows
toward the egg.

(view inside
an ovule)

**19**

**18**

**22**
(wind deposits
pollen grain
near ovule)

# 25.8. ANGIOSPERMS—THE FLOWERING, SEED-BEARING PLANTS (pp. 412–413)

*Selected Words:* angeion, Eucalyptus trees, Nymphaea, Arceuthobium, Monotropa uniflora, Lilium

*Boldfaced, Page-Referenced Terms*

(412) flowers _____

_____

(412) pollinators _____

_____

(412) dicots _____

_____

(412) monocots _____

_____

(413) fruits _____

_____

## Matching

1. ___examples of monocot plants
2. ___flowers
3. ___pollinators
4. ___double fertilization
5. ___fruits
6. ___endosperm
7. ___examples of dicot plants
8. ___seed

A. Nutritive tissue found within a seed
B. Palms, lilies, orchids, wheat, corn, rice, rye, and sugarcane
C. Unique angiosperm reproductive structures
D. mature ovaries; protect and help disperse plant embryos
E. Insects, bats, birds, and other animals that coevolved with the life cycles of flowering plants
F. Among all plants, unique to flowering plant life cycles; one sperm fertilizes the egg, the other sperm fertilizes a cell that gives rise to endosperm
G. Packaged in fruits; each covered by a protective tissue and containing an embryo and nutritive tissue
H. Most shrubs and trees, most nonwoody plants, cacti, and water lilies

## Complete the Table

9. Complete the table below to compare the plant groups studied in this chapter.

| Plant Group | Dominant Generation | Vascular Tissue Present? | Seeds Present? |
|---|---|---|---|
| a. Bryophytes | | | |
| b. Lycophytes | | | |
| c. Horsetails | | | |
| d. Ferns | | | |
| e. Gymnosperms | | | |
| f. Angiosperms | | | |

# Self-Quiz

___ 1. Members of the plant kingdom probably evolved from _____ more than 400 million years ago.
   a. unicellular brown algae
   b. multicellular green algae
   c. unicellular green algae
   d. multicellular red algae

___ 2. The _____ is *not* a trend in the evolution of plants.
   a. evolution of complex sporophytes
   b. shift from homospory to heterospory
   c. shift from diploid to haploid dominance
   d. development of xylem and phloem

___ 3. Existing nonvascular plants do *not* include _____ .
   a. horsetails
   b. mosses
   c. liverworts
   d. hornworts

___ 4. Plants possessing xylem and phloem are called _____ plants.
   a. gametophyte
   b. nonvascular
   c. vascular
   d. seedless

___ 5. Bryophytes _____ .
   a. have vascular systems that enable them to live on land
   b. include lycopods, horsetails, and ferns
   c. have true roots but not stems
   d. include mosses, liverworts, and hornworts

___ 6. _____ are not seedless vascular plants.
   a. Lycophytes
   b. Gymnosperms
   c. Horsetails
   d. Whisk Ferns
   e. Ferns

___ 7. In horsetails, lycopods, and ferns, _____ .
   a. spores give rise to gametophytes
   b. the main plant body is a gametophyte
   c. the sporophyte bears sperm- and egg-producing structures
   d. all of the above

___ 8. _____ are seed plants.
   a. Cycads and ginkgos
   b. Conifers
   c. Angiosperms
   d. all of the above

___ 9. In complex land plants, the diploid stage is resistant to adverse environmental conditions such as dwindling water supplies and cold weather. The diploid stage progresses through this sequence: _____ .
   a. gametophyte → male and female gametes
   b. spores → sporophyte
   c. zygote → sporophyte
   d. zygote → gametophyte

___10. Monocots and dicots are groups of _____ .
   a. gymnosperms
   b. club mosses
   c. angiosperms
   d. horsetails

# Chapter Objectives/Review Questions

*Page*        *Objectives/Questions*

(400)    1. Most of the members of the plant kingdom are _____ plants, with internal tissues that conduct and distribute water and solutes.
(400)    2. Distinguish between the vascular seed plants known as gymnosperms.
(400)    3. State the general functions of the root systems and shoot systems of vascular plants.
(400)    4. Explain the significance of "cells with lignified walls."
(400)    5. _____ tissue distributes water and dissolved ions through plant parts; _____ tissue distributes sugars and other photosynthetic parts.

(400)  6.  Be able to name the plant structures that protect leaves and young stems from water loss and also name the plant structures that serve as routes for absorbing carbon dioxide and controlling evaporative water loss.

(400)  7.  Give the reasons that diploid dominance allowed plants to successfully exploit the land environment.

(401)  8.  As plants evolved two spore types (heterospory), one spore type developed into pollen grains which become sperm-bearing male _____ ; the other spore type developed into female _____ , where eggs form and later become fertilized.

(401)  9.  The combination of a plant embryo, nutritive tissues, and protective tissues constitute a _____ .

(402)  10.  Mosses, liverworts, and hornworts belong to a plant group called the _____ .

(402)  11.  Describe and state the functions of rhizoids.

(402)  12.  What group of plants first displayed cuticles, cellular jackets around the parts that produce sperms and eggs, and large gametophytes that retain sporophytes?

(403)  13.  In mosses and their relatives, a _____ consists of a jacketed structure in which spores will develop, and a stalk.

(403)  14.  The remains of peat mosses accumulate into compressed, excessively moist mats called _____ _____ .

(404)  15.  Be able to list the four groups of seedless vascular plants.

(404)  16.  The _____ is the dominant phase in the seedless vascular plants.

(404–405)  17.  Be able to describe structural characteristics of *Psilotum*, *Lycopodium*, *Equisetum*, and a fern; be generally familiar with their life cycles.

(404)  18.  _____ are underground, branching, short, mostly horizontal stems that serve in absorption.

(404)  19.  Some sporophytes of club mosses have nonphotosynthetic, cone-shaped clusters of leaves known as _____ that bear spore sacs.

(405)  20.  Explain the meaning of the general term, *epiphyte*.

(406)  21.  Describe the sources of and the formation of coal.

(407)  22.  Microspores give rise to _____ _____ .

(407)  23.  Define the term *pollination*.

(407)  24.  Describe the relationship between seed ferns and progymnosperms.

(407)  25.  An _____ contains the female gametophyte, surrounded by nutritive tissue and a jacket of cell layers.

(408)  26.  Conifers, cycads, ginkgos, and gnetophytes are all members of the _____ lineage.

(408)  27.  Describe the structure of a typical conifer cone.

(408–409)  28.  Briefly characterize plants known as cycads, ginkgos, and gnetophytes.

(409)  29.  List reasons why *Ginkgo biloba* is a unique plant.

(409)  30.  *Gnetum*, *Ephedra*, and *Welwitschia* represent genera of _____ .

(410)  31.  Be generally familiar with the life cycle of *Pinus*, a somewhat typical gymnosperm.

(411)  32.  What is the largest threat to the existing conifers?

(412)  33.  Only angiosperms produce unique reproductive structures known as _____ .

(412)  34.  Most flowering plants evolved with _____ such as insects, bats, birds, and other animals.

(412)  35.  Name and cite examples of the two classes of flowering plants.

(413)  36.  Unlike gymnosperms, flowering plants provide their seeds with a unique, protective package of stored food, the _____ .

## Integrating and Applying Key Concepts

Explain why totally submerged aquatic plants that live in deep water never developed heterosporous life cycles.

# 26

# ANIMALS: THE INVERTEBRATES

---

## Interactive Exercises

---

*Madeleine's Limbs* (pp. 416–417)

### 26.1. OVERVIEW OF THE PLANT KINGDOM (pp. 418–419)
### 26.2. PUZZLES ABOUT ORIGINS (p. 420)
### 26.3. SPONGES—SUCCESS IN SIMPLICITY (pp. 420–421)

*Selected Words:* *anterior* end, *posterior* end, *dorsal* surface, *ventral* surface, <u>Paramecium</u>, <u>Volvox</u>, <u>Trichoplax</u> <u>adhaerens</u>, *plax*, *zoon* , <u>Euplectella</u>

## Boldfaced, Page-Referenced Terms

(418) animals _____

_____

(418) ectoderm _____

_____

(418) endoderm _____

_____

(418) mesoderm _____

_____

(418) vertebrates _____

_____

(418) invertebrates _____

_____

(418) bilateral symmetry _____

_____

(419) cephalization _____

_____

(419) gut _____

_____

(419) coelom _____

_____

(420) placozoan _____

_____

(420) sponges _____

_____

(420) collar cells _____

_____

(421) larva, -ae _____

_____

(421) adult _____

_____

## Matching

Choose the most appropriate answer for each.

1. ___ animals
2. ___ ventral surface
3. ___ ectoderm, endoderm, mesoderm
4. ___ anterior end
5. ___ vertebrates
6. ___ invertebrates
7. ___ radial symmetry
8. ___ bilateral symmetry
9. ___ dorsal surface
10. ___ gut
11. ___ coelom
12. ___ thoracic cavity
13. ___ abdominal cavity
14. ___ posterior end
15. ___ segmentation
16. ___ cephalization

A. All animals whose ancestors evolved before backbones did
B. An evolutionary process whereby sensory structures and nerve cells became concentrated in a head
C. The back surface
D. Animal body cavity lined with a peritoneum—found in most bilateral animals; some worms lack this cavity, other worms have a false cavity
E. head end
F. Animals having body parts arranged regularly around a central axis, like spokes of a bike wheel
G. Upper coelom cavity holding a heart and lungs
H. Primary tissue layers that give rise to all adult animal tissues and organs
I. Surface opposite the dorsal surface
J. Region inside animal body in which food is digested
K. Animals having right and left halves that are mirror images of each other
L. Series of animal body units that may or may not be similar to one another
M. End opposite the anterior end
N. Lower coelom cavity holding a stomach, intestines, and other organs
O. Multicellular organisms with tissues forming organs and organ systems, diploid body cells, heterotrophic, aerobic respiration, sexual reproduction, sometimes asexual, most are motile in some part of the life cycle, and the life cycle shows embryonic development
P. Animals with a "backbone"

## Complete the Table

17. Complete the table below by filling in the appropriate phylum or representative group name.

| Phylum | Some Representative Organisms | Number of Known Species |
|---|---|---|
| a. | *Trichoplax*; simplest animal | 1 |
| b. | Porifera | 8,000 |
| c. | Hydrozoans, jellyfishes, corals, sea anemones | 11,000 |
| d. Platyhelminthes | | 15,000 |
| e. | Pinworms, hookworms | 20,000 |
| f. | Tiny body with crown of cilia, great internal complexity; "wheel animals" | 1,800 |
| g. Mollusca | | 110,000 |
| h. | Leeches, earthworms, polychaetes | 15,000 |
| i. Arthropoda | | 1,000,000 |
| j. | Sea stars, sea urchins | 6,000 |
| k. | Invertebrate chordates: tunicates, lancelets | 2,100 |
| l. Chordata | | 45,000 |

## Choice

For questions 18–27, answer questions about animal origins by choosing from the following:

a. *Paramecium*   b. *Volvox*   c. *Trichoplax adhaerens*
d. different animal lineages arose from more than one group of protistanlike ancestors

18. ____ The only known placazoan

19. ____ Similar ciliate forerunners may have had multiple nuclei within a single cell

20. ____ Similar to colonies that became flattened and crept on the seafloor

21. ____ The answer to animal origins might require more than one answer

22. ____ As simple as an animal can get

23. ____ By another hypothesis, multicelled animals arose from flagellated cells that live in similar hollow, spherical colonies

24. ____ A soft-bodied marine animal, shaped a bit like a plate

25. ____ By one hypothesis, the animal forerunners were ciliates, much like this organism

26. ____ Has no symmetry, no tissues, and no mouth

27. ____ In a similar organism, the division of labor characterizing multicellularity might have begun

*Matching*

Choose the most appropriate answer for each.

28. ___oscula

29. ___fragmentation

30. ___sponge phylum

31. ___adult

32. ___larva

33. ___amoeboid cells

34. ___*Trichoplax*

35. ___sponge skeletal elements

36. ___microvilli

37. ___collar cells

38. ___water entering a sponge body

39. ___gemmules

A. Flagellated cells that absorb and move water through a sponge as well as engulf food
B. Reside in a gelatin-like substance between inner and outer cell linings
C. An organism whose two cell layers resemble those of a sponge
D. Sexually mature form of a species
E. Form the "collars" of collar cells
F. Clusters of sponge cells capable of germinating and establishing new colonies
G. Microscopic pores and chambers
H. Random chunks of sponge tissue break off and grow into more sponges
I. Spongin fibers, glasslike spicules of silica or calcium carbonate or both
J. Sexually immature form preceding the adult stage
K. Openings where water leaves a sponge
L. Porifera

## 26.4. CNIDARIANS—TISSUES EMERGE (pp. 422–423)

## 26.5. VARIATIONS ON THE CNIDARIAN BODY PLAN (pp. 424–425)

## 26.6. COMB JELLIES (p. 425)

*Selected Words:* <u>Hydra</u>, <u>Chrysaora</u>, <u>Obelia</u>, <u>Physalia</u>, <u>Cestum</u> <u>veneris</u>, <u>Mnemiopsis</u>

*Boldfaced, Page-Referenced Terms*

(422) Cnidaria _____

_____

(422) nematocysts _____

_____

(422) medusa _____

_____

(422) polyp _____

_____

(423) epithelium _____

_____

(423) nerve cells _____

_____

(423) contractile cells _____

_____

(423) hydrostatic skeleton _____

_____

(424) gonads _____

_____

(424) planulas _____

_____

(425) comb jellies _____

_____

## Fill-in-the-Blanks

All members of the phylum (1) _____ are tentacled, radial animals; they include jellyfishes, sea anemones, corals, and animals like *Hydra*. Most of these animals live in the sea and they alone produce (2) _____ that are capsules capable of discharging threads that entangle or pierce prey. Cnidarians have two common body plans, the (3) _____ that look like bells or upside down saucers, and the (4) _____ that has a tubelike body with a tentacle-fringed mouth at one end. The saclike cnidarian gut processes food with its (5) _____ , a sheetlike lining with glandular cells that secrete digestive enzymes. The (6) _____ covers the rest of the body's surfaces. Each of these linings is referred to as an (7) _____ . (8) _____ cells form a "nerve net" that coordinates responses to stimulation. Working together with (9) _____ cells and (10) _____ cells, the nerve net controls movement and shape changes. The (11) _____ is a layer of secreted material that lies between the epidermis and gastrodermis. Most polyps have little mesoglea and use water in their gut as a (12) _____ skeleton, a fluid-filled cavity or cell mass. Reef-forming (13) _____ consist of interconnected polyps that secrete calcium-reinforced external skeletons that, over time, build reefs. Many cnidarians have only a polyp or a medusa stage in the life cycle with the medusa being the sexual form. They have simple (14) _____ that rupture and release gametes. Zygotes formed at fertilization develop into (15) _____ , a type of swimming or creeping larva that usually has ciliated epidermal cells. Eventually a mouth opens at one end transforming the larva into a polyp or a medusa, and the cycle begins anew.

## Labeling

Identify each indicated part of the illustration below:

16. _____  _____

17. _____  _____

18. _____  _____

19. _____  _____

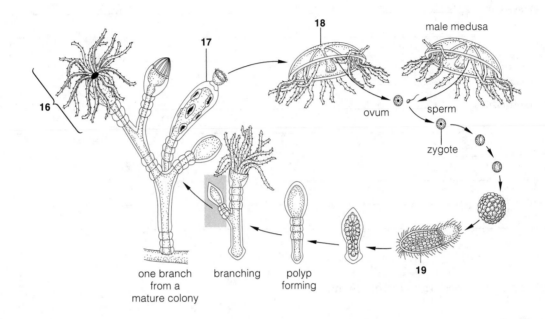

## Dichotomous Choice

Circle one of two possible answers given between parentheses in each statement.

20. Evolutionarily, comb jellies are of interest because they possess cells with (single/multiple) cilia as in many complex animals.
21. Comb jellies are classified in the phylum (Ctenophora/Cnidaria).
22. The comb jellies are the simplest animals with embryonic tissues that are like (mesoderm/endoderm).
23. If comb jellies are sliced into two equal halves and then the halves are sliced into quarters, two of the quarters will be (mirror/identical) images of the other two.
24. All comb jellies are weak-swimming predators in planktonic communities, and all show modified (bilateral/radial) symmetry.
25. A comb jelly has (eight/ten) rows of comblike structures made of thick, fused, cilia.
26. In a comb jelly, all the combs in a row beat in waves and propel the animal (backward/forward), usually mouth first.
27. Comb jellies (do/do not) produce nematocysts.
28. Some comb jellies use their sticky (lips/combs) to capture prey.
29. Some species of comb jellies have (four/two) muscular tentacles with branches that are equipped with sticky cells.

## 26.7. ACOELOMATE ANIMALS—AND THE SIMPLEST ORGAN SYSTEMS
(pp. 426–427)

*Selected Words:* organ-system level of construction, primary host, intermediate host

*Boldfaced, Page-Referenced Terms*

(426) organ _____

_____

(426) organ system _____

_____

(426) flatworms _____

_____

(426) hermaphrodites _____

_____

(427) proglottids _____

_____

(427) ribbon worms _____

_____

*Choice*

For questions 1–25, choose from the following:

a. turbellarians      b. flukes      c. tapeworms      d. ribbon worms

1. ____ Parasitic worms.

2. ____ Flame cells.

3. ____ Parasitize the intestines of vertebrates.

4. ____ Planarians that reproduce asexually by transverse fission.

5. ____ Possess a scolex.

6. ____ They swallow or suck body fluids from worms, mollusks, and crustaceans.

7. ____ Ancestral forms probably had a gut but later lost it during their evolution in animal intestines.

8. ____ Phylum Nemertea.

9. ____ Water-regulating systems have one or more tiny, branched tubes called protonephridia.

10. ____ Only planarians and a few others live in freshwater.

11. ____ Their life cycles have sexual and asexual phases and at least two kinds of hosts.

12. ____ May be related to flatworms but have a complete gut, circulatory system, and other traits that are notable departures from them.

13. ____ Flame cells, each with a tuft of cilia, that drive out excess water to the surroundings.

14. ____ Thrive in predigested food in vertebrate intestines.

15. ____ Resemble flatworms in their tissue organization.

16.____After division, each half regenerates the missing parts.

17.____Proglottids are new units of the body that bud just behind the head.

18.____These parasites attach to the intestinal wall by a scolex.

19.____They have a proboscis, which is a tubular, prey-piercing, venom-delivering device tucked inside the head end.

20.____Possess a structure equipped with suckers, hooks, or both.

21.____They reach sexual maturity in an animal that serves as their primary host; larval stages use an intermediate host, in which they either develop or become encysted.

22.____Older proglottids store fertilized eggs; they break off and leave the body in feces.

23.____They differ from flatworms in having a circulatory system, a complete gut, and separate sexes.

24.____Similar to flatworms in having ciliated surfaces, secreting mucus and move by beating their cilia through it.

25.____Intermediate hosts may meet up with their eggs.

## Labeling

Identify the parts of the animals shown dissected in the accompanying drawings.

26. _____ _____

27. _____

28. _____

29. _____ _____

30. _____

31. _____

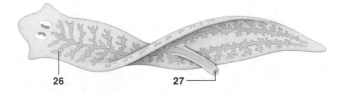

Answer exercises 32–35 for the drawings of the accompanying dissected animal.

32. What is the common name (or genus) of the animal dissected? _____
33. Is the animal parasitic? _____
34. Is the animal hermaphroditic? _____
35. Name the coelom type exhibited by this animal. _____

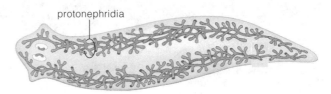

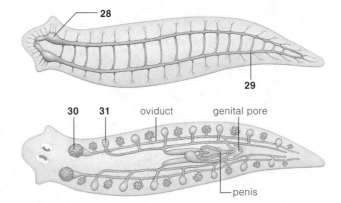

**26.8. ROUNDWORMS** (p. 428)

**26.9.** *Focus on Science:* **A ROGUE'S GALLERY OF WORMS** (pp. 428–429)

**26.10. ROTIFERS** (p. 430)

*Selected Words:* <u>Paratylenchus</u>, <u>Caenorhabditis</u> <u>elegans</u>, *schistosomiasis*, <u>Schistosoma</u> <u>japonicum</u>, <u>Enterobius</u> <u>vermicularis</u>, <u>Taenia</u> <u>saginata</u>, <u>Trichinella</u> <u>spiralis</u>, <u>Wuchereria</u> <u>bancrofti</u>, *elephantiasis*, <u>Philodina</u> <u>roseola</u>

*Boldfaced, Page-Referenced Terms*

(428) roundworms _____

_____

(430) rotifers _____

_____

Answer exercises 1–3 for the drawing of the dissected animal below.

1. What is the common name of the animal dissected? _____
2. Is the animal hermaphroditic? _____
3. Name the coelom type exhibited by this animal. _____

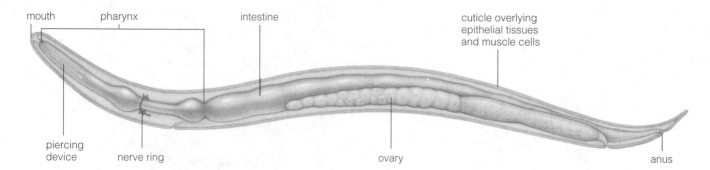

mouth   pharynx   intestine   cuticle overlying epithelial tissues and muscle cells

piercing device   nerve ring   ovary   anus

Answer exercises 4–8 for the drawing of the animal below.

4. What is the common name of this animal? _____

5. What type of symmetry does this animal possess? _____

6. Is this animal cephalized? _____

7. Name the type of coelom this animal possesses. _____

8. Why is this animal said to be a "wheel animal"? _____

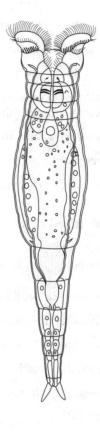

## Choice

For questions 9–26, choose from the following:

a. roundworms    b. rotifers

9. ____ All but about 5 percent live in freshwater, such as lakes, ponds, and even films of water on mosses and other plants.

10. ____ At night, the centimeter-long females migrate to the host's anal region to lay eggs; itching and scratching occurs that transfers eggs.

11. ____ Hookworms, pinworms, and other types parasitize plants and animals.

12. ____ Probably the most abundant of all multicelled animals alive today.

13. ____ All of them have a crown of cilia used in swimming and wafting food to the mouth.

14. ____ All of them have a bilateral, cylindrical body, usually tapered at both ends, and protected by a tough cuticle.

15. ____ Parasitic forms can do extensive damage to their hosts, which include humans, cats, dogs, cows, and sheep as well as valued crop plants.

16. ____ Most species are less than a millimeter long, yet they have a pharynx, an esophagus, digestive glands, a stomach, usually an intestine and anus, and protonephridia.

17. ____ Adult forms become lodged in the body's lymph nodes; elephantiasis occurs, an enlargement of legs and other body regions due to blockage of lymph flow.

18. ____ Humans become infected by walking barefoot where a juvenile may penetrate the skin; the parasite then travels the bloodstream to the lungs.

19. ____ Some have "eyes."

20.___Two "toes" exude substances that attach free-living species to substrates feeding time.

21.___A mosquito is *Wuchereria*'s intermediate host.

22.___Its rhythmic motions reminded early microscopists of a turning wheel.

23.___Humans become infected by *Trichinella spiralis* mostly by eating insufficiently cooked meat from pigs or certain game animals.

24.___They eat bacteria and microscopic algae.

25.___*Enterobius vermicularis* lives in the large intestine of humans.

26.___Thousands of scavenging types may occupy a handful of rich soil.

## Fill-in-the-Blanks

The numbered items on the illustration below (blood fluke life cycle) represent missing information; complete the corresponding numbered blanks in the narrative to supply missing information about flukes.

The life cycle of the Southeast Asian blood fluke (*Schistosoma japonicum*) requires a human primary host standing in water in which the fluke larvae can swim. This life cycle also requires an aquatic snail as an intermediate host. Flukes reproduce (27) _____ , and eggs mature in a human body. Eggs leave the body in feces, then hatch into ciliated, swimming (28) _____ that burrow into a (29) _____ and multiply asexually. In time, many fork-tailed (30) _____ develop. These leave the snail and swim until they contact (31) _____ skin. They bore inward and migrate to thin-walled intestinal veins, and the cycle begins anew. In infected humans, white blood cells that defend the body attack the masses of fluke eggs, and grainy masses form in tissues. In time, the liver, spleen, bladder, and kidneys deteriorate.

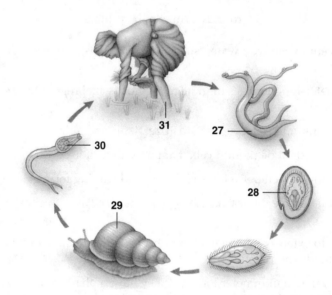

## Fill-in-the-Blanks

The numbered items on the illustration below (beef tapeworm life cycle) represent missing information; complete the corresponding numbered blanks in the narrative to supply missing information about the beef tapeworm.

(32) _____ , each with an inverted scolex of a future tapeworm, become encysted in (33) _____ host tissues (e.g., skeletal muscle).  A (34) _____ , a definitive host, eats infected and undercooked beef containing tapeworm cysts.  The scolex of a larva turns inside out, attaches to the wall of the host's (35) _____ _____ and begins to absorb host nutrients.  Many (36) _____ form, by budding; each of these segments becomes sexually mature and has both male and female reproductive (37) _____ . Ripe (38) _____ containing fertilized eggs leave the host in (39) _____ , which may contaminate water and vegetation.  Inside each fertilized egg, an embryonic (40) _____ form develops.  Cattle may ingest embryonated eggs or ripe proglottids and so become (41) _____ hosts.

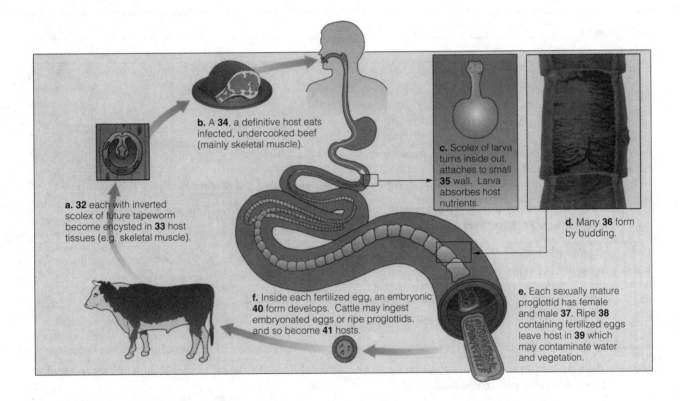

b. A **34**, a definitive host eats infected, undercooked beef (mainly skeletal muscle).

c. Scolex of larva turns inside out, attaches to small **35** wall.  Larva absorbes host nutrients.

a. **32** each with inverted scolex of future tapeworm become encysted in **33** host tissues (e.g. skeletal muscle).

d. Many **36** form by budding.

f. Inside each fertilized egg, an embryonic **40** form develops.  Cattle may ingest embryonated eggs or ripe proglottids. and so become **41** hosts.

e. Each sexually mature proglottid has female and male **37**. Ripe **38** containing fertilized eggs leave host in **39** which may contaminate water and vegetation.

## 26.11. TWO MAJOR DIVERGENCES (p. 430)

## 26.12. A SAMPLING OF MOLLUSCAN DIVERSITY (p. 431)

## 26.13. EVOLUTIONARY EXPERIMENTS WITH MOLLUSCAN BODY PLANS
(pp. 432–433)

*Selected Words:* molluscus, Aplysia, Dosidiscus, *jet propulsion*

## Boldfaced, Page-Referenced Terms

(430) protostomes _____

_____

(430) deuterostomes _____

_____

(430) spiral cleavage _____

_____

(430) radial cleavage _____

_____

(431) mollusks _____

_____

(431) mantle _____

_____

(432) torsion _____

_____

## Choice

For questions 1–10, choose from the following:

a. protostomes     b. deuterostomes

1. ____The first external opening in these embryos become the anus; the second becomes the mouth

2. ____Animals having a developmental pattern in which the early cell divisions are parallel and perpendicular to the axis

3. ____A coelom arises from spaces in the mesoderm

4. ____Radial cleavage

5. ____The first external opening in these embryos becomes the mouth

6. ____A coelom forms from outpouchings of the gut wall

7. ____Spiral cleavage

8. ____Animal having a developmental pattern in which early cell divisions are at oblique angles relative to the genetically prescribed body axis

9. ____Echinoderms and chordates

10. ____Mollusks, annelids, and arthropods

## Matching

Identify the animals pictured below by matching each with the appropriate description.

11. _____ Animal A     I. Bivalve

12. _____ Animal B     II. Cephalopod

13. _____ Animal C     III. Gastropod

## Labeling

Identify each numbered part in the drawings below by writing its name in the appropriate blank.

14. _____
15. _____
16. _____
17. _____
18. _____
19. _____
20. _____
21. _____

22. _____
23. _____
24. _____
25. _____
26. _____
27. _____
28. _____ _____
29. _____

30. _____ _____
31. _____
32. _____ _____
33. _____

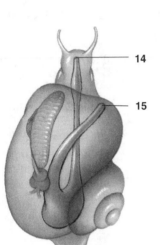

**Animal A**

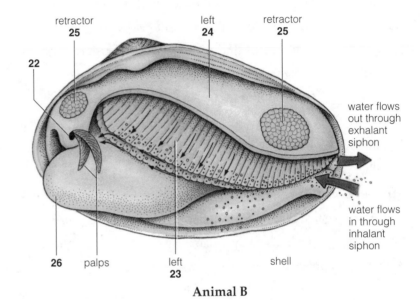

retractor **25**    left **24**    retractor **25**

water flows out through exhalant siphon

water flows in through inhalant siphon

**22**

**26**    palps    left **23**    shell

**Animal B**

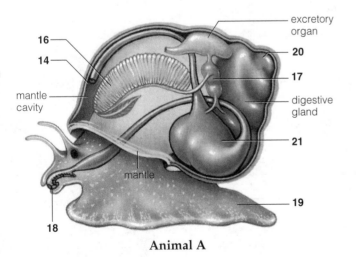

**16**
**14**
mantle cavity
mantle

excretory organ
**20**
**17**
digestive gland
**21**
**19**
**18**

**Animal A**

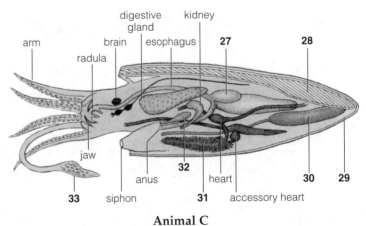

arm    radula    brain    digestive gland    esophagus    kidney    **27**    **28**

jaw    anus    **32**    heart    **30**    **29**

**33**    siphon    **31**    accessory heart

**Animal C**

Animals: The Invertebrates    **285**

## Fill-in-the-Blanks

The name of the phylum, Mollusca, means (34) _____ bodied. All mollusks have (35) _____ symmetry, a small coelom, and they alone have a (36) _____ , a tissue fold that hangs skirtlike over the body mass. Most have a (37) _____ of calcium carbonate (or a reduced version of one). They have (38) _____ that are the respiratory organs constructed of thin-walled leaflets for gas exchange. The vast majority of mollusks have a fleshy (39) _____ . Those with a well-developed (40) _____ have eyes and tentacles. Many have a (41) _____ , a tonguelike, toothed organ that preshreds food. The four classes of mollusks are the gastropods, chitons, bivalves, and the (42) _____ . With 90,000 species, (43) _____ or "belly foots" are the largest class. During the development of gastropods embryos, a process called (44) _____ twists the shell and most of the visceral mass 180° counterclockwise. This process places the anus very near the mouth. These are the shelled snails and (45) _____ that have no shell or only a remnant of one. Clams, scallops, oysters, and mussels are well-known (46) _____ . (47) _____ have been eating one type or another since prehistoric times. In nearly all mollusks, the gills function in collecting food and in (48) _____ . As water moves through the mantle cavity, (49) _____ on the gills traps bits of food. Cilia move the mucus to (50) _____ , which sort out the food before acceptable bits are driven to the mouth. Bivalves hide in mud or sand with an embedded large foot and have a pair of (51) _____ ; water is drawn into the mantle cavity by one and it leaves through the other carrying (52) _____ . (53) _____ include squids, cuttlefish, octopuses, and nautiluses. Some are the largest and (54) _____ of the known invertebrates. They have separate sexes, are footless, and are active marine predators. They also have prey-capturing (55) _____ . They move by (56) _____ propulsion by forcing water out of the mantle cavity through a funnel-shaped siphon. Active cephalopods have a great demand for (57) _____ and they alone among the mollusks have a (58) _____ circulatory system with two (59) _____ . Cephalopods have a well-developed (60) _____ system. Compared to other mollusks, they have the largest (61) _____ relative to body size. Cephalopod eyes resemble those of (62) _____ although they form in a different way. In terms of memory and learning, the cephalopods are the most complex (63) _____ .

## 26.14. ANNELIDS—SEGMENTS GALORE (pp. 434–435)

*Selected Words:* *setae* or *chaetae, oligo-, poly-,* <u>Hirudo</u> <u>medicinalis</u>

*Boldfaced, Page-Referenced Terms*

(434) annelids _____

_____

(435) nephridia _____

_____

(435) brain _____

_____

(435) nerve cords _____

_____

(435) ganglion _____

_____

## Matching

Choose the appropriate answer for each.

1. ___cuticle
2. ___annelids
3. ___earthworms
4. ___nephridia with cells similar to flame cells
5. ___marine polychaetes
6. ___brain
7. ___nephridia
8. ___nerve cords
9. ___setae or chaetae
10. ___advantage of segmentation
11. ___hydrostatic skeleton
12. ___leeches
13. ___earthworm locomotion
14. ___ganglion

A. Fluid-cushioned coelomic chambers
B. Oligochaete scavengers with a closed circulatory system and few bristles per segment
C. Paired, each a bundle of extensions of nerve cell bodies leading away from the brain
D. Possess many bristles per segment
E. Muscle contraction with protraction and retraction of segment bristles
F. Secreted wrapping around the body surface of most annelids that permits respiratory exchange
G. A rudimentary aggregation of nerve cell bodies that integrate sensory input and muscle responses for the whole body
H. Lack bristles
I. Implies an evolutionary link between flatworms and annelids
J. Different body parts can evolve separately and specialize in different tasks
K. Except for leeches, chitin-reinforced bristles on each side of the body on nearly all segments
L. The term means "ringed forms"
M. An enlargement of the nerve cord in each segment of an earthworm
N. Regulates volume and composition of body fluids; often begins with a funnel-shaped structure in each segment

## Labeling

Identify each indicated part of the illustration.

15. _____

16. _____

17. _____  _____

18. _____

19. _____  _____

20. _____

21. _____

22. Name this animal. _____

23. Name this animal's phylum. _____

24. Name two distinguishing characteristics of this group. _____

_____

25. Protostome (   ) or deuterostome (   )

26. Is this animal segmented? (   ) yes (   ) no

27. Symmetry of adult: (   ) radial (   ) bilateral

28. Does this animal have a true coelom? (   ) yes (   ) no

## 26.15. ARTHROPODS—THE MOST SUCCESSFUL ORGANISMS ON EARTH (p. 436)

**Selected Words:** *division of labor*

**Boldfaced, Page-Referenced Terms**

(436) arthropods _____

_____

(436) molting _____

_____

(436) metamorphosis _____

_____

## Choice

For questions 1–15, choose from the following six adaptations that contributed to the success of arthropods.

a. hardened exoskeletons      b. fused and modified segments      c. jointed appendages
d. specialized respiratory structures      e. efficient nervous system and sensory organs    f. division of labor

1. ____ Intricate eyes and other sensory organs.

2. ____ Sexually immature larvae that molt and change as they grow.

3. ____ Hardened and protective, due to protein, chitin, and surface waxes; an evolutionary innovation that led to specialized appendages.

4. ____ In most of their existing descendants, however, the serial repeats of the body wall and organs are masked.

5. ____ Metamorphosis from larvae into adult forms.

6. ____ Might have evolved as defenses against predators.

7. ____ Larval stages specialize in feeding and growth; the adult is concerned with dispersal and reproduction.

8. ____ Gills of aquatic arthropods.

9. ____ In insects, different segments became combined into a head, a thorax, and an abdomen.

10. ____ Evolved specializations that include diverse wings, antennae, and legs.

11. ____ Many species have a wide angle of vision and can process visual information from many directions.

12. ____ Tracheas, the air-conducting tubes of land-dwelling arthropods.

13. ____ In spiders, segments fused as a forebody and hindbody.

14. ____ Waxy surfaces restrict evaporative water loss and can support a body deprived of water's buoyancy.

15. ____ Crabs and lobsters are equipped with a calcium-stiffened armor plating.

## Short Answer

16. Why are the arthropods said to be the most biologically successful organisms on earth?

_____

_____

_____

_____

_____

## 26.16. A LOOK AT SPIDERS AND THEIR KIN (p. 437)

## 26.17. A LOOK AT THE CRUSTACEANS (p. 438–439)

## 26.18. HOW MANY LEGS? (p. 439)

*Selected Words:* *open* circulatory system, <u>Scutigera</u> <u>coleoptrata</u>

## Boldfaced, Page-Referenced Terms

(439) millipedes _____

_____

(439) centipedes _____

_____

## Matching

Select the most appropriate answer for each.

1. ___ticks
2. ___arachnid forebody appendages
3. ___arachnids
4. ___arachnid hindbody appendages
5. ___chelicerates
6. ___spider, internal organs
7. ___predatory arachnids

A. Scorpions and spiders that sting, bite and may subdue prey with venom
B. An open circulatory system and book lungs
C. A group including mites, horseshoe crabs, sea spiders, spiders, ticks, and chigger mites
D. Spin out silk thread for webs and egg cases
E. Some transmit bacterial agents of Rocky Mountain spotted fever or Lyme disease to humans
F. A familiar group including scorpions, spiders, ticks, and chigger mites
G. Four pairs of legs, a pair of pedipalps that have mainly sensory functions, and a pair of chelicerae that can inflict wounds and discharge venom

## Dichotomous Choice

Circle one of two possible answers given between parentheses in each statement.

8. Nearly all arthropods possess (strong claws/an exoskeleton).
9. The giant crustaceans are (lobsters and crabs/barnacles and pillbugs).
10. The simplest crustaceans have many pairs of (different/similar) appendages along their length.
11. (Barnacles/Lobsters and crabs) have strong claws that collect food, intimidate other animals, and sometimes dig burrows.
12. (Barnacles/Lobsters and crabs) have feathery appendages that comb microscopic bits of food from the water.
13. (Barnacles/Copepods) are the most numerous animals in aquatic habitats, maybe in the world.
14. Of all arthropods, only (lobsters and crabs/barnacles) have a calcified "shell."
15. Adult (barnacles/copepods) cement themselves to wharf pilings, rocks, and similar surfaces.
16. As is true of other arthropods, crustaceans undergo a series of (rapid feedings/molts) and so shed the exoskeleton during their life cycle.
17. Adult (millipedes/centipedes) have a rounded body, two pairs of legs per two earlier fused segments and move slowly.
18. Adult (millipedes/centipedes) have a flattened body, are fast-moving, and have a pair of walking legs on every segment except two.
19. (Millipedes/Centipedes) mainly scavenge for decaying vegetation in soil and forest litter.
20. (Millipedes/Centipedes), outfitted with fangs and venom glands, are aggressive predators of insects, earthworms, and snails.

## Labeling

Identify each numbered part of the animal pictured at right.

21. _____

22. _____

23. _____

24. _____

25. _____

26. _____

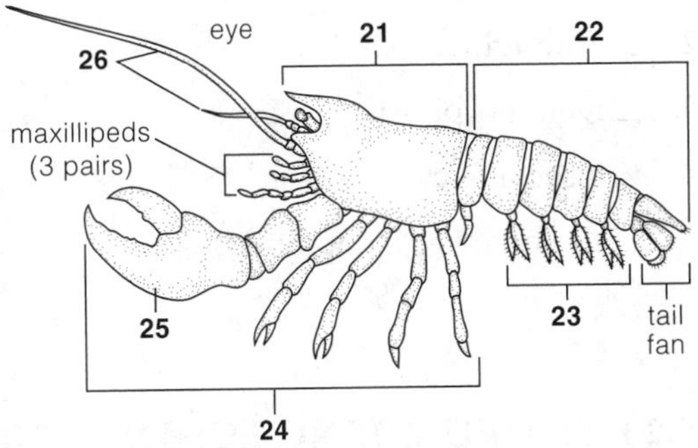

Answer exercises 27–33 for the animal pictured at the right.

27. Name the animal pictured.
_____

28. Name the subgroup of arthropods to which this animal belongs. _____

29. Name two distinguishing characteristics of this group. _____

30. Protostome (   ) or deuterostome (   )

31. Is this animal segmented? (   ) yes (   ) no

32. Symmetry of adult: (   ) radial (   ) bilateral

33. Does this animal have a true coelom? (   ) yes (   ) no

Identify each indicated body part in the illustration at the right.

34. _____ _____

35. _____

36. _____

37. _____

38. _____ _____

39. Name the subgroup of arthropods to which the animal belongs _____ .

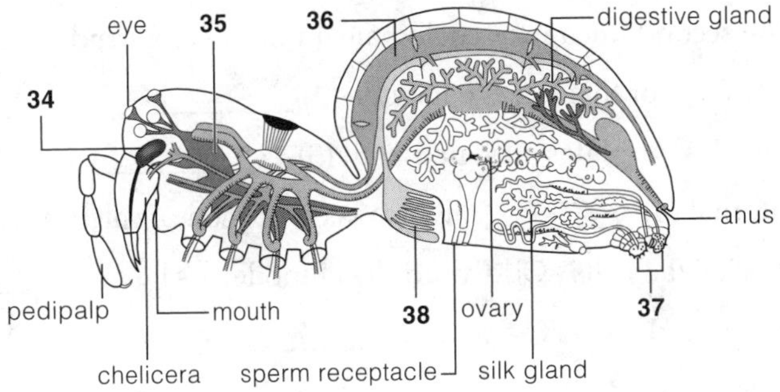

## 26.19. A LOOK AT INSECT DIVERSITY (pp. 440–441)

**Selected Words:** *incomplete* metamorphosis, *complete* metamorphosis

### Boldfaced, Page-Referenced Terms

(440) Malpighian tubules _____

_____

(440) nymphs _____

_____

(440) pupae _____

_____

## Matching

Choose the most appropriate answer for each.

1. ___ the most successful species of insects
2. ___ Malpighian tubules
3. ___ metamorphosis
4. ___ insect life cycle stages
5. ___ shared insect adaptations

A. Post-embryonic resumption of growth and transformation into an adult form
B. Head, thorax, and abdomen, paired antennae and mouthparts, three pairs of legs and two pairs of wings
C. Winged insects, also the only winged invertebrates
D. Larva—nymph—pupa—adult
E. Structures that collect nitrogen-containing wastes from blood and convert them to harmless crystals of uric acid that are eliminated with feces

## 26.20. THE PUZZLING ECHINODERMS (pp. 442–443)

### Boldfaced, Page-Referenced Terms

(442) echinoderms _____

_____

(443) water-vascular system _____

_____

### Fill-in-the Blanks

The second lineage of coelomate animals is referred to as the (1) _____ . The major invertebrate members of this lineage include (2) _____ . The body wall of all echinoderms has protective spines, spicules, or plates stiffened with (3) _____ _____ . Most echinoderms also have a well-developed internal (4) _____ , which is composed of calcium carbonate and other substances secreted from specialized cells. Oddly, adult echinoderms have (5) _____ symmetry with some bilateral features, but some produce larvae with (6) _____ symmetry. Adult echinoderms lack a (7) _____ but a decentralized (8) _____ system allows them to respond to information about food, predators, and so forth that is coming from different directions. For example, any (9) _____ of a sea star that senses the shell of a tasty scallop can become the leader, directing the remainder of the body to move in a direction suitable for prey capture. The (10) _____ feet of sea stars are used for walking, burrowing, clinging to a rock or gripping a meal of clam or snail. These "feet" are part of a (11) _____ vascular system unique to echinoderms. Each foot contains an (12) _____ that acts like a rubber bulb on a medicine dropper as it contracts and forces fluid into a foot that then lengthens. Tube feet change shape constantly as (13) _____ action redistributes fluid through the water vascular system. Some sea stars are able to swallow their prey (14) _____ ; some can push part of their stomach outside the mouth and around their prey, then start (15) _____ their meal even before swallowing it. Coarse, indigestible remnants are regurgitated back through the mouth. Their small (16) _____ is of no help in getting rid of empty clam or snail shells.

## Labeling

Identify each indicated part of the two illustrations below.

17. _____  _____

18. _____  _____

19. _____

20. _____

21. _____

22. _____  _____

23. _____

24. _____  _____

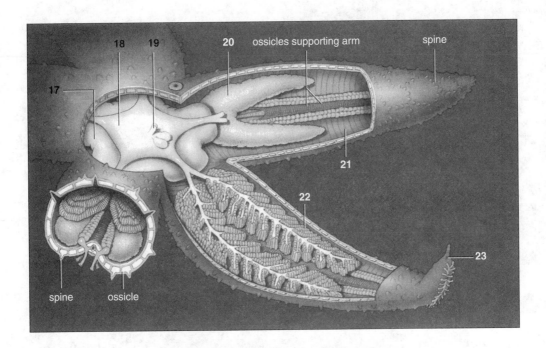

Answer exercises 25–27 for the animal pictured above.

25. Name the animal shown. _____
26. Does this animal have a true coelom? (   ) yes (   ) no
27. Protostome (   ) or deuterostome (   )

## Identifying

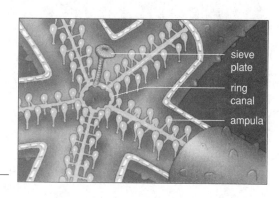

sieve
plate

ring
canal

ampula

28. Name the system shown at the right.

   _____ _____ system

29. Identify each creature in a through d below by its common name.

30. Name the phylum of the animals pictured.

   _____

31. Name two distinguishing characteristics of this group.

   _____

32. Symmetry of adults represented above: (   ) radial
   (   ) bilateral

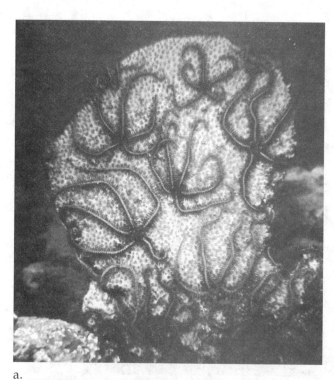

a. _____

b. _____

c. _____

d. _____

## Self-Quiz

___ 1. Which of the following is *not* true of sponges? They have no _____ .
a. distinct cell types
b. nerve cells
c. muscles
d. gut

___ 2. Which of the following is *not* a protostome?
a. earthworm
b. crayfish or lobster
c. sea star
d. squid

___ 3. Deuterostomes undergo _____ cleavage; protostomes undergo _____ cleavage.
a. radial; spiral
b. radial; radial
c. spiral; radial
d. spiral; spiral

___ 4. Bilateral symmetry is characteristic of _____ .
a. cnidarians
b. sponges
c. jellyfish
d. flatworms

___ 5. Flukes and tapeworms are parasitic _____ .
a. leeches
b. flatworms
c. jellyfish
d. flatworms

___ 6. Insects include _____ .
a. spiders, mites, ticks
b. centipedes and millipedes
c. termites, aphids, and beetles
d. all of the above

___ 7. Creeping behavior and a mouth located toward the "head" end of flatworms may have led, in some evolutionary lines, to _____ .
a. development of a circulatory system with blood
b. sexual reproduction
c. feeding on nutrients suspended in the water (filter feeding)
d. concentration of sensory structures and nerve cells in the head

___ 8. Which of the following is associated with the shift from radial to bilateral body form?
a. a circulatory system
b. a one-way gut
c. pairs of muscles and pairs of sensory structures, nerves and brain regions
d. the development of a water-vascular system

___ 9. The _____ body plan is characterized by bilateral symmetry, simple gas-exchange mechanisms, two-way traffic through a relatively unspecialized gut, and a thin cephalized body with all cells fairly close to the gut.
a. annelid
b. nematode
c. echinoderm
d. flatworm

___10. The _____ have bilateral symmetry, a tough cuticle, longitudinal muscles, and the simplest example of a complete digestive system.
a. nematodes
b. cnidarians
c. flatworms
d. echinoderms

___11. _____ with a peritoneum separates the gut and body wall of most bilateral animals.
a. A coelom
b. Mesoderm
c. A mantle
d. A water-vascular system

___12. In some annelid groups, a _____ is composed of cells similar to flatworm flame cells.
a. trachea
b. nephridium
c. mantle
d. parapodium

## Matching

Match each phylum below with the corresponding characteristics (a–l) and representatives (A–R). A phylum may match with more than one letter from the group of representatives.

____, ____13. Annelida

____, ____14. Arthropoda

____, ____15. Chordata

____, ____16. Cnidaria

____, ____17. Echinodermata

____, ____18. Mollusca

____, ____19. Nematoda

____, ____20. Ctenophora

____, ____21. Platyhelminthes

____, ____22. Porifera

____, ____23. Rotifera

____, ____24. Nemertea

a. choanocytes (= collar cells) + spicules
b. jointed legs + an exoskeleton
c. gill slits in pharynx + dorsal, tubular nerve cord + notochord
d. pseudocoelomate + wheel organ + soft body
e. soft body + mantle; may or may not have radula or shell
f. bilateral symmetry + blind-sac gut
g. radial symmetry + blind-sac gut; stinging cells
h. body compartmentalized into repetitive segments; coelom containing nephridia (= primitive kidneys)
i. bilateral, soft-bodied, ciliated, elongated predators; circulatory system, a complete gut, separate sexes, and a tubular prey-piercing proboscis
j. comb-bearing, weak-swimming predators, cells with multiple cilia
k. tube feet + calcium carbonate structures in skin
l. complete gut + bilateral symmetry + cuticle; includes many parasitic species, some of which are harmful to humans

A. Comb jellies
B. Corals, sea anemones, and *Hydra*
C. Salamanders and toads
D. Whales and opossums
E. Tapeworms and planaria
F. Insects
G. Jellyfish and the Portuguese man-of-war
H. Sand dollars and starfishes
I. Earthworms and leeches
J. Lobsters, shrimp, and crayfish

K. Organisms with spicules and choanocytes
L. Scorpions and millipedes
M. Octopuses and oysters
N. Ribbon worms
O. Flukes
P. Hookworm, trichina worm
Q. Small animals with a crown of cilia and two "exuding toes"
R. Dinosaurs

# Chapter Objectives/Review Questions

| Page | | Objectives/Questions |
|---|---|---|
| (418) | 1. | Be able to list the general characteristics that define an "animal." |
| (418) | 2. | _____ cells are the forerunners of the primary tissue layers, the ectoderm, endoderm, and, in most species, mesoderm. |
| (418) | 3. | List the primary characteristic that separates vertebrates from invertebrates. |
| (418–419) | 4. | Distinguish radial symmetry from bilateral symmetry and generally describe various animal gut types. |
| (419) | 5. | The _____ is a region inside the body in which food is digested, then absorbed into the internal environment. |
| (419) | 6. | List two benefits that the development of a coelom brings to an animal. |
| (419) | 7. | Describe the general structure of a segmented animal. |
| (420) | 8. | *Trichoplax* is as simple as an animal can get, having only _____ distinct layers of cells. |
| (420) | 9. | List the characteristics that distinguish sponges from other animal groups. |
| (420–421) | 10. | Describe the processes involved in sponge reproduction, both asexual and sexual. |
| (422) | 11. | State what nematocysts are used for and explain how they function. |
| (422) | 12. | Two cnidarian body types are the _____ and the _____ . |
| (424) | 13. | Describe the life cycle of the hydrozoan, *Obelia*. |

(425)    14. Describe a "comb jelly" in terms of its characteristics. Explain why the symmetry of a comb jelly is not truly bilateral.

(426)    15. Comb jellies are the simplest animals with embryonic tissues that are like _____ .

(426–427) 16. List the three main types of flatworms and briefly describe each; name the groups that are parasitic.

(427)    17. Members of the phylum Nemertea are known as _____ _____ ; describe them and the habitats they live in.

(428)    18. Describe the body plan of roundworms, comparing its various systems with those of the flatworm body plan.

(428–429) 19. Southeast Asian blood flukes, tapeworms, and pinworms are examples of _____ parasites.

(430)    20. Describe the size, structure, and the environment of the rotifers.

(430)    21. Define, by their characteristics, *protostome* and *deuterostome* and give examples of each group.

(431)    22. Define *mantle* and tell what role it plays in the molluscan body.

(432)    23. Describe the process of torsion that occurs only in gastropods.

(433)    24. Explain why cephalopods came to have such well-developed sensory and motor systems and are able to learn.

(434)    25. Describe the advantages of segmentation and tell how this relates to the development of specialized internal organs.

(436)    26. Be able to list the six arthropod adaptations that led to their success.

(436)    27. List four different lineages of arthropods; be able to briefly describe each.

(442)    28. The major invertebrate members of the _____ are the echinoderms; be able to list their major characteristics.

(442)    29. List five examples of echinoderms.

(443)    30. Describe how locomotion and eating occurs in sea stars.

---

## Integrating and Applying Key Concepts

Scan Table 26.1 to verify that most highly evolved animals have a complete gut, a closed blood vascular system, both central and peripheral nervous systems, and are dioecious. Why do you suppose having two sexes in separate individuals is considered to be more highly evolved than the monoecious condition utilized by many flatworms? Wouldn't it be more efficient if all individuals in a population could produce both kinds of gametes? Cross-fertilization would then result in both individuals being able to produce offspring.

# 27

# ANIMALS: THE VERTEBRATES

## Interactive Exercises

*Making Do (Rather Well) With What You've Got* (pp. 446–447)

### 27.1. THE CHORDATE HERITAGE (p. 448)

### 27.2. INVERTEBRATE CHORDATES (pp. 448–449)

### 27.3. EVOLUTIONARY TRENDS AMONG THE VERTEBRATES (pp. 450 -451)

### 27.4. EXISTING JAWLESS FISHES (p. 451)

### 27.5. EXISTING JAWED FISHES (pp. 452–453)

*Selected Words:* *"invertebrate chordates", "sea squirts", larvae.*

*Boldfaced, Page-Referenced Terms*

(448) chordates _____

(448) vertebrates _____

(448) notochord _____

(448) nerve cord _____

(448) pharynx _____

(448) tunicates _____

(448) filter feeders _____

(448) gill slits _____

(448) lancelets _____

(450) vertebrae _____

(450) jaws _____

(450) fins _____

(450) gills _____

(451) lungs _____

(451) ostracoderms _____

(451) placoderms _____

(451) hagfishes _____

(451) lampreys _____

(452) swim bladder _____

(452) cartilaginous fishes _____

(452) scales _____

_____

(452) bony fishes _____

_____

(453) lobe-finned fishes _____

_____

## Fill-in-the-Blanks

Four major features distinguish the embryos of chordates from those of all other animals: a hollow dorsal (1) _____ _____ , a (2) _____ with slits in its wall, a (3) _____ , and a tail that extends past the anus at least during part of its life. In some chordates, the (4) _____ chordates, the notochord is *not* divided into a skeletal column of separate, hard segments; in others, the (5) _____ , it is. Invertebrate chordates living today are represented by tunicates and (6) _____ , which obtain their food by (7) _____ - _____ ; they draw in plankton-laden water through the mouth and pass it over sheets of mucus, which trap the particulate food before the water exits through the (8) _____ _____ in the pharynx. (9) _____ are among the most primitive of all living chordates; when they are tiny, they look and swim like (10) _____ . A rod of stiffened tissue, the (11) _____ , runs the length of the larval body; it was the forerunner of the chordate's (12) _____ .

As the ancestors of land vertebrates began spending less time immersed and more of their lives exposed to air, use of gills declined and (13) _____ evolved; more elaborate and efficient (14) _____ systems evolved along with more complex and efficient lungs.

Even though the adult forms of hemichordates, echinoderms, and chordates look very different, they have similar embryonic developmental patterns. Chordates may have developed from a mutated ancestral deuterostome (15) _____ of a sessile, filter-feeding adult. A larva is an immature, motile form of an organism, but if a (16) _____ occurred that caused (17) _____ _____ to become functional in the larval body, then a motile larva that could reproduce would have been more successful in finding food and avoiding (18) _____ than a sessile adult; in time, the sessile stages in the species would be eliminated.

The ancestors of the vertebrate line may have been mutated forms of their closest relatives, the (19) _____ , in which the notochord became segmented and the segments became hardened (20) _____ . The vertebral column was the foundation for fast-moving (21) _____ , some of which were ancestral to all other vertebrates. The evolution of (22) _____ intensified the competition for prey and the competition to avoid being preyed upon; animals in which mutations expanded the nerve cord into a (23) _____ that enabled the animal to compete effectively survived more frequently than their duller-witted fellows and passed along their genes into the next generations. Fins became (24) _____ ; in some fishes, those became (25) _____ and equipped with skeletal supports; these forms set the stage for the development of legs, arms and wings in later groups.

Exercises 26–27 refer to the illustrations below.

26. Name the creature in illustration B. _____

27. Name the creature in illustration C. _____

## Labeling

Name the structures numbered in illustrations A, B and C.

28. _____

29. _____ with _____

    _____

30. _____

31. _____ _____

32. _____

33. _____

34. _____

35. _____ _____

36. _____ _____

37. _____ _____ _____

38. _____

39. _____

40. _____ _____ _____

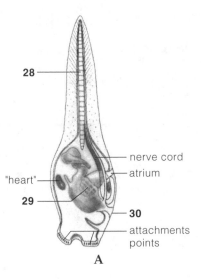

A

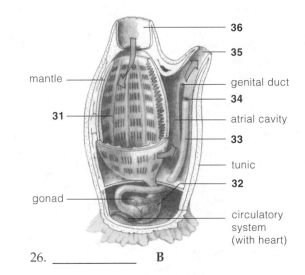

26. _____    B

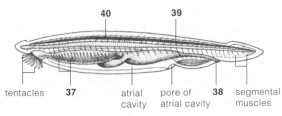

27. _____    C

41. _____    D

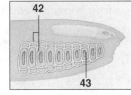

44. _____    E

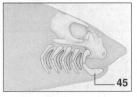

(Continued on p. 302)

41. Name the creature whose head is shown in illustration D on page 301.

_____

42. Name the structures. _____ _____
43. Name the structures. _____ _____
44. Name the creature shown in illustration E on page 301. _____
45. What structures occupy the front of its head space?

_____

## Complete the Table

Provide common names wherever (*) appears, and provide the class name wherever ☐ appears.

| Phylum Hemichordata | Phylum: Chordata | | |
|---|---|---|---|
| 46 acorn worms | Subphylum: Urochordata | Subphylum: Cephalochordata | Subphylum: Vertebrata |
| | 47* | 48* | Class: *Agnatha*        49* |
| | | | Class: *Placodermi*        50* |
| | | | Class: *Chondrichthyes*  51* |
| | | | Class: *Osteichthyes*      52* |
| | | | Class: 53 ☐        amphibians |
| | | | Class: 54 ☐        reptiles |
| | | | Class: 55 ☐        birds |
| | | | Class: 56 ☐        mammals |

## Fill-in-the-Blanks

Cartilaginous fishes include about 850 species of rays, skates, (57) _____ , and chimaeras. They have

conspicuous fins and five to seven (58) _____ _____ on both sides of the pharynx. Ninety-six

percent of the existing species of fishes are (59) _____ . Their ancestors arose during the Silurian period

(perhaps as early as 450 million years ago) and soon gave rise to three lineages: the (60) _____ -

_____ fishes, the lobe-finned fishes, and lungfishes.

## Matching

Match the numbered item with its letter. One letter is used twice.

61. ___
62. ___
63. ___
64. ___
65. ___
66. ___
67. ___
68. ___
69. ___
70. ___
71. ___
72. ___
73. ___
74. ___
75. ___
76. ___

A. Anal fin
B. Anus
C. Brain
D. Caudal fin
E. Dorsal fin
F. Gallbladder
G. Heart
H. Intestine
I. Kidney
J. Liver
K. Pectoral fin (paired)
L. Pelvic fin (paired)
M. Stomach
N. Swim bladder
O. Urinary bladder

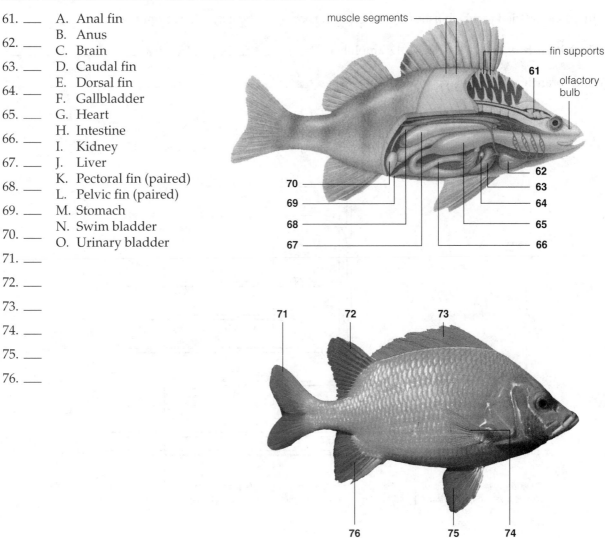

## Analysis and Short Answer

Figure 27.11 mentions features that distinguish the amphibian-lungfish lineage from the lineage that leads to sturgeons and other bony fishes. At least one of those features may have developed near ⑤ in this evolutionary tree.

Exercises 77–87 refer to the evolutionary diagram illustrated on p. 304.

77. What single feature do the lampreys, hagfishes, and extinct ostracoderms have in common that is different from the placoderms? _____

78. How did ostracoderms feed? _____

79. A mutation in ostracoderm stock led to the development of what feature in all organisms that descended from ①? _____

80. Mutation at ② led to the development of an endoskeleton made of what? _____

81. Mutations at ③ led to an endoskeleton of what? _____

82. Mutations at ④ led to which spectacularly diverse fishes that have delicate fins originating from the dermis? _____

83. Mutations at ⑤ led to which fishes whose fins incorporate fleshy extensions from the body? _____

84. Which branch, ④ or ⑤, gave rise to the amphibians? _____

85. Which branch gave rise to the modern bony fishes? _____

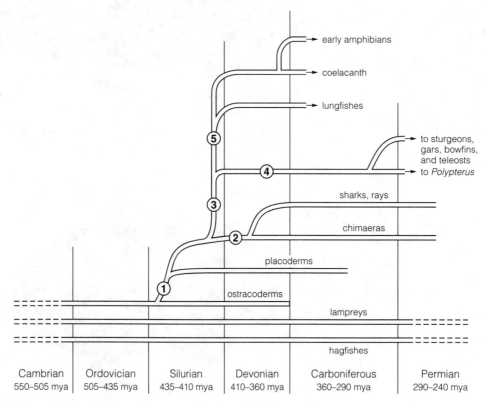

86. In which period did three distinctly different lineages of *bony* fishes appear in the fossil record?

_____

87. Approximately how many million years ago did the fork in the evolutionary path that led to the amphibians occur? _____

## 27.6. AMPHIBIANS (pp. 454–455)

## 27.7. THE RISE OF REPTILES (pp. 456–457)

## 27.8. MAJOR GROUPS OF EXISTING REPTILES (pp. 458–459)

*Selected Words:* Ichthyostega, Maiasaura, Velociraptor, *tuatara, stem reptile, synapsid, anapsid, archosaur, dinosaur, pterosaur.*

## Boldfaced, Page-Referenced Terms

(454) amphibian _____

_____

(455) salamander _____

_____

(455) frog _____

_____

(455) caecilian _____

_____

(456) reptiles _____

_____

(456) amniote egg _____

_____

(458) crocodilians _____

_____

(458) turtles _____

_____

(458) lizard _____

_____

(458) snake _____

_____

(459) tuatara _____

_____

## Fill-in-the-Blanks

Natural selection acting on lobe-finned fishes during the Devonian period favored the evolution of ever
more efficient (1) _____ used in gas exchange and stronger (2) _____ used in locomotion. Without
the buoyancy of water, an animal traveling over land must support its own weight against the pull of
gravity. The (3) _____ of early amphibians underwent dramatic modifications that involved evaluating
incoming signals related to vision, hearing, and (4) _____ . Although fish have (5) _____-
chambered hearts, amphibians have (6) _____-chambered hearts (see Fig. 39.4). Early in amphibian
evolution, mutations may have created the third chamber, which added a second (7) _____ in addition
to the already existing atrium and ventricle. The Devonian period brought humid, forested swamps with an
abundance of aquatic invertebrates and (8) _____—ideal prey for amphibians.

There are three groups of existing amphibians: (9) _____ , frogs and toads, and caecilians. Amphibians
require free-standing (10) _____ or at least a moist habitat to (11) _____ . Amphibian skin
generally lacks scales but contains many glands, some of which produce (12) _____ .

In the late Carboniferous, (13) _____ began a major adaptive radiation into the lush habitats on land, and only amphibians that mutated and developed certain (14) _____ features were able to follow them and exploit an abundant food supply. Several features helped: modification of (15) _____ bones favored swiftness, modification of teeth and jaws enabled them to feed efficiently on a variety of prey items, and the development of a (16) _____ egg protected the embryo inside from drying out, even in dry habitats.

(For questions 17–28, consult Figure 27.14 of the main text, and the time line below, and the figure for the Analysis and Short Answer on page 307 of this Study Guide.)

Today's reptiles include (17) _____ , crocodilians, snakes, and (18) _____ . All rely on (19) _____ fertilization, and most lay leathery-shelled eggs. Although amphibians originated during Devonian times, ancestral "stem" reptiles appeared during the (20) _____ period, about 340 million years ago. Reptilian groups living today that have existed on Earth longest are the (21) _____ ; their ancestral path diverged from that of the "stem" reptiles during the (22) _____ period. Crocodilian ancestors appeared in the early (23) _____ period, about 220 million years ago. Snake ancestry diverged from lizard stocks during the late (24) _____ period, about 140 million years ago. (25) _____ are more closely related to extinct dinosaurs and crocodiles than to any other existing vertebrates; they, too, have a (26) _____-chambered heart. Mammals have descended from therapsids, which in turn are descended from the (27) _____ group of reptiles, which diverged earlier from the stem reptile group during the (28) _____ period, approximately 320 million years ago.

In blanks 29–35, arrange the following groups in sequence, from earliest to latest, according to their appearance in the fossil record:

A. Birds    B. Crocodilians    C. Dinosaurs    D. Early ancestors of mammals (= synapsid reptiles)
    E. Early ancestors of turtles (= anapsid reptiles)    F. Lizards and snakes    G. "Stem" reptiles

29. ____  30. ____  31. ____  32. ____  33. ____  34. ____  35. ____

In blanks 36–42, select, from the choices immediately below, the geologic period in which each group (from 29–35) first appeared and write it in its corresponding space. For example, blank 29 contains the letter of the earliest group to appear on Earth; blank 36 should contain the letter of its geologic period chosen from below.

| Paleozoic Era | | Mesozoic Era | | |
|---|---|---|---|---|
| A. Carboniferous 360–290 mya | B. Permian 290–240 mya | C. Triassic 240–205 mya | D. Jurassic 205–138 mya | E. Cretaceous 138–65 mya |

36. ____  37. ____  38. ____  39. ____  40. ____  41. ____  42. ____

# Analysis and Short Answer

43. This chapter, on p. 462, mentions the features that distinguish, say, the mammals from all of the remaining groups on the right-hand side of the diagram; therefore, at some time during the evolution of mammals, mutations that produced those features appeared. Consult the evolutionary diagram below and imagine what sort of mutation(s) occurred at the numbered places.
    a. What may have occurred at ①?
    b. What may have occurred at ②?
    c. What may have occurred at ③?
    d. What may have occurred at ④?
    e. What may have occurred at ⑤?

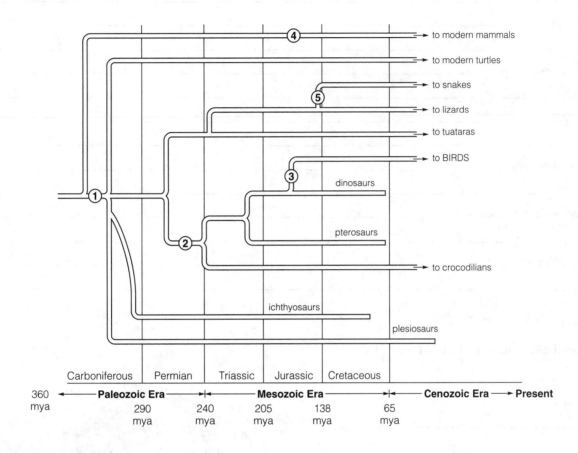

## 27.9. BIRDS (pp. 460–461)

## 27.10. THE RISE OF MAMMALS (pp. 462–463)

## 27.11. A PORTFOLIO OF EXISTING MAMMALS (pp. 464–465)

*Selected Words:* <u>Archaeopteryx</u>, *animal migration, underhair, guard hairs.*

*Boldfaced, Page-Referenced Terms*

(460) birds _____

_____

(460) feathers _____

_____

(462) mammals _____

_____

(462) behavioral flexibility _____

_____

(462) dentition _____

_____

(463) therapsids _____

_____

(463) therians _____

_____

(463) monotremes _____

_____

(463) marsupials _____

_____

(463) eutherians _____

_____

(464) convergent evolution _____

_____

(464) placenta _____

_____

## Labeling

Label the structures pictured at the right.

1. _____

2. _____

3. _____ _____

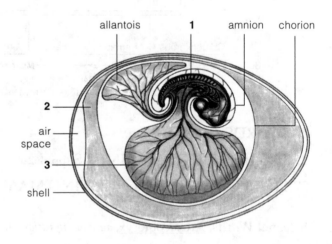

## Fill-in-the-Blanks

Birds descended from (4) _____ that ran around on two legs some 160 million years ago. All birds have (5) _____ that insulate and help get the bird aloft. Generally, birds have a greatly enlarged (6) _____ to which flight muscles are attached. Bird bones contain (7) _____ _____ , and air flows through sacs, not into and out of them, for gas exchange in the lungs. Birds also lay (8) _____ _____ eggs, have complex courtship behaviors, and generally nurture their offspring. Birds are able to regulate their body (9) _____ , which is generally higher than that of mammals. Existing birds do not have socketed (10) _____ in their beaks.

Flight demands (11) ☐ high ☐ low metabolic rates, which require an abundant supply of (12) _____ pumped to all body parts by means of a large, durable (13) ☐ 2 ☐ 3 ☐ 4 ☐ 6 - chambered heart. Almost 9,000 species of birds show amazing variation in body structure. The smallest adult bird is a(n) (14) _____ that weighs 2.25 grams. The largest existing bird is the (15) _____ , which weighs about 150 kilograms, can sprint fast but cannot fly.

During the Cenozoic Era (65 million years ago to the present), birds, (16) _____ , and flowering plants evolved to dominate Earth's assemblage of organisms. (16) are warm-blooded vertebrates with (17) _____ that began their evolution more than 200 million years ago.

There are three groups of existing mammals: those that lay eggs (examples are the (18) _____ and the spiny anteater), those that are (19) _____ (examples are the opossum and the kangaroo), and those that are (20) _____ mammals (there are more than 4,500 species of these). Mammals regulate their body temperature, have a (21) _____-chambered heart, and show a high degree of parental nurture. Most mammals have (22) _____ as a means of insulation, and mammalian mothers generally suckle their young with milk. Mammals differ from reptiles and birds in (23) _____ (the type, number and size of teeth). There are many groups, or (24) _____ , within the class Mammalia; each order has its distinctive array of characteristics.

The ancestors of today's mammals diverged from small, hairless reptiles called (25) _____ more than 200 million years ago during the Triassic. Through mutation and natural selection during Jurassic times they had developed hair and major changes in the body form, jaws and teeth; now they were called (26) _____ , which coexisted with the diverse groups of dominant (27) _____ through the Cretaceous. When the (27)s became extinct, diverse adaptive zones awaited exploitation by the three principal lineages, which, during the subsequent 65 million years have blossomed into 4,500 known mammalian species. (28) _____ _____ has occurred as evolutionarily distinct, geographically isolated lineages evolved in similar ways in similar habitats.

# Self-Quiz

___ 1. Filter-feeding chordates rely on _____, which have cilia that create water currents and mucous sheets that capture nutrients suspended in the water.
   a. notochords
   b. differentially permeable membranes
   c. filiform tongues
   d. gill slits

___ 2. In true fishes, the gills serve primarily _____ function.
   a. a gas-exchange
   b. a feeding
   c. a water-elimination
   d. both a feeding and a gas-exchange

___ 3. The heart in amphibians _____.
   a. pumps blood more rapidly than the heart of fish
   b. is efficient enough for amphibians but would not be efficient for birds and mammals
   c. has three chambers (ventricle and two atria)
   d. all of the above

___ 4. The feeding behavior of true fishes selected for highly developed _____.
   a. parapodia
   b. notochords
   c. sense organs
   d. gill slits

## Identification

Provide the common name and the major chordate group to which each creature pictured below and on pages 311–312 belongs.

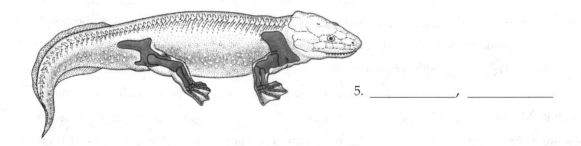

5. _____, _____

6. _____, _____

7. _____, _____

8. _____, _____

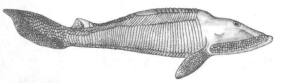

9. _____, _____

10. _____, _____

11. _____, _____

12. _____, _____

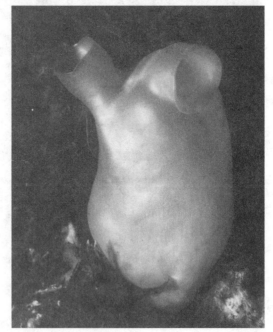

13. _____, _____

14. _____, _____

15. _____, _____

Match the following groups and classes with the corresponding characteristics (a - i) and representatives (A - I).

16. _____ , _____ Amphibians

17. _____ , _____ Birds

18. _____ , _____ Bony fishes

19. _____ , _____ Cartilaginous fishes

20. _____ , _____ Cephalochordates

21. _____ , _____ Jawless fishes

22. _____ , _____ Mammals

23. _____ , _____ Reptiles

24. _____ , _____ Urochordates

a. hair + vertebrae
b. feathers + hollow bones
c. jawless + cartilaginous skeleton (in existing species)
d. two pairs of limbs (usually) + glandular skin + "jelly"-covered eggs
e. amniote eggs + scaly skin + bony skeleton
f. invertebrate + sessile adult that cannot swim
g. jaws + cartilaginous skeleton + vertebrae
h. in adult, notochord stretches from head to tail; mostly burrowed-in, adult can swim
i. bony skeleton + skin covered with scales and mucus

A. lancelet
B. loons, penguins, and eagles
C. tunicates, sea squirts
D. sharks and manta rays
E. lampreys and hagfishes (and ostracoderms)
F. true eels and sea horses
G. lizards and turtles
H. caecilians and salamanders
I. platypuses and opossums

# Chapter Objectives/Review Questions

This section lists general and detailed chapter objectives that can be used as review questions. You can make maximum use of these items by writing answers on a separate sheet of paper. Fill in answers where blanks are provided. To check for accuracy, compare your answers with information given in the chapter or glossary.

*Page*      *Objectives/Questions*

(446–466)  1.  State how each of the groups of vertebrates obtains oxygen from the environment. List the principal skin structures that each kind of vertebrate produces.
(447–448)  2.  List three characteristics found only in chordates.
(448–449)  3.  Describe the adaptations that sustain the sessile or sedentary life-style seen in primitive chordates such as tunicates and lancelets.
(450–451)  4.  State what sort of changes occurred in the primitive chordate body plan that could have promoted the emergence of vertebrates.
(450–454)  5.  Describe the differences between primitive and advanced fishes in terms of skeleton, jaws, special senses, and brain.
(454–455)  6.  Describe the changes that enabled aquatic fishes to give rise to land dwellers.
(460–465)  7.  Discuss the effects that increased parental nurture of offspring in birds and mammals has had on courtship behavior and reproductive physiology.

# Integrating and Applying Key Concepts

Birds and mammals both have four-chambered hearts, high metabolic rates, and regulate their body temperatures efficiently. Both groups evolved from reptiles, so one would think that those same traits would have developed in ancestral reptiles. Data suggest that most reptiles have a heart intermediate between three and four chambers, lower metabolic rates, and body temperatures that are not well regulated and tend to rise and fall in accord with the environmental temperature. If the three traits mentioned in the first sentence had developed in reptilian groups, how might their lives have been different?

# 28

# HUMAN EVOLUTION: A CASE STUDY

*The Cave at Lascaux and the Hands of Gargas*

**EVOLUTIONARY TRENDS AMONG THE PRIMATES**
    Primate Classification
    Key Evolutionary Trends

**FROM PRIMATES TO HOMINIDS**
    Origins and Early Divergences
    The First Hominids

**EMERGENCE OF EARLY HUMANS**
    Defining "Human"
    *Homo habilis* and the First Stone Tools
    From *Homo erectus* to *Homo sapiens*

*Focus on Science:* OUT OF AFRICA—ONCE, TWICE, OR . . .

---

## Interactive Exercises

---

*The Cave at Lascaux and the Hands of Gargas* (pp. 468–469)

## 28.1. EVOLUTIONARY TRENDS AMONG THE PRIMATES (pp. 470–471)

## 28.2. FROM PRIMATES TO HOMINIDS (pp. 472–473)

*Selected Words:* "humanness", prosimian, tarsioid, prehensile, opposable, Plesiadapis, Aegyptopithecus, "Lucy", Australopithecus, dryopiths.

### Boldfaced, Page-Referenced Terms

(470) Primates _____

_____

(470) arboreal _____

_____

(470) anthropoids _____

_____

(470) hominoids _____

_____

(470) hominids _____

_____

(470) savannas _____

_____

(471) bipedalism _____

_____

(471) culture _____

_____

(473) australopiths _____

_____

## Fill-in-the-Blanks

During the Cenozoic Era (65 million years ago to the present), birds, (1) _____ , and flowering plants evolved to dominate Earth's assemblage of organisms. (1) are warm-blooded vertebrates with (2) _____ that began their evolution more than 200 million years ago. There are many groups ((3) _____) within the class Mammalia; each order has its distinctive array of characteristics. Humans, apes, monkeys, and prosimians are all (4) _____ ; members of this <u>order</u> have excellent (5) _____ perception as a result of their forward-directed eyes, hands that are (6) _____ ; they are adapted for (7) _____ instead of running, and primates rely less on their sense of (8) _____ and more on daytime vision. Their (9) _____ became larger and more complex; this trend was accompanied by refined technologies and the development of (10) _____ : the collection of behavior patterns of a social group, passed from generation to generation by learning and by symbolic behavior (language). Modification in the (11) _____ led to increased dexterity and manipulative skills. Changes in primate (12) _____ indicate that there was a shift from eating insects to fruit and leaves and on to a mixed diet. Primates began to evolve from ancestral mammals more than (13) _____ million years ago, during the Paleocene. The first primates resembled small (14) _____ or tree shrews; they foraged at night for (15) _____ , seeds, buds, and eggs on the forest floor, and they could clumsily climb trees searching for safety and sleep.

During the (16) _____ (25 → 5 million years ago), the continents began to assume their current positions, and climates became cooler and (17) _____ . Under these conditions, forests began to give way to mixed woodlands and (18) _____ ; an adaptive radiation of apelike forms—the first hominoids—took place as subpopulations of apes became reproductively isolated within the shrinking stands of trees. The chimpanzee-sized "tree apes," or (19) _____ , originated during this time; eventually, they ranged throughout Africa, Europe, and Southern (20) _____ . Between (21) _____ million and (22) _____ million years ago, three divergences occurred that gave rise to the ancestors of modern gorillas, chimpanzees, and humans.

## Matching

Fill in the blanks of the evolutionary diagram below with the appropriate letter from the choice list. Consult Table 28.1. Choose from the bold-faced letters to match boldfaced numbers.

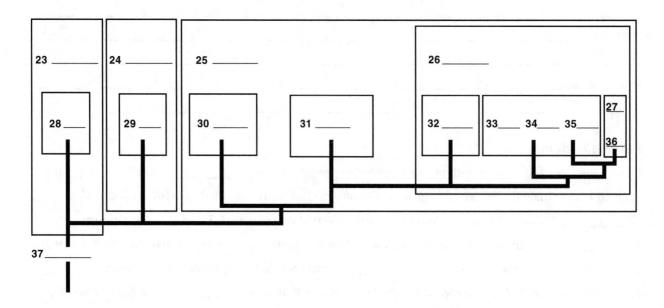

A. Anthropoids
B. Chimpanzee
C. Gibbon, siamang
D. Gorilla
E. Hominids
F. Hominoids
G. Humans and their most recent ancestors
H. Lemurs, lorises

I. New World monkeys (spider monkeys etc.)
J. Old World monkeys (baboons, etc.)
K. Orangutan
L. Prosimians
M. Tarsiers
N. Tarsioids
O. Rodentlike primate of the Paleocene

## More Fill-in-the-Blanks

The family Hominidae (hominids) (which includes all species on the genetic path leading to humans since the time when that path diverged from the path leading to the (38) _____ _____) emerged between (39) _____ million and (40) _____ million years ago, during the late Miocene. (41) _____-million-year-old fossils of humanlike forms have been discovered in Africa, and they all were bipedal and omnivorous and had an expanded brain. Lucy was one of the earliest (42) _____ , a collection of forms that combined ape and human features; they were fully two-legged, or (43) _____ , with essentially human bodies and ape-shaped heads.

## Choice

Choose from the choices below for 44–53. Choose the single *smallest* group to which each belongs.

a. Anthropoids      b. Hominids      c. Hominoids      d. Prosimians      e. Tarsioids

44.___australopiths

45.___chimpanzees

46.___dryopiths

47.___gibbons

48.___gorillas

49.___humans

50.___lemurs

51.___New World monkeys

52.___Old World monkeys

53.___orangutans

## Matching

Suppose you are a student who wants to be chosen to accompany a paleontologist who has spent forty years teaching and roaming the world in search of human ancestors. Above the desk in her office, she keeps reconstructions of many skull types. Five are shown on the opposite page, and you have heard that she chooses the graduate students who accompany her on her summer safaris on the basis of their ability on a short matching quiz. The quiz is presented below. *Without consulting your text (or any other)* while you are taking the quiz, match up all five skulls with *all* applicable letters. Sixteen correctly placed letters win you a place on the expedition. Fourteen correct answers put you on a waiting list. Thirteen or fewer and she suggests that you may wish to investigate dinosaur fossils instead of human ancestry. Which will it be for you?

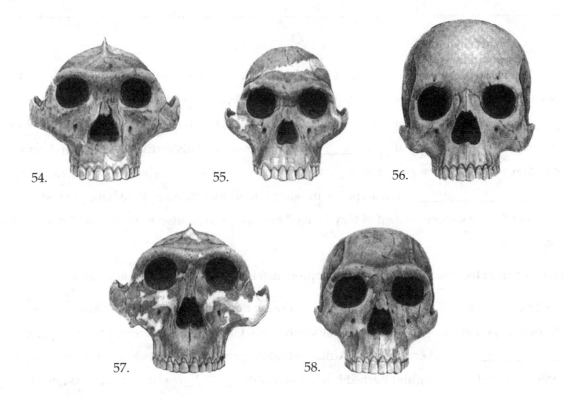

54.      55.      56.

57.      58.

| | |
|---|---|
| 54. ___ | A. Gracile form(s) of australopiths |
| 55. ___ | B. Robust form(s) of australopiths |
| | C. *Homo sapiens sapiens* |
| 56. ___ | D. Has a chin |
| | E. First controlled use of fire |
| 57. ___ | F. *Homo erectus* |
| 58. ___ | G. From a lineage that lived some time during the time span from approximately 3 million years ago until about 1.25 million years ago |
| | H. Lived 2 million years ago |
| | I. Evolved by 2 million years ago; survived until at least 53,000 years ago |
| | J. The most recent hominid to appear |
| | K. A species of *Australopithecus* |

## 28.3. EMERGENCE OF EARLY HUMANS (pp. 474–475)

## 28.4. *Focus on Science:* OUT OF AFRICA—ONCE, TWICE, OR . . . (p. 476)

*Selected Words:* Australopithecus afarensis, *cultural, "race".*

### Boldfaced, Page-Referenced Terms

(474) humans _____

_____

(474) *Homo habilis* _____

_____

(475) *Homo erectus* _____

_____

(475) *Homo sapiens* _____

_____

### Fill-in-the-Blanks

(1) _____ include all lineages that descended from the ancestral groups that had diverged from the

ancestors of chimpanzees approximately (2) _____ million years ago during the Miocene epoch.

(3) _____ such as Lucy were apelike in many skeletal details, but they walked upright like humans.

Other human features include: a large (4) _____ that provides for distinctive analytical and verbal

skills, complex social behavior, and technological innovation. Changes in the bones of the hand led to

greater (5) _____ _____ to manipulate the simple tools that the earliest humans, (6) *Homo*

_____ of Olduvai George, created as they foraged for fruits, leaves, insects, roots, and the remains of

animal carcasses.

The oldest fossils of the genus *Homo* date from approximately (7) _____ million years ago.

Between 2 million and until at least 53,000 years ago, during the Pleistocene periods of glaciation, there were

also intermittent periods of warming; during these interglacial times, a larger-brained human species,

(8) _____ _____ , migrated out of Africa and into China, Southeast Asia, and Europe. Over time,

(8) became better at toolmaking and learned how to control (9) _____ as their people adapted to a wide

range of habitats and were associated with abundant (10) _____ artifacts. The (11) _____ were a distinct hominid population that appeared 200,000 years ago in Southern France, Central Europe, and the Near East; their cranial capacity was indistinguishable from our own, they were massively built and they had a complex culture; their disappearance coincided with the appearance of anatomically modern humans.

Anatomically modern humans evolved from (12) _____ _____ . Analysis of (13) _____ _____ _____ studies and (14) _____ comparisons and specimens from the fossil record support the model of human origins which hypothesizes that *Homo erectus* migrated out of Africa approximately two million years ago, then formed distinctive subpopulations of modern humans as an outcome of genetic divergence in different geographic regions.

The oldest known fossils of early modern humans (*H. sapiens*) are from (15) _____ ; they are (16) _____ years old. From (17) _____ years ago to the present, human evolution has been almost entirely cultural rather than biological. By 30,000 years ago, there was only one remaining hominid species: (18) _____ _____ .

## Labeling

Identify each indicated part of the accompanying illustrations.

19._____          23._____

20._____          24._____ _____

21._____          25._____ _____

22._____

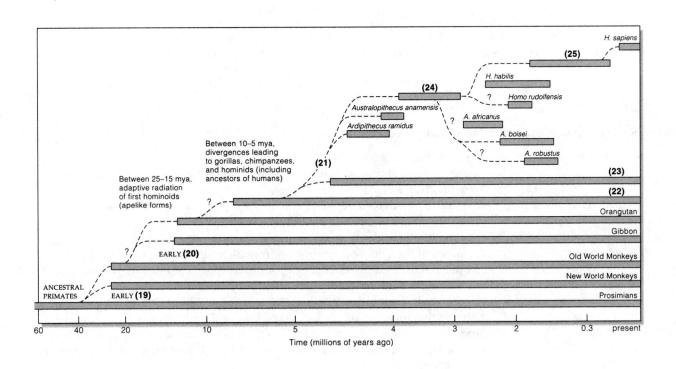

# Self-Quiz

___ 1. Which of the following is *not* considered to have been a key character in early primate evolution?
   a. eyes adapted for discerning color and shape in a three-dimensional field
   b. body and limbs adapted for tree climbing
   c. bipedalism and increased cranial capacity
   d. eyes adapted for discerning movement in a three-dimensional field

___ 2. Primitive primates generally live _____ .
   a. in tropical and subtropical forest canopies
   b. in temperate savanna and grassland habitats
   c. near rivers, lakes, and streams in the East African Rift Valley
   d. in caves where there are abundant supplies of insects

___ 3. The earliest fossils of *Homo habilis*, *Homo erectus*, and *Homo sapiens* date from approximately _____ , _____ , and _____ years ago respectively.
   a. 5, 4, and 2 million
   b. 2.5, 2, and 0.1 million
   c. 25,000, 2,000, and 100 thousand
   d. 5,000, 4,000, and 20 thousand

___ 4. The hominid evolutionary line stems from a divergence (fork in a phylogenetic tree) from the ape line that apparently occurred _____ .
   a. somewhere between 6 million and 4 million years ago
   b. about 3 million years ago
   c. during the Pliocene epoch
   d. less than 2 million years ago

___ 5. _____ was an Oligocene anthropoid that probably predated the divergence leading to Old World monkeys and the apes, with dentition more like that of dryopiths and less like that of the Paleocene primates with rodentlike teeth.
   a. *Aegyptopithecus*
   b. *Australopithecus*
   c. *Homo erectus*
   d. *Plesiadapis*

___ 6. Donald Johanson, from the University of California at Berkeley, discovered Lucy (named for the Beatles tune), who was a(n) _____ .
   a. dryopith
   b. australopith
   c. member of *Homo*
   d. prosimian

___ 7. A hominid of Europe and Asia that became extinct nearly 30,000 years ago was _____ .
   a. a dryopith
   b. *Australopithecus*
   c. *Homo erectus*
   d. Neanderthal

## Matching

Choose the one most appropriate answer for each.

8. ___anthropoids
9. ___australopiths
10. ___Cenozoic
11. ___hominids
12. ___hominoids
13. ___Miocene
14. ___primates
15. ___prosimians

A. A group that includes apes and humans
B. Organisms in a suborder that includes New World and Old World monkeys, apes, and humans
C. An era that began 65 to 63 million years ago; characterized by the evolution of birds, mammals, and flowering plants
D. A group that includes humans and their most recent ancestors
E. An epoch of the Cenozoic era lasting from 25 million to 5 million years ago; characterized by the appearance of primitive apes, whales, and grazing animals of the grasslands
F. Organisms in a suborder that includes tree shrews, lemurs, and others
G. A group that includes prosimians, tarsioids, and anthropoids
H. Bipedal organisms living from about 4 million to 1 million years ago, with essentially human bodies and ape-shaped heads; brains no larger than those of chimpanzees

*Crossword Puzzle: Primate Evolution*

## ACROSS

1. A group that includes prosimians, tarsioids, and anthropoids
3. A group that includes apes and humans
7. _____ evolution is a time of rapid branchings and adaptive radiation
10. A kind of tall, showy flower; usually lavender or purple
11. A cheek tooth that crushes and grinds.
12. All the behavior patterns of a social group passed by learning and language from generation to generation
13. Pay great homage to; worship
14. Forest apes that lived 13 million years ago in Africa, Europe and southern Asia
17. Pointed teeth that enable the tearing of flesh
18. *Homo* _____ was the first out-of-Africa hominid to control and use fire for heating and cooking
19. A group of primates intermediate between lemurs and monkeys

## DOWN*

2. Southern ape-humans, the fossils of which have dates from 3.7 to 1.25 million years ago
4. A vertebrate with hair
5. A flat chisel or conelike tooth that nips or cuts food
6. A group that includes modern humans and their direct-line ancestors since divergence from the ape line
8. The specific (species) name of modern humans
9. Habitual two-legged method of locomotion
15. _____ *Homo* used simple stone tools
16. Hard structures that can provide clues about what an animal typically eats

# Chapter Objectives/Review Questions

This section lists general and detailed chapter objectives that can be used as review questions. You can make maximum use of these items by writing answers on a separate sheet of paper. Fill in answers where blanks are provided. To check for accuracy, compare your answers with information given in the chapter or glossary.

*Page*        *Objectives/Questions*

(470)        1.  Where do tarsier survivors dwell today on Earth?

(470)        2.  Beginning with the primates most closely related to humans, list the main groups of primates in order by decreasing closeness of relationship to humans.

(470–471)    3.  Five key characters of primate evolution are _____ , _____ , _____ , _____ and _____ .

(472)        4.  Describe the general physical features and behavioral patterns attributed to early primates.

(472)        5.  Trace primate evolutionary development through the Cenozoic era. Describe how Earth's climates were changing as primates changed and adapted. Be specific about times of major divergence.

(472–474)    6.  State which anatomical features underwent the greatest changes along the evolutionary line from early anthropoids to humans.

(475–477)    7.  Explain how you think *Homo sapiens sapiens* arose. Make sure your theory incorporates existing paleontological (fossil), biochemical, and morphological data.

# Integrating and Applying Key Concepts

Suppose someone told you that some time between 12 million and 6 million years ago dryopiths were forced by larger predatory members of the cat family to flee the forests and take up residence in estuarine, riverine, and sea coastal habitats where they could take refuge in the nearby water to evade the tigers. Those that, through mutations, became naked, developed an upright stance, developed subcutaneous fat deposits as insulation, and developed a bridged nose that had advantages in watery habitats (features that other dryopiths that remained inland never developed) survived and expanded their populations. As time went on, predation by the big cats and competition with other animals for available food caused most of the terrestrial dryopiths to become extinct, but the water-habitat varieties survived as scattered remnant populations, adapting to easily available shellfish and fish, wild rice and oats, and various tubers, nuts, and fruits. It was in these aquatic habitats that the first food-getting tools (baskets, nets, and pebble tools) were developed, as well as the first words that signified different kinds of food. How does such a story fit with current speculations about and evidence of human origins? How could such a story be shown to be true or false?

# 29

# PLANT TISSUES

## Interactive Exercises

***Plants Versus the Volcano*** (pp. 480–481)

## 29.1. OVERVIEW OF THE PLANT BODY (pp. 482–483)

*Selected Words:* <u>Solanum</u> <u>esculentum</u>, *transitional* meristems, *lengthening* of stems, *thickening* of stems

*Boldfaced, Page-Referenced Terms*

(482) shoots _____

_____

(482) roots _____

_____

(482) ground tissue system _____

_____

(482) vascular tissue system _____

_____

(482) dermal tissue system _____

_____

(483) meristems _____

_____

(483) apical meristem _____

_____

(483) vascular cambium _____

_____

(483) cork cambium _____

_____

## Label-Match

Identify each part of the accompanying illustration. Choose from dermal tissues, root system, ground tissues, shoot system, and vascular tissues. Complete the exercise by matching and entering the letter of the proper description in the parentheses following each label.

1. _____ _____ (   )
2. _____ _____ (   )
3. _____ _____ (   )
4. _____ _____ (   )
5. _____ _____ (   )

A. Typically consists of stems, leaves, flowers (reproductive shoots)
B. Typically grows below ground, anchors aboveground parts, absorbs soil water and minerals, stores and releases food, and anchors the aboveground parts
C. Tissues that make up the bulk of the plant body
D. Covers and protects the plant's surfaces
E. Two conducting tissues that distribute water and solutes throughout the plant body

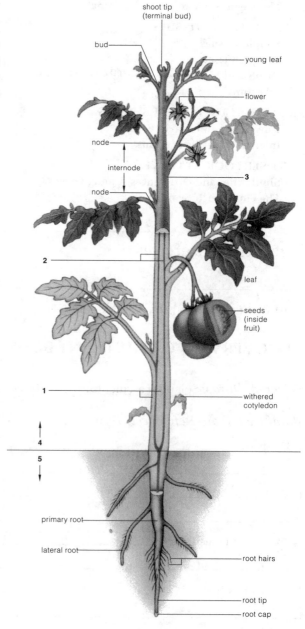

## Matching

Choose the most appropriate answer for each.

6. ___ secondary growth

7. ___ vascular cambium and cork cambium

8. ___ meristems

9. ___ apical meristems

10. ___ primary growth

11. ___ transitional meristems

A. Represented by a lengthening of stems and roots originating from cell divisions at apical meristems
B. Localized regions of embryonic, self-perpetuating cells
C. Cell populations forming from apical meristems; include protoderm, ground meristem, and procambium
D. Lateral meristems giving rise to, respectively, secondary vascular tissues, and a sturdier plant covering that replaces epidermis
E. Represented by a thickening of stems and roots due to activity of the lateral meristems
F. Located in the dome-shaped tips of all shoots and roots; responsible for lengthening those organs

## Labeling

It is important to understand the terms that identify the thin sections (slices) of plant organs and tissues that are prepared for study. Label each of the three diagrams below. Choose from radial section (cut along the radius of the organ), transverse or cross section (cuts perpendicular to the long axis of the organ), or tangential section (cuts made at right angles to the radius of the organ). Note: the white area indicates the slice.

12. _____ section

13. _____ section

14. _____ section

## Label and Match

Identify each part of the accompanying illustration. Choose from shoot apical meristem, root apical meristem, root transitional meristems, shoot transitional meristems, and lateral meristems. Complete the exercise by matching and entering the letter of the proper description in the parentheses following each label.

15. _____ _____ _____ (  )

16. _____ _____ _____ (  )

17. _____ _____ _____ (  )

18. _____ _____ _____ (  )

19. _____ _____ _____ (  )

A. Near all root tips, also gives rise to three transitional meristems from which the root's primary tissue systems develop (lengthening)

B. Embryonic descendants of the shoot apical meristem that divide, grow, and differentiate into a shoot's primary tissue systems

C. Found only inside the older stems and roots of woody plants; sources of secondary growth (increases in diameter)

D. A region of embryonic cells near the dome-shaped tip of all shoots; the source of primary growth (lengthening)

E. Embryonic descendants of the root apical meristem that divide, grow and differentiate into the root's primary tissue systems

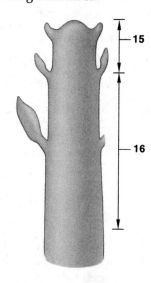

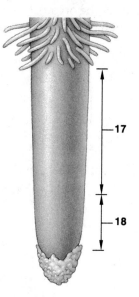

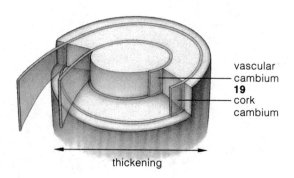

vascular cambium
**19**
cork cambium

thickening

## Complete the Table

20. The cells of the transitional meristems and the lateral meristems give rise to tissues, either primary or secondary. Complete the table below, which summarizes the activity of these meristems.

| Meristem | Tissue(s) | Primary or Secondary |
|---|---|---|
| a. Protoderm | | |
| b. Ground meristem | | |
| c. Procambium | | |
| d. Vascular cambium | | |
| e. Cork cambium | | |

## 29.2. TYPES OF PLANT TISSUES (pp. 484–485)

**Selected Words:** *fibers, sclereids, vessel members, tracheids, sieve-tube members, companion cells*

## Boldfaced, Page-Referenced Terms

(484) parenchyma _____

_____

(484) mesophyll _____

_____

(484) collenchyma _____

_____

(484) sclerenchyma _____

_____

(484) xylem _____

_____

(485) phloem _____

_____

(485) epidermis _____

_____

(485) cuticle _____

_____

(485) stoma _____

_____

(485) periderm _____

_____

(485) dicots _____

_____

(485) monocots _____

_____

## Labeling

Label the following illustrations as either collenchyma, sclerenchyma, or parenchyma. The following are helpful details and descriptions of these illustrations: collenchyma (uneven wall thickenings, often appearing like thickened corners); parenchyma (thin-walled with intercellular spaces); sclerenchyma (very thick walls in cross section, includes fibers and sclereids such as stone cells).

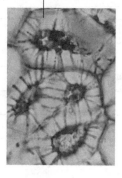

1. _____

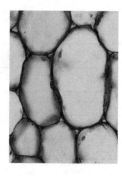

2. _____

3. _____

thick secondary wall

4. _____

5. _____

6. _____

## Choice

For questions 7–17, choose from the following:

a. parenchyma     b. collenchyma     c. sclerenchyma

7.____Patches or cylinders of this tissue are found near the surface of lengthening stems

8.____The cells have thick, lignin-impregnated walls.

9.____Cells are alive at maturity and retain the capacity to divide.

10.____Some types specialize in storage, secretion, and other tasks.

11.____Cells are mostly elongated, with unevenly thickened walls.

12.____Provides flexible support for primary tissues.

13.____Abundant air spaces are found around the cells.

14.____Supports mature plant parts and often protects seeds.

15.____Cells are thin-walled, pliable, and many-sided.

16.____Mesophyll is specialized for photosynthesis.

17.____Fibers and sclereids.

## Labeling

Identify the cell types and cellular structures in the accompanying illustrations by entering the correct name in the blanks. Complete the exercise by entering the name of the complex tissue (xylem or phloem) in which that cell or structure is found.

18. _____ (_____)

19. _____ (_____)

20. _____ (_____)

21. _____ (_____)

22. _____ (_____)

23. _____ (_____)

24. _____ (_____)

25. _____ (_____)

26. _____ (_____)

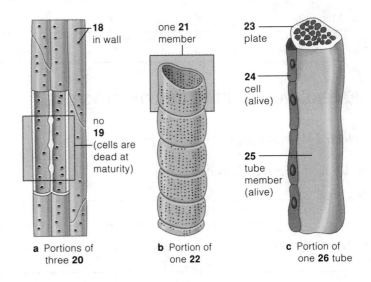

**18** in wall

no **19** (cells are dead at maturity)

**a** Portions of three **20**

one **21** member

**b** Portion of one **22**

**23** plate

**24** cell (alive)

**25** tube member (alive)

**c** Portion of one **26** tube

*Complete the Table*

27. The two types of vascular tissues, xylem and phloem, occur in strands called vascular bundles within the plant body (each of these complex tissues also has fibers and parenchyma). Complete the table below, which summarizes important information about these vascular tissues.

| Vascular Tissue | Major Conducting Cells | Cells Alive? | Function(s) |
|---|---|---|---|
| a. Xylem | | | |
| b. Phloem | | | |

28. Two types of dermal tissues cover the plant body. Complete the table below, which summarizes information about the dermal tissues.

| Dermal Tissue | Primary or Secondary Plant Body | Function(s) |
|---|---|---|
| a. Epidermis | | |
| b. Periderm | | |

29. There are two classes of flowering plants, monocots and dicots. Complete the table below, which summarizes information about the two groups of flowering plants.

| Class | Number of Cotyledons | Number of Floral Parts | Leaf Venation | Pollen Grains | Vascular Bundles |
|---|---|---|---|---|---|
| a. Monocots | | | | | |
| b. Dicots | | | | | |

## 29.3. PRIMARY STRUCTURE OF SHOOTS (pp. 486–487)

*Selected Words:* Coleus, Medicago, Zea mays

*Boldfaced, Page-Referenced Terms*

(486) bud _____

_____

(486) vascular bundles _____

_____

(486) cortex _____

_____

(486) pith _____

_____

## Labeling

Name the structures numbered in the illustrations of leaf development and leaf forms below.

1. _____
2. _____ _____
3. _____
4. _____
5. _____
6. _____
7. _____
8. _____
9. _____ _____
10. _____
11. _____
12. _____
13. _____
14. _____ _____
15. _____ _____

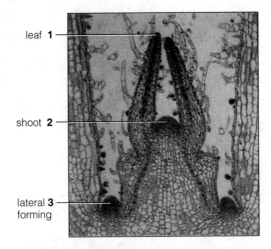

leaf **1**

shoot **2**

lateral **3** forming

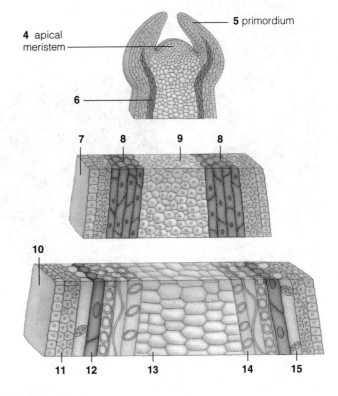

**5** primordium

**4** apical meristem

**6**

7  8  9  8

10

11  12  13  14  15

## Labeling

Name the structures numbered in the illustrations of the monocot stem below. Choose from air space, epidermis, sclerenchyma cells, vessel, vascular bundle, sieve-tube, ground tissue, and companion cell.

16. _____

17. _____  _____

18. _____  _____

19. _____  _____

20. _____  _____

21. _____  _____

22. _____  _____

23. _____  _____

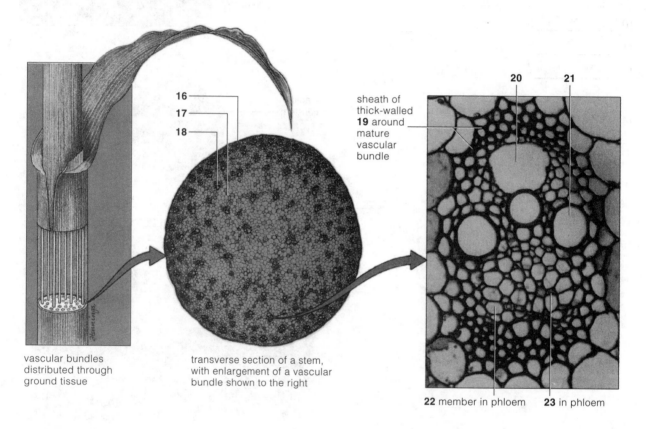

vascular bundles distributed through ground tissue

transverse section of a stem, with enlargement of a vascular bundle shown to the right

sheath of thick-walled **19** around mature vascular bundle

**20**   **21**

**22** member in phloem   **23** in phloem

## Labeling

Name the structures numbered in illustrations of the herbaceous dicot stem below.

Choose from sieve tube and companion cells in phloem, vascular bundle, xylem vessel, cortex, vascular cambium, pith, phloem fibers, and epidermis.

24. _____

25. _____

26. _____ _____

27. _____

28. _____

29. _____ _____

30. _____ _____

31. _____

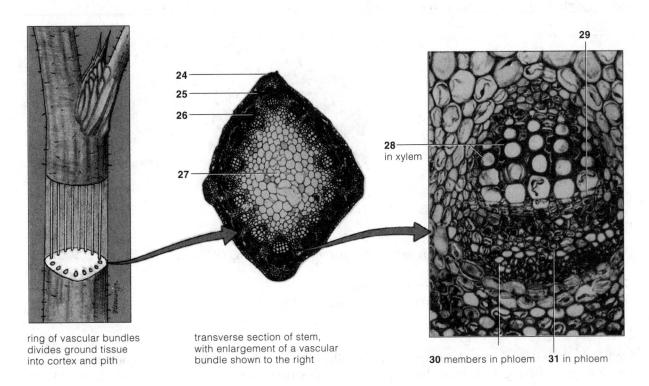

ring of vascular bundles
divides ground tissue
into cortex and pith

transverse section of stem,
with enlargement of a vascular
bundle shown to the right

**30** members in phloem    **31** in phloem

## 29.4. A CLOSER LOOK AT LEAVES (pp. 488–489)

## 29.5. *Commentary:* USES AND ABUSES OF SHOOTS (pp. 490–491)

**Selected Words:** *deciduous, simple* leaves, *compound* leaves, <u>Coleus</u>, <u>Phaseolus</u>, *palisade* mesophyll, *spongy* mesophyll

### Boldfaced, Page-Referenced Terms

(488) leaf _____

_____

(489) veins _____

_____

### Labeling

Name the structures numbered in the illustrations of leaf development and leaf forms.

1. _____
2. _____
3. _____ _____
4. _____
5. _____
6. _____

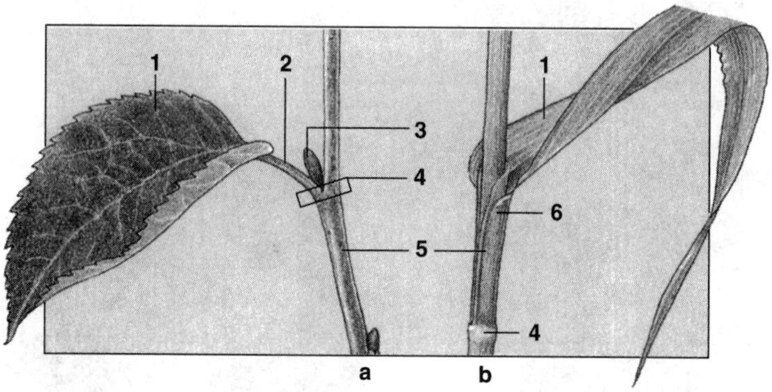

### Dichotomous Choice

Circle one of two possible answers given between parentheses in each statement.

7. The leaf illustrated on the left above is a (monocot/dicot).
8. The leaf illustrated on the right above is a (monocot/dicot).

### Label-Match

Identify each indicated part of the accompanying illustration (Figure 29.15, text). Complete the exercise by matching and entering the letter of the proper function description in the parentheses following each label.

9. _____ _____ ( )

10. _____ _____ ( )

11. _____ _____ ( )

12. _____ ( )

13. _____ _____ ( )

A. Lowermost cuticle-covered cell layer
B. Loosely packed photosynthetic parenchyma cells just above the lower epidermal layer
C. Allows movement of oxygen and water vapor out of leaves and allows carbon dioxide to enter
D. Photosynthetic parenchyma cells just beneath the upper epidermis
E. Move water and solutes to photosynthetic cells and carry products away from them

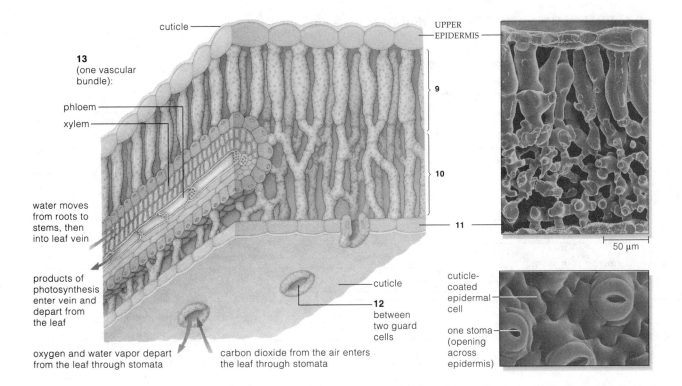

cuticle

**13**
(one vascular bundle):

phloem

xylem

water moves from roots to stems, then into leaf vein

products of photosynthesis enter vein and depart from the leaf

oxygen and water vapor depart from the leaf through stomata

carbon dioxide from the air enters the leaf through stomata

UPPER EPIDERMIS

9

10

11

cuticle

**12**
between two guard cells

50 µm

cuticle-coated epidermal cell

one stoma (opening across epidermis)

## Matching

Choose the most appropriate answer for each.

14. ___ nightshade leaves

15. ___ digitalis from foxglove leaves

16. ___ *Aloe vera* leaves

17. ___ periwinkle leaves

18. ___ century plant

19. ___ Manila hemp leaf fibers

20. ___ Panamanian palm fronds

21. ___ Mexican cockroach plants

22. ___ neem tree leaves

23. ___ tobacco plants

24. ___ marijuana

25. ___ coca leaves

26. ___ henbane

27. ___ flax fibers

A. Smoked by Mayan priests to carry their priestly thoughts to the gods
B. Kills a variety of insects, mites, and nematodes without killing natural predators
C. Soothe sun-damaged skin
D. Used to produce hats
E. Source of twine and rope
F. Produces mind-altering products; linked with low sperm counts
G. Extracts used to treat Parkinson's disease
H. Caused Hamlet's death
I. Cords and textiles
J. Kills fleas, lice, flies, and cockroaches
K. Helps stabilize heartbeat and circulation
L. Used to treat Hodgkin's disease
M. Source of cocaine that is used medicinally and as a mind-altering drug
N. Used to make linen

## 29.6. PRIMARY STRUCTURE OF ROOTS (pp. 492–493)

*Selected Words:* *adventitious* structures, <u>Salix</u>, <u>Ranunculus</u>

### Boldfaced, Page-Referenced Terms

(492) primary root _____

_____

(492) lateral roots _____

_____

(492) taproot system _____

_____

(492) fibrous root system _____

_____

(492) root hairs _____

_____

(493) vascular cylinder _____

_____

### Label-Match

Identify each indicated part of the accompanying illustration. Complete the exercise by matching the letter of the proper description in the parentheses following each label. Some choices are used more than once.

1. _____ _____ (  )

2. _____ (  )

3. _____ (  )

4. _____ (  )

5. _____ (  )

6. _____ _____ _____ (  )

7. _____ _____ (  )

8. _____ (  )

9. _____ (  )

10. _____ _____ (  )

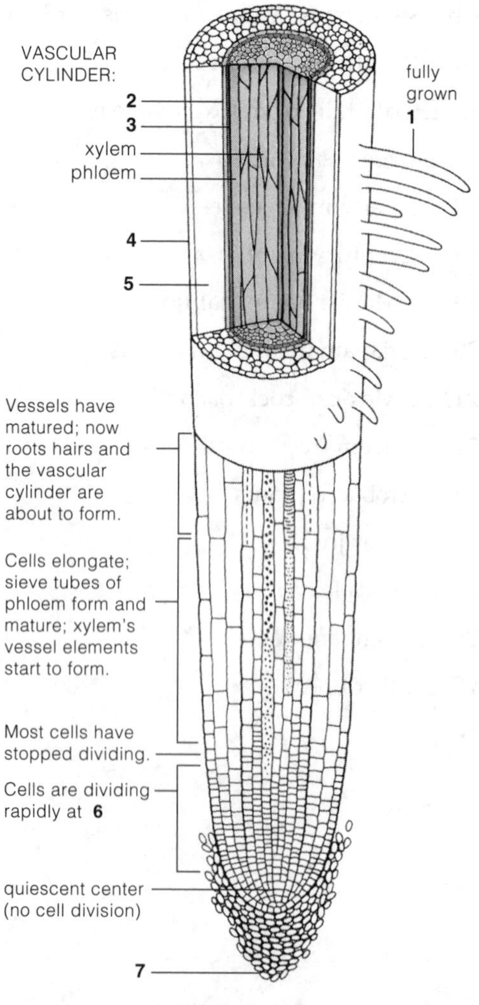

VASCULAR CYLINDER:
2
3
xylem
phloem
4
5

fully grown
1

Vessels have matured; now roots hairs and the vascular cylinder are about to form.

Cells elongate; sieve tubes of phloem form and mature; xylem's vessel elements start to form.

Most cells have stopped dividing.

Cells are dividing rapidly at **6**

quiescent center (no cell division)

7

A. Dome-shaped cell mass produced by the apical meristem
B. Part of the vascular cylinder; gives rise to lateral roots
C. Part of the vascular cylinder; transports photosynthetic products
D. Ground tissue region surrounding the vascular cylinder
E. The absorptive interface with the root's environment
F. The region of dividing cells
G. Innermost part of the root cortex; helps control water and mineral movement into the vascular column
H. Epidermal cell extensions; greatly increases the surface available for taking up water and solutes

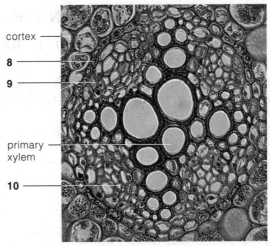

cortex
8
9

primary
xylem

10

Vascular Cylinder

## Fill-in-the-Blanks

The first part of a seedling to emerge from the seed coat is the (11) _____ root. In most dicot seedlings, the primary root increases in diameter while it grows downward. Later, (12) _____ roots begin forming in internal tissues at an angle perpendicular to the primary root's axis, then they erupt through epidermis. Oak trees, carrots, and dandelions are examples of plants whose primary root and its lateral branchings represent a(n) (13) _____ system. In monocots such as grasses, the primary root is short-lived; in its place, numerous (14) _____ roots arise from the stem of the young plant. Such roots and their branches are somewhat alike in length and diameter and form a (15) _____ root system. Some root epidermal cells send out absorptive extensions called root (16) _____ . Vascular tissues form a (17) _____ cylinder, a central column inside the root consisting of primary xylem and phloem and one or more layers of parenchyma called the (18) _____ . Ground tissues surrounding the cylinder are called the root (19) _____ . A monocot's vascular cylinder divides the ground tissue system into cortex and (20) _____ regions. There are many (21) _____ spaces in between cells of the ground tissue system, and (22) _____ can easily diffuse through them. Water entering the root moves from cell to cell until it reaches the (23) _____ , the innermost cell layer of the root cortex. Abutting walls of its cells are waterproof, so they force incoming water to pass through the cytoplasm. This arrangement helps (24) _____ the movement of water and dissolved substances into the vascular cylinder. Just inside the endodermis is the (25) _____ . This part of the vascular cylinder gives rise to (26) _____ roots. (27) _____ roots provide a plant with a tremendous surface area for absorbing water and solutes.

## 29.7. ACCUMULATED SECONDARY GROWTH—THE WOODY PLANTS (pp. 494–495)
## 29.8. A LOOK AT WOOD AND BARK (pp. 496–497)

*Selected Words:* *nonwoody* plants, *woody* plants, <u>Castanea</u>, *fusiform initials*, *ray initials*, *inner* face, *outer* face, *girdling*, <u>Acer</u> <u>saccharum</u>, *early* wood, *late* wood

### Boldfaced, Page-Referenced Terms

(494) annuals _____

_____

(494) biennials _____

_____

(494) perennials _____

_____

(496) bark _____

_____

(496) cork _____

_____

(496) heartwood _____

_____

(496) sapwood _____

_____

(497) growth rings _____

_____

(497) hardwood _____

_____

(497) softwood _____

_____

(497) compartmentalization _____

_____

*Matching*

Choose the most appropriate answer for each.

1. ___growth rings
2. ___heartwood
3. ___sapwood
4. ___compartmentalization
5. ___early wood
6. ___nonwoody plants
7. ___biennial
8. ___woody plants
9. ___softwood
10. ___girdling
11. ___bark
12. ___cork
13. ___fusiform initials
14. ___late wood
15. ___annual
16. ___hardwood
17. ___ray initials
18. ___perennials

A. Produced by cells of the cork cambium
B. Constituted by periderm and secondary phloem; includes all tissues external to the vascular cambium
C. Alternating bands of early and late wood, which reflect light differently; show secondary growth during two or more growing seasons
D. Complete the life cycle in a single growing season and are generally herbaceous
E. Produce many rays of parenchyma cells
F. Another term for herbaceous plants
G. The first xylem cells produced at the start of the growing season; tend to have large diameters and thin walls
H. Produced by conifers that lack fibers and vessels in their xylem
I. Give rise to secondary xylem and phloem which extend longitudinally through the stem
J. Term applied to the wood of dicot trees possessing vessels, tracheids, and fibers in their xylem
K. Plants that continue vegetative growth and seed formation year after year; secondary tissues are often formed
L. Xylem cells produced during the drier days of summer; cells have smaller diameters and thicker walls
M. A dumping ground at the center of older stems and roots for resins, oils, gums, and tannins
N. The total number of responses that a plant has to various attacks
O. A deliberate stripping away of a band of secondary phloem around a trunk's circumference
P. Plants that live for two growing seasons
Q. Secondary growth located between heartwood and the vascular cambium; wet, usually pale
R. Plants that add secondary growth during two or more growing seasons

## Label-Match

Identify each indicated part of the accompanying illustration. Complete the exercise by matching and entering the letter of the proper description in the parentheses following each label.

19. _____ (   )

20. _____ _____ (   )

21. _____ _____ (   )

22. _____ (   )

23. year _____ (   )

24. years _____ and _____ (   )

25. _____ _____ (   )

A. All secondary growth
B. Produced later in the growing season; vessels have smaller diameters with thick walls
C. Includes the primary growth and some secondary growth
D. Includes all tissues external to the vascular cambium
E. Large-diameter conducting cell in the xylem
F. Produced early in the growing season; vessels have large diameters and thin walls
G. Produces secondary vascular tissues, xylem and phloem

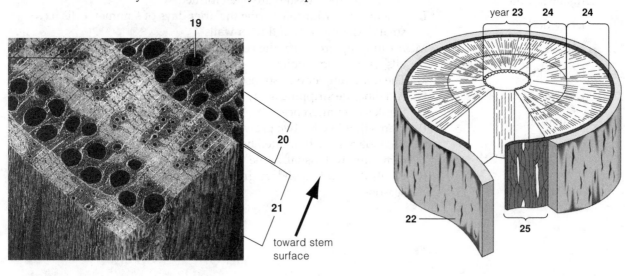

toward stem surface

---

# Self-Quiz

___ 1. _____ develops into the plant's surface layers.
   a. Ground tissue
   b. Dermal tissue
   c. Vascular tissue
   d. Pericycle

___ 2. Which of the following is *not* considered a ground cell type?
   a. Epidermis
   b. Parenchyma
   c. Collenchyma
   d. Sclerenchyma

___ 3. The _____ produces secondary xylem growth.
   a. apical meristem
   b. lateral meristem
   c. cork cambium
   d. endodermis

___ 4. The _____ is a leaflike structure that is part of the embryo; monocot embryos have one, dicot embryos have two.
   a. shoot tip
   b. root tip
   c. cotyledon
   d. apical meristem

___ 5. Leaves are differentiated and buds develop at specific points along the stem called _____ .
   a. nodes
   b. internodes
   c. vascular bundles
   d. cotyledons

___ 6. Which of the following structures is *not* considered to be meristematic?
   a. vascular cambium
   b. lateral meristem
   c. cork cambium
   d. endodermis

___ 7. Which of the following statements about monocots is *false*?
   a. They are usually herbaceous.
   b. They develop one cotyledon in their seeds.
   c. Their vascular bundles are scattered throughout the ground tissue of their stems.
   d. They have a single central vascular cylinder in their stems.

___ 8. New plants grow and older plant parts lengthen through cell divisions at _____ meristems present at root and shoot tips; older roots and stems of woody plants increase in diameter through cell divisions at _____ meristems.
   a. lateral; lateral
   b. lateral; apical
   c. apical; apical
   d. apical; lateral

___ 9. Vascular bundles called _____ form a network through a leaf blade.
   a. xylem
   b. phloem
   c. veins
   d. stomata

___10. A primary root and its lateral branchings represent a _____ system.
   a. lateral root
   b. adventitious root
   c. taproot
   d. branch root

___11. Plants whose vegetative growth and seed formation continue year after year are _____ plants.
   a. annual
   b. perennial
   c. biennial
   d. herbaceous

___12. The _____ layer of a root divides to produce lateral roots.
   a. endodermis
   b. pericycle
   c. xylem
   d. cortex

## Chapter Objectives/Review Questions

| Page | Objectives/Questions |
|---|---|
| (482) | 1. The aboveground parts of flowering plants are called _____ ; the plants' descending parts are called _____ . |
| (482) | 2. Distinguish between the ground tissue system, the vascular tissue system, and the dermal tissue system. |
| (483) | 3. Plants grow at localized regions of self-perpetuating embryonic cells called _____ . |
| (483) | 4. Lengthening of stems and roots originates at _____ meristems and all dividing tissues derived from them; this is called _____ growth. |
| (483) | 5. Cell populations of protoderm, ground meristem, and procambium are derived from the apical meristem and are known as _____ meristems. |
| (483) | 6. Increases in the diameter of a plant originate at _____ meristems. |
| (483) | 7. Describe the role of vascular cambium and cork cambium in producing secondary tissues of the plant body. |

(484)    8.  Be able to visually identify and generally describe the simple tissues called parenchyma, collenchyma, and sclerenchyma.

(484)    9.  Fibers and sclereids are both types of _____ cells.

(484)   10.  Name the cell wall compound that was necessary for the evolution of rigid and erect land plants.

(484–485) 11.  _____ tissue conducts soil water and dissolved minerals, and it mechanically supports the plant.

(485)   12.  _____ tissue transports sugars and other solutes.

(485)   13.  Name and describe the functions of the conducting cells in xylem and phloem.

(485)   14.  All surfaces of primary plant parts are covered and protected by a dermal tissue system called _____ and a surface coating called a _____ .

(485)   15.  What is the function of guard cells and stomata found within the epidermis of young stems and leaves?

(485)   16.  The cork cells of _____ replace the epidermis of stems and roots showing secondary growth.

(485)   17.  Distinguish between monocots and dicots by listing their characteristics and citing examples of each group.

(486)   18.  Leaves develop from leaf _____ located on the flanks of the apical meristem.

(486)   19.  A _____ is an undeveloped shoot of mostly meristematic tissue, often protected by scales.

(486)   20.  The primary xylem and phloem develop as vascular _____ .

(486)   21.  Distinguish between the stem's cortex and its pith.

(487)   22.  Be able to visually distinguish between monocot stems and dicot stems, as seen in cross section.

(488)   23.  Describe the principal difference between deciduous and evergreen plants.

(488)   24.  How does the "simple" leaf type differ from "compound" leaves?

(488–489) 25.  Describe the structure (cells and layers) and major functions of leaf epidermis, mesophyll, and vein tissue.

(490–491) 26.  Be able to list the names and uses of at least six plants described in the Commentary, "Uses and Abuses of Shoots."

(492)   27.  The _____ root is the first to poke through the coat of a germinating seed; later, _____ roots erupt through the epidermis.

(492)   28.  How does a taproot system differ from a fibrous root system?

(492)   29.  Define the term, *adventitious*.

(492)   30.  Describe the origin and function of root hairs.

(493)   31.  A _____ _____ consists of primary xylem and primary phloem and one or more layers of parenchyma cells called the pericycle.

(493)   32.  Describe the passage of soil water through root epidermis to the xylem of the vascular cylinder; include the role of the endodermis.

(494)   33.  Define these three categories of flowering plants: *annuals, biennials,* and *perennials*.

(494)   34.  Distinguish a woody plant from a nonwoody plant in terms of secondary growth.

(494)   35.  Each growing season, new tissues that increase the girth of woody plants originate at their _____ meristems.

(494–495) 36.  State the functions of fusiform and ray initials of the vascular cambium.

(496)   37.  Describe the formation of cork and bark.

(496–497) 38.  Distinguish early wood from late wood; heartwood from sapwood.

(497)   39.  Explain the origin of the annual growth layers (tree rings) seen in a cross-section of a tree trunk.

(497)   40.  Hardwood trees possess _____ and _____ but softwood trees lack these cells.

(497)   41.  Define the term *compartmentalization*.

## Integrating and Applying Key Concepts

Try to imagine the specific behavioral restrictions that might be imposed if the human body resembled the plant body in having (1) open growth with apical meristematic regions, (2) stomata in the epidermis, (3) cells with chloroplasts, (4) excess carbohydrates stored primarily as starch rather than as fat, and (5) dependence on the soil as a source of water and inorganic compounds.

# 30

# PLANT NUTRITION AND TRANSPORT

## Interactive Exercises

*Flies for Dinner* (pp. 500–501)

### 30.1. SOIL AND ITS NUTRIENTS (pp. 502–503)

*Selected Words:* Dionaea muscipula, Utricularia, Darlingtonia californica, *profile* properties, *macro*nutrients, *micro*nutrients

### Boldfaced, Page-Referenced Terms

(500) carnivorous plants _____

_____

(500) plant physiology _____

_____

(502) soil _____

_____

(502) humus _____

_____

(502) loams _____

_____

(502) topsoil _____

_____

(502) nutrients _____

_____

(503) leaching _____

_____

(503) erosion _____

_____

## Matching

Choose the most appropriate answer for each.

1. ___soil
2. ___humus
3. ___loams
4. ___profile properties
5. ___topsoil
6. ___nutrients
7. ___macronutrients
8. ___micronutrients
9. ___leaching
10. ___erosion

A. Elements essential for a given organism because, directly or indirectly, they have roles in metabolism that are unable to be fulfilled by any other element
B. Refers to the layered characteristics of soils, which are in different stages of development in different places
C. Refers to the organic material in soil
D. The movement of land under the force of wind, running water, and ice
E. Elements other than the macronutrients that are essential for plant growth
F. Consists of weather-formed mineral particles mixed with variable amounts of humus
G. Refers to the removal of some of the nutrients in soil as water percolates through it
H. Uppermost part of the soil that is highly variable in depth (A horizon); the most essential layer for plant growth
I. Soils having more or less equal proportions of sand, silt, and clay; best for plant growth
J. Six of the essential elements required for plant growth

## Fill-in-the-Blanks

The three essential elements that plants use as their main metabolic building blocks are oxygen, carbon, and

(11) _____ . Plant survival depends on the uptake of at least (number) (12) _____ other elements.

These are typically available to plants as dissolved (13) _____ _____ . Of these, six are present in

easily detectable concentrations in plant tissues and are known as (14) _____ . The remainder occur in

very small amounts in plant tissues and are known as (15) _____ .

## Complete the Table

16. Thirteen essential elements are available to plants as mineral ions. Complete the table below, which summarizes information about these important plant nutrients. Refer to Table 30.1 (p. 503), in the text.

| Mineral Element | Macronutrient or Micronutrient | Known Functions |
| --- | --- | --- |
| a. | | Roles in chlorophyll synthesis, electron transport |
| b. | | Activation of enzymes, role in maintaining water-solute balance |
| c. | | Role in chlorophyll synthesis; coenzyme activity |
| d. | | Role in root, shoot growth; role in photolysis |
| e. | | Role in chlorophyll synthesis; coenzyme activity |
| f. | | Component of enzyme used in nitrogen metabolism |
| g. | | Component of proteins, nucleic acids, coenzymes, chlorophyll |
| h. | | Component of most proteins, two vitamins |
| i. | | Roles in flowering, germination, fruiting, cell division, nitrogen metabolism |
| j. | | Role in formation of auxin, chloroplasts, and starch; enzyme component |
| k. | | Roles in cementing cell walls, regulation of many cell functions |
| l. | | Component of several enzymes |
| m. | | Component of nucleic acids, phospholipids, ATP |

## 30.2. ABSORPTION OF WATER AND MINERAL IONS INTO ROOTS (pp. 504–505)

*Selected Words:* Rhizobium, Bradyrhizobium

*Boldfaced, Page-Referenced Terms*

(504) vascular cylinder _____

_____

(504) endodermis _____

_____

(504) Casparian strip _____

_____

(504) exodermis _____

_____

(504) root hairs _____

_____

(505) mutualism _____

_____

(505) nitrogen fixation _____

_____

(505) root nodules _____

_____

(505) mycorrhizae _____

_____

## *Label-Match*

Identify each indicated part of the illustration below that deals with the control of nutrient uptake by plant roots. Choose from the following: epidermis, water movement, cytoplasm, root hair, vascular cylinder, endo-dermal cell wall, exodermis, endodermis, cortex, and Casparian strip. Complete the exercise by matching from the list below and entering the correct letter in the parentheses following each label.

1. _____ (   )
2. _____ _____ (   )
3. _____ (   )
4. _____ _____ (   )
5. _____ (   )
6. _____ (   )
7. _____ (   )
8. _____ _____ (   )
9. _____ _____ (   )
10. _____ _____ _____ (   )

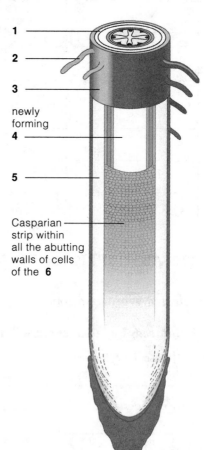

1
2
3
newly forming
4
5
Casparian strip within all the abutting walls of cells of the **6**

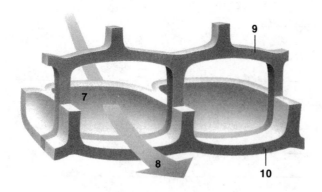

7
8
9
10

A. Cellular area through which water and dissolved nutrients must move due to Casparian strips
B. A layer of cortex cells just inside the epidermis; also equipped with Casparian strips
C. Cellular area of a root between the exodermis (if present) and the endodermis.
D. Waxy band acting as an impermeable barrier between the walls of abutting endodermal cells; forces water and dissolved nutrients through the cytoplasm of endodermal cells
E. Tiny extensions of the root epidermal cells that greatly increase absorptive capacity of a root.
F. Specific location of the waxy strips known as Casparian strips
G. Substance whose diffusion occurs through the cytoplasm of endodermal cells due to Casparian strips
H. Outermost layer of root cells.
I. A cylindrical layer of cells that wraps around the vascular column
J. Tissues include the xylem, phloem, and pericycle

## Sequence

Arrange in correct chronological sequence the path that water and nutrients take from the soil to cells in living plant tissues. Write the letter of the first step next to 11, the letter of the second step next to 12, and so on.

11. ___      A. Cortex cells lacking Casparian strips
             B. Endodermis cells with Casparian strips
12. ___      C. Root epidermis and root hairs
13. ___      D. Exodermis cells with Casparian strips (in some plants)
             E. Vascular cylinder
14. ___

15. ___

## Dichotomous Choice

Circle one of two possible answers given between parentheses in each statement.

16. A form of symbiosis, (mutualism/parasitism), refers to permanent and intimate interactions between species in which two-way benefits pass between species.
17. (Gaseous nitrogen/Nitrogen "fixed" by bacteria) represents the chemical form of nitrogen plants can use in their metabolism.
18. Nitrogen-fixing, mutualistic bacteria reside in localized swellings on legume plants known as (root hairs/root nodules).
19. Mycorrhizae represent symbiotic relationships in which fungi and the roots they cover both benefit; the roots receive (sugars and nitrogen-containing compounds/scarce minerals that the fungus is better able to absorb).
20. In a mycorrhizal interaction between a young root and a fungus, the fungus benefits by (obtaining sugars and nitrogen-containing compounds/scarce minerals that the fungus is better able to absorb).

## 30.3. A THEORY OF WATER TRANSPORT THROUGH PLANTS (pp. 506–507)

*Selected Words:* cohesion, tension

*Boldfaced, Page-Referenced Terms*

(506) transpiration _____

_____

(506) xylem _____

_____

(506) tracheids _____

_____

(506) vessel members _____

_____

(506) cohesion-tension theory _____

_____

## Fill-in-the-Blanks

The cohesion-tension theory explains (1) _____ transport to the tops of, often very high, plants. It travels pipelines in the (2) _____ , which are formed by hollow, dead, cells called (3) _____ and (4) _____ members. The process begins with the drying power of air, which causes (5) _____ , the evaporation of water from plant parts exposed to air. The collective strength of (6) _____ bonds between water molecules in the narrow, tubular xylem cells, imparts (7) _____ . This provides unbroken, fluid columns of water. The xylem water columns, as they are pulled upward, are under (8) _____ . This force extends from the veins inside leaves, down through the stems, and on into the young (9) _____ where water is being absorbed. As long as water molecules continue to escape from the plant, the continuous tension in the (10) _____ permits more molecules to be pulled upward from the roots to replace them.

## 30.4. CONSERVATION OF WATER IN STEMS AND LEAVES (pp. 508–509)

*Selected Words:* *inward* diffusion, *outward* diffusion, <u>Opuntia</u>, <u>Commelina communis</u>

### Boldfaced, Page-Referenced Terms

(508) cuticle _____

_____

(508) cutin _____

_____

(508) stomata _____

_____

(508) guard cells _____

_____

(509) turgor pressure _____

_____

(509) CAM plants _____

_____

## True-False

If the statement is true, write a T in the blank. If the statement is false, make it correct by changing the underlined word(s) and writing the correct word(s) in the answer blank.

_____ 1. Of the water moving into a leaf, <u>2 percent</u> or more is lost by evaporation into the surrounding air.

_____ 2. When evaporation exceeds water uptake by roots, plant tissues <u>wilt</u> and water-dependent activities are seriously disrupted.

_____ 3. The surfaces of all plant epidermal cell walls have an outer layer of waxes embedded in cutin, the <u>cuticle</u>, which restricts water loss, restricts inward diffusion of carbon dioxide, and limits outward diffusion of oxygen by-products.

_____ 4. Plant epidermal layers are peppered with tiny openings called <u>guard cells</u> through which water leaves the plant and carbon dioxide enters.

_____ 5. When a pair of guard cells swells with turgor pressure, the opening between them <u>closes</u>.

_____ 6. In most plants, stomata remain <u>open</u> during the daylight photosynthetic period; water is lost from plants but they gain carbon dioxide.

_____ 7. Stomata stay <u>closed</u> at night in most plants.

_____ 8. Photosynthesis starts when the sun comes up; as the morning progresses, carbon dioxide levels <u>increase</u> in cells, including guard cells.

_____ 9. As the morning progresses, a drop in carbon dioxide level within guard cells triggers an inward active transport of potassium ions; water follows by osmosis and the fluid pressure <u>closes</u> the stoma.

_____ 10. When the sun goes down and photosynthesis stops, carbon dioxide levels rise; stomata <u>close</u> when potassium, then water, moves out of the guard cells.

_____ 11. CAM plants such as cacti and other succulents open stomata during the <u>day</u> when they fix carbon dioxide by way of a special C4 metabolic pathway.

_____ 12. CAM plants use carbon dioxide in photosynthesis the following <u>night</u> when stomata are closed.

## 30.5. DISTRIBUTION OF ORGANIC COMPOUNDS THROUGH THE PLANT
(pp. 510–511)

**Selected Words:** _source, sink,_ <u>Sonchus</u>

**Boldfaced, Page-Referenced Terms**

(510) phloem _____

_____

(510) sieve tubes _____

_____

(510) companion cells _____

_____

(510) translocation _____

_____

(510) pressure flow theory _____

_____

## Fill-in-the-Blanks

Sucrose and other organic compounds resulting from (1) _____ are used throughout the plant. Most plant cells store their carbohydrates as (2) _____ in plastids. Quantities of (3) _____ and sometimes fats become stored in some fruits; (4) _____ store proteins and fats. (5) _____ molecules are too large to cross cell membranes and too insoluble to be transported to other regions of the plant body. (6) _____ are largely insoluble in water and cannot be transported from storage sites. (7) _____ proteins do not lend themselves to transport. Plant cells convert storage forms of organic compounds to (8) _____ of smaller size that are more easily transported through the phloem. For example, the cells degrade starch to glucose monomers. When one of these monomers combines with fructose, the result is (9) _____ , an easily transportable sugar. Experiments with phloem-embedded aphid mouthparts revealed that (10) _____ , an easily transportable sugar, was the most abundant carbohydrate being forced out of those tubes.

## Matching

Choose the most appropriate answer for each.

11. ___translocation
12. ___sieve tube members
13. ___companion cells
14. ___aphids
15. ___source
16. ___sink
17. ___pressure flow theory

A. Any region where organic compounds are being loaded into the sieve tubes
B. Nonconducting cells adjacent to sieve tube members that supply energy to load sucrose at the source
C. Any region of the plant where organic compounds are being unloaded from the sieve tube system and used or stored
D. Process occurring in phloem that distributes sucrose and other organic compounds through the plant (apparently under pressure)
E. Internal pressure builds up at the source end of a sieve tube system and pushes the solute-rich solution toward a sink, where they are removed
F. Passive conduits for translocation within vascular bundles; water and organic compounds flow rapidly through large pores on their end walls
G. Insects used to verify that in most plant species sucrose is the main carbohydrate translocated under pressure

# Self-Quiz

___ 1. The three elements that are present in carbohydrates, lipids, proteins, and nucleic acids are _____ .
a. oxygen, carbon, and nitrogen
b. oxygen, hydrogen, and nitrogen
c. oxygen, carbon, and hydrogen
d. carbon, nitrogen, and hydrogen

___ 2. Macronutrients are the six dissolved mineral ions that _____ .
a. play vital roles in photosynthesis and other metabolic events
b. occur in only small traces in plant tissues
c. become detectable in plant tissues
d. can function only without the presence of micronutrients
e. both a and c

___ 3. Gaseous nitrogen is converted to a plant-usable form by _____ .
a. root nodules
b. mycorrhizae
c. nitrogen-fixing bacteria
d. Venus flytraps

___ 4. _____ prevent(s) inward-moving water from moving past the abutting walls of the root endodermal cells.
a. Cytoplasm
b. Plasma membranes
c. Osmosis
d. Casparian strips

___ 5. Most of the water moving into a leaf is lost through _____ .
a. osmotic gradients being established
b. evaporation of water from plant parts exposed to air
c. pressure-flow forces
d. translocation

___ 6. Stomata remain _____ during daylight, when photosynthesis occurs, but remain _____ during the night when carbon dioxide accumulates through aerobic respiration.
a. open; open
b. closed; open
c. closed; closed
d. open; closed

___ 7. By control of _____ levels inside the guard cells of stomata, the activity of stomata is controlled when leaves are losing more water than roots can absorb.
a. oxygen
b. potassium
c. carbon dioxide
d. ATP

___ 8. Without _____ , plants would rapidly wilt and die during hot, dry spells.
a. a cuticle
b. mycorrhizae
c. phloem
d. cotyledons

___ 9. The _____ theory of water transport states that hydrogen bonding allows water molecules to maintain a continuous fluid column as water is pulled from roots to leaves.
a. pressure flow
b. cohesion-tension
c. evaporation
d. abscission

___10. Leaves represent _____ regions; growing leaves, stems, fruits, seeds, and roots represent _____ regions.
a. source; source
b. sink; source
c. source; sink
d. sink; sink

# Chapter Objectives/Review Questions

| Page | Objectives/Questions |
|---|---|
| (500) | 1. Explain how carnivorous plants such as Venus flytraps and bladderworts accomplish their nutritional needs. |
| (500) | 2. _____ _____ is the study of adaptations by which plants function in their environment. |
| (502) | 3. Define *soil*. |
| (502) | 4. Plants do best in _____ , which are the soils having more or less equal proportions of sand, silt, and clay. |
| (502) | 5. Name the soil layer that is the most essential for plant growth. |
| (502) | 6. Define *nutrients* in terms of plant nutrition. |
| (502) | 7. Name the three elements considered essential for plant nutrition. |
| (503) | 8. Distinguish between macronutrients and micronutrients in relation to their role in plant nutrition. |
| (503) | 9. _____ refers to the removal of some of the nutrients in soil as water percolates through it. |
| (503) | 10. The movement of land under the force of wind, running water, and ice is called _____ . |

(504)  11.  Be able to trace the path of water and mineral ions into roots; name the structure and function of plant structures involved.

(504)  12.  Differentiate between the endodermis and the exodermis of the root cortex.

(504)  13.  Due to the presence of the _____ strip in the walls of endodermal cells, water can move into the vascular cylinder only by crossing the plasma membrane and diffusing through the cytoplasm.

(504)  14.  Explain why root hairs are so valuable in root absorption.

(505)  15.  Define *mutualism*.

(505)  16.  Describe the roles of root nodules and mycorrhizae in plant nutrition.

(506)  17.  The evaporation of water from leaves as well as from stems and other plant parts is known as _____ .

(506)  18.  Henry Dixon's _____-_____ theory explains how water moves upward in an unbroken column through xylem to the tops of tall trees.

(506)  19.  In a plant's vascular tissues, water moves through pipelines called _____ .

(506–507)) 20.  Be able to give the key points of Dixon's explanation of upward water transport in plants.

(508)  21.  Even mildly stressed plants would rapidly wilt and die if it were not for the _____ covering their parts; describe additional functions of this structure.

(508)  22.  Describe the chemical compounds found in cutin.

(508)  23.  Evaporation from plant parts occurs mostly at _____ , tiny epidermal passageways of leaves and stems.

(509)  24.  Explain the mechanism by which stomata open during daylight and close during the night.

(509)  25.  Define *turgor pressure* and explain its role in controlling water loss at stomata.

(509)  26.  Describe the mechanisms by which CAM plants conserve water.

(510)  27.  Describe the role of phloem sieve tube members and companion cells in translocation.

(510)  28.  _____ is the main form in which sugars are transported through most plants.

(510)  29.  Define *translocation*.

(510)  30.  According to the _____ _____ theory, pressure builds up at the source end of a sieve tube system and pushes solutes toward a sink, where they are removed.

(510)  31.  Companion cells supply the _____ that loads sucrose at the source.

(511)  32.  Name the gradients responsible for continuous flow of organic compounds through phloem.

---

# Integrating and Applying Key Concepts

How do you think maple syrup is made from maple trees? Which specific systems of the plant are involved, and why are maple trees tapped only at certain times of the year?

# 31

# PLANT REPRODUCTION

## Interactive Exercises

*A Coevolutionary Tale* (pp. 514–515)

## 31.1. REPRODUCTIVE STRUCTURES OF FLOWERING PLANTS (pp. 516–517)

## 31.2. *Focus on The Environment:* POLLEN SETS ME SNEEZING (p. 517)

*Selected Words:* <u>Angraecum</u> <u>sesquipedale</u>, <u>Prunus</u>, <u>Rosa</u>, *sperma*, *perfect* flowers, *imperfect* flowers, *allergic rhinitis*

*Boldfaced, Page-Referenced Terms*

(514) flower _____

_____

(514) coevolution _____

_____

(514) pollinator _____

_____

(516) sporophyte _____

_____

(516) gametophytes _____

_____

(517) stamens _____

_____

(517) pollen grains _____

_____

(517) carpels _____

_____

(517) ovary _____

_____

## Label-Match

Identify each indicated part of the accompanying illustration.
Complete the exercise by matching and entering the letter of
the proper description in the parentheses following each label.

1. _____ (   )

2. _____ (   )

3. _____ (   )

4. _____ (   )

5. _____ (   )

6. _____ (   )

A. An event that produces a young sporophyte
B. A reproductive shoot produced by the sporophyte
C. The "plant"; a vegetative body that develops from a zygote
D. Cellular division event occurring within flowers to produce
   spores
E. Produces haploid eggs by mitosis
F. Produces haploid sperm by mitosis

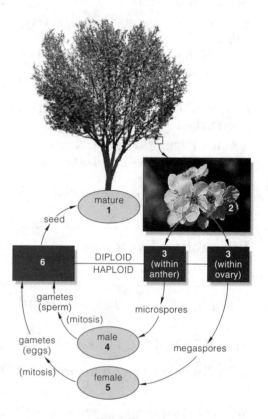

## Labeling

Identify each indicated part of the accompanying illustration.

7. _____

8. _____

9. _____

10. _____

11. _____

12. _____

13. _____

14. _____

15. _____

16. _____

17. _____

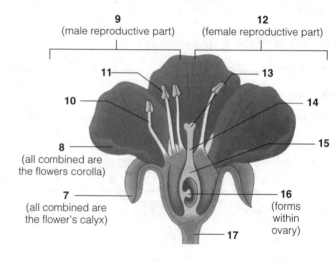

## Matching

Choose the most appropriate answer for each.

18. ___sepals
19. ___petals
20. ___stamens
21. ___ovule
22. ___pollen grain
23. ___carpels
24. ___ovary
25. ___perfect flowers
26. ___imperfect flowers

A. Have both male and female parts
B. Mature haploid spore whose contents develop into a male gametophyte
C. Collectively, the flower's "corolla"
D. Have male or female parts, but not both
E. Female reproductive part; includes stigma, style, and ovary
F. Structure that matures to become a seed
G. Found just inside the flower's corolla, the male reproductive parts
H. Lower portion of the carpel where egg formation, fertilization, and seed development occur
I. Outermost leaflike whorl of floral organs; collectively, the calyx

## 31.3. A NEW GENERATION BEGINS (pp. 518–519)

*Selected Words:* Prunus

*Boldfaced, Page-Referenced Terms*

(518) microspores _____

_____

(518) ovule _____

_____

(518) megaspores _____

_____

(518) endosperm _____

_____

(518) pollination _____

_____

(519) double fertilization _____

_____

## Fill-in-the-Blanks

The numbered items on the illustration below represent missing information; complete the blanks of the following narrative to supply that information.

Within each (1) _____ , mitotic divisions produce four masses of spore-forming cells, each mass forming within a (2) _____ _____ . Each one of these diploid cells is known as a (3) _____ _____ cell and undergoes (4) _____ to produce four haploid (5) _____ . Mitosis within each haploid microspore results in a two-celled haploid body, the immature male gametophyte. One of these cells will give rise to a (6) _____ _____ ; the other cell will develop into a (7) _____-_____ cell. Mature microspores are eventually released from the pollen sacs of the anther as (8) _____ . Pollination occurs and after the pollen lands on a (9) _____ of a carpel, the pollen tube develops from one of the cells in the pollen grain; the other cell within the pollen grain divides to form two sperm cells. As the pollen tube grows through the carpel tissues, it contains the two sperm cells and a tube nucleus. The pollen tube with its included two sperm cells and the tube nucleus is known as the mature (10) _____ _____ .

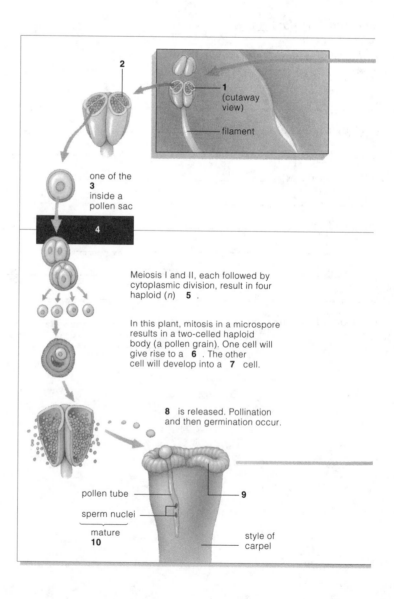

## Fill-in-the-Blanks

The numbered items on the illustration below represent missing information; complete the numbered blanks in the narrative below to supply that information.

In the carpel of the flower, one or more dome-shaped, diploid tissue masses develop on the inner wall of the ovary. Each mass is the beginning of an (11) _____ . A tissue forms inside a domed mass as it grows, and one or two protective layers called (12) _____ form around it. Inside each mass, a diploid cell (the megaspore mother cell) divides by (13) _____ to form four haploid spores known as (14) _____ . Commonly, all but one (15) _____ disintegrates. The remaining (16) _____ undergoes (17) _____ three times without cytoplasmic division. At first, this structure is a cell with (18) _____ haploid nuclei. Cytoplasmic division results in a seven-cell (19) _____ _____ which represents the mature (20) _____ _____ . Six of those cells have a single nucleus, but one cell has (21) _____ (number) nuclei (2n) and represents the (22) _____ mother cell (n + n). Another

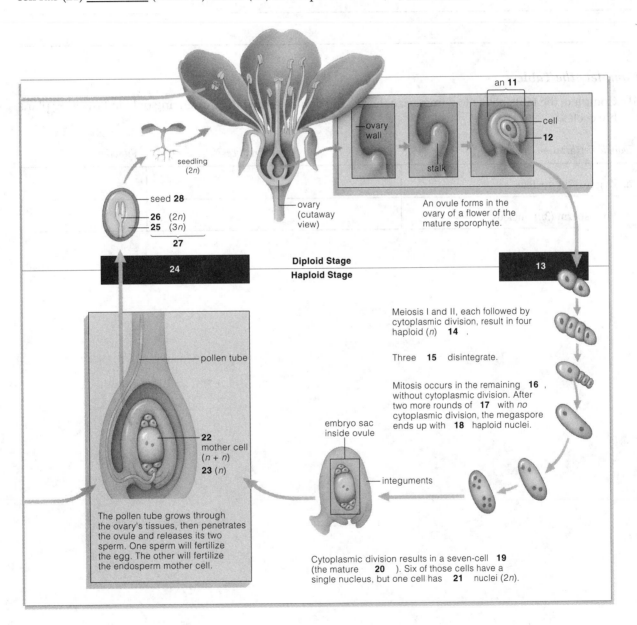

an **11**

ovary wall

cell

**12**

seedling (2n)

seed **28**

**26** (2n)
**25** (3n)

**27**

ovary (cutaway view)

An ovule forms in the ovary of a flower of the mature sporophyte.

**Diploid Stage**

**24**

**Haploid Stage**

**13**

Meiosis I and II, each followed by cytoplasmic division, result in four haploid (n) **14** .

Three **15** disintegrate.

pollen tube

Mitosis occurs in the remaining **16** , without cytoplasmic division. After two more rounds of **17** with *no* cytoplasmic division, the megaspore ends up with **18** haploid nuclei.

embryo sac inside ovule

**22**
mother cell (n + n)
**23** (n)

integuments

The pollen tube grows through the ovary's tissues, then penetrates the ovule and releases its two sperm. One sperm will fertilize the egg. The other will fertilize the endosperm mother cell.

Cytoplasmic division results in a seven-cell **19** (the mature **20** ). Six of those cells have a single nucleus, but one cell has **21** nuclei (2n).

haploid cell within the embryo sac is the (23) _____ . Following (24) _____ _____ with one sperm, the n + n cell will help form the 3$n$ (25) _____ , a nutritive tissue for the forthcoming embryo. The other sperm involved in this unique fertilization process fertilizes the haploid egg; this combination forms the diploid (26) _____ . Thus, the ovule is transformed to a (27) _____ that is composed of three parts, a seed (28) _____ , an embryo, and nourishment for the embryo, the endosperm.

## Short Answer

29. What guides the growth of the pollen tube down through the female floral tissues toward the chamber holding the egg? _____

_____

30. Describe the site of double fertilization known only in flowering plants. _____

_____

_____

## Complete the Table

31. Complete the table below to summarize the unique double fertilization occurring only in flowering plant life cycles.

| Double Fertilization Products | Origin | Produces? | Function |
|---|---|---|---|
| a. Zygote (2$n$) nucleus | | | |
| b. Endosperm (3$n$) nucleus | | | |

## 31.4. FROM ZYGOTE TO SEEDS AND FRUITS (pp. 520–521)

*Selected Words:* <u>Capsella</u>, <u>Fragaria</u>, <u>Malus</u>, <u>Ananas</u> <u>comosus</u>, *endo*carp, *meso*carp, *exo*carp

### Boldfaced, Page-Referenced Terms

(520) fruit _____

_____

(520) cotyledons _____

_____

(521) seed _____

_____

(521) pericarp _____

_____

### Complete the Table

1. Complete the following table, which summarizes concepts associated with seeds and fruits.

| Structure | Origin |
|---|---|
| a. Cotyledons | |
| b. Seeds | |
| c. Seed Coat | |
| d. Fruit | |

## Labeling

Identify each indicated part of the accompanying illustration.

2. _____

3. _____

4. _____

5. _____

6. _____

7. _____  _____

8. _____  _____

9. _____

10. _____

11. _____

12. _____  _____

13. _____

14. _____

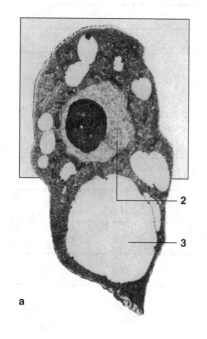

a

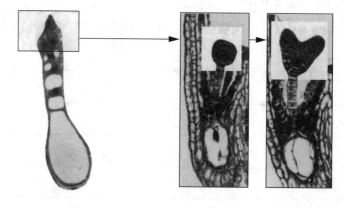

b Upper part of **4** gives rise to embryo

c Globular **5** stage

d Heart-shaped stage of **6**

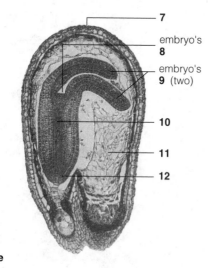

**7**

embryo's **8**

embryo's **9** (two)

**10**

**11**

**12**

e

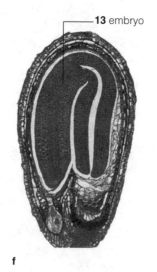

**13** embryo

f

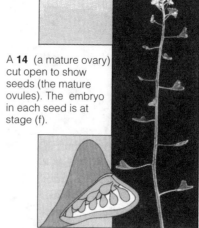

A **14** (a mature ovary) cut open to show seeds (the mature ovules). The embryo in each seed is at stage (f).

g

## Fill-in-the-Blanks

Following fertilization, the newly formed zygote initiates a course of (15) _____ cell divisions that lead to a mature embryo (16) _____ . The embryo develops as part of an (17) _____ and is accompanied by the formation of a (18) _____ , a mature ovary.

By the time a *Capsella* embryo is near maturity, two (19) _____ , or seed leaves have begun to develop from two lobes of meristematic tissue. Dicot embryos, such as *Capsella*, have (20) _____ (number) cotyledons and monocot embryos have (21) _____ (number) cotyledon. Like most dicots, the *Capsella* embryo has rather thick cotyledons that absorb nutrients from the (22) _____ of the seed and stores them in its cotyledons. In corn, wheat, and most other monocots, endosperm is not tapped until the seed (23) _____ . Digestive enzymes become stockpiled inside the (24) _____ cotyledons of monocot embryos. When the (25) _____ do become active, nutrients stored in the endosperm will be released and transferred to the growing (26) _____ .

From the time a zygote forms until a mature embryo develops, a parent sporophyte plant transfers nutrients to tissues of the (27) _____ . Food reserves accumulate in (28) _____ or cotyledons. Eventually, the ovule separates from the (29) _____ wall, and its integuments thicken and harden into a seed (30) _____ . The embryo, food reserves, and seed coat are a self-contained package—a (31) _____ , which is defined as a mature (32) _____ . While seeds are forming, changes in other parts of the flower are forming (33) _____ . They may be fleshy or (34) _____ , they may consist of one or more (35) _____ , and they may incorporate tissues besides those of the ovary, such as tissues of the (36) _____ . Botanists commonly refer to three divisions of a (37) _____ fruit, although sometimes they are not clearly visible. The innermost portion of a fruit that surrounds the seed is the (38) _____ . The (39) _____ is the fleshy portion, and the (40) _____ is the skin. Together, these three regions of the fruit are called a (41) _____ .

## Choice

For questions 42–48, refer to Table 31.1 on page 520 of the text, and choose from the following:

a. simple, wall dry; splits at maturity    b. simple, wall dry; intact at maturity
c. simple, fleshy; often leathery    d. aggregate
e. multiple    f. accessory, simple    g. accessory, aggregate

42. ____ apple, pear

43. ____ pea, magnolia, mustard

44. ____ strawberry

45. ____ sunflower, wheat, rice, maple

46. ____ pineapple, fig, mulberry

47. ____ grape, banana, lemon, cherry

48. ____ blackberry, raspberry

## 31.5. DISPERSAL OF FRUITS AND SEEDS (p. 522)

## 31.6. *Focus on Science:* WHY SO MANY FLOWERS AND SO FEW FRUITS? (p. 523)

*Selected Words:* seed dispersal, Acer

### Choice

For questions 1–11, choose from the following:

a. wind-dispersed fruit       b. fruits dispersed by animals       c. water-dispersed fruits

1. ____heavy wax coats
2. ____coconut palms
3. ____seed coats assaulted by digestive enzymes to assist in releasing embryos
4. ____maples
5. ____orchids
6. ____air sacs
7. ____saguaro
8. ____hooks, spines, hairs, and sticky surfaces
9. ____cacao
10. ____dandelions
11. ____cockleburs, bur clover, and bedstraw

## 31.7. ASEXUAL REPRODUCTION OF FLOWERING PLANTS (pp. 524–525)

*Selected Words:* Populus tremuloides, Larrea divaricata, Daucus carota

### Boldfaced, Page-Referenced Terms

(524) vegetative growth _____

_____

(524) parthenogenesis _____

_____

(524) tissue culture propagation _____

_____

## Matching

Using Table 31.2 on page 524 of the text as a reference, match the following asexual reproductive modes of flowering plants.

1. ___corm
2. ___bulb
3. ___parthenogenesis
4. ___runner
5. ___vegetative propagation on modified stems
6. ___rhizome
7. ___tuber
8. ___tissue culture propagation (induced propagation)
9. ___vegetative growth

A. In a general sense, new plants develop from tissues or organs that drop or separate from parent plants
B. New shoots arise from axillary buds (enlarged tips of slender underground rhizomes)
C. New plants arise from cells in parent plant that were not irreversibly differentiated; a laboratory technique
D. New plants arise at nodes of underground horizontal stem
E. New plant arises from axillary bud on short, thick vertical underground stem
F. New plants arise at nodes on an aboveground horizontal stem
G. New bulb arises from an axillary bud on a short underground stem
H. Involves asexual reproduction utilizing runners, rhizomes, corms, tubers, and bulbs
I. Embryo develops without nucleus or cellular fusion

## Matching

Choose the most appropriate example for each modified stem.

10. ___bulb
11. ___rhizome
12. ___tuber
13. ___runner
14. ___corm

A. Potato
B. Strawberry
C. Gladiolus
D. Onion, lily
E. Bermuda grass

---

# Self-Quiz

___ 1. The joint evolution of flowers and their pollinators is known as _____ .
a. adaptation
b. coevolution
c. joint evolution
d. covert evolution

___ 2. A stamen is _____ .
a. composed of a stigma
b. the mature male gametophyte
c. the site where microspores are produced
d. part of the vegetative phase of an angiosperm

___ 3. The portion of the carpel that contains an ovule is the _____ .
a. stigma
b. anther
c. style
d. ovary

___ 4. The phase in the life cycle of plants that gives rise to spores is known as the _____ .
a. gametophyte
b. embryo
c. sporophyte
d. seed

___ 5. A gametophyte is _____ .
a. a gamete-producing plant
b. haploid
c. both a and b
d. the plant produced by the fusion of gametes

___ 6. A characteristic of a seed is that it _____ .
a. contains an embryo sporophyte
b. represents an arrested growth stage
c. is covered by hardened and thickened integuments
d. all of these

___ 7. An immature fruit is a(n) _____ and an immature seed is a(n) _____ .
  a. ovary; megaspore
  b. ovary; ovule
  c. megaspore; ovule
  d. ovule; ovary

___ 8. In flowering plants, one sperm nucleus fuses with that of an egg, and a zygote forms that develops into an embryo. Another sperm fuses with _____ .
  a. a primary endosperm cell to produce three cells, each with one nucleus
  b. a primary endosperm cell to produce one cell with one triploid nucleus
  c. both nuclei of the endosperm mother cell, forming a primary endosperm cell with a single triploid nucleus
  d. one of the smaller megaspores to produce what will eventually become the seed coat

___ 9. "Simple, aggregate, multiple, and accessory," refer to types of _____ .
  a. carpels
  b. seeds
  c. fruits
  d. ovaries

___10. "When a leaf falls or is torn away from a jade plant, a new plant can develop from the leaf, from meristematic tissue." This statement refers to _____ .
  a. parthenogenesis
  b. runners
  c. tissue culture propagation
  d. vegetative propagation

# Chapter Objectives/Review Questions

| Page | | Objectives/Questions |
|---|---|---|
| (514) | 1. | _____ refers to two or more species jointly evolving as an outcome of close ecological interactions. |
| (514) | 2. | Describe the role of a pollinator. |
| (516) | 3. | Be able to distinguish between sporophytes and gametophytes. |
| (516) | 4. | _____ are shoots on sporophytes that are specialized for reproduction. |
| (516) | 5. | Be able to identify the various parts of a typical flower and state their functions. |
| (517) | 6. | ____ reproduction requires formation of gametes, followed by fertilization. |
| (517) | 7. | Walled microspores form in pollen sacs and develop into sperm-producing bodies called _____ _____ . |
| (517) | 8. | A _____ is another name for a pistil. |
| (517) | 9. | The lower portion of either single or fused carpels is the _____ . |
| (517) | 10. | Distinguish between a flower that is *perfect* and one that is *imperfect*. |
| (517) | 11. | Describe the condition called *allergic rhinitis*. |
| (518–519) | 12. | Relate the sequence of events and structures involved that give rise to microspores and megaspores. |
| (518) | 13. | _____ is the transfer of pollen grains to a receptive stigma. |
| (518–519) | 14. | What structures represent the male gametophyte and female gametophyte in flowering plants? List the contents of each. |
| (519) | 15. | The endospore mother cell in the embryo sac is composed of two _____ . |
| (519) | 16. | Describe the *double fertilization* that occurs uniquely in the flowering plant life cycle. |
| (519) | 17. | How is endosperm formed? What is the function of endosperm? |
| (520–521) | 18. | Describe the formation of the embryo sporophyte; give the origin and formation of seeds and fruits. |
| (520) | 19. | Review the general types of fruits produced by flowering plants (text, Table 31.1). |
| (521) | 20. | Name and describe the three regions of a fruit; together, these three regions are called the _____ . |
| (522) | 21. | Seeds and fruits are structurally adapted for _____ by air currents, water currents, and many kinds of animals. |

(524–525) 22. Distinguish between parthenogenesis, vegetative propagation, and tissue culture propagation; cite an example for each.

(524)    23. Be able to list representative plant examples of a runner, a rhizome, a corm, a tuber, and a bulb.

## Integrating and Applying Key Concepts

In terms of botanical morphology, a flower is interpreted as "a reproductive shoot bearing organs." Try to list structural evidence, not discussed in the chapter, that botanists might have discovered that led them to this view.

# 32

# PLANT GROWTH AND DEVELOPMENT

## Interactive Exercises

*Foolish Seedlings and Gorgeous Grapes* (pp. 528–529)

## 32.1. PATTERNS OF EARLY GROWTH AND DEVELOPMENT—AN OVERVIEW
(pp. 530–531)

***Selected Words:*** Gibberella fujikuroi, Eschscholtzia californica, Vitus, Phaseolus vulgaris, Zea mays

***Boldfaced, Page-Referenced Terms***

(528) gibberellin _____

_____

(528) hormones _____

_____

(530) germination _____

_____

(530) imbibition _____

_____

## Fill-in-the-Blanks

Before or after seed dispersal, the growth of the (1) _____ sporophyte idles. For seeds, (2) _____ is the resumption of growth by an immature stage in the life cycle after a time of arrested development. Germination depends on (3) _____ factors, such as soil temperature, moisture and oxygen levels, and the number of seasonal daylight hours available. By a process known as (4) _____ , water molecules move into the seed. As more water moves in, the seed swells, and its coat (5) _____ . Once the seed coat splits, more oxygen reaches the embryo, and (6) _____ respiration moves into high gear. The embryo's (7) _____ cells begin to divide rapidly. In general, the (8) _____ meristem is the first to be activated. Its meristematic descendants divide, elongate, and give rise to the (9) _____ _____ . When this structure breaks through the seed coat, (10) _____ is over. For both monocots and dicots, the patterns of germination, growth, and development that unfold have a (11) _____ basis; they are dictated by the plant's (12) _____ . All cells in the new plant arise from the same cell, the (13) _____ . Thus, all cells inherit the same (14) _____ instructions. Unequal (15) _____ divisions between daughter cells lead to differences in their (16) _____ and output. Activities in daughter cells start to vary as a result of (17) _____ gene expression. As an example, genes governing the synthesis of growth-stimulating (18) _____ are activated in some cells but not others. (19) _____ among genes, hormones, and the environment govern how an individual plant grows and develops.

## Labeling

Identify each numbered part of the accompanying illustration.

20. _____    27. _____    34. _____

21. _____    28. _____    35. _____

22. _____    29. _____    36. _____

23. _____    30. _____    37. _____

24. _____    31. _____    38. _____

25. _____    32. _____    39. _____

26. _____    33. _____

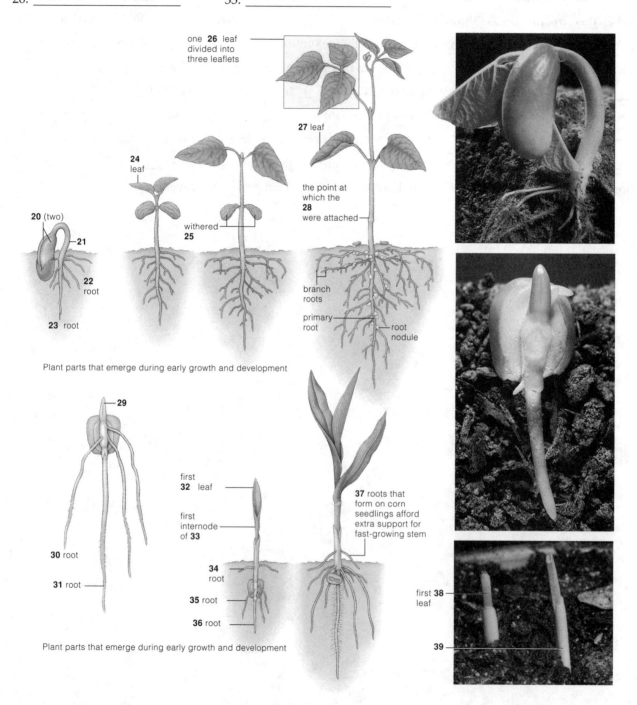

one **26** leaf divided into three leaflets

**27** leaf

the point at which the **28** were attached

**24** leaf

**20** (two)

**21**

withered **25**

**22** root

branch roots

primary root

root nodule

**23** root

Plant parts that emerge during early growth and development

**29**

first **32** leaf

first internode of **33**

**37** roots that form on corn seedlings afford extra support for fast-growing stem

**30** root

**31** root

**34** root

**35** root

**36** root

first **38** leaf

**39**

Plant parts that emerge during early growth and development

## 32.2. HORMONAL EFFECTS ON PLANT GROWTH AND DEVELOPMENT (pp. 532–533)

*Selected Words:* *quantitative* terms, *qualitative* terms

### Boldfaced, Page-Referenced Terms

(532) growth _____

_____

(532) development _____

_____

(532) auxins _____

_____

(532) cytokinins _____

_____

(532) ethylene _____

_____

(532) abscisic acid _____

_____

(532) coleoptile _____

_____

(533) herbicides _____

_____

(533) apical dominance _____

_____

### Choice

For questions 1–20, choose from the following; also refer to Table 32.1 on page 533 of the text.

a. auxins      b. gibberellins      c. cytokinins      d. abscisic acid      e. ethylene

1. ____ Ancient Chinese burned incense to hurry fruit ripening

2. ____ Natural and synthetic versions are used to prolong the shelf life of cut flowers, lettuces, mushrooms, and other vegetables

3. ____ Orchardists spray trees with IAA to thin out overcrowded seedlings in the spring

4. ____ Inhibits cell growth, promotes bud dormancy, and prevents seeds from germinating prematurely; causes stomata to close and help conserve plant water

5. ____ IAA, the most important naturally occurring compound of its type

6. ____ Exposed to oranges and other citrus fruits to brighten their color of their rind before being displayed in the market

7. ____ Stimulates stem lengthening; influence plant responses to gravity and light and promote coleoptile elongation

8. ___Causes stems to lengthen; help buds and seeds break dormancy and resume growth in the spring

9. ___Used to prevent premature fruit drop—all fruit can be picked at the same time

10. ___Stimulate cell division

11. ___Used by food distributors to ripen green fruit after shipment

12. ___Most abundant in root and shoot meristems and in the tissues of maturing fruit

13. ___Under its influence, fruit ripens and flowers, fruits, and leaves drop away from plants at prescribed times of the year

14. ___Promote leaf expansion and retard leaf aging

## Matching

Choose the most appropriate answer for each.

15. ___2,4-D

16. ___hormone

17. ___apical dominance

18. ___herbicide

19 ___growth

20. ___"foolish seedling" effect on rice plants

21. ___IAA

22. ___development

23. ___2,4,5-T

24. ___coleoptile

25. ___target cell

A. Mixed with 2,4-D to produce Agent Orange for defoliation in Viet Nam
B. The emergence of specialized, morphologically different body parts; measured in qualitative terms
C. Gibberellin from fungal extracts
D. A cell with receptors to bind a given signaling molecule
E. Hormonal effect that inhibits lateral bud growth, which promotes stem elongation
F. Synthetic auxin used as an herbicide to selectively kill broadleaf weeds that compete with valuable crop plants
G. An increase in the number, size, and volume of cells; measured in quantitative cells
H. A signaling molecule released from one cell that changes the activity of target cells
I. The most important naturally occurring auxin
J. Any synthetic auxin compound used to selectively kill plants
K. A sheath around the primary shoot of grass seedlings, such as corn plants

## 32.3. ADJUSTMENTS IN THE RATE AND DIRECTION OF GROWTH (pp. 534–535)

*Selected Words:* trope, auxein, thigma

*Boldfaced, Page-Referenced Terms*

(534) plant tropism _____

_____

(534) gravitropism _____

_____

(534) statoliths _____

_____

(534) phototropism _____

_____

(535) flavoprotein _____

_____

(535) thigmotropism _____

_____

(535) vines _____

_____

(535) tendrils _____

_____

## Choice

For questions 1–10, choose from the following:

a. phototropism      b. gravitropism      c. thigmotropism      d. mechanical stress

  1.____More intense sunlight on one side of a plant—stems curve toward the light

  2.____Vines climbing around a fencepost as they grow upward

  3.____Plants grown outdoors have shorter stems than plants grown in a greenhouse

  4.____A root turned on its side will curve downward

  5.____Briefly shaking a plant daily inhibits the growth of the entire plant

  6.____A potted seedling turned on its side—the growing stem curves upward

  7.____Leaves turn until their flat surfaces face light

  8.____Flavoprotein may be a central component

  9.____Tendrils are sometimes involved

10.____Generally, statoliths are involved

## 32.4. BIOLOGICAL CLOCKS AND THEIR EFFECTS (pp. 536–537)

*Selected Words:* circadian

*Boldfaced, Page-Referenced Terms*

(536) biological clocks _____

_____

(536) phytochrome _____

_____

(536) circadian rhythm _____

_____

(536) photoperiodism _____

_____

(536) long-day plants _____

_____

(536) short-day plants _____

(536) day-neutral plants _____

## Matching

Choose the most appropriate response for each.

1. ___photoperiodism
2. ___Pr
3. ___day-neutral plants
4. ___phytochrome activation
5. ___"long-day" plants
6. ___circadian rhythms
7. ___Pfr
8. ___biological clocks
9. ___rhythmic leaf movements
10. ___phytochrome

A. Biological activities that recur in cycles of twenty-four hours or so
B. Flower in spring when daylength exceeds a critical value
C. Internal time-measuring mechanisms with roles in adjusting daily activities
D. Any biological response to a change in the relative length of daylight and darkness in the 24-hour cycle; active Pfr may be an alarm button for this process
E. A blue-green pigment that absorbs red or far-red wavelengths, with different results
F. An example of a circadian rhythm
G. Active form of phytochrome
H. Flower when mature enough to do so
I. Inactive form of phytochrome
J. May induce plant cells to take up free calcium ions or induce certain plant cell organelles to release them

## Labeling

Identify each indicated part of the accompanying illustration.

11.___
12.___
13.___
14.___
15.___

red light

| 11 |  | 12 |  | 15 |
(inactive)  far-red light  (active)  Growth of plant part is promoted or inhibited.

**13** reverts to **14** in the dark.

## Fill-in-the-Blanks

Photoperiodism is especially apparent in the (16) _____ process, which is often keyed to day length changes throughout the year. "Long-day plants" flower in spring when day length becomes (17) [choose one] □ shorter, □ longer than some critical value. "Short-day plants" flower in late summer or early autumn when daylength becomes (18) [choose one] □ shorter, □ longer than some critical value. "Day-neutral plants" flower whenever they become (19) _____ enough to do so without regard to daylength. Spinach is a (20) _____ - _____ plant because it will not flower and produce seeds unless it is exposed to fourteen hours of light every day for two weeks. Cocklebur is termed a (21) _____ - _____ plant because it flowers after a single night that is longer than 8-1/2 hours. (22) _____ - _____ plants flower when mature enough to do so.

## 32.5. LIFE CYCLES END, AND TURN AGAIN (pp. 538–539)

## 32.6. *Focus on Science:* THE RISE AND FALL OF A GIANT (p. 540)

*Selected Words:* <u>Acer</u>, *vernalis,* <u>Secale</u> <u>cereale</u>, <u>Quercus</u> <u>agrifolia</u>, <u>Armillaria</u>

### Boldfaced, Page-Referenced Terms

(538) abscission _____

_____

(538) senescence _____

_____

(538) dormancy _____

_____

(539) vernalization _____

_____

### Choice

For questions 1–12 choose from the following:

a. senescence     b. abscission     c. vernalization     d. entering dormancy     e. breaking dormancy

1. ____ Dropping of leaves or other parts from a plant

2. ____ A process at work between fall and spring; temperatures become milder, and rain and nutrients become available again

3. ____ Strong cues are short days, long, cold nights, and dry, nitrogen-deficient soil

4. ____ The process proceeds at tissues in the base of leaves, flowers, fruits, or other plant parts; a special zone is involved

5. ____ The sum total of processes leading to the death of plant parts or the whole plant

6. ____ A recurring cue for this process is a decrease in daylength

7. ____ When a plant stops growing under conditions that appear quite suitable for growth

8. ____ Unless buds of some biennials and perennials are exposed to low winter temperatures, flowers will not form on their stems when spring rolls around

9. ____ The forming of ethylene in cells near the breakpoints may trigger the process

10. ____ Stopping nutrient drain to reproductive parts blocks this process

11. ____ Keeping germinating seeds of winter rye at near-freezing temperature to induce flowering the following summer

12. ____ Many perennial and biennial plants start to shut down growth as autumn approaches and days grow shorter.

13. ____ Postponement can be effected when gardeners remove flower buds from plants to maintain vegetative growth

# Self-Quiz

___ 1. Promoting fruit ripening and abscission of leaves, flowers, and fruits is a function ascribed to _____ .
a. gibberellins
b. ethylene
c. abscisic acid
d. auxins

___ 2. Auxins _____ .
a. cause flowering
b. promote stomatal closure
c. promote cell division
d. promote cell elongation in coleoptiles and stems

___ 3. _____ is demonstrated by a germinating seed whose first root always curves down while the stem always curves up.
a. Phototropism
b. Photoperiodism
c. Gravitropism
d. Thigmotropism

___ 4. Light of _____ wavelengths is the main stimulus for phototropism.
a. blue
b. yellow
c. red
d. green

___ 5. Plants whose leaves are open during the day but fold them at night are exhibiting a(n) _____ .
a. growth movement
b. circadian rhythm
c. biological clock
d. both b and c are correct

___ 6. 2,4-D, a potent dicot weed killer, is a synthetic _____ .
a. auxin
b. gibberellin
c. cytokinin
d. phytochrome

___ 7. All the processes that lead to the death of a plant or any of its organs are called _____ .
a. dormancy
b. vernalization
c. abscission
d. senescence

___ 8. Phytochrome is converted to an active form, _____ , at sunrise and reverts to an inactive form, _____ , at sunset, night or in the shade.
a. Pr; Pfr
b. Pfr; Pfr
c. Pr; Pr
d. Pfr; Pr

___ 9. Which of the following is not promoted by the active form of phytochrome?
a. Seed germination
b. Flowering
c. Leaf expansion
d. Stem elongation

___10. When a perennial or biennial plant stops growing under conditions suitable for growth, it has entered a state of _____ .
a. senescence
b. vernalization
c. dormancy
d. abscission

# Chapter Objectives/Review Questions

| Page | | Objectives/Questions |
|---|---|---|
| (528) | 1. | _____ are signaling molecules. |
| (530) | 2. | _____ is a resumption of growth after a time of arrested embryonic development. |
| (530) | 3. | Define *imbibition* and describe its role in germination. |
| (530) | 4. | List the environmental factors that influence germination. |
| (530) | 5. | The primary _____ breaks through the seed coat first. |
| (530) | 6. | The basic patterns of growth and development are heritable, dictated by the plant's _____ . |
| (531) | 7. | Compare and contrast the major features of the growth and development of a monocot plant and a dicot plant. |
| (532) | 8. | _____ of a multicelled organism means the number, size, and volume of cells increase. |
| (532) | 9. | Explain why plant growth is measured in quantitative terms and plant development in qualitative terms. |
| (532) | 10. | Describe the general role of plant hormones. |
| (532) | 11. | _____ promote cell elongation in coleoptiles and stems; involved in phototropism and gravitropism. |
| (532) | 12. | _____ promote stem elongation; help break dormancy of seeds and buds that resume growth in the spring. |
| (533) | 13. | Synthetic auxins are used as _____ , compounds that kill some plants but not others. |
| (533) | 14. | _____ promote cell division; promote leaf expansion and retard leaf aging. |
| (533) | 15. | _____ _____ promotes stomatal closure; promotes bud and seed dormancy. |
| (533) | 16. | _____ promotes fruit ripening; promotes abscission of leaves, flowers, and fruits. |
| (533) | 17. | Describe the form of growth inhibition known as apical dominance. |
| (534) | 18. | Define phototropism, gravitropism, and thigmotropism and cite examples of each. |
| (534) | 19. | Explain how statoliths form the basis for gravity-sensing mechanisms. |
| (535) | 20. | Plants make the strongest phototropic response to light of _____ wavelengths; _____ is a yellow pigment molecule that absorbs blue wavelengths. |
| (535) | 21. | Provide an example of how mechanical stress can affect plants. |
| (536) | 22. | Plants have internal time-measuring mechanisms called biological _____ . |
| (536) | 23. | The alarm button for some biological clocks in plants is the blue-green pigment molecule _____ . |
| (536) | 24. | What are circadian rhythms? Give an example. |
| (536) | 25. | Phytochrome is converted to an active form, _____ , at sunrise, when red wavelengths dominate the sky. It reverts to an inactive form, _____ , at sunset, at night, or even in shade, where far-red wavelengths predominate. |
| (536) | 26. | _____ is a biological response to a change in the relative length of daylight and darkness in a 24-hour cycle. |
| (536) | 27. | _____ serves as a switching mechanism in the biological clock governing flowering responses. |
| (536) | 28. | Describe the photoperiodic responses of "long-day," "short-day," and "day-neutral" plants. |
| (538) | 29. | _____ is the dropping of leaves, flowers, fruits, or other plant parts. |
| (538) | 30. | Describe the events that signal plant senescence. |
| (538) | 31. | List environmental cues that send a plant into dormancy? |
| (539) | 32. | The low-temperature stimulation of flowering is called _____ . |
| (539) | 33. | What conditions are instrumental in the dormancy-breaking process? |

# Integrating and Applying Key Concepts

An oak tree has grown up in the middle of a forest. A lumber company has just cut down all of the surrounding trees except for a narrow strip of woods that includes the oak. How will the oak be likely to respond as it adjusts to its changed environment? To what new stresses will it be exposed? Which hormones will most probably be involved in the adjustment?

# 33

# TISSUES, ORGAN SYSTEMS, AND HOMEOSTASIS

*Meerkats, Humans, It's All the Same*

**EPITHELIAL TISSUE**
General Characteristics
Cell-to-Cell Contacts
Glandular Epithelium

**CONNECTIVE TISSUE**
Connective Tissue Proper
Specialized Connective Tissue

**MUSCLE TISSUE**

**NERVOUS TISSUE**

*Focus on Science:* FRONTIERS IN TISSUE RESEARCH

**ORGAN SYSTEMS**
Overview of Major Organ Systems
Tissue and Organ Formation

**HOMEOSTASIS AND SYSTEMS CONTROL**
Concerning the Internal Environment
Mechanisms of Homeostasis

---

## Interactive Exercises

---

*Meerkats, Humans, It's All the Same* (pp. 544–545)

### 33.1. EPITHELIAL TISSUE (pp. 546–547)

In addition to the boldfaced terms, the text features other important terms essential to understanding the assigned material. "Selected Words" is a list of these terms, which appear in the text in italics, in quotation marks, and occasionally in roman type. Latin binomials found in this section are underlined and in roman type to distinguish them from other italicized words.

*Selected Words: anatomy, physiology,* Suricata suricatta, *simple epithelium, stratified epithelium,* Dendrobates

### Boldfaced, Page-Referenced Terms

The page-referenced terms are important; they are in boldface type in the chapter. Write a definition for each term in your own words without looking at the text. Next, compare your definition with that given in the chapter or in the text glossary. If your definition seems inaccurate, allow some time to pass and repeat this procedure until you can define each term rather quickly (how fast you can answer is a gauge of your learning effectiveness).

(544) internal environment _____

_____

(544) homeostasis _____

_____

(544) tissue _____

_____

(544) organ _____

_____

(545) organ system _____

_____

(545) division of labor _____

_____

(546) epithelium _____

_____

(546) tight junctions _____

_____

(546) adhering junctions _____

_____

(546) gap junctions _____

_____

(547) exocrine glands _____

_____

(547) endocrine glands _____

_____

## Matching

Choose the most appropriate answer for each.

1. ___ internal environment
2. ___ anatomy
3. ___ physiology
4. ___ homeostasis
5. ___ tissue
6. ___ organ
7. ___ organ system
8. ___ division of labor

A. Consists of two or more organs that are interacting physically, chemically, or both in a common task
B. How the body functions
C. Consists of interstitial fluid (tissue fluids) and blood that bathes the living cells of any complex animal
D. Consists of different tissues that are organized in specific proportions and patterns
E. Cells, tissues, organs, and organ systems, split up the work in ways that contribute to the survival of the animal as a whole
F. How the animal body is structurally put together
G. An interactive group of cells and intercellular substances that take part in one or more particular tasks
H. With respect to the animal body, refers to stable operating conditions in the internal environment

## Fill-in-the-Blanks

(9) _____ tissue has a free surface, which faces either a body fluid or the outside environment.

(10) _____ _____ has a single layer of cells and functions as a lining for body cavities, ducts, and tubes. (11) _____ _____ has two or more layers and typically functions in protection, as it does in

the skin. (12) _____ junctions are strands of proteins that help stop substances from leaking across a tissue. (13) _____ junctions cement cells together. (14) _____ junctions help communicate by promoting the rapid transfer of ions and small molecules among them. (15) _____ junctions in the epithelium of your stomach help prevent a condition called (16) _____ ulcer. (17) _____ glands secrete mucus, saliva, earwax, milk, oil, digestive enzymes, and other cell products. These products are usually released onto a free (18) _____ surface through ducts or tubes. (19) _____ glands lack ducts; their products are (20) _____ , which are secreted directly into the fluid bathing the gland. Typically, the (21) _____ picks up the hormone molecules and distributes them to target cells elsewhere in the body.

## 33.2. CONNECTIVE TISSUE (pp. 548–549)
## 33.3. MUSCLE TISSUE (p. 550)
## 33.4. NERVOUS TISSUE (p. 551)
## 33.5. *Focus on Science:* FRONTIERS IN TISSUE RESEARCH (p. 551)

*Selected Words:* plasma, contract, striated, laboratory-grown epidermis, designer organs, type I diabetes mellitus

*Boldfaced, Page-Referenced Terms*

(548) loose connective tissue _____

_____

(548) dense, irregular connective tissue _____

_____

(548) dense, regular connective tissue _____

_____

(548) cartilage _____

_____

(549) bone _____

_____

(549) adipose tissue _____

_____

(549) blood _____

_____

(550) skeletal muscle tissue _____

_____

(550) smooth muscle tissue _____

_____

(550) cardiac muscle tissue _____

_____

(551) nervous tissue _____

_____

(551) neurons _____

_____

(551) neuroglia _____

_____

## Choice

For questions 1–10, choose from the following types of connective tissue proper:

a. loose     b. dense, irregular     c. dense regular

1. ___ Contains fibers, mostly collagen-containing ones, and a few fibroblasts.

2. ___ Rows of fibroblasts often intervene between the bundles.

3. ___ Has its fibers and cells loosely arranged in a semifluid ground substance.

4. ___ Has parallel bundles of many collagen fibers and resists being torn apart.

5. ___ Forms protective capsules around organs that do not stretch much.

6. ___ Often serves as a support framework for epithelium.

7. ___ Found in tendons, which attach skeletal muscle to bones.

8. ___ Besides fibroblasts, it contains infection-fighting white blood cells.

9. ___ Found in elastic ligaments, which attach bones to each other.

10. ___ It is also present in the deeper part of skin.

## Complete the Table

11. Supply the name of the specialized connective tissue after reading each description.

| Specialized Connective | Description |
|---|---|
| a. | Chockfull of large fat cells; stores excess carbohydrates and proteins; richly supplied with blood |
| b. | Intercellular material, solid yet pliable, resists compression; structural models for vertebrate embryo bones; maintains shape of nose, outer ear, and other body parts; cushions joints |
| c. | Derived mainly from connective tissue, has transport functions; circulating within plasma are a great many red blood cells, white blood cells, and platelets |
| d. | The weight-bearing tissue of vertebrate skeletons, which support or protect softer tissues and organs; mineral-hardened with calcium-salt laden collagen fibers and ground substance; interact with skeletal muscles attached to them |

## Dichotomous Choice

Circle one of two possible answers given between parentheses in each statement.

12. Contractile cells of (skeletal/smooth) muscle tissue taper at both ends.
13. The contractile walls of the heart are composed of (striated/cardiac) muscle tissue.
14. Walls of the stomach and intestine contain (smooth/skeletal) muscle tissue.
15. The only muscle tissue attached to bones is (skeletal/smooth).
16. (Smooth/Skeletal) muscle cells are bundled together in parallel.
17. "Involuntary" muscle action is associated with (smooth/skeletal) muscle tissue.
18. Striated means (bundled/striped).
19. (Smooth/Skeletal) muscle tissue has sheath of tough connective tissue enclosing several bundles of muscle cells.
20. The function of smooth muscle tissue is to (pump blood/move internal organs).
21. Cell junctions fuse together the plasma membranes of (smooth/cardiac) muscle cells.
22. (Muscle/Nervous) tissue exerts the greatest control over the body's responsiveness to changing conditions.
23. Excitable cells are the (neuroglia/neurons).
24. (Neuroglia/Muscle) cells protect and structurally and metabolically support the neurons.
25. When a (neuron/muscle cell) is suitably stimulated, an electrical "message" travels over its plasma membrane that may result in stimulation of other cells of the same type or of other types.
26. Different types of (neuroglia/neurons) detect specific stimuli, integrate information, and issue or relay commands for response.
27. The lives of people with type I diabetes mellitus might, in the future, be made more normal with (a designer organ/a sheet of laboratory-grown epidermis).

## Label and Match

Label each of the illustrations below with one of the following: connective, epithelial, muscle, nervous, or gametes. Complete the exercise by writing *all* appropriate letters and numbers from each group below in the parentheses following each label.

28. _____ (   )
29. _____ (   )
30. _____ (   )
31. _____ (   )
32. _____ (   )
33. _____ (   )
34. _____ (   )
35. _____ (   )
36. _____ (   )
37. _____ (   )
38. _____ (   )
39. _____ (   )
40. _____ (   )

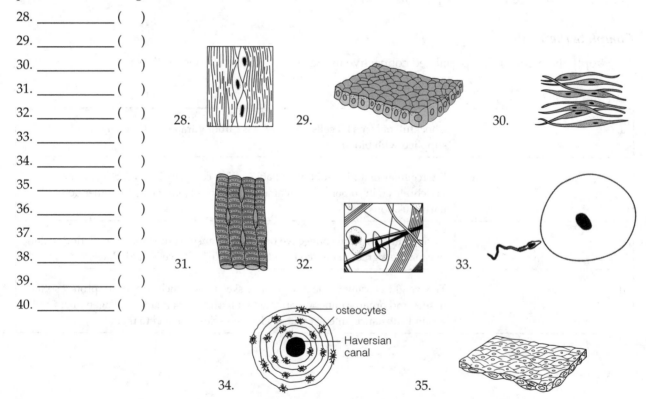

28.     29.     30.

31.     32.     33.

— osteocytes

— Haversian canal

34.     35.

A. Adipose
B. Bone
C. Cardiac
D. Dense, regular
E. Loose
F. Simple columnar
G. Simple cuboidal
H. Simple squamous
I. Smooth

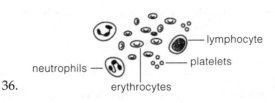

36.

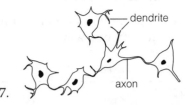

37.

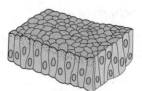

38.

1. Absorption
2. Maintain diploid number of chromosomes in sexually reproducing populations
3. Communication by means of electrical signals
4. Energy reserve
5. Contraction for voluntary movements
6. Diffusion
7. Padding
8. Contract to propel substances along internal passageways; not striated
9. Attaches muscle to bone and bone to bone
10. In vertebrates, provides the strongest internal framework of the organism
11. Elasticity
12. Secretion
13. Pumps circulatory fluid; striated
14. Insulation
15. Transport of nutrients and waste products to and from body cells

39.

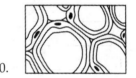

40.

## 33.6. ORGAN SYSTEMS (pp. 552–553)

*Selected Words: midsagittal* plane, *frontal* plane, *transverse* plane, *anterior, posterior, superior, inferior, distal, proximal*

### Boldfaced, Page-Referenced Terms

(553) ectoderm _____

_____

(553) mesoderm _____

_____

(553) endoderm _____

_____

### Fill-in-the-Blanks

The brain is housed in the (1) _____ cavity.  The (2) _____ cavity contains the spinal cord and the

beginnings of spinal nerves.  The heart and lungs are found within the (3) _____ cavity.  The stomach,

spleen, liver, gallbladder, pancreas, small intestine, most of the large intestine, the kidneys, and ureters lie

inside the (4) _____ cavity.  The (5) _____ cavity contains the urinary bladder, sigmoid colon,

rectum, and reproductive organs

## Complete the Table

6. Supply the name of the primary tissue of the embryo that does the job indicated by becoming specialized in particular ways.

| Primary Tissue | Functions |
|---|---|
| a. | Forms internal skeleton and muscle, circulatory, reproductive, and urinary systems |
| b. | Forms inner lining of gut and linings of major organs formed from the embryonic gut |
| c. | Forms outer layer of skin and the tissues of the nervous system |

## Labeling

Identify each numbered part of the accompanying illustration that reviews the directional terms and planes of symmetry for the human body.

7. _____

8. _____

9. _____

10. _____

11. _____

12. _____

13. _____

14. _____

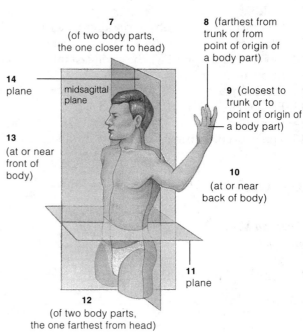

7 (of two body parts, the one closer to head)

8 (farthest from trunk or from point of origin of a body part)

9 (closest to trunk or to point of origin of a body part)

14 plane

midsagittal plane

13 (at or near front of body)

10 (at or near back of body)

11 plane

12 (of two body parts, the one farthest from head)

## Labeling

Label each organ system described.

15. _____ Picks up nutrients absorbed from gut and transports them to cells throughout body

16. _____ Helps cells use nutrients by supplying them with oxygen and relieving them of $CO_2$ wastes

17. _____ Helps maintain the volume and composition of body fluids that bathe the body's cells

18. _____ Provides basic framework for the animal and supports other organs of the body

19. _____ Uses chemical messengers to control and guide body functions

20. _____ Protects the body from viruses, bacteria, and other foreign agents

21. _____ Produces younger, temporarily smaller versions of the animal

22. _____ Breaks down larger food molecules into smaller nutrient molecules that can be absorbed by body fluids and transported to body cells

23. _____ Consists of contractile parts that move the body through the environment and propel substances about in the animal

24. _____ Serves as an electrochemical communications system in the animal's body

25. _____ In the meerkat, served as a heat catcher in the morning and protective insulation at night

## Matching

Match the most appropriate function with each system shown in the illustrations below.

26. ___Male: production and transfer of sperm to the female. Female: production of eggs; provision of a protected nutritive environment for developing embryo and fetus. Both systems have hormonal influences on other organ systems.

27. ___Ingestion of food, water; preparation of food molecules for absorption; elimination of food residues from the body

28. ___Movement of internal body parts; movement of whole body; maintenance of posture; heat production

29. ___Detection of external and internal stimuli; control and coordination of responses to stimuli; integration of activities of all organ systems

30. ___Protection from injury and dehydration; body temperature control; excretion of some wastes; reception of external stimuli; defense against microbes

31. ___Provisioning of cells with oxygen; removal of carbon dioxide wastes produced by cells; pH regulation

32. ___Support, protection of body parts; sites for muscle attachment, blood cell production, and calcium and phosphate storage

33. ___Hormonal control of body functioning; works with nervous system in integrative tasks

34. ___Maintenance of the volume and composition of extracellular fluid

35. ___Rapid internal transport of many materials to and from cells; helps stabilize internal temperature and pH

36. ___Return of some extracellular fluid to blood; roles in immunity (defense against specific invaders of the body)

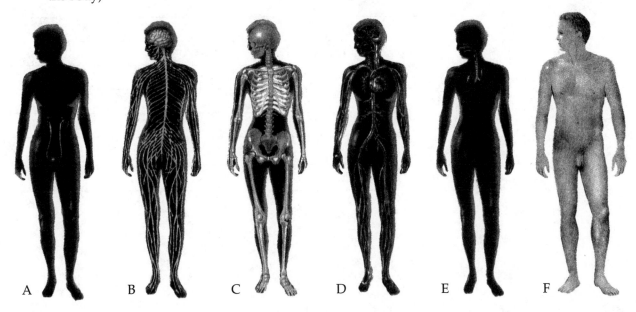

A           B           C           D           E           F

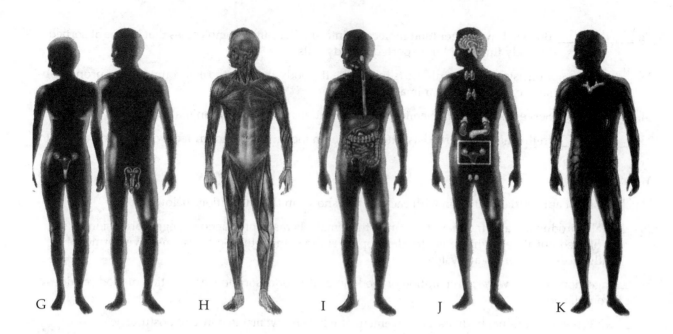

G          H         I         J         K

## 33.7. HOMEOSTASIS AND SYSTEMS CONTROL (pp. 554–555)

*Selected Words:* interstitial, plasma, intensify

### Boldfaced, Page-Referenced Terms

(554) extracellular fluid _____

_____

(554) sensory receptors _____

_____

(554) stimulus _____

_____

(554) integrator _____

_____

(554) effectors _____

_____

(554) negative feedback mechanism _____

_____

(555) positive feedback mechanism _____

_____

## Matching

Choose the most appropriate answer for each.

1. ___extracellular fluid
2. ___interstitial fluid
3. ___plasma
4. ___sensory receptors
5. ___stimulus
6. ___integrator
7. ___effectors
8. ___negative feedback mechanism
9. ___positive feedback mechanism

A. An activity alters a condition in the internal environment, and this triggers a response that reverses the altered condition
B. The fluid portion of blood
C. An example is the brain, a central command post where different bits of information are pulled together in the selection of a response
D. The fluid not inside cells
E. A specific change in the environment
F. Examples are muscles and glands
G. Fluid that occupies spaces between cells and tissues
H. Sets in motion a chain of events that intensify a change from an original condition—and after a limited time, the intensification reverses the change
I. Cells or cell parts that can detect a stimulus

## Fill-in-the-Blanks

The component parts of any animal work together to maintain the (10) _____ fluid environment required by all of its living cells. (11) _____ refers to stable operating conditions in the internal environment. Feedback (12) _____ mechanisms help maintain physical and chemical aspects of the body's internal environment within ranges that are most favorable for individual cell activities. In a (13) _____ feedback mechanism, a chain of events is set in motion that intensifies the original condition before returning to a set point; sexual arousal and childbirth are two examples. Generally, physiological controls work by means of (14) _____ feedback, in which an activity changes some condition in the internal environment, and the change causes the condition to be reversed; the maintenance of body temperature close to a "set point" is an example. Your brain is a(n) (15) _____ a control point where different bits of information are pulled together in the selection of a response. Muscles and (16) _____ are examples of effectors. Internal temperature control of a husky running around on a hot day is achieved by (17) _____ in the skin and elsewhere sensing an increasing temperature change in the dog's body and relaying the neural information to an integrator. In this example, the (18) _____ is the integrator. The husky searches for shade and rests under a tree. Many effectors carry out the specific responses. One example is the increased salivary gland secretions that increases evaporation from the dog's tongue. This has a (19) _____ effect.

# Self-Quiz

___ 1. Which of the following is not included in connective tissues?
   a. Bone
   b. Blood
   c. Cartilage
   d. Skeletal muscle

___ 2. A surrounding material within which something originates, develops, or is contained is known as a _____
   a. lamella
   b. ground substance
   c. plasma
   d. lymph

___ 3. Blood is considered to be a(n) _____ tissue.
   a. epithelial
   b. muscular
   c. connective
   d. none of these

___ 4. _____ are abundant in tissues of the heart and liver where they promote diffusion of ions and small molecules from cell to cell.
   a. Adhesion junctions
   b. Filter junctions
   c. Gap junctions
   d. Tight junctions

___ 5. Muscle that is not striped and is involuntary is _____
   a. cardiac
   b. skeletal
   c. striated
   d. smooth

___ 6. Chemical and structural bridges link groups or layers of like cells, uniting them in structure and function as a cohesive _____ .
   a. organ
   b. organ system
   c. tissue
   d. cuticle

___ 7. A fish embryo was accidentally stabbed by a graduate student in developmental biology. Later, the embryo developed into a creature that could not move and had no supportive or circulatory systems. Which embryonic tissue had suffered the damage?
   a. ectoderm
   b. endoderm
   c. mesoderm
   d. protoderm

___ 8. A tissue whose cells are striated and fused at the ends by cell junctions so that the cells contract as a unit is called _____ tissue.
   a. smooth muscle
   b. dense fibrous connective
   c. supportive connective
   d. cardiac muscle

___ 9. The secretion of tears, milk, sweat, and oil are functions of _____ tissues.
   a. epithelial
   b. loose connective
   c. lymphoid
   d. nervous

___10. Memory, decision making, and issuing commands to effectors are functions of _____ tissue.
   a. connective
   b. epithelial
   c. muscle
   d. nervous

___11. An animal that feels heated from the sun moves to an environment that tends to cool its body. This is an example of _____ .
   a. intensifying an original condition
   b. positive feedback mechanism
   c. positive phototropic response
   d. negative feedback mechanism

___12. Which group is arranged correctly from smallest structure to largest?
   a. muscle cells, muscle bundle, muscle
   b. muscle cells, muscle, muscle bundle
   c. muscle bundle, muscle cells, muscle
   d. none of the above

## Matching

Choose the most appropriate answer for each term.

13. ___circulatory system

14. ___digestive system

15. ___endocrine system

16. ___immune system

17. ___integumentary system

18. ___muscular system

19. ___nervous system

20. ___reproductive system

21. ___respiratory system

22. ___skeletal system

23. ___urinary system

A. Picks up nutrients absorbed from gut and transports them to cells throughout body

B. Helps cells use nutrients by supplying them with oxygen and relieving them of $CO_2$ wastes

C. Helps maintain the volume and composition of body fluids that bathe the body's cells

D. Provides basic framework for the animal and supports other organs of the body

E. Uses chemical messengers to control and guide body functions

F. Protects the body from viruses, bacteria, and other foreign agents

G. Produces younger, temporarily smaller versions of the animal

H. Breaks down larger food molecules into smaller nutrient molecules that can be absorbed by body fluids and transported to body cells

I. Consists of contractile parts that move the body through the environment and propel substances about in the animal

J. Serves as an electrochemical communications system in the animal's body

K. In the meerkat, served as a heat catcher in the morning and protective insulation at night

This section lists general and detailed chapter objectives that can be used as review questions. You can make maximum use of these items by writing answers on a separate sheet of paper. Fill in answers where blanks are provided. To check for accuracy, compare your answers with information given in the chapter or glossary.

# Chapter Objectives/Review Questions

| Page(s) | | Objectives/Questions |
|---|---|---|
| (544–545) | 1. | Explain how the meerkat maintains a rather constant internal environment in spite of changing external conditions. |
| (544–545) | 2. | Cells are the basic units of life; in a multicellular animal, like cells are grouped into a(n) _____ , and these are organized in specific proportions and patterns that compose a(n) _____ . |
| (544, 552) | 3. | Explain how, if each cell can perform all its basic activities, organ systems contribute to cell survival. |
| (546) | 4. | _____ tissues cover the body surface of all animals and line internal organs from gut cavities to vertebrate lungs; this tissue always has one _____ surface that faces a body fluid or the outside _____ . |
| (546) | 5. | Distinguish simple epithelium from stratified epithelium. |
| (546) | 6. | Name and describe three kinds of cell junctions that occur in epithelia and other tissues. |
| (546–547) | 7. | Be able to name and describe the various types of epithelial tissues as well as their location and general functions. |
| (547) | 8. | Define the term *gland*. |
| (547) | 9. | _____ glands usually secrete their products onto a free epithelial surface through ducts or tubes; cite examples of their products. |
| (547) | 10. | _____ glands lack ducts; their products are _____ which are secreted directly into the fluid bathing the gland. |
| (548) | 11. | Distinguish between loose connective, dense, irregular, and dense, regular, connective tissues on the basis of their structures and functions. |

(548–549) 12. Cartilage, bone, adipose tissue, and blood are known as the specialized connective tissues; describe their structures and various functions.

(550)      13. Distinguish between skeletal, cardiac, and smooth muscle tissues in terms of location, structure, and function.

(550)      14. Muscle tissues contain specialized cells that can _____ .

(551)      15. Neurons are organized as lines of _____ .

(552–553) 16. List each of the eleven principal organ systems in humans, and list the main task of each.

(553)      17. Be able to list the major cavities in the human body and the organs they house.

(553)      18. Know the directional terms and planes of symmetry used for description of the human body.

(553)      19. _____ gives rise to the skin's outer layer and tissues of the nervous system; _____ gives rise to muscles, bones, and most of the circulatory, reproductive, and urinary systems; _____ gives rise to the lining of the digestive tract and to organs derived from it.

(554)      20. Describe the ways by which extracellular fluid helps cells survive.

(554)      21. Distinguish between interstitial fluid and plasma.

(554)      22. Be able to prepare a diagram that illustrates the components necessary for negative feedback at the organ level.

(554–555) 23. Describe the relationships among receptors, integrators, and effectors in a negative feedback system.

(555)      24. Explain the mechanisms involved in a positive feedback mechanism.

(555)      25. Tell why negative and positive feedback mechanisms are viewed as homeostatic controls.

---

## Interpreting and Applying Key Concepts

Explain why, of all places in the body, marrow is located on the interior of long bones. Explain why your bones are remodeled after you reach maturity. Why does your body not keep the same mature skeleton throughout life?

# 34

# INFORMATION FLOW AND THE NEURON

---

## Interactive Exercises

---

*TORNADO!* (pp. 558–559)

## 34.1. NEURONS—THE COMMUNICATION SPECIALISTS (pp. 560–561)

## 34.2. A CLOSER LOOK AT ACTION POTENTIALS (pp. 562–563)

**Selected Words:** neuroglia, *input* zones, *conducting* zone, *trigger* zone, *output* zones, *unipolar* cells, *bipolar,*
*multipolar, graded* signals, *local* signals, *all-or-nothing event,* <u>Loligo</u> *"giant" axons.*

### Boldfaced, Page-Referenced Terms

(559) neurons _____

_____

(559) sensory neuron _____

_____

(559) stimulus _____

_____

(559) interneurons _____

_____

(559) motor neuron _____

_____

(560) dendrites _____

_____

(560) axon _____

_____

(560) resting membrane potential _____

_____

(560) action potential _____

_____

(561) sodium-potassium pumps _____

_____

(562) threshold level _____

_____

(562) positive feedback _____

_____

## Fill-in-the-Blanks

Nerve cells that conduct messages are called (1) _____ . (2) _____ cells, which support and nurture the activities of neurons, make up more than half the volume of the nervous system. (3) _____ neurons are receptors for environmental stimuli, (4) _____ connect different neurons in the central nervous system, and (5) _____ neurons are linked with muscles or glands. All neurons have a (6) _____ _____ that contains the nucleus and the metabolic means to carry out protein synthesis. (7) _____ are short, slender extensions of (6), and together these two neuronal parts are the neurons' "input zone" for receiving (8) _____ . The (9) _____ is a single long cylindrical extension of the (10) _____ _____ ; in motor neurons, the (11) _____ has finely branched (12) _____ that terminate on muscle or gland cells and are "output zones," where messages are sent on to other cells.

A neuron at rest establishes unequal electric charges across its plasma membrane, and a (13) _____ _____ is maintained. Another name for (13) is the (14) _____ _____ _____ ; it represents a tendency for activity to happen along the membrane. Weak disturbances of the neuronal membrane might set off only slight changes across a small patch, but strong disturbances can cause a(n) (15) _____ _____ , which is an abrupt, short-lived reversal in the polarity of charge across the plasma membrane of the neuron. For a fraction of a second, the cytoplasmic side of a bit of membrane becomes positive with respect to the outside. The (16) _____ that travels along the neuronal membrane is nothing more than short-lived changes in the membrane potential. When action potentials reach the end of a motor neuron, they cause (17) _____ to be released that serve as chemical signals to adjacent muscle cells. Muscles (18) _____ in response to the signals.

How is the resting membrane potential established, and what restores it between action potentials? The concentrations of (19) _____ ions (K⁺), sodium ions (20) (___⁺), and other charged substances are not the same on the inside and outside of the neuronal membrane. (21) _____ proteins that span the membrane affect the diffusion of specific types of ions across it. (22) _____ proteins that span the membrane pump sodium and potassium ions against their concentration gradients across it by using energy stored in ATP. A neuronal membrane has many more positively charged (23) _____ ions inside than out and many more positively charged (24) _____ ions outside than inside. An electrical gradient also exists across the neuronal membrane; compared with the outside, the inside of a neuron at rest has an overall (25) _____ charge. For many neurons in most animals, the difference in charge across the neuronal membrane is about 70 (26) _____ . There are about (27) _____ times more potassium ions on the cytoplasmic side as outside, and there are about (28) _____ times more sodium ions outside as inside. These ions can cross the membrane only by traveling along passages through (29) _____ proteins. Some channel proteins leak ions through them all the time; others have (30) _____ that open only when stimulated. Transport proteins called (31) _____-_____ _____ counter the leakage of ions across the neuronal membrane and maintain the resting membrane potential.

## Labeling

Identify the parts of the neuron illustrated below.

32. _____  _____

33. _____  _____

34. _____

35. _____-_____  _____

36. _____  _____

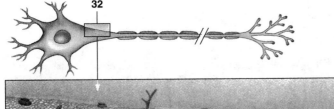

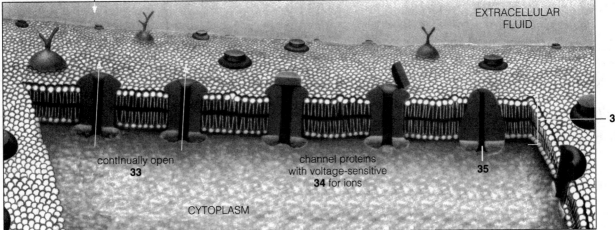

continually open
**33**

channel proteins
with voltage-sensitive
**34** for ions

**36**

**35**

EXTRACELLULAR
FLUID

CYTOPLASM

## Fill-in-the-Blanks

In all neurons, stimulation at an input zone produces (37) _____ signals that do not spread very far (half a millimeter or less). (38) _____ means that signals can vary in magnitude—small or large—depending on the intensity and (39) _____ of the stimulus. When stimulation is intense or prolonged, graded signals can spread into an adjacent (40) _____ _____ of the membrane—the site where action potentials can be initiated.

A(n) (41) _____ _____ is an all-or-nothing, brief reversal in membrane potential. Once an action potential has been achieved, it is an (42) _____-_____-_____ event; its amplitude will not change even if the strength of the stimulus changes. The minimum change in membrane potential needed to achieve an action potential is the (43) _____ value. Each action potential is followed by a time of insensitivity to stimulation.

## Labeling

Identify the parts of the neuron illustrated below.

44. _____ _____

45. _____

46. _____ _____ _____

47. _____

48. _____

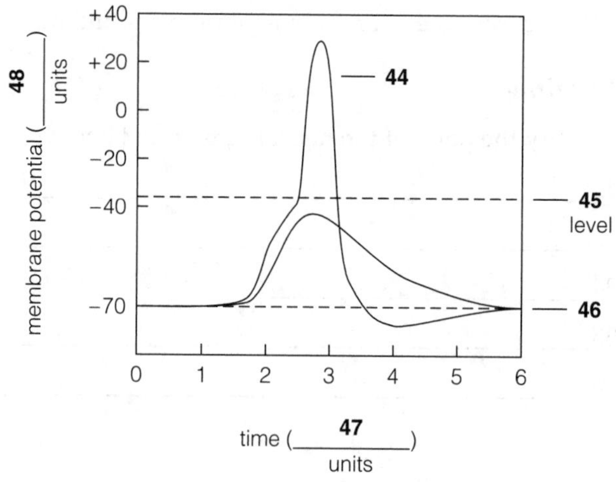

## 34.3. CHEMICAL SYNAPSES (pp. 564–565)

## 34.4. PATHS OF INFORMATION FLOW (pp. 566–567)

## 34.5. *Focus on Health:* SKEWED INFORMATION FLOW (p. 568)

*Selected Words:* *excitatory* effect, *inhibitory* effect, *GABA, depolarizing* effect, *hyperpolarizing* effect, *summation, divergent, convergent and reverberating* circuits, *multiple sclerosis,* Clostridium botulinum, *botulism,* C. tetani, *tetanus, muscle spindle.*

## Boldfaced, Page-Referenced Terms

(564) neurotransmitters _____

_____

(564) chemical synapse _____

_____

(564) acetylcholine (ACh) _____ —

_____

(564) neuromodulators _____

_____

(565) EPSP (excitatory postsynaptic potentials) _____

_____

(565) IPSP (inhibitory postsynaptic potentials) _____

_____

(565) synaptic integration _____

_____

(566) nerves _____

_____

(566) myelin sheath _____

_____

(566) Schwann cells _____

_____

(566) reflexes _____

_____

## Fill-in-the-Blanks

The junction specialized for transmission between a neuron and another cell is called a (1) _____

_____ . Usually, the signal being sent to the receiving cell is carried by chemical messengers called

(2) _____ . (3) _____ is an example of this type of chemical messenger that diffuses across the

synaptic cleft, combines with protein receptor molecules on the muscle cell membrane, and soon thereafter

is rapidly broken down by enzymes. At an (4) _____ synapse, the membrane potential is driven

toward the threshold value and increases the likelihood that an action potential will occur. At an

(5) _____ synapse, the membrane potential is driven away from the threshold value, and the receiving

neuron is less likely to achieve an action potential. A specific neurotransmitter can have either excitatory or

inhibitory effects depending on which type of protein channel it opens up in the (6) _____ membrane.

A (7) _____ _____ is a synapse between a motor neuron and muscle cells.

(8) _____ acts on brain cells that govern sleeping, sensory perception, temperature regulation, and

emotional states. (9) _____ are neuromodulators that inhibit perceptions of pain and may have roles in

memory and learning, emotional depression, and sexual behavior. (10) _____ _____ at the cellular

level is the moment-by-moment tallying of all excitatory and inhibitory signals acting on a neuron.

Incoming information is (11) _____ by cell bodies, and the charge differences across the membranes are

either enhanced or inhibited. An (12) _____ postsynaptic potential (EPSP) brings the membrane closer

to threshold and has a depolarizing effect. An inhibitory postsynaptic potential (IPSP) drives the membrane

away from threshold and either has a (13) _____ effect or maintains the membrane at its resting level.

Some narrow-diameter neurons are wrapped in lipid-rich (14) _____ produced by specialized neuroglial cells called Schwann cells; each of these is separated from the next by a(n) (15) _____ _____—a small gap where the axon is exposed to extracellular fluid. An action potential jumps from one node to the next in line and in the largest myelinated axon, signals travel (16) _____ meters per second.

## Labeling

Label the parts of the nerve illustrated at the right and below.

17. _____
18. _____ _____
19. _____ _____
20. _____
21. _____ _____
22. _____ _____

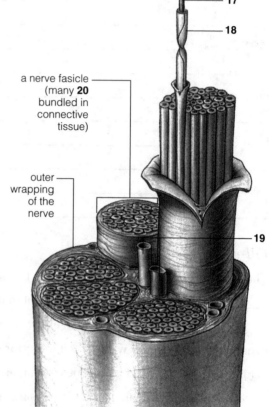

a nerve fasicle
(many **20**
bundled in
connective
tissue)

outer
wrapping
of the
nerve

17

18

19

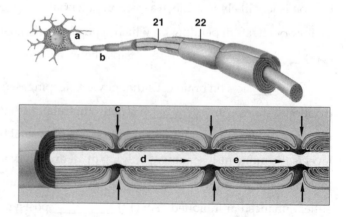

21    22

a
b

c

d →    e →

## Labeling

Label the parts of neurons and types of neurons in the illustration.

23. _____

24. _____ -

25. _____

26. _____ _____ or _____

27. _____ _____

28. _____ _____

29. _____

30. _____ _____

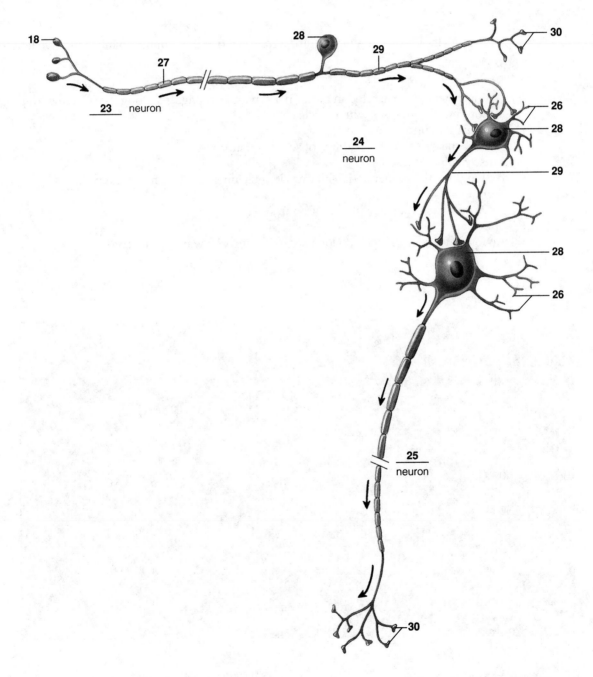

## Fill-in-the-Blanks

A (31) _____ is an involuntary sequence of events elicited by a stimulus. During a (32) _____ _____ , a muscle contracts involuntarily whenever conditions cause a stretch in length; many of these help you maintain an upright posture despite small shifts in balance. Located within skeletal muscles are length-sensitive organs called (33) _____ _____ , which generate action potentials when stretched beyond a critical point; these potentials are conducted rapidly to the (34) _____ _____ , where they are communicated to motor neurons leading right back to the muscle that was stretched.

Imbalances can occur at chemical synapses; a neurotoxin produced by *Clostridium tetani* interferes with the effect of (35) _____ on motor neurons, which may cause tetanus—a prolonged, spastic paralysis that can lead to death.

## Matching

Match the choices below with the correct number in the diagram. Two of the numbers match with two lettered choices.

36. ___
37. ___
38. ___
39. ___
40. ___
41. ___
42. ___
43. ___
44. ___

A. Response
B. Action potentials generated in motor neuron and propagated along its axon toward muscle
C. Motor neuron synapses with muscle cells
D. Muscle cells contract
E. Local signals in receptor endings of sensory neuron
F. Muscle spindle stretches
G. Action potentials generated in all muscle cells innervated by motor neuron
H. Stimulus
I. Axon endings synapse with motor neuron
J. Spinal cord
K. Action potential propagated along sensory neuron toward spinal cord

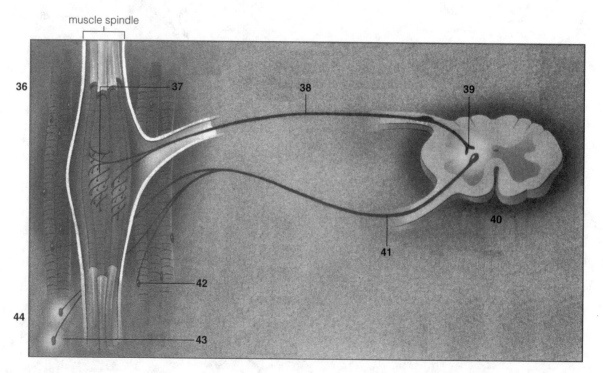

muscle spindle

# Self-Quiz

___ 1. Which of the following is *not* true of an action potential?
   a. It is a short-range message that can vary in size.
   b. It is an all-or-none brief reversal in membrane potential.
   c. It doesn't decay with distance.
   d. It is self-propagating.

___ 2. The conducting zone of a neuron is the _____ .
   a. axon
   b. axonal terminals
   c. cell body
   d. dendrite

___ 3. The integrative zone of a neuron is the _____ .
   a. axon
   b. axonal terminals
   c. cell body
   d. dendrite

___ 4. In the nervous system, a cotransport mechanism involves _____ .
   a. the inactivation of signals along the parasympathetic nerves by signals from the sympathetic division
   b. acetylcholine being reciprocally inactivated by cholinesterase in the synaptic zone
   c. the exchange of signals between the peripheral and central nervous systems
   d. two different substances being exchanged across a membrane by the same enzyme system

___ 5. An action potential is brought about by _____ .
   a. a sudden membrane impermeability
   b. the movement of negatively charged proteins through the neuronal membrane
   c. the movement of lipoproteins to the outer membrane
   d. a local change in membrane permeability caused by a greater-than-threshold stimulus

___ 6. The resting membrane potential _____ .
   a. exists as long as a charge difference sufficient to do work exists across a membrane
   b. occurs because there are more potassium ions outside the neuronal membrane than there are inside
   c. occurs because of the unique distribution of receptor proteins located on the dendrite exterior
   d. is brought about by a local change in membrane permeability caused by a greater-than-threshold stimulus

___ 7. The phrase "all or none" used in conjunction with discussion about an action potential means that _____ .
   a. a resting membrane potential has been received by the cell
   b. an impulse does not decay or dissipate as it travels away from the stimulus point
   c. the membrane either achieves total equilibrium or remains as far from equilibrium as possible
   d. propagation along the neuron is much faster than in other neurons

___ 8. Endorphins _____ .
   a. are neuromodulators
   b. block perceptions of pain
   c. may play a role in causing emotional depression
   d. are involved in all of the above roles

___ 9. An action potential passes from neuron to neuron across a synaptic cleft by _____ .
   a. myelin bridges
   b. the resting membrane potential
   c. neurotransmitter substances
   d. neuromodulator substances

___10. _____ are responsible for integration in the nervous system.
   a. Interneurons
   b. Schwann cells
   c. Motor neurons
   d. Sensory neurons

# Chapter Objectives/Review Questions

This section lists general and detailed chapter objectives that can be used as review questions. You can make maximum use of these items by writing answers on a separate sheet of paper. Fill in answers where blanks are provided. To check for accuracy, compare your answers with information given in the chapter or glossary.

| Page | | Objectives/Questions |
|------|------|------|
| (560) | 1. | Draw a neuron and label it according to its three general zones, its specific structures, and the specific function(s) of each structure. |
| (560) | 2. | Define *resting membrane potential*; explain what establishes it and how it is used by the cell neuron. |
| (560) | 3. | Define *action potential* by stating its three main characteristics. |
| (560–561) | 4. | Define *sodium-potassium pump* and state how it helps maintain the resting membrane potential. |
| (560–563) | 5. | Describe the distribution of the invisible array of large proteins, ions, and other molecules in a neuron, both at rest and as a neuron experiences a change in potential. |
| (560–563) | 6. | Explain the chemical basis of the action potential. Look at Figure 34.6 in your text and determine which part of the curve represents the following:<br>a. the point at which the stimulus was applied;<br>b. the events prior to achievement of the threshold value;<br>c. the opening of the ion gates and the diffusing of the ions;<br>d. the change from net negative charge inside the neuron to net positive charge and back again to net negative charge; and<br>e. the active transport of sodium ions out of and potassium ions into the neuron. |
| (562) | 7. | Explain how graded signals differ from action potentials. |
| (563) | 8. | Define *period of insensitivity* and state what causes it. |
| (564–567) | 9. | Understand how a nerve impulse is received by a neuron, conducted along a neuron, and transmitted across a synapse to a neighboring neuron, muscle, or gland. |
| (564–568) | 10. | Outline some of the ways by which information flow is regulated and integrated in the human body. |
| (565) | 11. | Distinguish the way excitatory synapses function from the way inhibitory synapses function. |
| (566) | 12. | Define *Schwann cell, unsheathed nodes,* and *myelin sheath* and explain how each helps narrow-diameter neurons conduct nerve impulses quickly. |
| (566–567) | 13. | Explain what a reflex is by drawing and labeling a diagram and telling how it functions. |
| (567) | 14. | Explain what the stretch reflex is and tell how it helps an animal survive. |

# Integrating and Applying Key Concepts

What do you think might happen to human behavior if inhibitory postsynaptic potentials did not exist and if the threshold stimulus necessary to provoke an EPSP were much higher?

# 35

# INTEGRATION AND CONTROL: NERVOUS SYSTEMS

---

## Interactive Exercises

---

### 35.1. INVERTEBRATE NERVOUS SYSTEMS (pp. 572–573)

*Selected Words: drug, dealer, radial* symmetry, *bilateral* symmetry

*Boldfaced, Page-Referenced Terms*

(572) nervous system _____

_____

(572) nerve _____

_____

(572) nerve net _____

_____

(572) reflex pathways _____

_____

(572) ganglia _____

_____

(573) planula _____

_____

## Fill-in-the-Blanks

All animals except sponges have some type of (1) _____ system in which nerve cells, such as (2) _____ , are oriented in signal-conducting and information-processing pathways. At the least, the animal has communication lines of its component cells that receive information about changing conditions outside and inside the (3) _____ , then elicit suitable responses from muscle and gland cells. Nerves in a sea urchin deal with sampling the environment, moving through it, and getting food into the gut. The nerves interconnect with one another by a nerve (4) _____ . Information flow through this nervous system is diffuse, not highly (5) _____ . Animals with (6) _____ symmetry have the simplest nervous systems. The nerve (7) _____ of cnidarians interacts with sensory cells and contractile cells of the epithelium in (8) _____ _____ . In such pathways, (9) _____ stimulation directly triggers simple, stereotyped movements. Flatworms are the simplest animals having a (10) _____ nervous system. There are equivalent body parts on the left and right sides of the body's (11) _____ plane. The ladderlike nervous system of the flatworm has (12) _____ (number) cordlike nerves running longitudinally through the body. Besides these nerve cords, some nervous systems have (13) _____ , which are clusters of nerve cell bodies that serve as integrating centers. At the head end of a flatworm, ganglia form a two-part (14) _____ structure that coordinates signals from paired (15) _____ organs, including two eyespots, and provide a degree of control over the nerve cords.

Perhaps bilateral nervous systems evolved from (16) _____ _____ . Ancient (17) _____ crawled about on the seafloor in Cambrian times. Chance mutations of (18) _____ genes in such animals may have (19) _____ cells concentrated at their leading end. This would have made possible much more rapid, effective responses to stimuli. Probably natural (20) _____ favored animals with concentrations of (21) _____ cells at their leading ends. (22) _____ (formation of a head) and bilateral symmetry may have started this way. Both of these features do occur in most (23) _____ lineages.

## 35.2. VERTEBRATE NERVOUS SYSTEMS—AN OVERVIEW (pp. 574–575)

*Selected Words:* afferent, efferent

### Boldfaced, Page-Referenced Terms

(574) neural tube _____

_____

(575) central nervous system _____

_____

(575) peripheral nervous system _____

_____

(575) tracts _____

_____

(575) white matter _____

_____

(575) gray matter _____

_____

(575) neuroglial cells _____

_____

### Complete the Table

1. A comparison of the brains of some existing vertebrates (Fig. 35.5, p. 574) suggests an evolutionary trend toward an expanded, more complex brain. Complete the following table by entering forebrain, midbrain, and hindbrain in the correct blanks to understand how the anterior end of the dorsal, hollow nerve cord expanded into functionally distinct regions, which also increased in complexity in certain lineages.

| Brain Region | Basic Functions |
|---|---|
| a. | Coordinates reflex responses to sight, sounds |
| b. | Receives, integrates sensory information from nose, eyes, and ears; in land-dwelling vertebrates, contains the highest integrating centers |
| c. | Reflex control of respiration, blood circulation, other basic tasks; in complex vertebrates, coordination of sensory input, motor dexterity, and possibly mental dexterity |

## Matching

Choose the most appropriate answer for each.

2. ___neural tube

3. ___central nervous system

4. ___peripheral nervous system

5. ___tracts

6. ___white matter tracts

7. ___gray matter tracts

8. ___neuroglial cells

9. ___afferent

10. ___efferent

A. Refers to nerves carrying motor output away from the central nervous system to muscles and glands

B. Term applied to the nerve cord that persists in all vertebrate embryos

C. Remember, protect or structurally and functionally support neurons

D. The spinal cord and brain

E. Refers to nerves carrying sensory input to the central nervous system

F. Consists of unmyelinated axons, dendrites, and nerve cell bodies and neuroglial cells

G. Consists mainly of nerves that thread through the rest of the body and carry signals into and out of the central nervous system

H. The communication lines inside the brain and spinal cord

I. Contain axons with glistening white myelin sheaths and specialize in rapid signal transmission

## 35.3. THE MAJOR EXPRESSWAYS (pp. 576–577)

*Selected Words:* spinal nerves, *cranial* nerves, *meningitis*

### Boldfaced, Page-Referenced Terms

(576) somatic nerves _____

_____

(576) autonomic nerves _____

_____

(576) parasympathetic nerves _____

_____

(577) sympathetic nerves _____

_____

(577) fight-flight response _____

_____

(577) spinal cord _____

_____

(577) meninges _____

_____

## Labeling

Label each numbered part of the accompanying illustration.

1. _____

2. _____

3. _____

4. _____

5. _____ _____

6. _____

7. _____

8. _____

9. _____

### THE 1 NERVOUS SYSTEM

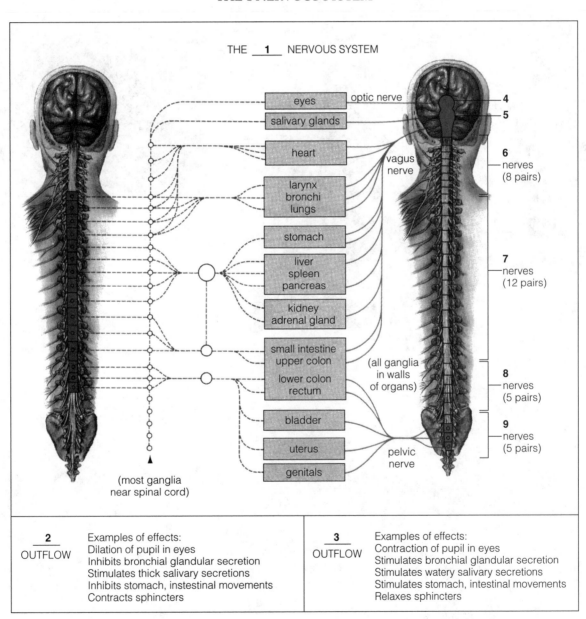

THE __1__ NERVOUS SYSTEM

eyes — optic nerve — 4

salivary glands — 5

heart

vagus nerve — 6 nerves (8 pairs)

larynx bronchi lungs

stomach

liver spleen pancreas — 7 nerves (12 pairs)

kidney adrenal gland

small intestine upper colon

(all ganglia in walls of organs)

lower colon rectum — 8 nerves (5 pairs)

bladder

uterus — 9 nerves (5 pairs)

genitals — pelvic nerve

(most ganglia near spinal cord)

__2__ OUTFLOW

Examples of effects:
Dilation of pupil in eyes
Inhibits bronchial glandular secretion
Stimulates thick salivary secretions
Inhibits stomach, instestinal movements
Contracts sphincters

__3__ OUTFLOW

Examples of effects:
Contraction of pupil in eyes
Stimulates bronchial glandular secretion
Stimulates watery salivary secretions
Stimulates stomach, intestinal movements
Relaxes sphincters

## Labeling

Identify the numbered parts of the accompanying illustrations.

10. _____ _____

11. _____

12. _____

13. _____

14. _____ _____

15. _____

16. _____ _____

17. _____ _____

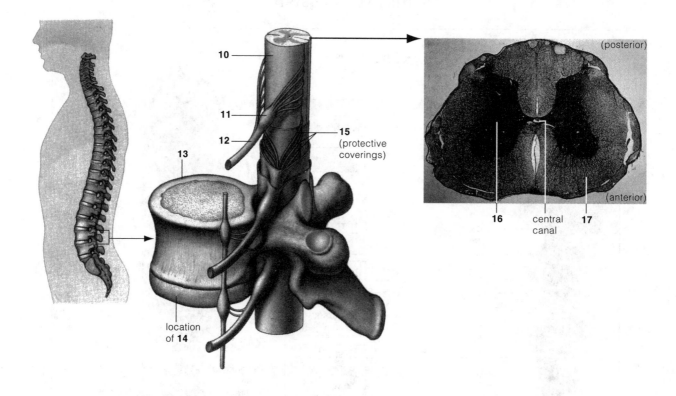

*Choice*

For questions 18–34, choose from the following:

       a. peripheral—somatic nerves     b. peripheral—autonomic sympathetic nerves
      c. peripheral—autonomic parasympathetic nerves     d. spinal cord nerves

18.____Dominate when the body is not receiving much outside stimulation.

19.____Carry signals about moving your head, trunk, and limbs.

20.____Dominate in times of sharpened awareness.

21.____Can be attacked by meningitis.

22.____Sensory axons inside these nerves deliver information from receptors in the skin, skeletal muscles, and tendons to the central nervous system.

23.____Tend to slow down the body overall and divert energy to basic "housekeeping" tasks, such as digestion.

24.____The meninges, three tough, tubelike coverings are part of the structure.

25.____A vital expressway for signals between the peripheral nervous system and the brain.

26.____The fight-flight response.

27.____Threads through a canal formed by bones of the vertebral column.

28.____Commands one's heart to beat faster.

29.____Tend to shelve housekeeping tasks.

30.____Their motor axons deliver commands from the brain and spinal cord to the body's skeletal muscles.

31.____Signals cause the release of epinephrine.

32.____Some of its interneurons exert direct control over certain reflex pathways.

33.____Commands your heart to beat a little slower.

34.____Gray matter that plays an important role in controlling reflexes for limb movement and organ activity.

## 35.4. FUNCTIONAL DIVISIONS OF THE VERTEBRATE BRAIN (pp. 578–579)

### *Boldfaced, Page-Referenced Terms*

(578) brain _____

_____

(578) brain stem _____

_____

(578) medulla oblongata _____

_____

(578) cerebellum _____

_____

(578) pons _____

_____

(578) tectum _____

_____

(579) cerebrum _____

_____

(579) thalamus _____

_____

(579) hypothalamus _____

_____

(579) reticular formation _____

_____

(579) cerebrospinal fluid _____

_____

(579) blood-brain barrier _____

_____

## Fill-in-the-Blanks

The (1) _____ is the body's master control center and merges with the anterior end of the spinal cord. This organ receives, integrates, stores, and retrieves (2) _____ . It also (3) _____ responses to information by adjusting activities throughout the body. Like the spinal cord, it is protected by (4) _____ and membranes.

The forebrain, midbrain, and hindbrain of vertebrates form from three successive portions of the (5) _____ _____ . The nervous tissue that evolved first in all three regions is called the brain (6)_____ . It still contains many simple, basic (7) _____ centers. Over evolutionary time, expanded layers of (8) _____ matter developed from the brain stem. The more recent additions have been correlated with an increasing reliance on three major (9) _____ organs: the nose, ears, and eyes. The (10) _____ and possibly parts of the (11) _____ , which is part of the hindbrain region, contains the newest additions of gray matter.

The medulla oblongata, cerebellum, and pons are all components of the (12) _____ . The roof of the (13) _____ is formed by the (14) _____ , a more recent layering of gray matter. The midbrain includes a pair of brain centers called (15) _____ lobes. The cerebrum, olfactory lobes, thalamus, hypothalamus, limbic system, pituitary gland, and pineal gland are all parts of the (16) _____ . An evolutionary ancient mesh of interneurons, the (17) _____ formation, still extends from the uppermost part of the spinal cord, on through the brain stem, and on into higher integrative centers of the cerebral cortex. The (18) _____ fluid is a clear extracellular fluid that cushions the brain and spinal cord from sudden, jarring movements. A mechanism called the (19) _____-_____ barrier exerts some control over which solutes enter the cerebrospinal fluid and thereby helps to protect the brain and spinal cord. In most complex vertebrates, the highest integrative centers reside in the forebrain, especially in its (20) _____ hemispheres.

## Matching

Choose the most appropriate answer for each.

21. ___medulla oblongata

22. ___limbic system

23. ___cerebellum

24. ___pons

25. ___pituitary gland

26. ___tectum

27. ___cerebrum

28. ___thalamus

29. ___hypothalamus

30. ___pineal gland

31. ___olfactory lobes

A. Forebrain: control of some circadian rhythms; has a role in mammalian reproductive physiology
B. Midbrain: in mammals, mainly reflex centers that rapidly relay sensory input to the forebrain
C. Hindbrain: coordinates motor activity for limb movements, maintaining posture, and spatial orientation
D. Forebrain: relaying of sensory input from the nose to olfactory centers of the cerebrum
E. Hindbrain: contains tracts between pons and spinal cord; has reflex centers for vital tasks, such as respiration, heart rate, and blood circulation
F. Forebrain: localizes and processes sensory inputs; initiates and controls skeletal muscle activity; in the most complex vertebrates, roles in memory, mediating emotions, and abstract thought
G. Hindbrain: "bridge" of tracts between cerebrum and cerebellum; its other tracts connect forebrain and spinal cord; works with medulla oblongata to control rate and depth of respiration
H. Forebrain: main center for homeostatic control over the internal environment; became central to behaviors related to internal organ activities, such as thirst, hunger, and sex; and to emotional expression, such as sweating with fear
I. Forebrain: a complex of structures that govern emotions and that have roles in memory
J. Forebrain: with hypothalamus, endocrine control of metabolism, growth, and development
K. Forebrain: relay stations for conducting sensory signals to and from cerebral cortex; role in memory

## 35.5. A CLOSER LOOK AT THE HUMAN CEREBRUM (pp. 580–581)

## 35.6. *Focus on Science:* SPERRY'S SPLIT-BRAIN EXPERIMENTS (pp. 582–583)

*Selected Words:* occipital lobe, temporal lobe, parietal lobe, frontal lobe, epilepsy

### Boldfaced, Page-Referenced Terms

(580) cerebral hemispheres _____

_____

(580) cerebral cortex _____

_____

(581) limbic system _____

_____

## Labeling

Identify each numbered part of the accompanying illustration.

1. _____
2. _____
3. _____
4. _____
5. _____  _____
6. _____
7. _____
8. _____
9. _____
10. _____  _____

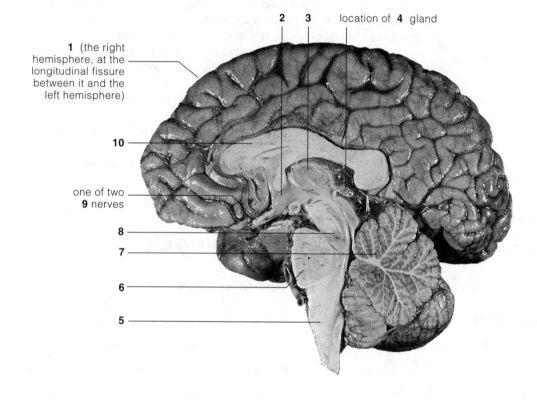

**2**  **3**  location of **4** gland

**1** (the right hemisphere, at the longitudinal fissure between it and the left hemisphere)

10

one of two **9** nerves

8

7

6

5

## Complete the Table

11. Complete the table below by identifying the area of the brain in the left column whose functions are described in the right column. Choose from: cerebral cortex, occipital lobe; cerebral cortex, temporal lobe; cerebral cortex, parietal lobe; cerebral cortex, frontal lobe; left cerebral hemisphere, right cerebral hemisphere, cerebral cortex, limbic system, and corpus callosum.

| Brain Area | Functions |
|---|---|
| a. | Located at the middle of the cerebral hemispheres; governs our emotions and has roles in memory; distantly related to olfactory lobes and still deals with the sense of smell |
| b. | Near each temple, has centers that deal with hearing and visual associations |
| c. | Affords most control over spatial abilities, music, and other abstract, nonverbal skills |
| d. | Motor cortex deals with signals for motor responses; thumb, finger, and tongue muscles get much of the brain's attention |
| e. | A nerve tract that affords two-way communication between both cerebral hemispheres |
| f. | At the back of each cerebral hemisphere; has centers for vision |
| g. | Affords most of the control over speech, mathematics, and analytical skills |
| h. | The somatosensory cortex, the main receiving area for sensory input from the skin and joints |
| i. | A thin layer of gray matter above the brain's axons with localized activities in four tissue lobes in each cerebral hemisphere |

## Short Answer

12. In an effort to relieve the frequent seizures of severe epilepsy, neural surgeon Dr. Roger Sperry cut the neural bridge of the corpus callosum of several of these patients. The seizures did subside in frequency and intensity. Summarize the subsequent findings of Dr. Sperry regarding the function of the corpus callosum. _____

_____

_____

_____

_____

_____

## 35.7. MEMORY (p. 583)

*Selected Words:* *short-term* storage, *long-term* storage, *facts, skills, amnesia, Parkinson's disease, Alzheimer's disease*

### Boldfaced, Page-Referenced Terms

(583) memory _____

_____

### Labeling

Identify each numbered part of the accompanying illustration.

1. _____
2. _____
3. _____
4. _____
5. _____
6. _____
7. _____ _____
8. _____ _____
9. _____ and _____
10. _____ _____
11. _____ _____

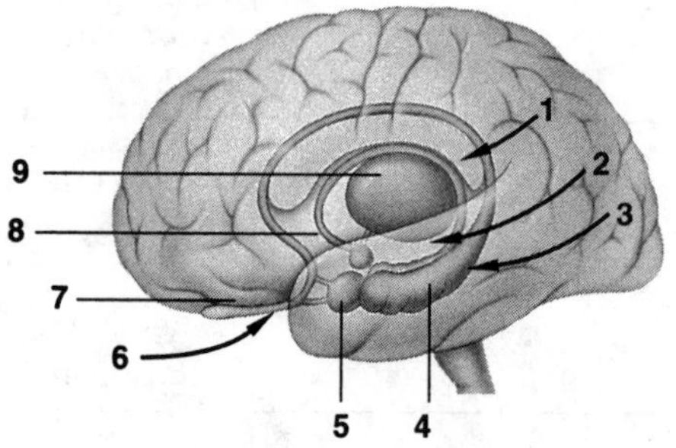

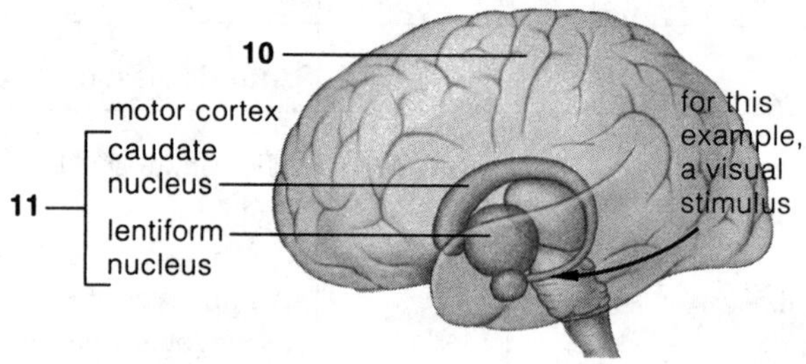

(illustration labels: 9, 8, 7, 6, 5, 4, 1, 2, 3, 10, 11; motor cortex; caudate nucleus; lentiform nucleus; for this example, a visual stimulus)

### Fill-in-the-Blanks

(12) _____ is the capacity of an individual's brain to store and retrieve information about past sensory experience. (13) _____ and adaptive modifications of our behavior would be impossible without it. Information is stored in stages. (14) _____-_____ storage is a stage of neural excitation that lasts a few seconds to a few hours; it is limited to a few bits of sensory information. In (15) _____-_____ storage, seemingly unlimited amounts of information get tucked away more or less permanently.

Only some (16) _____ input is selected for transfer to brain structures involved in short-term memory. If information is (17) _____ , it is forgotten, otherwise it is consolidated with banks of information in long-term storage structures.

The human brain processes facts separately from (18) _____ . Explicit bits of information are soon forgotten or filed away in (19) _____-_____ storage, along with the circumstance in which they were learned. (20) _____ are gained by practicing specific motor activities and are best recalled by actually performing the motor activity involved.

Separate memory circuits handle different kinds of (21) _____ . A circuit leading to fact memory starts with inputs at the sensory cortex that flow to the (22) _____ and (23) _____ , structures in the limbic system. The (24) _____ is the gatekeeper and connects the sensory cortex with parts of the thalamus and hypothalamus that govern emotional states. The (25) _____ mediates learning and spatial relations. Information flows to the prefrontal cortex, where multiple banks of (26) _____ memories are retrieved and used to stimulate or inhibit other parts of the brain. New input also flows to (27) _____ ganglia, which send it back to the cortex in a feedback loop that reinforces the input until it can be consolidated in (28) _____-_____ storage.

(29) _____ memory also begins at the (30) _____ cortex, but this circuit routes sensory input to the (31) _____ _____ , which promotes motor responses. The circuit for motor skills extends to the (32) _____ , the brain region that coordinates motor activity.

(33) _____ is a loss of memory, the severity of which depends on whether the hippocampus, amygdala, or both are damaged, as by a severe head blow; this does not affect capacity to learn new (34) _____ . By contrast, basal ganglia are destroyed and learning ability is lost during (35) _____ disease, yet skill memory is retained. With a usual onset during later life, (36) _____ disease is linked to structural changes in the cerebral cortex and (37) _____ . Affected people often can remember long-standing (38) _____ but they have difficulty remembering what has just happened to them. In time they become confused, depressed, and unable to complete a train of thought.

## 35.8. STATES OF CONSCIOUSNESS (p. 584)

## 35.9. *Focus on Health:* DRUGGING THE BRAIN (pp. 584–585)

*Selected Words:* *alpha rhythm, slow-wave sleep, REM sleep, rapid eye movements, EEG arousal,* Cannabis

### Boldfaced, Page-Referenced Terms

(584) consciousness _____

_____

(584) EEGs _____

_____

(584) drug _____

_____

(584) drug addiction _____

_____

## Matching

Choose the most appropriate answer for each.

1. ___consciousness

2. ___EEGs

3. ___alpha rhythm

4. ___slow-wave sleep

5. ___REM sleep

6. ___EEG arousal

A. Briefly punctuates slow-wave sleep; accompanied by REM, irregular breathing, faster heartbeat, and twitching fingers; experience vivid dreams

B. The prominent EEG wave pattern for someone who is relaxed, with eyes closed

C. The spectrum includes sleeping and aroused states, during which neural chattering shows up as wavelike patterns in EEGs

D. A transition that occurs when conscious effort is made to focus on external stimuli or even on one's own thoughts

E. Electrical recordings of the frequency and strength of membrane potentials at the surface of the brain

F. A pattern that dominates when sensory input is low and the mind is more or less idling; subjects seemed to be mulling over recent, ordinary events

## Choice

For questions 7–15, choose from the following:

a. stimulants          b. depressants, hypnotics          c. analgesics          d. psychedelics, hallucinogens

7. ___amphetamines

8. ___caffeine

9. ___cocaine

10. ___codeine

11. ___ethyl alcohol

12. ___heroin

13. ___lysergic acid diethylkamide or LSD

14. ___marijuana

15. ___nicotine

# Self-Quiz

___ 1. All nerves that lead away from the central nervous system are _____
   a. efferent nerves
   b. sensory nerves
   c. afferent nerves
   d. spinal nerves
   e. peripheral nerves

___ 2. _____ nerves generally dominate internal events when environmental conditions permit normal body functioning.
   a. Ganglia
   b. Pacemaker
   c. Sympathetic
   d. Parasympathetic
   e. All of the above

___ 3. The center of consciousness and intelligence is the _____
   a. medulla
   b. thalamus
   c. hypothalamus
   d. cerebellum
   e. cerebrum

___ 4. The _____ are the protective coverings of the brain.
   a. ventricles
   b. meninges
   c. tectums
   d. olfactory bulbs
   e. pineal glands

___ 5. The _____ monitors internal organs; acts as gatekeeper to the limbic system; helps the reasoning centers of the brain to dampen rage and hatred; and governs hunger, thirst, and sex drives.
   a. medulla
   b. pons
   c. thalamus
   d. hypothalamus
   e. reticular formation

___ 6. The left hemisphere of the brain is responsible for _____ .
   a. music
   b. artistic ability and spatial relationships
   c. language skills
   d. abstract abilities

___ 7. The part of the brain that controls the basic responses necessary to maintain life processes (breathing, heartbeat) is _____ .
   a. the cerebral cortex
   b. the cerebellum
   c. the corpus callosum
   d. the medulla

___ 8. To produce a split-brain individual, an operation would be required to sever the _____ .
   a. pons
   b. fissure of Rolando
   c. hypothalamus
   d. reticular formation
   e. corpus callosum

___ 9. The center for balance and coordination in the human brain is the _____ .
   a. cerebrum
   b. pons
   c. cerebellum
   d. hypothalamus
   e. thalamus

___10. The sleep center of the human brain is the _____ .
   a. medulla
   b. pons
   c. thalamus
   d. hypothalamus
   e. reticular formation

## Chapter Objectives/Review Questions

This section lists general and detailed chapter objectives that can be used as review questions. You can make maximum use of these items by writing answers on a separate sheet of paper. To check for accuracy, compare your answers with information given in the chapter or glossary.

*Page*      *Objectives/Questions*

(572)      1. Describe a "nerve net."
(572–573)  2. Fully explain how the shift from radial to bilateral symmetry within invertebrate animals influenced the complexity of nervous systems.
(574)      3. The nerve cord that persists in all vertebrate embryos is called the _____ _____ .
(575)      4. Define and contrast the vertebrate central and peripheral nervous systems.
(575)      5. The communication lines inside the brain and spinal cord are called _____ , not nerves.
(575)      6. Compare the structures of the spinal cord and brain with respect to white matter and gray matter.
(575)      7. _____ cells protect or structurally and functionally support neurons.
(575)      8. In terms of sensory input and motor output, _____ means "to bring to," and _____ means to carry outward.
(576)      9. Distinguish between somatic nerves and autonomic nerves.
(576–577) 10. Explain how parasympathetic nerve activity balances sympathetic nerve activity. List activities of the sympathetic and parasympathetic nerves in regulating pupil diameter, rate of heartbeat, activities of the gut, and elimination of urine.
(577)     11. Describe the basic structural and functional organization of the spinal cord. In your answer, distinguish spinal cord from vertebral column.
(578–579) 12. List the parts of the brain found in the hindbrain, midbrain, and forebrain and tell the basic functions of each.
(579)     13. The _____ formation is an ancient mesh of interneurons that still persists as a low-level pathway to motor centers of the medulla oblongata and spinal cord.

(579)     14.  Describe the function of cerebrospinal fluid.
(579)     15.  In terms of structure and function, explain how the mechanism called the blood-brain barrier protects the brain and spinal cord.
(580)     16.  The _____ cerebral hemisphere affords most of the control over speech, mathematics, and analytical skills; the _____ hemisphere affords most control over spatial abilities, music, and other abstract, nonverbal skills.
(580)     17.  Be able to list the four tissue lobes of the cerebral cortex and state the function of each.
(580–581) 18.  Describe how the cerebral hemispheres are structurally and functionally related to the other parts of the forebrain.
(582)     19.  State what the results of the "split-brain" experiments suggest about the functioning of the cerebral hemispheres.
(583)     20.  Distinguish between short-term information storage and long-term storage.
(583)     21.  The human brain processes _____ separately from _____ .
(583)     22.  Define amnesia, Parkinson's disease, and Alzheimer's disease; list the characteristics of each.
(584)     23.  Explain what an electroencephalogram is and what EEGs can tell us about the levels of conscious experience. Describe three typical EEG patterns and tell which level of consciousness each characterizes.
(584–585) 24.  List the major classes of psychoactive drugs and provide an example of each class.

## Integrating and Applying Key Concepts

Suppose that anger is eventually determined to be caused by excessive amounts of specific transmitter substances in the brains of angry people. Also suppose that an inexpensive antidote to anger that neutralizes these anger-producing transmitter substances is readily available. Can violent murderers now argue that they have been wrongfully punished because they were victimized by their brain's transmitter substances and could not have acted in any other way? Suppose an antidote is prescribed to curb violent tempers in an easily angered person. Suppose also that the person forgets to take the pill and subsequently murders a family member. Can the murderer still claim to be victimized by transmitter substances?

# 36

# SENSORY RECEPTION

## Interactive Exercises

*Different Stokes for Different Folks* (pp. 588–589)

## 36.1. SENSORY RECEPTORS AND PATHWAYS—AN OVERVIEW (pp. 590–591)

*Selected Words:* *"ultrasounds", compound sensations, amplitude, frequency*

*Boldfaced, Page-Referenced Terms*

(588) echolocation _____

_____

(590) sensory systems _____

_____

(590) sensation _____

_____

(590) perception _____

_____

(590) mechanoreceptors _____

_____

(590) thermoreceptors _____

_____

(590) pain receptors (nociceptors) _____

_____

(590) chemoreceptors _____

_____

(590) osmoreceptors _____

_____

(590) photoreceptors _____

_____

(591) sensory adaptation _____

_____

(591) somatic sensations _____

_____

(591) special senses _____

_____

## Fill-in-the-Blanks

A sensory system consists of sensory receptors for specific stimuli, (1) _____ _____ that conduct information from those receptors to the brain, and (2) _____ _____ where information is evaluated.  A(n) (3) _____ is conscious awareness of change in internal or external conditions; this is not to be confused with (4) _____ , which is an understanding of what sensation means.  The specialized peripheral endings of sensory neurons that detect specific kinds of stimuli are (5) _____ . A (6) _____ is any form of energy that activates a specific type of sensory receptor.  (7) _____ detect the chemical energy of specific substances dissolved in the fluid surrounding them; (8) _____ detect mechanical energy associated with changes in pressure, position, or acceleration; (9) _____ detect the energy of visible and ultraviolet wavelengths of light; and (10) _____ detect radiant energy associated with temperature changes.

Every type of sensation is caused by (11) _____ _____ arriving from particular nerve pathways activating specific neurons in the (12) _____ .  Besides sensing a stimulus, the brain also interprets variations in (13) _____ _____ .  Interpretation is based on the (14) _____ of action potentials propagated along single axons and the (15) _____ of axons carrying action potentials from a given tissue.

Sometimes, the (16) _____ of action potentials decreases or stops even when a stimulus is being maintained at constant strength; such a decrease is known as (17) _____ _____ .

Some mechanoreceptors only signal a (18) _____ in a stimulus; if a stimulus is constant, but deserves no response, these receptors will quit responding. But there are other types of receptors that adapt slowly or not at all. (19) _____ receptors that continually inform the brain about the (20) _____ of particular muscles help maintain balance and posture.

Sensory receptors that are present at more than one body location generally inform the brain about (21) _____ _____ , that is, how different parts of the body feel at any particular time. Special senses are restricted to specific locations, such as inside the eyes or ears.

## Matching

Select the best match for each item below.

22. ___ Vision is associated with _____ .

23. ___ Pain is associated with _____ .

24. ___ Odors are detected by _____ .

25. ___ Hearing is detected by _____ .

26. ___ $CO_2$ concentration in the blood is detected by _____ .

27. ___ Environmental temperature is detected by _____ .

28. ___ Internal body temperature is detected by _____ .

29. ___ Touch is detected by _____ .

30. ___ Rods and cones

31. ___ Hair cells in the ear's organ of Corti

32. ___ Pacinian corpuscles in the skin

33. ___ Olfactory receptors in the nose

34. ___ Any stimulus that causes tissue damage

35. ___ The movement of fluid in the inner ear is associated with _____ .

A. chemoreceptors
B. mechanoreceptors
C. nociceptors
D. photoreceptors
E. thermoreceptors

## 36.2. SOMATIC SENSATIONS (pp. 592–593)
## 36.3. SENSES OF TASTE AND SMELL (p. 594)

*Selected Words:* *somatic pain, visceral pain, Pacinian corpuscles, Ruffini endings, bulb of Krause, Meissner's corpuscle, referred pain, phantom pain, chemical senses, vomeronasal organ*

## Boldfaced, Page-Referenced Terms

(592) somatosensory cortex _____

_____

(592) free nerve endings _____

_____

(592) encapsulated receptors _____

_____

(592) pain _____

_____

(594) taste receptors _____

_____

(594) olfactory receptors _____

_____

(594) pheromones _____

_____

## Fill-in-the-Blanks

Signals from receptors in the skin and joints travel to the
primary (1) _____ _____ _____ , which is a
strip little more than an inch wide running from the top of
the (2) _____ to just above the (3) _____ on the
surface of each (4) _____ _____. The largest
portion of the somatosensory cortex is A (see figure at
right), which receives signals coming from the
(5) _____ . The second largest region is C, which
receives signals coming from the (6) _____ .

The somatic sensations (awareness of (7) _____ ,
pressure, heat, (8) _____ , and pain) start with
receptor endings that are embedded in (9) _____ and
other tissues at the body's surfaces, in (10) _____
muscles, and in the walls of internal organs. All skin (11) _____ are easily deformed by pressure on the
skin's surface; these make you aware of touch, vibrations, and pressure. (12) _____ nerve endings
serve as "heat" receptors, and their firing of action potentials increases with increases in temperature.
(13) _____ is the perception of injury to some body region; the perception begins when (14) _____ ,
which include free nerve endings, send signals via the thalamus to the parietal lobe of the brain, where they
are interpreted. When the (15) _____ _____ pass a certain threshold, the signals generated are
translated into sensations of pain. The brain sometimes gets confused and may associate perceived pain
with a tissue some distance from the damaged area; this phenomenon is called (16) _____ _____ .
Usually the nerve pathways to both the injured and the mistaken area pass through the same segment of
spinal cord. (17) _____ in skeletal muscle, joints, tendons, ligaments, and (18) _____ are
responsible for awareness of the body's position in space and of limb movements.

## Label-Match

Identify each indicated part of the illustration at the right. Complete the exercise by entering the appropriate letter in the parentheses that follow the labels.

19. _____ _____

    _____ ( )

20. _____ _____ ( )

21. _____ _____ ( )

22. _____

23. _____

24. _____ _____ ( )

A. React continually to ongoing stimuli
B. Contribute to sensations of vibrations
C. Involved in sensing heat, light pressure, and pain
D. Stimulated at the beginning and end of sustained pressure

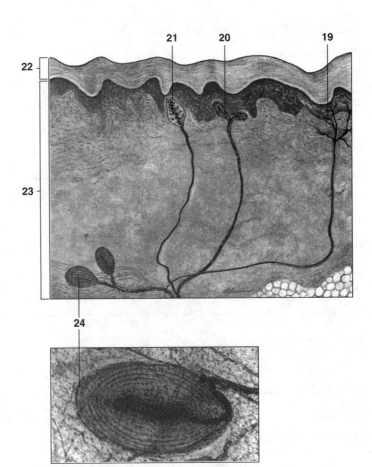

## Fill-in-the-Blanks

(25) _____ receptors detect molecules that become dissolved in fluid next to some body surface. Receptors are the modified dendrites of (26) _____ neurons. In the case of taste, these receptors, when located on animal tongues, are often part of sensory organs, (27) _____ _____ , which are enclosed by circular papillae. Animals smell substances by means of (28) _____ receptors, such as the ones in your (29) _____ ; humans have about (30) _____ million of these in a nose. Sensory nerve pathways lead from the nasal cavity to the region of the brain where odors are identified and associated with their sources—the (31) _____ bulb and nerve tract. (32) _____ receptors sampling odors from food in the mouth are important for our sense of taste. When we suffer from the common cold and have a "runny" nose, odor molecules from food have difficulty reaching the olfactory receptors and contributing their information signals. Our senses of taste and (33) _____ are both dulled.

## 36.4. SENSE OF BALANCE (p. 595)
## 36.5. SENSE OF HEARING (pp. 596–597)

**Selected Words:** *equilibrium position, dynamic equilibrium, "semicircular" canals, static equilibrium, motion sickness, amplitude, frequency, intensity, pitch, pinna, oval window, basilar membrane, organ of Corti, tectorial membrane*

## Boldfaced, Page-Referenced Terms

(595) inner ears _____

_____

(595) vestibular apparatus _____

_____

(595) hair cells _____

_____

(595) otoliths _____

_____

(596) hearing _____

_____

(596) middle ear _____

_____

(596) external ear _____

_____

(596) cochlea _____

_____

(596) acoustical receptors _____

_____

## Fill-in-the-Blanks

The (1) _____ (perceived loudness) of sound depends on the height of the sound wave. The
(2) _____ (perceived pitch) of sound depends on how fast the wave changes occur. The faster the
vibrations, the (3) [choose one] ☐ higher, ☐ lower the sound. Hair cells are (4) [choose one] ☐ nociceptors,
☐ mechanoreceptors, ☐ thermoreceptors that detect vibrations. The hammer, anvil, and stirrup are located
in the (5) [choose one] ☐ inner, ☐ middle ear. The (6) _____ is a coiled tube that resembles a snail shell
and contains the (7) _____ _____ _____—the organ that changes vibrations into
electrochemical impulses. Structures that detect rotational acceleration in humans are (8) _____

_____ .

## *Labeling*

Identify each indicated part of the accompanying illustrations.

9. _____ _____

10. _____

11. _____ _____

12. _____ _____

13. _____ _____

14. _____ _____

15. _____ _____

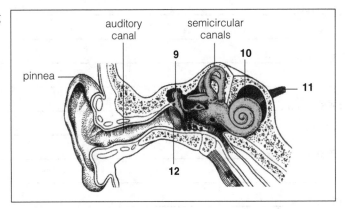

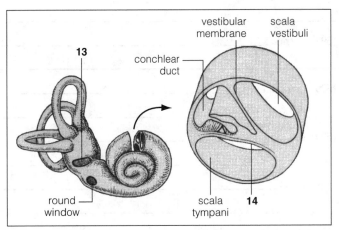

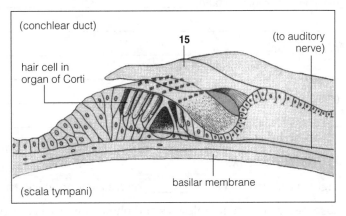

**36.6. SENSE OF VISION** (pp. 598–599)

**36.7. STRUCTURE AND FUNCTION OF VERTEBRATE EYES** (pp. 600–601)

**36.8. CASE STUDY: FROM SIGNALING TO VISUAL PERCEPTION** (pp. 602–603)

**36.9. *Focus on Science*: DISORDERS OF THE HUMAN EYE** (pp. 604–605)

*Selected Words: rhabdomeric and ciliary photoreceptors, simple eye, iris, vitreous body, sclera, choroid, pupil, ciliary muscle, fovea, optic disk, color-blind, red-green color blindness, astigmatism, nearsightedness, farsightedness, histoplasmosis,* Herpes simplex*, trachoma, cataracts, glaucoma, retinal detachment, corneal transplant surgery, radial keratotomy, laser coagulation*

## Boldfaced, Page-Referenced Terms

(598) vision _____

_____

(598) eyes _____

_____

(598) ocellus _____

_____

(598) visual field _____

_____

(598) lens _____

_____

(598) cornea _____

_____

(599) camera eyes _____

_____

(599) compound eye _____

_____

(599) retina _____

_____

(601) visual accommodation _____

_____

(602) rod cells _____

_____

(602) cone cells _____

_____

## Fill-in-the-Blanks

Light is a stream of (1) _____ –discrete energy packets. (2) _____ is a process in which photons are absorbed by pigment molecules and photon energy is transformed into the electrochemical energy of a nerve signal. (3) _____ requires precise light focusing onto a layer of photoreceptive cells that are dense enough to sample details of the light stimulus, followed by image formation in the brain.

(4) _____ are simple clusters of photosensitive cells, usually arranged in a cuplike depression in the epidermis. (5) _____ are well-developed photoreceptor organs that allow at least some degree of image formation. The (6) _____ is a transparent cover of the lens area, and the (7) _____ consists of tissue containing densely packed photoreceptors. Compound eyes contain several thousand photosensitive units known as (8) _____ . In the vertebrate eye, lens adjustments assure that the (9) _____ _____ for a specific group of light rays lands on the retina. (10) _____ refers to the lens adjustments that bring

about precise focusing onto the retina. (11) _____ people focus light from nearby objects posterior to the retina. (12) _____ cells are concerned with daytime vision and, usually, color perception. A (13) _____ is a funnel-shaped pit on the retina that provides the greatest visual acuity.

## Labeling

Identify each indicated part of the accompanying illustration.

14. _____ _____
15. _____
16. _____
17. _____
18. _____ _____
19. _____
20. _____
21. _____
22. _____ _____
23. _____ _____
24. _____

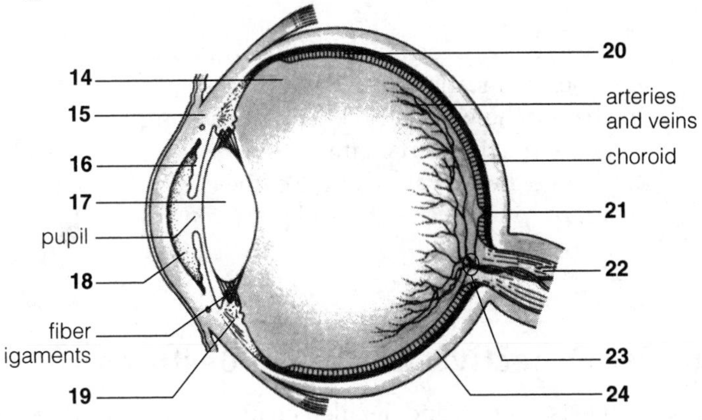

Labels on illustration:

20
arteries and veins
choroid
21
22
23
24
14
15
16
17
pupil
18
fiber ligaments
19

# Self-Quiz

___ 1. According to the mosaic theory, _____ .
  a. the basement membrane's pigment molecules prevent the scattering of light
  b. light falling on the inner area of an "on-center" field activates firing of the cells
  c. hair cells in the semicircular canals cooperate to detect rotational acceleration
  d. each ommatidium detects information about only one small region of the visual field; many ommatidia contribute "bits" to the total image
  e. all of the above

___ 2. The principal place in the human ear where sound waves are amplified is _____ .
  a. the pinna
  b. the ear canal
  c. the middle ear
  d. the organ of Corti
  e. none of the above

___ 3. The place where vibrations are translated into patterns of nerve impulses is _____.
  a. the pinna
  b. the ear canal
  c. the middle ear
  d. the organ of Corti
  e. none of the above

For questions 4–8, choose from the following answers:
  a. fovea
  b. cornea
  c. iris
  d. retina
  e. sclera

___ 4. The white protective fibrous tissue of the eye is the _____ .

___ 5. Rods and cones are located in the _____.

___ 6. The highest concentration of cones is in the _____ .

___ 7. The adjustable ring of contractile and connective tissues that controls the amount of light entering the eye is the _____ .

___ 8. The outer transparent protective covering of part of the eyeball is the _____ .

___ 9. Accommodation involves the ability to _____ .
   a. change the sensitivity of the rods and cones by means of transmitters
   b. change the width of the lens by relaxing or contracting certain muscles
   c. change the curvature of the cornea
   d. adapt to large changes in light intensity
   e. all of the above

___10. Nearsightedness is caused by _____ .
   a. eye structure that focuses an image in front of the retina
   b. uneven curvature of the lens
   c. eye structure that focuses an image posterior to the retina
   d. uneven curvature of the cornea
   e. none of the above

## Chapter Objectives/Review Questions

This section lists general and detailed chapter objectives that can be used as review questions. You can make maximum use of these items by writing answers on a separate sheet of paper. Fill in answers where blanks are provided. To check for accuracy, compare your answers with information given in the chapter or glossary.

*Page*       *Objectives/Questions*

(590)     1.  Define and distinguish among *chemoreceptors, mechanoreceptors, photoreceptors,* and *thermoreceptors.* Name at least one example of each type that appears in an animal.

(590)     2.  Distinguish the types of stimuli detected by tactile and stretch receptors from those detected by hearing and equilibrium receptors.

(590–595)  3.  Explain how a taste bud works and distinguish the types of stimuli it detects from those detected by touch or stretch receptors.

(595)     4.  Explain how the three semicircular canals of the human ear detect changes of position and acceleration in a variety of directions.

(596–597)  5.  Follow a sound wave from pinna to organ of Corti; mention the name of each structure it passes and state where the sound wave is amplified and where the pattern of pressure waves is translated into electrochemical impulses.

(596–597)  6.  State how low- and high-amplitude sounds affect the organ of Corti.

(597)     7.  State how low- and high-pitch sounds affect the organ of Corti.

(598)     8.  Explain what a visual system is and list four of the five aspects of a visual stimulus that are detected by different components of a visual system.

(598–600)  9.  Contrast the structure of compound eyes with the structures of invertebrate eyespots and of the human eye.

(600–601;  10. Define *nearsightedness* and *farsightedness* and relate each to eyeball structure.
605)

(602–603) 11. Describe how the human eye perceives color and black-and-white.

(602–603) 12. Explain the general principles that affect how light is detected by photoreceptors and changed into electrochemical messages.

## Integrating and Applying Key Concepts

How might human behavior be changed if human eyes were compound eyes composed of ommatidia and if humans perceived only vibrations—as fish do—rather than sounds?

# 37

# ENDOCRINE CONTROL

## Interactive Exercises

*Hormone Jamboree* (pp. 608–609)

## 37.1. THE ENDOCRINE SYSTEM (pp. 610–611)

*Selected Words:* *vomeronasal* organ, *hormon, endon, krinein*

*Boldfaced, Page-Referenced Terms*

(610) hormones _____

_____

(610) neurotransmitters _____

_____

(610) local signaling molecules _____

_____

(610) pheromones _____

_____

(611) endocrine system _____

_____

## Matching

Choose the most appropriate answer for each.

1. ___hormones
2. ___neurotransmitters
3. ___vomeronasal organ
4. ___target cells
5. ___local signaling molecules
6. ___pheromones

A. Signaling molecules released from axon endings of neurons that act swiftly on target cells
B. A pheromone detector discovered in humans
C. Released by many types of body cells and alter conditions within localized regions of tissues
D. Nearly odorless secretions of particular exocrine gland; common signaling molecules that act on cells of other animals of the same species and help integrate social behavior
E. Secretions from endocrine glands, endocrine cells, and some neurons that the bloodstream distributes to nonadjacent target cells
F. Cells that have receptors for any given type of signaling molecule

## Complete the Table

7. Complete the table below by identifying the numbered components of the endocrine system shown in the illustration on the facing page as well as the hormones produced by each gland (see text Figure 37.2, pp. 610–611).

| Gland Name | Number | Hormones Produced |
|---|---|---|
| a. Hypothalamus | | |
| b. Pituitary, anterior lobe | | |
| c. Pituitary, posterior lobe | | |
| d. Adrenal glands (cortex) | | |
| e. Adrenal glands (medulla) | | |
| f. Ovaries (two) | | |
| g. Testes (two) | | |
| h. Pineal | | |
| i. Thyroid | | |
| j. Parathyroids (four) | | |
| k. Thymus | | |
| l. Pancreatic islets | | |

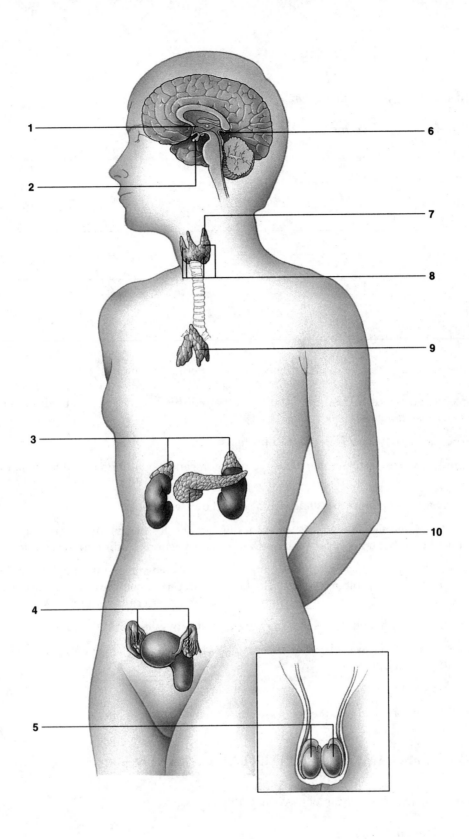

## 37.2. SIGNALING MECHANISMS (pp. 612–613)

*Selected Words:* *testicular feminization syndrome*

### Boldfaced, Page-Referenced Terms

(612) steroid hormones _____

_____

(613) peptide hormones _____

_____

(613) second messenger _____

_____

### Choice

For questions 1–10, choose from the following.

         a. steroid hormones      b. peptide hormones

1. ____ Lipid-soluble molecules derived from cholesterol; can diffuse directly across the lipid bilayer of a target cell's plasma membrane.

2. ____ Various peptides, polypeptides, and glycoproteins.

3. ____ One example involves testosterone, defective receptors, and a condition called testicular feminization syndrome.

4. ____ Hormones that often require assistance from second messengers.

5. ____ Hormones that issue signals right at a receptor of a target cell's plasma membrane; signals activate specific membrane-bound enzyme systems that initiate reactions leading to the cellular response.

6. ____ Lipid-soluble hormones that move through the target cell's plasma membrane to the nucleus, where it binds to some type of protein receptor. The hormone-receptor complex moves into the nucleus and interacts with specific DNA regions to stimulate or inhibit transcription of mRNA.

7. ____ Water-soluble signaling molecules that may incorporate anywhere from 3 to 180 amino acids.

8. ____ Involves molecules such as cyclic AMP that activates many enzymes in cytoplasm that, in turn, cause alteration in some cell activity.

9. ____ Glucagon is an example.

10. ____ Cyclic AMP relays a signal into the cell's interior to activate protein kinase A.

## 37.3. THE HYPOTHALAMUS AND PITUITARY GLAND (pp. 614–615)

*Selected Words:* *posterior* lobe, *anterior* lobe, *second* capillary bed

### Boldfaced, Page-Referenced Terms

(614) hypothalamus _____

_____

(614) pituitary gland _____

_____

(615) releasers _____

_____

(615) inhibitors _____

_____

## Choice-Match

Label each hormone given below with an "A" if it is secreted by the anterior lobe of the pituitary, a "P" if it is released from the posterior pituitary, or an "I" if it is released from intermediate tissue. Complete the exercise by entering the letter of the corresponding action in the parentheses following each label.

1. ___ ( ) ACTH
2. ___ ( ) ADH
3. ___ ( ) FSH
4. ___ ( ) STH (GH)
5. ___ ( ) LH
6. ___ ( ) MSH
7. ___ ( ) OCT
8. ___ ( ) PRL
9. ___ ( ) TSH

A. Stimulates egg and sperm formation in ovaries and testes
B. Targets are pigmented cells in skin and other surface coverings; induces color changes in response to external stimuli and affects some behaviors
C. Stimulates and sustains milk production in mammary glands
D. Stimulates progesterone secretion, ovulation and corpus luteum formation in females; promotes testosterone secretion and sperm release in males
E. Induces uterine contractions and milk movement into secretory ducts of the mammary glands
F. Stimulates release of thyroid hormones from the thyroid gland
G. Acts on the kidneys to conserve water required in control of extracellular fluid volume
H. Stimulates release of adrenal steroid hormones from the adrenal cortex
I. Promotes growth in young; induces protein synthesis and cell division; has roles in adult glucose and protein metabolism

## Dichotomous Choice

Circle one of two possible answers given between parentheses in each statement.

10. The (hypothalamus/pituitary gland) region of the brain monitors internal organs and activities related to their functioning, such as eating and sexual behavior; it also secretes some hormones.
11. The (posterior/anterior) lobe of the pituitary stores and secretes two hormones, ADH and OCT, that are produced by the hypothalamus.
12. The (posterior/anterior) lobe of the pituitary produces and secretes its own hormones that govern the release of hormones from other endocrine glands.
13. Humans lack the (posterior/intermediate) lobe of the pituitary gland, one that is possessed by many other vertebrates.
14. Most hypothalamic hormones acting in the anterior pituitary lobe are (releasers/inhibitors) and cause target cells there to secrete hormones of their own.
15. Some hypothalamic hormones slow down secretion from their targets in the anterior pituitary; these are classed as (releasers/inhibitors).

## 37.4. EXAMPLES OF ABNORMAL PITUITARY OUTPUT (p. 616)

**Selected Words:** *pituitary dwarfism, acromegaly, diabetes insipidus*

## Complete the Table

1. Complete the table below to summarize examples of abnormal pituitary output.

| Condition | Hormone/Abnormality | Characteristics |
|---|---|---|
| a. | Excessive somatotropin produced during childhood | Affected adults are proportionally similar to a normal person but larger |
| b. | Insufficient somatotropin produced during childhood | Affected adults are proportionally similar to a normal person but much smaller |
| c. | Diminished ADH secretion by a damaged posterior pituitary lobe | Large volumes of dilute urine are secreted causing life-threatening dehydration |
| d. | Excessive somatotropin output during adulthood when long bones can no longer lengthen | Abnormal thickening of bone, cartilage, and other connective tissues in the hands, feet, and jaws |

## 37.5. SOURCES AND EFFECTS OF OTHER HORMONES (p. 617)

1. Complete the table below by matching the gland/organ and the hormone(s) produced by it to the descriptions of hormone action. Refer to Table 37.3 and p. 617 in the text.

Gland/Organ

A. adrenal cortex
B. adrenal medulla
C. thyroid
D. parathyroids
E. testes
F. ovaries
G. pancreas (alpha cells)
H. pancreas (beta cells)
I. pancreas (delta cells)
J. thymus
K. pineal

Hormones

(a)  thyroxine and triiodothyronine
(b)  glucagon
(c)  PTH
(d)  androgens (includes testosterone)
(e)  somatostatin
(f)  thymosins
(g)  glucocorticoids
(h)  estrogens, general (includes progesterone)
(i)  epinephrine
(j)  melatonin
(k)  insulin
(l)  progesterone
(m)  mineralocorticoids (including aldosterone)
(n)  calcitonin
(o)  norepinephrine

| Gland/Organ | Hormone | Hormone Action |
|---|---|---|
| a. | | Elevates calcium levels in blood |
| b. | | Influences carbohydrate metabolism of insulin-secreting cells |
| c. | | Required in egg maturation and release; preparation of uterine lining for pregnancy and its maintenance in pregnancy; influences growth and development; genital development; maintains sexual traits |
| d. | | Promote protein breakdown and conversion to glucose in most cells |
| e. | | Lowers blood sugar level in muscle and adipose tissue |
| f. | | In general, required in sperm formation; genital development and maintenance of sexual traits; influences growth and development |
| g. | | In most cells, regulates metabolism, plays roles in growth and development |
| h. | | In the gonads, influences daily biorhythms and influences gonad development and reproductive cycles |
| i. | | In liver, muscle, and adipose tissues; raises blood sugar level, fatty acids; increases heart rate and force of contraction |
| j. | | Targets lymphocytes, has roles in immunity |
| k. | | Raises blood sugar level |
| l. | | Prepares, maintains uterine lining for pregnancy; stimulates breast development |
| m. | | In kidneys, promotes sodium reabsorption and control of salt-water balance |
| n. | | In smooth muscle cells of blood vessels, promotes constriction or dilation of blood vessels |
| o. | | In bone, lowers calcium levels in blood |

## 37.6. FEEDBACK CONTROL OF HORMONAL SECRETIONS (pp. 618–619)

**Selected Words:** *inhibit, hypoglycemia, stress response, fight-flight response, goiter, hypothyroidism, hyperthyroidism, primary* reproductive organs

### Boldfaced, Page-Referenced Terms

(618) negative feedback _____

_____

(618) positive feedback _____

_____

(618) adrenal cortex _____

_____

(618 adrenal medulla _____

_____

(618) thyroid gland _____

_____

(619) gonads _____

_____

## Matching

Choose the most appropriate answer for each.

1. ___adrenal medulla
2. ___ACTH
3. ___nervous system
4. ___glucocorticoids
5. ___CRH
6. ___cortisol
7. ___hypoglycemia
8. ___cortisol-like drugs
9. ___adrenal cortex
10. ___feedback mechanisms

A. Occurs when the glucose blood level falls below a set point, and a negative feedback mechanism kicks in
B. Outer portion of each adrenal gland; some of its cells secrete hormones such as glucocorticoids
C. Negative ultimately inhibits hormone secretion, positive stimulates further hormone secretion
D. Initiates a stress response during severe stress, painful injury, or prolonged illness; cortisol helps to suppress inflammation
E. Inner portion of the adrenal gland; neurons located here release epinephrine and norepinephrine
F. Stimulates the adrenal cortex to secrete cortisol; this helps raise the level of glucose by preventing muscle cells from taking up more blood glucose
G. Used to counter asthma and other chronic inflammatory disorders
H. Help maintain blood glucose concentration and help suppress inflammatory responses
I. Secreted by the hypothalamus in response to falling glucose blood level; stimulates the anterior pituitary to secrete ACTH
J. Secreted by the adrenal cortex; blocks the uptake and use of blood glucose by muscle cells; also stimulates liver cells to form glucose from amino acids

## Fill-in-the-Blanks

Thyroxine and triiodothyronine are the main hormones secreted by the human (11) _____ gland. They are critical for normal development of many tissues, and they control overall (12) _____ rates in humans and other warm-blooded animals. The synthesis of thyroid hormones requires (13) _____ , which is obtained from food. When iodine is absorbed from the gut, iodine is converted to (14) _____ . In the absence of that element, blood levels of these hormones decrease. The anterior pituitary responds by secreting (15) _____ . When thyroid hormones cannot be synthesized, the feedback signal continues— and so does TSH secretion. TSH secretion continues and overstimulates the thyroid gland and causes (16) _____ , an enlargement of the thyroid gland. (17) _____ results from insufficient blood level concentrations of thyroid hormones. Such (18) _____ adults are often overweight, sluggish, dry-skinned, intolerant of cold, and sometimes confused and depressed. (19) _____ results from excess concentrations of thyroid hormones. Affected adults show an increased heart rate, elevated blood pressure, weight loss despite normal caloric intake, heat intolerance, and profuse sweating. They are also typically nervous, agitated, and have trouble sleeping.

The primary reproductive organs are known as (20) _____ . These organs produce and secrete sex (21) _____ essential to reproduction. These organs are known as the (22) _____ in human males and (23) _____ in females. Testes secrete (24) _____ . Ovaries secrete estrogens and progesterone. All these hormones influence (25) _____ sexual traits. Both types of organs produce (26) _____ or sex cells.

## 37.7. RESPONSES TO LOCAL CHEMICAL CHANGES (pp. 620–621)

*Selected Words:* rickets, exocrine cells, endocrine cells, alpha cells, glucagon, beta cells, insulin, delta cells, diabetes mellitus, type 1 diabetes, type 2 diabetes

### Boldfaced, Page-Referenced Terms

(620) parathyroid glands _____

_____

(620) pancreatic islet _____

_____

### Fill-in-the-Blanks

Humans have four (1) _____ glands positioned next to the posterior (or back) of the human thyroid; they secrete (2) _____ in response to a low (3) _____ level in blood. This hormone induces living bone cells to secrete enzymes that digest bone tissue and thereby release (4) _____ and other minerals to interstitial fluid, then to the blood. It enhances calcium (5) _____ from the filtrate flowing from the nephrons of the kidneys. PTH induces some kidney cells to secrete enzymes that act on blood-borne precursors of the active form of vitamin (6) _____ , a hormone. This hormone stimulates (7) _____ cells to increase calcium absorption from the gut lumen. In a child with vitamin D deficiency, too little calcium and phosphorus are absorbed and so rapidly growing bones develop improperly. This ailment is called (8) _____ and is characterized by bowed legs, a malformed pelvis, and in many cases a malformed skull and rib cage.

### Complete the Table

9. Complete the table below to summarize function of the pancreatic islets.

| Pancreatic Islet Cells | Hormone Secreted | Hormone Action |
| --- | --- | --- |
| a. Alpha cells | | |
| b. Beta cells | | |
| c. Delta cells | | |

## Dichotomous Choice

Circle one of two possible answers given between parentheses in each statement.

10. Insulin deficiency can lead to diabetes mellitus, a disorder in which the glucose level (rises/decreases) in the blood, then in the urine.
11. In a person with diabetes mellitus, urination becomes (reduced/excessive), so the body's water-solute balance becomes disrupted; people become abnormally dehydrated and thirsty.
12. Lacking a steady glucose supply, body cells of persons with diabetes mellitus begin breaking down their own fats and proteins for (energy/water).
13. Weight loss occurs and (amino acids/ketones) accumulate in blood and urine; this promotes excessive water loss with a life-threatening disruption of brain function.
14. After a meal, blood glucose rises; pancreatic beta cells secrete (glucagon/insulin); targets use glucose or store it as glycogen.
15. Blood glucose levels decrease between meals. (Glucagon/Insulin) is secreted by stimulated pancreas alpha cells; targets convert glycogen back to glucose, which then enters the blood.
16. In ("type 1 diabetes"/"type 2 diabetes") the body mistakenly mounts an autoimmune response against its own insulin-secreting beta cells and destroys them.
17. Juvenile-onset diabetes is also known as ("type 1 diabetes"/"type 2 diabetes"); these patients survive with insulin injections.
18. In ("type 1 diabetes"/"type 2 diabetes"), insulin levels are close to or above normal, but target cells fail to respond to insulin.
19. ("Type 1 diabetes"/"Type 2 diabetes") usually is manifested during middle age and is less dramatically dangerous than the other type; beta cells produce less insulin as a person ages.

## 37.8. HORMONAL RESPONSES TO ENVIRONMENTAL CUES (pp. 622–623)

*Selected Words: winter blues*

### Boldfaced, Page-Referenced Terms

(622) pineal gland _____

_____

(622) puberty _____

_____

(623) molting _____

_____

## Choice

For questions 1–10, choose from the following.

a. melatonin       b. ecdysone

1. ___ The hormone that controls molting

2. ___ Hormone secreted by the pineal gland

3. ___ High blood levels of this hormone in winter (long nights) suppresses sexual activity in hamsters

4. ___ Winter blues

5. ___ Chemical interactions that cause an old cuticle to detach from the epidermis and muscles

6. ___ Triggers puberty

7.___Suppresses growth of a bird's gonads in fall and winter

8.___Jet lag

9.___Waking up at sunrise

10.___Stored by insects and crustaceans

---

# Self-Quiz

___ 1. The ____ governs the release of hormones from other endocrine glands; it is controlled by the _____ .
a. pituitary, hypothalamus
b. pancreas, hypothalamus
c. thyroid, parathyroid glands
d. hypothalamus, pituitary
e. pituitary, thalamus

___ 2. Neurons of the _____ produce ADH and oxytocin that are stored within axon endings of the _____ .
a. anterior pituitary, posterior pituitary
b. adrenal cortex, adrenal medulla
c. posterior pituitary, hypothalamus
d. posterior pituitary, thyroid
e. hypothalamus, posterior pituitary

___ 3. If you were lost in the desert and had no fresh water to drink, the level of _____ in your blood would increase as a means to conserve water.
a. insulin
b. corticotropin
c. oxytocin
d. antidiuretic hormone
e. salt

For questions 4–6, choose from the following answers:
a. estrogen
b. PTH
c. FSH
d. somatotropin
e. prolactin

___ 4. _____ stimulates bone cells to release calcium and phosphate and the kidneys to conserve it.

___ 5. _____ stimulates and sustains milk production in mammary glands.

___ 6. Protein synthesis and cell division are activities stimulated by _____.

For questions 7–9, choose from the following answers:
a. adrenal medulla
b. adrenal cortex
c. thyroid
d. anterior pituitary
e. posterior pituitary

___ 7. The _____ produces glucocorticoids that help maintain the blood level of glucose and suppress inflammatory responses.

___ 8. The gland that is most closely associated with emergency situations is the _____ .

___ 9. The _____ gland regulates the basic metabolic rate.

___10. If all sources of calcium were eliminated from your diet, your body would secrete more _____ in an effort to release calcium stored in your body and send it to the tissues that require it.
a. parathyroid hormone
b. aldosterone
c. calcitonin
d. mineralocorticoids
e. none of the above

## Matching

11. ___ACTH

12. ___ADH

13. ___thymosins

14. ___oxytocin

15. ___cortisol

16. ___epinephrine and norepinephrine

17. ___estrogens

18. ___glucagon

19. ___insulin

20. ___melatonin

21. ___parathyroid hormone

22. ___STH (GH)

23. ___calcitonin

24. ___testosterone

25. ___thyroxine

26. ___progesterone

27. ___TSH

A. Raises the glucose level in the blood

B. Influences daily biorhythms, gonad development, and reproductive cycles

C. Affects development and maintenance of male sexual traits; required in sperm formation, genital development, growth, development

D. Increases heart rate and contraction, controls blood volume, dilates lung airways; controls features of the "fight-flight" response

E. Produced by ovaries; genital development, essential for egg maturation and release, prepares and maintains uterine lining for pregnancy, maintenance of secondary sex characteristics in the female, growth, development

F. The water conservation hormone; released from posterior pituitary

G. Lowers blood sugar by encouraging muscle and adipose cells to take in glucose; promotes the synthesis of proteins and fats; inhibits protein conversion to glucose

H. Stimulates adrenal cortex to secrete cortisol

I. Elevates calcium levels in blood by stimulating calcium reabsorption from bone and kidneys and calcium absorption from gut

J. Influences overall metabolic rate, growth and development

K. Roles in immunity

L. Triggers uterine contractions during labor and causes milk release during nursing

M. Prepares, maintains uterine lining for pregnancy; stimulates breast development

N. Inhibits uptake and use of blood glucose by muscle cells

O. Lowers calcium levels in blood; bone is the target

P. Secreted by anterior pituitary; stimulates release of thyroid hormones

Q. Secreted by anterior pituitary; enhances growth in young animals, especially of cartilage and bone; induces protein synthesis, cell division, roles in glucose, protein metabolism in adults

# Chapter Objectives/Review Questions

*Page*      *Objectives/Questions*

(610)      1.  Hormones, neurotransmitters, local signaling molecules, and pheromones are all known as _____ molecules that carry out integration.

(610)      2.  _____ are the secretory products of endocrine glands, endocrine cells, and some neurons.

(610)      3.  Define *neurotransmitters, local signaling molecules,* and *pheromones.*

(611)      4.  Collectively, sources of hormones came to be viewed as the _____ system.

(611)      5.  Be able to locate and name the components of the human endocrine systems on a diagram such as text Figure 37.2b.

(611)      6.  _____ cells are any cells that have receptors for a specific signaling molecule and that may alter their activities in response to it.

(612–613)  7.  Contrast the proposed mechanisms of hormonal action on target cell activities by (a) steroid hormones and (b) peptide hormones that are proteins or are derived from proteins.

(614)      8.  The _____ and the pituitary gland interact as a major neural-endocrine control center.

(614)  9. Explain how, even though the anterior and posterior lobes of the pituitary are compounded as one gland, the tissues of each part differ in character.

(614–615) 10. Identify the hormones released from the posterior lobe of the pituitary and state their target tissues.

(614–615) 11. Identify the hormones produced by the anterior lobe of the pituitary and tell which target tissues or organs each acts on.

(615) 12. Most hypothalamic hormones acting in the anterior lobe are _____ ; they cause target cells to secrete hormones of their own but others are _____ that slow down secretion from their targets.

(616) 13. Pituitary dwarfism, gigantism, and acromegaly are all associated with abnormal secretion of _____ by the pituitary gland.

(616) 14. One cause of _____ _____ is damage to the pituitary's posterior lobe and the diminished secretion or lack of ADH.

(617) 15. Be familiar with the major human hormone sources, their secretions, main targets, and primary actions as shown on text Table 37.3.

(618) 16. With _____ feedback, an increase or decrease in the concentration of a secreted hormone triggers events that inhibit further secretion.

(618) 17. With _____ feedback, an increase in the concentration of a secreted hormone triggers events that stimulate further secretion.

(618) 18. The adrenal _____ secretes glucocorticoids.

(618) 19. Define *hypoglycemia*; relate the role of the hypothalamus and the anterior pituitary in this condition.

(618) 20. The adrenal _____ contains neurons that secrete epinephrine and norepinephrine.

(618) 21. The _____ system initiates the stress response.

(618) 22. Describe the role of cortisol in a stress response.

(618) 23. List the features of the fight-flight response.

(619) 24. Excess _____ in the blood overstimulates the thyroid gland; this causes an enlargement known as a form of _____ .

(619) 25. Describe the characteristics of hypothyroidism and hyperthyroidism.

(619) 26. The _____ are primary reproductive organs that produce and secrete hormones with essential roles in reproduction.

(620) 27. Name the glands that secrete PTH and state the function of this hormone.

(620) 28. Describe an ailment called rickets and state its cause.

(620) 29. Give two examples that illustrate the effects of *local signaling molecules.*

(620) 30. Be able to name the hormones secreted by alpha, beta, and delta pancreatic cells; list the effect of each.

(620–621) 31. Describe the symptoms of diabetes mellitus and distinguish between type 1 and type 2 diabetes.

(622) 32. The pineal gland secretes the hormone _____ ; relate two examples of the action of this hormone.

(622) 33. Explain the cause of "jet lag" and "winter blues."

(623) 34. The invertebrate hormone _____ is related to the control of a phenomenon known as _____ that occurs among crustaceans and insects.

---

# Integrating and Applying Key Concepts

Suppose you suddenly quadruple your already high daily consumption of calcium. State which body organs would be affected and tell how they would be affected. Name two hormones whose levels would most probably be affected and tell whether your body's production of them would increase or decrease. Suppose you continue this high rate of calcium consumption for ten years. Can you predict the organs that would be subject to the most stress as a result?

# 38

# PROTECTION, SUPPORT, AND MOVEMENT

---

## Interactive Exercises

---

*Of Men, Women, and Polar Huskies* (pp. 626–627)

**38.1. INTEGUMENTARY SYSTEM** (pp. 628–629)

**38.2. A LOOK AT HUMAN SKIN** (p. 630)

**38.3.** *Focus on Health:* **SUNLIGHT AND SKIN** (p. 631)

*Selected Words:* collagen, melanin, keratin, *stratified* epithelium, *cold sweats*, *acne*, "split ends", *hirsutism*, vitamin D, *aging*, elastin, *cold sores*

*Boldfaced, Page-Referenced Terms*

(628) integument _____

_____

(628) cuticle _____

_____

(628) skin _____

_____

(628) epidermis _____

_____

(628) dermis _____

_____

(630) keratinocytes _____

_____

(630) melanocytes _____

_____

(630) sweat glands _____

_____

(630) oil glands _____

_____

(630) hair _____

_____

(631) Langerhans cells _____

_____

(631) Granstein cells _____

_____

## Fill-in-the-Blanks

Human skin is an organ system that consists of two layers: the outermost (1) _____ , which contains mostly dead cells, and the (2) _____ , which contains hair follicles, nerves, tiny muscles associated with the hairs, and various types of glands.  The (3) _____ layer, with its loose connective tissue and store of fat in (4) _____ tissue, lies beneath the skin.

## Labeling

Label the numbered parts of the illustration below.

5. _____

6. _____ _____
   _____

7. _____ _____

8. _____ _____

9. _____ _____

10. _____ _____

11. _____ _____

12. _____

13. _____

14. _____

## Choice

Match the following proteins (or protein derivatives) with the particular ability that each substance lends to the skin. For questions 15–21, choose from these letters:

   a. collagen        b. elastin        c. keratin        d. melanin        e. hemoglobin

15.\_\_\_\_Fibers that run through the dermis and lend a flexible, but substantive structure to it

16.\_\_\_\_Protects against loss of moisture

17.\_\_\_\_Protects against ultraviolet radiation and sunburn

18.\_\_\_\_Helps skin to be stretchable, yet return to its previous shape

19.\_\_\_\_Beaks, hooves, hair, fingernails and claws contain a lot of this

20.\_\_\_\_Helps ward off bacterial attack by making the skin surface rather impermeable

21.\_\_\_\_Located in red blood cells; binds with $O_2$

## Fill-in-the-Blanks

The sun's ultraviolet wavelengths stimulate (22) _____ production in melanocytes. (23) _____ glands lubricate and soften hair and the skin; (24) _____ is a skin inflammation occurring after bacteria have successfully infected the ducts leading from these glands. (25) _____ , excessive hairiness, may result when the body produces abnormal amounts of testosterone.

Cholecalciferol, otherwise known as (26) _____ _____ , is a steroid-like compound that helps the human body absorb (27) _____ from food. (26) is produced by special cells in the skin upon exposure to (28) _____ from a precursor molecule that is related to (29) _____ . (26) is released into the blood stream and it travels to absorptive cells in the lining of the (30) _____ where it acts.

Sunlight also damages two types of cells that mobilize the body's (31) _____ system. (32) _____ cells are phagocytes produced in the bone marrow but later reside in the skin where they envelope and digest viruses and bacteria; they then signal other parts of the immune system to join the battle and round up the foreign invaders. Exposure to excess ultraviolet radiation can damage (32) cells and render the skin susceptible to viral infections such as (33) _____ _____ . (34) _____ cells issue suppressor signals that dampen immune responses and keep them under control.

## 38.4. TYPES OF SKELETONS (pp. 632–633)
## 38.5. CHARACTERISTICS OF BONE (pp. 634–635)
## 38.6. HUMAN SKELETAL SYSTEM (pp. 636–637)

*Selected Words:* *hydraulic,* pressure, cuticle, cartilage, *compact* bone tissue, Haversian system, *spongy* bone tissue, osteoblast, shaft, *osteoporosis, appendicular* and *axial* portions, *herniated disks, fibrous, cartilaginous* and *synovial* joints, *strain, sprain, osteoarthritis, rheumatoid arthritis,* vertebral column, cranium, clavicle, sternum, scapula, radius, carpal bones, femur, tibia, tarsal bones, metatarsals

### *Boldfaced, Page-Referenced Terms*

(632) hydrostatic skeleton _____

_____

(632) exoskeleton _____

_____

(632) endoskeleton _____

_____

(634) bones _____

_____

(635) red marrow _____

_____

(635) yellow marrow _____

_____

(635) osteocytes _____

_____

(635) bone tissue turnover _____

_____

(636) vertebrae _____

_____

(636) intervertebral disks _____

_____

(636) joints _____

_____

(636) ligaments _____

_____

## Fill-in-the-Blanks

All motor systems are based on muscle cells that are able to (1) _____ and (2) _____ , and on the presence of a medium against which the (3) _____ force can be applied. Longitudinal and radial muscle layers work as an (4) _____ muscle system, in which the action of one motor element opposes the action of another. A membrane filled with fluid resists compression and can act as a (5) _____ skeleton. Arthropods have muscles attached to an (6) [choose one] ☐ endoskeleton, ☐ exoskeleton.

(7) _____ are hard compartments that enclose and protect the brain, spinal cord, heart, lungs, and other vital organs of vertebrates. Bones support and anchor (8) _____ and soft organs, such as eyes.

(9) _____ systems are found in the long bones of mammals and contain living bone cells that receive their nutrients from the blood. (10) _____ _____ is a major site of blood cell formation. Bone tissue serves as a "bank" for (11) _____ , (12) _____ , and other mineral ions; depending on metabolic needs, the body deposits ions into and withdraws ions from this "bank."

Bones develop from (13) _____ secreting material inside the shaft and on the surface of the cartilage model. Bone can also give ions back to interstitial fluid as bone cells dissolve out component minerals and remodel bone in response to (14) _____ signals, lack of exercise, and calcium deficiencies in the diet. Extreme decreases in bone density result in (15) _____ , particularly among older women.

The (16) _____ portion of the human skeleton includes the skull, vertebral column, ribs, and breastbone; the (17) _____ portion includes the pectoral and pelvic girdles and the forelimbs and hindlimbs, when they exist in vertebrates.

(18) _____ joints are freely movable and are lubricated by a fluid secreted into the capsule of dense connective tissue that surrounds the bones of the joint. Bones are often tipped with (19) _____ ; as a person ages, the cartilage at (20) _____ joints may simply wear away, a condition called (21) _____ . By contrast, in (22) _____ _____ , the synovial membrane becomes inflamed, cartilage degenerates, and bone becomes deposited in the joint.

## Labeling

Identify each indicated part of the accompanying illustrations.

23. _____  _____
24. _____  _____  _____
25. _____  _____
26. _____  _____
27. _____  _____  _____
28. _____  _____
29. _____  _____
30. _____  _____
31. _____

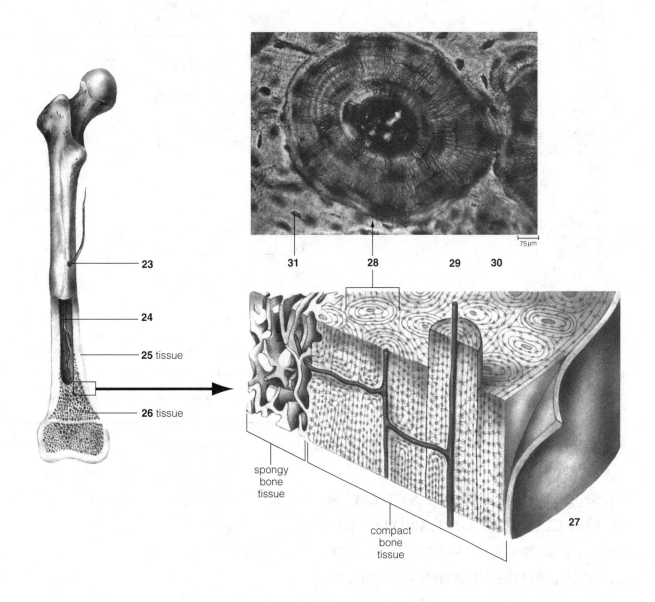

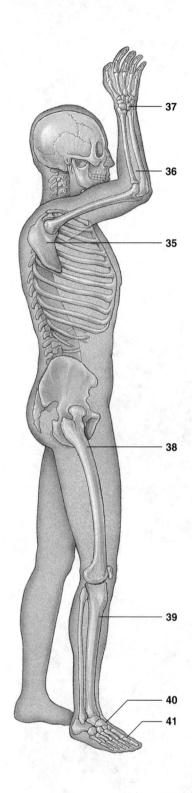

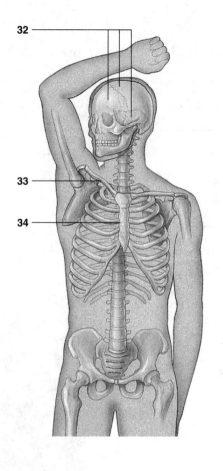

32. _____
33. _____
34. _____
35. _____
36. _____
37. _____   _____
38. _____
39. _____
40. _____   _____
41. _____

**Selected Words:** *skeletal* muscle, *biceps brachii, deltoid, gluteus maximus, biceps femoris, gastrocnemius, pectoralis major, rectus abdominis, quadriceps femoris, latissimus dorsi, trapezius, myofibril, T tubules, isometric, isotonic and legthening contractions, muscle fatigue, aerobic exercise, strength training*

## Boldfaced, Page-Referenced Terms

(638) skeletal muscles _____

_____

(638) tendons _____

_____

(640) sarcomeres _____

_____

(640) actin _____

_____

(640) myosin _____

_____

(641) sliding-filament model _____

_____

(641) cross-bridge formation _____

_____

(642) action potential _____

_____

(642) sarcoplasmic reticulum _____

_____

(643) creatine phosphate _____

_____

(643) oxygen debt _____

_____

(644) muscle tension _____

_____

(644) motor unit _____

_____

(644) muscle twitch _____

_____

(644) tetanus _____

_____

(644) exercise _____

_____

(645) anabolic steroids _____

_____

## Sequence

Arrange in order of decreasing size.

1. ___
2. ___
3. ___
4. ___
5. ___
6. ___

A. Muscle fiber (muscle cell)
B. Myosin filament
C. Muscle bundle
D. Muscle
E. Myofibril
F. Actin filament

## Labeling

Identify each indicated part of the accompanying illustration.

7. _____ _____
8. _____ _____
9. _____ _____
10. _____ _____
11. _____ _____
12. _____ _____
13. _____
14. _____
15. _____
16. _____ _____
17. _____ _____
18. _____ _____
19. _____ _____
20. _____ _____

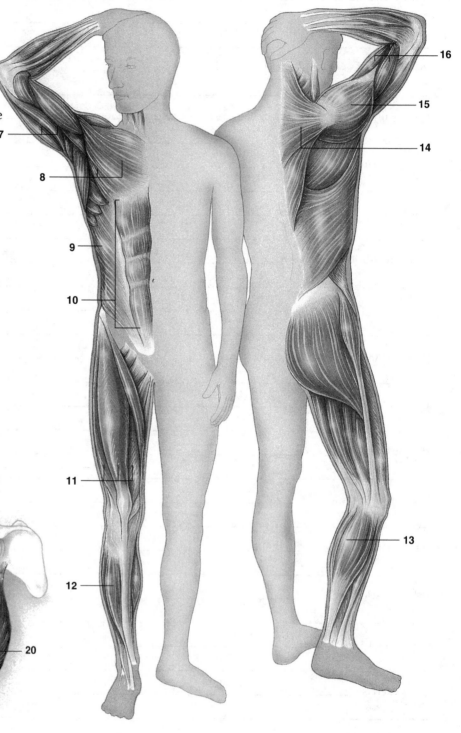

## Fill-in-the-Blanks

There are three types of muscle tissue: (21) _____ , which is striated, involuntary, and located in the heart; smooth, which is involuntary, not striated, and mostly located in the walls of internal organs in vertebrates, and (22) _____ which is striated, largely voluntary, and generally attached to (23) _____ or cartilage by means of (24) _____ . (25) _____ muscle interacts with the skeleton to bring positional changes of body parts and to move the animal through its environment.

Together, the skeleton and its attached muscles are like a system of levers in which rigid rods, (26) _____ , move about at fixed points, called (27) _____ . Most attachments are close to joints, so a muscle has to shorten only a small distance to produce a large movement of some body part.

When the (28) _____ _____ contracts, the elbow joint bends (flexes). As it relaxes and as its partner, the (29) _____ _____ , contracts, the forelimb extends and straightens.

Muscle cells have three properties in common: contractility, elasticity, and (30) _____ . Like a (31) _____ , a muscle cell is "excited" when a wave of electrical disturbance, a (an) (32) _____ _____ , travels along its cell membrane. Neurons that send signals to muscles are (33) _____ neurons.

Each muscle cell contains (34) _____ : threadlike structures packed together in parallel array. Every (34) is functionally divided into (35) _____ , which appear to be striped and are arranged one after another along its length. Each myofibril contains (36) _____ filaments, which have cross-bridges and (37) _____ filaments, which are thin and lack cross-bridges. According to the (38) _____ -_____ model, (39) _____ filaments physically slide along actin filaments and pull them toward the center of a (40) _____ during a contraction. The energy that drives the forming and breaking of the cross-bridges comes immediately from (41) _____ , which obtained its phosphate group from (42) _____ _____ which is stored in muscle cells.

(43) _____ of an entire muscle is brought about by the combined decreases in length of the individual sarcomeres that make up the myofibrils of the muscle cells.

The parallel orientation of muscle's component parts directs the force of contraction toward a (44) _____ that must be pulled in some direction.

When excited by incoming signals from motor neurons, muscle cell membranes cause (45) _____ ions to be released from the (46) _____ _____ . These ions remove all obstacles that might interfere with myosin heads binding to the sites along (47) _____ filaments. When muscle cells relax/rest, calcium ions are sent back to the sarcoplasmic reticulum by (48) _____ _____ .

By controlling the (49) _____ _____ that reach the (50) _____ _____ in the first place, the nervous system controls muscle contraction by controlling calcium ion levels in muscle tissue.

## Complete the Table

For each item in the table below, state its specific role in muscle contraction.

| | |
|---|---|
| Aerobic respiration | 51. |
| ATP | 52. |
| Calcium ions | 53. |
| Creatine phosphate | 54. |
| Glycogen | 55. |
| Glycolysis alone | 56. |
| Motor neuron | 57. |
| Myosin heads | 58. |
| Sarcomere | 59. |
| Sarcoplasmic reticulum | 60. |

## Fill-in-the-Blanks

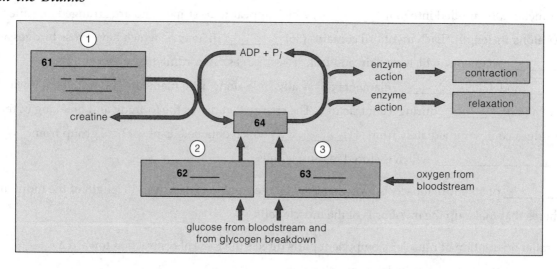

## Fill-in-the-Blanks

The larger the (65) _____ of a muscle, the greater its strength. A (66) _____ neuron and the muscle cells under its control are called a (67) _____ _____ . A (68) _____ _____ is a response in which a muscle contracts briefly when reacting to a single, brief stimulus and then relaxes. The strength of the muscular contraction depends on how far the (69) _____ response has proceeded by the time another signal arrives. (70) _____ is the state of contraction in which a motor unit that is being stimulated repeatedly is maintained. In a (71) _____ contraction, the nervous system activates only a small number of motor units; in a stronger contraction, a larger number are activated at a high (72) _____ of stimulation.

## Labeling

Label the numbered parts of the accompanying illustrations.

73. _____ bundle

74. _____ _____

75. _____

76. _____

77. _____ _____

78. _____ _____

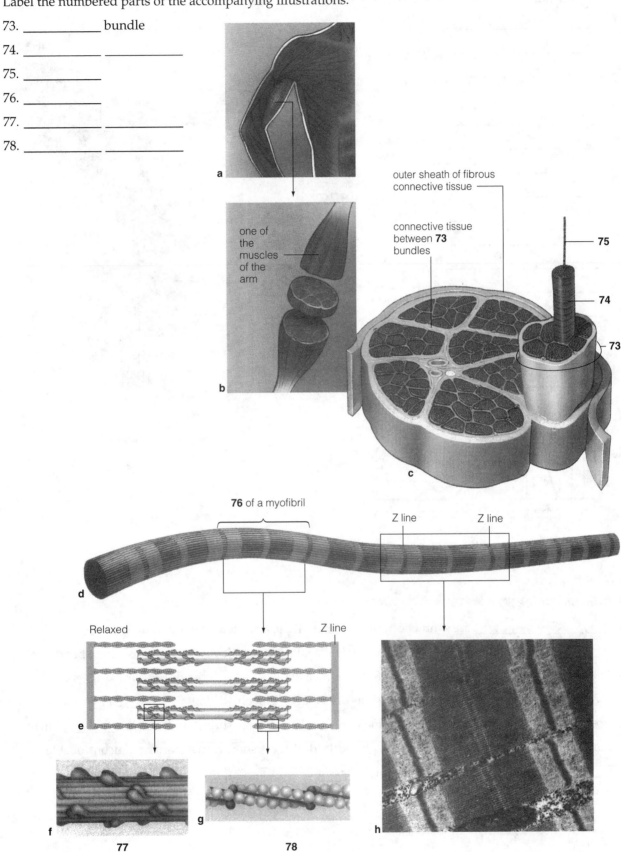

outer sheath of fibrous connective tissue

connective tissue between **73** bundles

one of the muscles of the arm

**76** of a myofibril

Z line            Z line

Relaxed                                    Z line

**77**                        **78**

## Labeling

Identify the numbered parts of the accompanying illustrations.

79. _____

_____

80. _____

_____

81. _____

_____

82. _____

_____

83. _____

_____

84. _____

85. _____

86. _____ (_____)

87. _____

_____

spinal cord
(section)

**79**

**81**

**80**

**82**

**87**
(Y-axis label)

latent
period

**84**    **85**

**83**    **86**
(X-axis label)

## Fill-in-the-Blanks

(88) _____ _____ are synthetic hormones that mimic testosterone in building greater

(89) _____ mass in both men and women.  It is illegal for competitive athletes to use them because of

unfair advantage and because of the side effects.  In men, (90) _____ , baldness, shrinking

(91) _____ , and infertility are the first signs of damage.  In women, (92) _____ hair becomes more

noticeable, (93) _____ _____ become irregular, breasts may shrink, and the (94) _____ may

become grossly enlarged.  Aside from these physical side effects, some men experience uncontrollable

(95) _____ , delusions, and wildly manic behavior.

# Self-Quiz

For questions 1–7, choose from the following answers:

    a. bone
    b. cartilage
    c. epidermis
    d. dermis
    e. hypodermis

____ 1. Fat cells in adipose tissue are most likely to be located in this.

____ 2. Keratinized squamous cells are most likely to be located in this.

____ 3. Melanin in melanocytes is most likely to be here.

____ 4. Smooth muscles attached to hairs are probably here.

____ 5. This makes up the original "model" of the skeletal framework.

____ 6. The receiving ends of sensory receptors are most likely here.

____ 7. This serves as a "bank" for withdrawing and depositing calcium and phosphate ions.

For questions 8–11, choose from the following answers:

    a. ligaments
    b. osteoblasts
    c. mature bone cells
    d. red marrow
    e. tendons

____ 8. _____ secrete bone-dissolving enzymes.

____ 9. Major site of blood cell formation.

____ 10. _____ remove $Ca^{++}$ and $PO_4^{\equiv}$ ions from blood and build bone.

____ 11. _____ attach muscles to bone.

For questions 12–16, choose from the following answers:

    a. an action potential
    b. cross-bridge formation
    c. the sliding-filament model
    d. tension
    e. tetanus

____ 12. _____ is a mechanical force that causes muscle cells to shorten if it is not exceeded by opposing forces.

____ 13. A wave of electrical disturbance that moves along a neuron or muscle cell in response to a threshold stimulus.

____ 14. _____ is a large contraction caused by repeated stimulation of motor units that are not allowed to relax.

____ 15. _____ is assisted by calcium ions and ATP.

____ 16. _____ explains how myosin filaments move to the centers of sarcomeres and back.

For questions 17–20, choose from the following answers:

    a. actin
    b. myofibril
    c. myosin
    d. sarcomere
    e. sarcoplasmic reticulum

____ 17. A(n) _____ contains many repetitive units of muscle contraction.

____ 18. The repetitive unit of muscle contraction.

____ 19. Thin filaments that depend upon calcium ions to clear their binding sites so that they can attach to parts of thick filaments.

____ 20. _____ stores calcium ions and releases them in response to an action potential.

# Chapter Objectives/Review Questions

This section lists general and detailed chapter objectives that can be used as review questions. You can make maximum use of these items by writing answers on a separate sheet of paper. Fill in answers where blanks are provided. To check for accuracy, compare your answers with information given in the chapter or glossary.

*Page*        *Objectives/Questions*

(627)      1.  Name the four functions of human skin.
(628)      2.  Describe the two-layered structure of human skin and identify the items located in each layer.
(631)      3.  Describe several ways in which sunlight affects human skin.
(632–633)  4.  Compare invertebrate and vertebrate motor systems in terms of skeletal and muscular components and their interactions.
(634–635)  5.  Explain the various roles of osteoblasts, mature bone cells, cartilage models, and long bones in the development of human bones.
(637)      6.  Identify human bones by name and location.
(639)      7.  Refer to Figure 38.18 of your main text and indicate (a) a muscle used in sit-ups, (b) another used in dorsally flexing and inverting the foot, and (c) another used in flexing the elbow joint.
(640–642)  8.  Explain in detail the structure of muscles, from the molecular level to the organ systems level. Then explain how biochemical events occur in muscle contractions and how antagonistic muscle action refines movements.
(640–641)  9.  Describe the fine structure of a muscle fiber; use terms such as *myofibril, sarcomere, motor unit, actin,* and *myosin.*
(640–643) 10.  List, in sequence, the biochemical and fine structural events that occur during the contraction of a skeletal muscle fiber and explain how the fiber relaxes.
(644)     11.  Distinguish twitch contractions from tetanic contractions.

# Integrating and Applying Key Concepts

If humans had an exoskeleton rather than an endoskeleton, would they move differently from the way they do now? Name any advantages or disadvantages that having an exoskeleton instead of an endoskeleton would present in human locomotion.

# 39

# CIRCULATION

## Interactive Exercises

*Heartworks* (pp. 648–649)

## 39.1. CIRCULATORY SYSTEMS—AN OVERVIEW (pp. 650–651)

*Selected Words:* "internal environment", *closed* circulatory system, *open* circulatory system, blood flow *velocity, gills, lungs*

### Boldfaced, Page-Referenced Terms

(648) electrocardiograms _____

_____

(650) circulatory system _____

_____

(650) interstitial fluid _____

_____

(650) blood _____

_____

(650) heart _____

_____

(650) capillary beds _____

_____

(650) capillaries _____

_____

(651) pulmonary circuit _____

_____

(651) systemic circuit _____

_____

(651) lymphatic system _____

_____

## Fill in the Blanks

Cells survive by taking in from their surroundings what they need, (1) _____ , and giving back to their surroundings materials that they don't need: (2) _____ . In most animals, substances move rapidly to and from living cells by way of a (3) _____ circulatory system. (4) _____ , a fluid connective tissue within the (5) _____ and blood vessels, is the transport medium.

Most of the cells of animals are bathed in a(n) (6) _____ _____ ; blood is constantly delivering nutrients and removing wastes from that fluid. The (7) _____ generates the pressure that keeps blood flowing. Blood flows (8) (choose one) ☐ rapidly ☐ slowly through large diameter vessels to and from the heart, but where the exchange of nutrients and wastes occurs, in the (9) _____ beds, the blood is divided up into vast numbers of smaller-diameter vessels with tremendous surface area that enables the exchange to occur by diffusion. An elaborate network of drainage vessels attracts excess interstitial fluid and reclaimable (10) _____ and returns them to the circulatory system. This network is part of the (11) _____ system, which also helps clean the blood of disease agents.

Nutrients are absorbed into the blood from the (12) _____ and (13) _____ systems. Carbon dioxide is given to the (14) _____ system for elimination, and excess water, solutes and wastes are eliminated by the (15) _____ system.

## Labeling

Label the numbered parts in the illustrations below.

16. _____

17. _____  _____

18. _____

Describe the kind of circulatory system in:

19. Creature A. _____

20. Creature B. _____

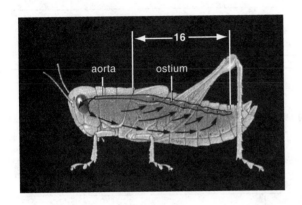

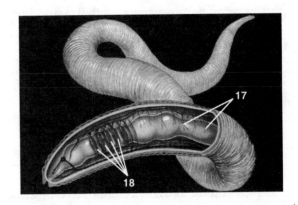

## 39.2. CHARACTERISTICS OF BLOOD (pp. 652–653)

## 39.3. *Focus on Health:* BLOOD DISORDERS (p. 654)

## 39.4. BLOOD TRANSFUSION AND TYPING (pp. 654–655)

***Selected Words:*** *macrophages, plasma proteins, neutrophils, lymphocytes, monocytes, eosinophils, basophils, erythrocytes, leukocytes, megakaryocytes, hemorrhagic anemias, chronic anemias, hemolytic anemias, iron defficiency anemia, $B_{12}$ deficiency, sickle-cell anemia, thalassemias, polycythemias, "blood doping", infectious mononucleosis, leukemias, "self" markers, erythroblastosis fetalis*

### Boldfaced, Page-Referenced Terms

(652) plasma _____

_____

(652) red blood cells _____

_____

(652) white blood cells _____

_____

(652) platelets _____

_____

(653) stem cells _____

_____

(653) cell count _____

_____

(654) anemias _____

_____

(654) blood transfusions _____

_____

(654) agglutination _____

_____

(655) ABO blood typing _____

_____

(655) Rh blood typing _____

_____

## Complete the Table

1. Complete the following table, which describes the components of blood.

| Components | Relative Amounts | Functions |
|---|---|---|
| Plasma Portion (50%–60% of total volume): | | |
| Water | 91%–92% of plasma volume | Solvent |
| a. (albumin, globulins, fibrinogen, etc.) | 7%–8% | Defense, clotting, lipid transport, roles in extracellular fluid volume, and so forth. |
| Ions, sugars, lipids, amino acids, hormones, vitamins, dissolved gases | 1%–2% | Roles in extracellular fluid volume, pH, etc. |
| Cellular Portion (40%–50% of total volume): | | |
| b. | 4,800,000–5,400,000 per microliter | $O_2$, $CO_2$ transport |
| White blood cells: | | |
| c. | 3,000–6,750 | Phagocytosis |
| d. | 1,000–2,700 | Immunity |
| Monocytes (macrophages) | 150–720 | Phagocytosis |
| Eosinophils | 100–360 | Roles in inflammatory response, immunity |
| Basophils | 25–90 | Roles in inflammatory response, anticlotting |
| e. | 250,000–300,000 | Roles in clotting |

## Fill-in-the-Blanks

Blood is a highly specialized fluid (2) _____ tissue that helps stabilize internal (3) _____ and

equalize internal temperature throughout an animal's body. Oxygen binds with the (4) _____ atom in a

hemoglobin molecule. The red blood (5) _____ _____ in males is about 5.4 million cells per

microliter of blood; in females, it is 4.8 million per microliter. The plasma portion constitutes approximately

(6) _____ to _____ percent of the total blood volume. Erythrocytes are produced in the

(7) _____ _____ . (8) _____ _____ are immature cells not yet fully differentiated.

(9) _____ and monocytes are highly mobile and phagocytic; they chemically detect, ingest, and destroy

bacteria, foreign matter, and dead cells. (10) _____ (thrombocytes) are membrane-bound cell fragments

that aid in forming blood clots by releasing substances that initiate the process.

In humans, red blood cells lack their (11) _____ , but they contain enough resources to sustain them for

about (12) _____ months. Platelets also have no (13) _____ , but they last a maximum of

(14) _____ days in the human bloodstream.

If you are blood type (15) _____ , you have no antibodies against A or B markers in your plasma.

(16) _____ is a response in which antibodies act against "foreign" cells bearing specific markers and

cause them to clump together.

## Label-Match

Identify the numbered cell types in the illustration below. Complete the exercise by matching and entering the
letter of the appropriate function in the parentheses following the given cell types. A letter may be used more
than once.

17. _____

18. _____

_____ ( )

19. _____ ( )

20. _____ ( )

21. _____ ( )

22. _____ ( )

23. _____

24. _____ ( )

A. Phagocytosis
B. Plays a role in the
   inflammatory
   response
C. Plays a role in clotting
D. Immunity
E. $O_2$, $CO_2$ transport
F. Play a role in
   producing blood cells

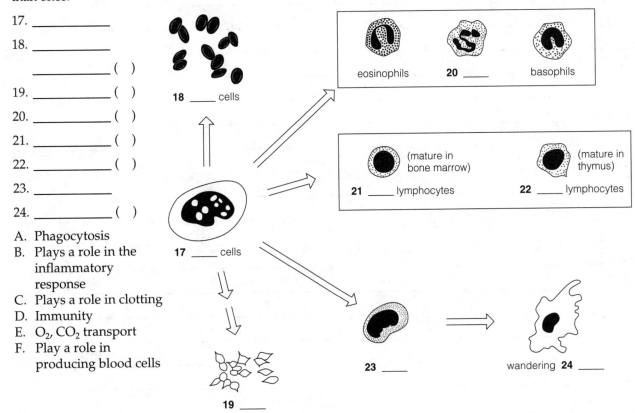

## Matching

25. ___ B$_{12}$ deficiency
26. ___ chronic anemias
27. ___ erythroblastosis fetalis
28. ___ hemolytic anemias
29. ___ hemorrhagic anemias
30. ___ infectious mononucleosis
31. ___ leukemias
32. ___ polycythemias
33. ___ sickle-cell anemia
34. ___ thalassemias

A. a potential hazard for strict vegetarians and alcoholics
B. a category of cancers that suppress or impair white blood cell formation in bone marrow
C. disorder caused by sluggish blood flow caused by far too many red blood cells; "blood doping" and some bone marrow cancers
D. caused by mixing Rh⁺ and Rh⁻ blood types; RhoGam can prevent this
E. abnormal forms of hemoglobin caused by a gene mutation
F. results from a sudden blood loss as from a severe wound
G. disorders caused by specific infectious bacteria and parasites as they replicate inside red blood cells and then lyse them
H. an Epstein-Barr virus causes this highly contagious disease which results from too many monocytes and lymphocytes
I. these disorders result from ongoing but slight blood loss; hemorrhoids, a bleeding ulcer or monthly blood loss by premenopausal women could be the cause

## 39.5. HUMAN CARDIOVASCULAR SYSTEM (pp. 656–657)

## 39.6. THE HEART IS A LONELY PUMPER (pp. 658–659)

**Selected Words:** *"cardiovascular", pulmonary circuit, systemic circuit, coronary circulation, systole, diastole, striated*

## Boldfaced, Page-Referenced Terms

(656) arteries _____

_____

(656) arterioles _____

_____

(656) venules _____

_____

(656) veins _____

_____

(656) aorta _____

_____

(658) endothelium _____

_____

(658) atrium _____

_____

(658) ventricle _____

_____

(658) cardiac cycle _____

_____

(659) cardiac conduction system _____

_____

(659) cardiac pacemaker _____

_____

## Fill-in-the-Blanks

A(n) (1) _____ carries blood away from the heart.  A(n) (2) _____ is a blood vessel with such a small diameter that red blood cells must flow through in single-file; its wall consists of no more than a single layer of (3) _____ cells resting on a basement membrane.  In each (4) _____ _____ , small molecules move between the bloodstream and the (5) _____ fluid.  (6) _____ are in the walls of veins and prevent backflow.  Both (7) _____ and (8) _____ serve as temporary reservoirs for blood volume.

(9) _____ are pressure reservoirs that keep blood flowing smoothly away from the heart while the (10) _____ are relaxing.  (11) _____ are control points where adjustments can be made in the volume of blood flow to be delivered to different capillary beds.  They offer great resistance to flow, so there is a major drop in (12) _____ in these tubes.

In the pulmonary circuit, the heart pumps (13) _____-poor blood to the lungs; then the (14) _____-enriched blood flows back to the (15) _____ . The (16) _____ _____ is the cardiac pacemaker.  During a cardiac cycle, contraction of the (17) _____ is the driving force for blood circulation; (18) _____ contraction helps fill the ventricles.

A receiving zone of a vertebrate heart is called a(n) (19) _____ ; a departure zone is called a(n) (20) _____ .  Each contraction period is called (21) _____ ; each relaxation period is (22) _____ . The heart is a pumping station for two major blood transport routes: the (23) _____ circulation to and from the lungs, and the (24) _____ circulation to and from the rest of the body.

## Fill-in-the-Blanks

Look at Fig. 39.11 in your text and use a red pen to redden all tubes in this illustration (except the pulmonary artery) that are indicated by "aorta" or "artery." Memorize the names, then fill in all of the blanks. Also, redden the parts of the heart that contain oxygen-rich blood.

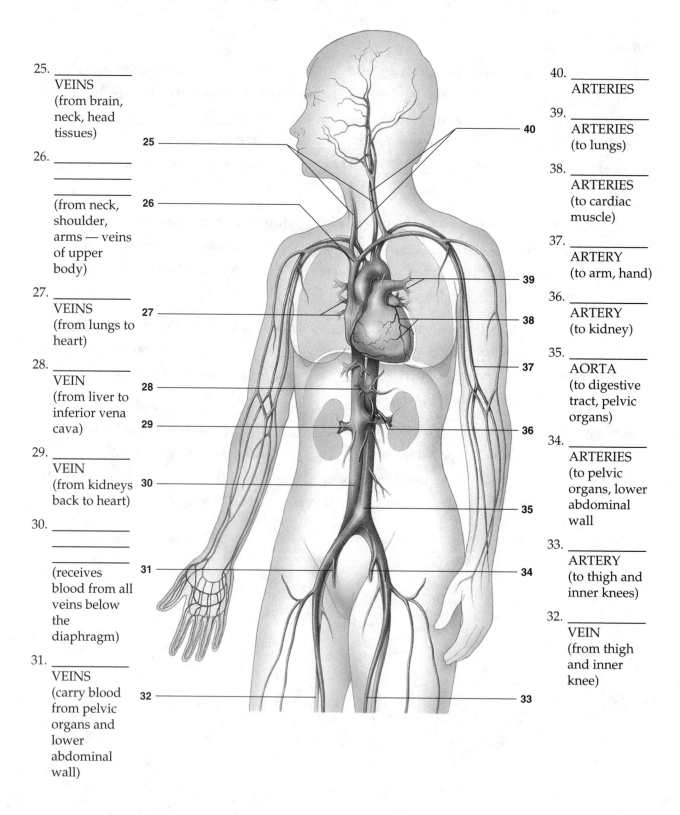

25. _____
VEINS
(from brain,
neck, head
tissues)

26. _____
_____
_____
(from neck,
shoulder,
arms — veins
of upper
body)

27. _____
VEINS
(from lungs to
heart)

28. _____
VEIN
(from liver to
inferior vena
cava)

29. _____
VEIN
(from kidneys
back to heart)

30. _____
_____
_____
(receives
blood from all
veins below
the
diaphragm)

31. _____
VEINS
(carry blood
from pelvic
organs and
lower
abdominal
wall)

40. _____
ARTERIES

39. _____
ARTERIES
(to lungs)

38. _____
ARTERIES
(to cardiac
muscle)

37. _____
ARTERY
(to arm, hand)

36. _____
ARTERY
(to kidney)

35. _____
AORTA
(to digestive
tract, pelvic
organs)

34. _____
ARTERIES
(to pelvic
organs, lower
abdominal
wall

33. _____
ARTERY
(to thigh and
inner knees)

32. _____
VEIN
(from thigh
and inner
knee)

## Labeling

Identify each indicated part of the accompanying illustration.

41. _____

42. _____ _____

     _____ _____

43. _____ _____

44. _____ _____

45. _____ _____

     _____

46. _____ _____

47. _____ _____

     _____

48. _____ _____

     _____

arteries

48

47

41

right pulmonary veins

trunk of pulmonary artery

42

right atrium

left atrium

semilunar valve

43

atrioventricular valve

46

44

right ventricle

septum

45

(apex of heart)

## 39.7. BLOOD PRESSURE IN THE CARDIOVASCULAR SYSTEM (pp. 660–661)

## 39.8. FROM CAPILLARY BEDS BACK TO THE HEART (pp. 662–663)

## 39.9. *Focus On Health*: CARDIOVASCULAR DISORDERS (pp. 664–665)

***Selected Words:*** *systolic and diastolic pressure, pulse pressure, epinephrine, angiotensin, carotid arteries, aortic arch, mean arterial pressure, edema, elephantiasis, hypertension, atherosclerosis, heart attack, strokes, arteriosclerosis, atherosclerotic plaque, thrombus, embolus, stress electrocardiograms, angiography, coronary bypass surgery, laser angioplasty, balloon angioplasty, arrhythmias, bradycardia, tachycardia, atrial fibrillation, ventricular fibrillation*

### Boldfaced, Page-Referenced Terms

(660) blood pressure _____

_____

(661) vasodilation _____

_____

(661) vasoconstriction _____

_____

(661) baroreceptor reflex _____

_____

(662) ultrafiltration _____

_____

(662) reabsorption _____

_____

(665) LDLs, low-density lipoproteins _____

_____

(665) HDLs, high-density lipoproteins _____

_____

## *Fill-in-the-Blanks*

Blood pressure is normally high in the (1) _____ immediately after leaving the heart, but then it drops as the fluid passes along the circuit through different kinds of blood vessels. Energy in the form of (2) _____ is lost as it overcomes (3) _____ to the flow of blood. Arterial walls are thick, muscular and (4) ____ and have large diameters. Arteries present [choose one] (5) □ much □ little resistance to blood flow, so pressure [choose one] (6) □ drops a lot □ does not drop much in the arterial portion of the systemic and pulmonary circuits.

The greatest drop in pressure occurs at [choose one] (7) □ capillaries □ arterioles □ veins. With this slowdown, blood flow can now be alloted in different amounts to different regions of the body in response to signals from the (8) _____ system and endocrine system or even changes in local chemical conditions.

When a person is resting, blood pressure is influenced most by reflex centers in the (9) _____ _____ . When the resting level of blood pressure increases, reflex centers command the heart to [choose one] (10) □ beat more slowly □ beat faster, and command smooth muscle cells in arteriole walls to [choose one] (11) □ contract □ relax, which results in [ choose one] (12) □ vasodilation □ vasoconstriction. Capillary beds are (13) _____ zones where substances are exchanged between blood and interstitial fluid. Capillaries have the thinnest walls across which (14) _____ _____ drives fluid forcing molecules of water and solutes to move in the same direction. Veins contain (15) _____ that prevent blood from flowing backward; they contain [choose one] (16) □ 20–30 □ 35–45 □ 50–60 percent of the total blood volume.

## Labeling

Identify each indicated part of the accompanying illustrations.

17. _____    20. _____

18. _____    21. _____ _____

19. _____    22. _____

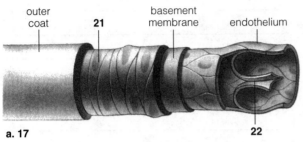

a. 17

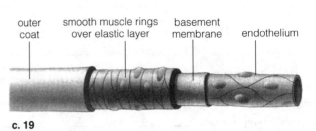

c. 19

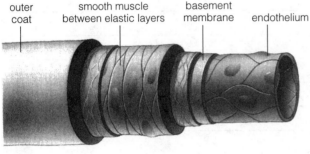

b. 18

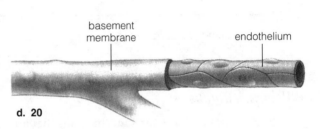

d. 20

## True-False

If the statement is true, write a T in the blank. If false, make it true by changing the underlined word.

_____23. The pulse <u>rate</u> is the difference between the systolic and the diastolic pressure readings.

_____24. Because the total volume of blood remains constant in the human body, blood pressure must <u>also</u> <u>remain</u> <u>constant</u> throughout the circuit.

## Fill-in-the-Blanks

One cause of (25) _____ is the rupture of one or more blood vessels in the brain.  A (26) _____

_____ blocks a coronary artery.  (27) _____ is a term for a formation that can include cholesterol, calcium salts, and fibrous tissue.  It is not healthful to have a high concentration of (28) _____-density lipoproteins in the bloodstream.  A clot that stays in place is a (29) _____ , but a clot that travels in the bloodstream is an embolus.

## 39.10. HEMOSTASIS (p. 666)
## 39.11. LYMPHATIC SYSTEM (pp. 666–667)

*Selected Words:* "valves", *tonsils.*

## Boldfaced, Page-Referenced Terms

(666) hemostasis _____

_____

(666) lymph vascular system _____

_____

(666) lymph capillaries _____

_____

(666) lymph vessels _____

_____

(667) lymphoid organs and tissues _____

_____

(667) lymph nodes _____

_____

(667) spleen _____

_____

(667) thymus gland _____

_____

## Fill-in-the-Blanks

Bleeding is stopped by several mechanisms that are referred to as (1) _____ ; the mechanisms include
blood vessel spasm, (2) _____ _____ _____ , and blood (3) _____ . Once the platelets
reach a damaged vessel, through chemical recognition they adhere to exposed (4) _____ fibers in
damaged vessel walls. Reactions cause rod-shaped proteins to assemble into long (5) _____ fibers.
These trap blood cells and components of plasma. Under normal conditions, a clot eventually forms at the
damaged site.

## Choice

Choose from these possibilities for numbers 6–10.

a. bone marrow      b. lymph nodes      c. lymph vascular system      d. spleen      e. thymus gland

6. ____ A huge reservoir of red blood cells and a filter of pathogens and used-up blood cells from the blood;
   contains red pulp.

7. ____ Delivers water and plasma proteins from capillary beds to the blood circulation, delivers fats from
   the small intestine to the blood, and delivers pathogens, foreign cells and material and cellular debris
   to the organized disposal centers.

8. ____ Immature T lymphocytes become mature here and hormones are produced here.

9. ____ Contain white blood cells that destroy invading bacteria and viruses as they are filtered from the
   lymph.

10. ____ White blood cells are produced here before they go on to other parts of the lymphatic system.

## Identification/Fill-in-the-Blanks

11. _____

12. _____ gland

13. _____ duct

14. _____

15. _____ _____

16. organized arrays of _____ and

_____

17. _____ _____

18. _____ vessels reclaim fluid lost
from the bloodstream, purify the blood
of microorganisms, and transport
(19) _____ from the
(20) _____ _____
to the bloodstream.

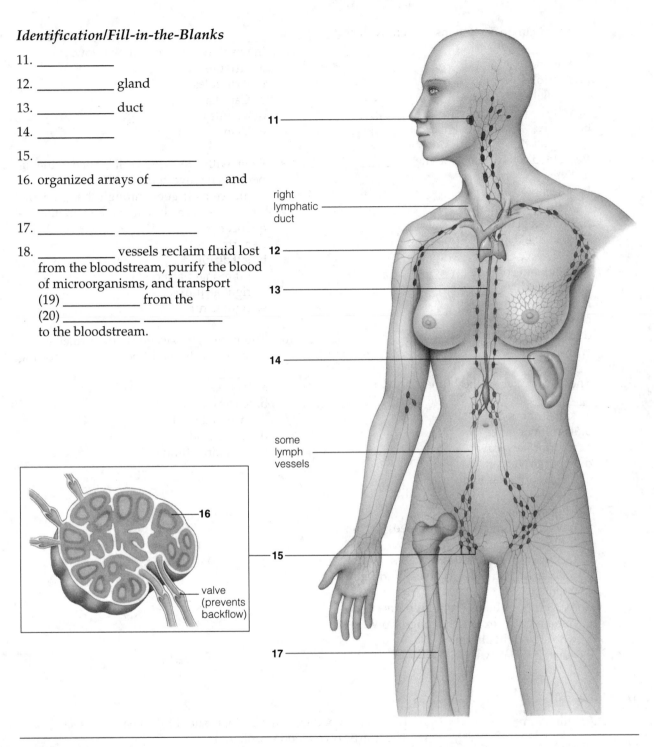

right
lymphatic
duct

some
lymph
vessels

valve
(prevents
backflow)

# Self-Quiz

____ 1. Most of the oxygen in human blood is transported by _____ .
   a. plasma
   b. serum
   c. platelets
   d. hemoglobin
   e. leukocytes

____ 2. Of all the different kinds of white blood
cells, two classes of _____ are the ones
that respond to specific invaders and confer
immunity to a variety of disorders.
   a. basophils
   b. eosinophils
   c. monocytes
   d. neutrophils
   e. lymphocytes

___ 3. Open circulatory systems generally lack _____ .
  a. a heart
  b. arterioles
  c. capillaries
  d. veins
  e. arteries

___ 4. Red blood cells originate in the _____ .
  a. liver
  b. spleen
  c. yellow bone marrow
  d. thymus gland
  e. red bone marrow

___ 5. Hemoglobin contains _____ .
  a. copper
  b. magnesium
  c. sodium
  d. calcium
  e. iron

___ 6. The pacemaker of the human heart is the _____ .
  a. sinoatrial node
  b. semilunar valve
  c. inferior vena cava
  d. superior vena cava
  e. atrioventricular node

___ 7. During systole, _____ .
  a. oxygen-rich blood is pumped to the lungs
  b. the heart muscle tissues contract
  c. the atrioventricular valves suddenly open
  d. oxygen-poor blood from all parts of the human body, except the lungs, flows toward the right atrium
  e. none of the above

___ 8. _____ are reservoirs of blood pressure in which resistance to flow is low.
  a. Arteries
  b. Arterioles
  c. Capillaries
  d. Venules
  e. Veins

___ 9. Begin with a red blood cell located in the superior vena cava and travel with it in proper sequence as it goes through the following structures. Which will be *last* in the sequence?
  a. aorta
  b. left atrium
  c. pulmonary artery
  d. right atrium
  e. right ventricle

___10. The lymphatic system is the principal avenue in the human body for transporting _____ .
  a. fats
  b. wastes
  c. carbon dioxide
  d. amino acids
  e. interstitial fluids

## Matching

11. ___agglutination

12. ___angiogram, angiography

13. ___atherosclerosis

14. ___carotid arteries

15. ___coronary arteries

16. ___edema

17. ___embolism, embolus

18. ___erythroblastosis fetalis

A. Receive(s) blood from the brain, tissues of the head and neck
B. The heart's own blood supplier(s)
C. Excessively fast rate of heart beat; occurs during heavy exercising
D. Damage to brain caused by damaged blood vessels
E. High blood pressure
F. A blood clot that is on the move from one place to another
G. Diagnostic technique that uses opaque dyes and x-rays to discover blocked blood vessels
H. Deliver(s) blood to the head, neck, brain
I. The clumping of red blood cells, or of antibodies with antigens
J. Delivers blood to kidneys, where its composition and volume are adjusted

19. ___hypertension

20. ___inferior vena cava

21. ___jugular veins

22. ___renal arteries

23. ___stroke

24. ___tachycardia

25. ___thrombosis, thrombus

K. A blood clot that is lodged in a blood vessel and is blocking it

L. Receive(s) blood from all veins below the diaphragm

M. Progressive thickening of the arterial wall and narrowing of the arterial lumen (space)

N. Blood cell destruction caused by mis-matched Rh blood types (Rh⁻ mother and Rh⁺ fetus)

O. Accumulation of excess fluid in interstitial spaces; extreme in elephantiasis

## Chapter Objectives/Review Questions

This section lists general and detailed chapter objectives that can be used as review questions. You can make maximum use of these items by writing answers on a separate sheet of paper. Fill in answers where blanks are provided. To check for accuracy, compare your answers with information given in the chapter or glossary.

| Page | | Objectives/Questions |
|---|---|---|
| (650–651) | 1. | Distinguish between open and closed circulatory systems. |
| (652) | 2. | Describe the composition of human blood, using percentages of volume. |
| (652–653) | 3. | Distinguish the five types of leukocytes from each other in terms of structure and functions. |
| (653) | 4. | State where erythrocytes, leukocytes, and platelets are produced. |
| (654–655) | 5. | Describe how blood is typed for the ABO blood group and for the Rh factor. |
| (655–661) | 6. | Describe the composition and function of the lymphatic system. |
| (656–657) | 7. | Trace the path of blood in the human body. Begin with the aorta and name all major components of the circulatory system through which the blood passes before it returns to the aorta. |
| (658–659) | 8. | Explain what causes a heart to beat. Then describe how the rate of heartbeat can be slowed down or speeded up. |
| (660) | 9. | Describe how the structures of arteries, capillaries, and veins differ. |
| (660–661) | 10. | Explain what causes high pressure and low pressure in the human circulatory system. Then show where major drops in blood pressure occur in humans. |
| (660–665) | 11. | List the factors that cause blood to leave the heart and the factors that cooperate to return blood to the heart. |
| (662–663) | 12. | Explain how veins and venules can act as reservoirs of blood volume. |
| (664) | 13. | Describe how hypertension develops, how it is detected, and whether it can be corrected. |
| (664–665) | 14. | Distinguish a stroke from a coronary artery blockage, occlusion. |
| (665) | 15. | State the significance of high- and low-density lipoproteins to cardiovascular disorders. |
| (666) | 16. | List in sequence the events that occur in the formation of a blood clot. |

## Integrating and Applying Key Concepts

You observe that some people appear as though fluid had accumulated in their lower legs and feet. Their lower extremities resemble those of elephants. You inquire about what is wrong and are told that the condition is caused by the bite of a mosquito that is active at night. Construct a testable hypothesis that would explain (1) why the fluid was not being returned to the torso, as normal, and (2) what the mosquito did to its victims.

# 40

# IMMUNITY



## Interactive Exercises

### *Russian Roulette, Immunological Style* (pp. 670–671)

### 40.1. THREE LINES OF DEFENSE (p. 672)

### 40.2. COMPLEMENT PROTEINS (p. 673)

### 40.3. INFLAMMATION (pp. 674–675)

***Selected Words:*** "immune", Jenner, *smallpox*, Koch, Bacillus anthracis, *athlete's foot*, Lactobacillus, *nonspecific* and *specific* responses, *complement "coat"*, *chemotaxins*, *interleukin*, *endogenous pyrogen*, *"set point"*, *interleukin-1*

***Boldfaced, Page-Referenced Terms***

(671) vaccination ______________________________

______________________________

(672) pathogens ______________________________

______________________________

(672) lysozyme _____

_____

(673) complement system _____

_____

(673) lysis _____

_____

(674) neutrophils _____

_____

(674) eosinophils _____

_____

(674) basophils _____

_____

(674) macrophages _____

_____

(674) acute inflammation _____

_____

(674) mast cells _____

_____

(674) histamine _____

_____

(675) fever _____

_____

## Fill-in-the-Blanks

The best way to deal with damaging foreign agents is to prevent their entry. Several barriers prevent pathogens from crossing the boundaries of your body. Intact skin and (1) _____ membranes are effective barriers. (2) _____ is an enzyme that destroys the cell wall of many bacteria. (3) _____ fluid destroys many food-borne pathogens in the gut. Normal (4) _____ residents of the skin, gut, and vagina outcompete pathogens for resources and help keep their numbers under control.

Even simple aquatic invertebrates defend themselves with (5) _____ cells and antimicrobial substances including lysozymes that cause bacteria to burst open. In most animals, when a sharp object cuts through the skin and foreign microbes enter, some plasma proteins come to the rescue and seal the wound with a (6) _____ mechanism. (7) _____ white blood cells engulf bacteria soon thereafter.

Also, in vertebrates, there are about twenty plasma proteins [collectively referred to as the (8) _____ _____ ] that are activated one after another in a "cascade" of reactions to help destroy invading microorganisms. When the complement system is activated, circulating basophils and mast cells in tissues release (9) _____ , which increases the permeability of (10) _____ and makes them "leaky," so fluid seeps out and causes the inflamed area to become swollen and warm.

Vertebrates have (11) _____ , general defenses such as those mentioned above that isolate and destroy threatening agents identified as "nonself." (11) defenses do not identify <u>which</u> <u>specific</u> foreign agent needs to be destroyed; they only identify that there is <u>some</u> foreign agent that needs to be destroyed. If foreign agents penetrate the boundaries of your body, (11) defenses constitute a fast way to exterminate them. If (11) defenses fail to control invading pathogens, (12) _____ defenses come to the rescue, in which well-coordinated armies of (13) _____ blood cells make highly-focused counterattacks against pathogens whose specific identity has been identified. These armies and the organs of our body that house them constitute our (14) _____ system.

### Complete the Table

Complete the table by providing the specific functions carried out by each of the four different kinds of white blood cells listed below.

| Type of Cells | Functions |
|---|---|
| Basophils | 15. |
| Eosinophils | 16. |
| Neutrophils | 17. |
| Macrophages | 18. |

## 40.4. THE IMMUNE SYSTEM (pp. 676–677)
## 40.5. LYMPHOCYTE BATTLEGROUNDS (p. 678)
## 40.6. CELL-MEDIATED RESPONSES (pp. 678–679)
## 40.7. ANTIBODY—MEDIATED RESPONSES (pp. 680–681)

**Selected Words:** *immunological specificity, immunological memory, self, nonself, effector* cells, *memory* cells, *differentiation, "touch killing", processed* antigen, *antibody-mediated* responses, *cell-mediated responses, primary* immune response, *secondary* immune response, *clone, "double signal", IgM, IgD, IgG, IgA, IgE*

### Boldfaced, Page-Referenced Terms

(676) B lymphocytes (=B cells) _____

_____

(676) T lymphocytes (=Tcells) _____

_____

(676) immune system _____

_____

(676) antigen _____

_____

(676) MHC markers _____

_____

(676) antigen-MHC complexes _____

_____

(676) antigen-presenting cell _____

_____

(676) helper T cells _____

_____

(676) cytotoxic T cells _____

_____

(677) antibodies _____

_____

(678) TCRs, T-cell Receptors _____

_____

(678) interleukins _____

_____

(678) perforins _____

_____

(678) apoptosis _____

_____

(679) natural killer cells (NK cells) _____

_____

(681) immunoglobulins, Igs _____

_____

## Fill-in-the-Blanks

If the (1) _____ defenses (fast-acting white blood cells such as neutrophils, eosinophils and basophils, slower-acting macrophages and the plasma proteins involved in clotting and complement) fail to repel the microbial invaders, then the body calls on its (2) _____ _____ , which identifies *specific* targets to kill and *remembers* the identities of its targets. Your own unique (3) _____ _____ patterns identify your cells as "self" cells. Any other surface pattern is, by definition, (4) _____ , and doesn't belong in your body.

The principal actors of the immune system are (5) _____ descended from stem cells (consult Fig. 39.7) in the bone marrow which have two different strategies of action to deal with their different kinds of enemies. (6) _____ _____ clones secrete antibodies and act principally against the extracellular (ones that stay outside the body cells) enemies that are pathogens in blood or on the cell surfaces of body tissues. (7) _____ _____ clones descend from lymphocytes that matured in the (8) _____ where they acquired specific markers on their cell surfaces; they defend principally against intracellular pathogens such as (9) _____ , and against any cells ((10) _____ cells and grafts of foreign tissue) that are perceived as abnormal or foreign.

Each kind of cell, virus or substance bears unique molecular configurations (patterns) that give it a unique (11) _____ . A(n) (12) _____ is any molecular configuration that causes the formation of lymphocyte armies. Any cell that processes and displays (12) together with a suitable MHC molecule is known as an (13) _____-_____ cell that can activate lymphocytes to undergo rapid cell divisions. Lymphocyte subpopulations that fight and destroy enemies are known as (14) _____ cells; among these are (15) _____ _____ cells that promote the formation of large populations of effector cells, memory cells, and (16) _____ _____ cells that eliminate infected body cells and tumor cells by a "touch killing." Together with other "killers," they execute the (17) _____-_____ immune responses. By contrast, (6) produce Y-shaped antigen-binding receptor molecules called (18) _____ . (6) and (18) carry out the (19) _____-_____ immune responses. Other lymphocyte subpopulations, (20) _____ cells enter a resting phase, but they "remember" the specific agent that was conquered and will undertake a larger, more rapid response if it shows up again.

## Labeling

Identify each numbered part in the accompanying illustration.

21. _____ - _____ _____

22. _____ _____ _____ _____

23. _____ _____ _____

24. _____

25. _____ _____ _____

26. _____ _____ _____

27. _____

28. _____

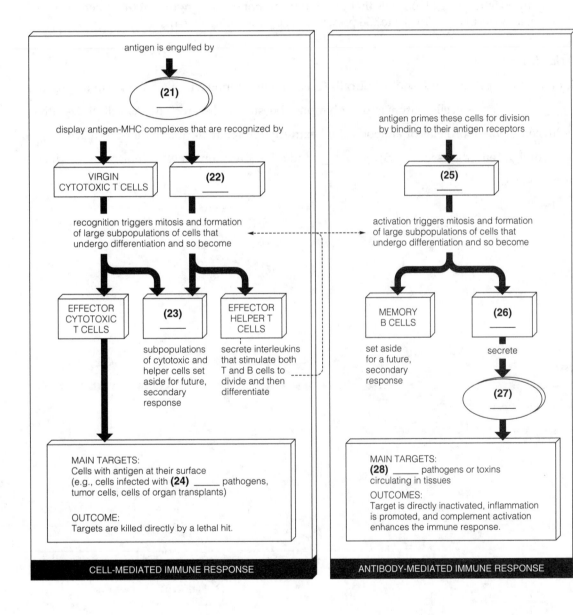

## Choice

Match each of the white blood cell types with the appropriate function (29–33)

a. effector cytotoxic T cells    b. effector helper T cells    c. macrophages
d. memory cells    e. effector B cells

29. ___Lymphocytes that directly destroy by "touch killing" body cells already infected by certain viruses or by parasitic fungi

30. ___A portion of B and T cell populations that were set aside as a result of a first encounter, now circulate freely and respond rapidly to any later attacks by the same type of invader

31. ___Lymphocytes and their progeny that produce antibodies

32. ___Lymphocytes that serve as master switches of the immune system; stimulate the rapid division of B cells and cytotoxic T cells

33. ___Nonlymphocytic white blood cells that develop from monocytes, engulf anything perceived as foreign, and alert helper T cells to the presence of specific foreign agents

## Fill-in-the-Blanks

In addition to effector B cells and cytotoxic T cells (executioner lymphocytes that mature in the thymus), (34) _____ _____ cells mature in other lymphoid tissues and search out any cell that is either coated with complement proteins or antibodies, *or* bears any foreign molecular pattern. When cytotoxic T cells find foreign cells, they secrete (35) _____ and other toxic substances to poison and induce cell death by apoptosis.

## Labeling

Identify each numbered part in the accompanying illustrations.

36. _____

37. _____ _____

38. _____ _____

39. _____ - _____

40. _____ _____ cell

41. _____

42. _____ _____

43. _____ _____

Example of an (36) _____-mediated immune response to a bacterial invasion.

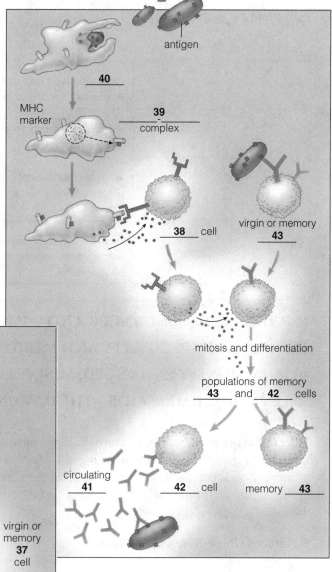

antigen

**40**

MHC marker

**39** - complex

**38** cell

virgin or memory **43**

mitosis and differentiation

populations of memory **43** and **42** cells

circulating **41**

**42** cell

memory **43**

MHC marker

**39** - complex

**40**

**38** cell

virgin or memory **37** cell

mitosis and differentiation

population of effector and memory **37** cells

infected cell

disengaged effector **37** cell

Example of a(n) (41) _____-mediated immune response to a bacterial invasion.

*Note*: If the same number is used more than once, that is because the label is the same in all such situations. If you have labeled it correctly on the lefthand illustration then its identification should also be correct on the one above.

The (44) _____ _____ _____ is the route taken during a first-time contact with an antigen; the secondary immune response is more rapid because patrolling battalions of (45) _____ _____ are in the bloodstream on the lookout for enemies they have conquered before. When these meet up with recognizable (46) _____ , they divide at once, producing large clones of B or T cells within two to three days.

| Immunoglobulin Type | Function |
|---|---|
| 47. | enter mucus-coated surfaces of the respiratory, digestive and reproductive tracts where they neutralize pathogens. |
| 48. | triggers inflammation when parasitic worms attack the body; play a role in allergic responses. |
| 49. | activate complement proteins; neutralize many toxins; long-lasting; can cross placenta and protect developing fetus; also present in colostrum from mammary glands. |
| 50. | first to be secreted during immune responses; after binding to antigen, trigger complement cascade; also tag invaders and bind them in clumps for later phagocytosis. |

## 40.8. *Focus on Health:* CANCER AND IMMUNOTHERAPY (p. 681)

## 40.9. IMMUNE SPECIFICITY AND MEMORY (pp. 682–683)

## 40.10. IMMUNITY ENHANCED, MISDIRECTED, OR COMPROMISED (pp. 684–685)

## 40.11. *Focus on Health:* AIDS—THE IMMUNE SYSTEM COMPROMISED (pp. 686–687)

***Selected Words:*** *carcinomas, sarcomas, leukemia, immunotherapy, Milstein and Kohler, monoclonal antibodies, LAK cells, therapeutic vaccines, variable regions, Macfarlane Burnet, clonal selection hypothesis, active immunization, passive immunization, asthma, hay fever, anaphylactic shock, antihistamine, Grave's disorder, myasthenia gravis, rheumatoid arthritis, severe combined immunodeficiencies (SCIDs),* Pneumocystis carinii, *AZT, ddi*

### Boldfaced, Page-Referenced Terms

(684) immunization _____

_____

(684) vaccine _____

_____

(684) allergens _____

_____

(684) allergy, allergies _____

_____

(685) autoimmune response _____

_____

## Fill-in-the-Blanks

Various strategems have been developed that enhance immunological defenses against tumors as well as against certain pathogens; collectively they are referred to as (1) _____ .

(2) _____ refers to cells that have lost control over cell division. Milstein and Kohler developed a means of producing large amounts of (3) _____ _____ , which are produced by clones of proliferating (4) _____ cells: Mouse B cells fused with tumor cells that divide nonstop. A different therapy involves researchers extracting (5) _____ from tumors, activating them by exposing them to a (6) _____ , and injecting these *LAK* cells back into the patient, where they actively kill tumor cells.

Antibodies are plasma proteins that are part of the (7) _____ group of proteins. Some of these circulate in blood; others are present in other body fluids or bound to B cells. (8) _____ are Y-shaped proteins that lock onto specific foreign targets and thereby tag them for destruction by phagocytes or by activating the complement system. While B or T cells are maturing, different regions of the genes that code for antigen receptors are shuffled at random into one of millions of possible combinations; this process of (9) _____ produces a gene that codes for just one of millions of possible antigen receptors.

## Labeling

Identify each numbered part in the accompanying illustration and its legend.

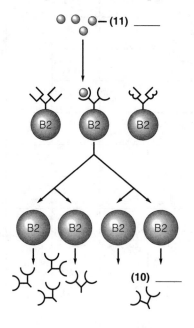

10. _____

11. _____

12. _____ _____

13. _____ _____

14. _____ _____

Clonal selection of a (12) _____ _____ cell, the descendants of which produced the specific (10) _____ that can combine with a specific antigen.

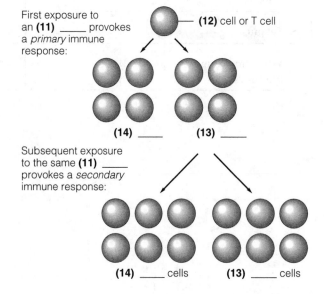

First exposure to an **(11)** _____ provokes a *primary* immune response:

**(12)** cell or T cell

**(14)** _____    **(13)** _____

Subsequent exposure to the same **(11)** _____ provokes a *secondary* immune response:

**(14)** _____ cells    **(13)** _____ cells

## Fill-in-the-Blanks

The (15) _____ _____ _____ explains how an individual has immunological memory, which is the basis of a secondary immune response; the hypothesis also explains in part how *self* cells are distinguished from (16) _____ cells in the vertebrate immune response.

Deliberately provoking the production of memory lymphocytes is known as (17) _____ . In a(n) (18) _____ immunization, a vaccine containing antigens is injected into the body or taken orally. The first one elicits a (19) _____ _____ _____ , and the second one (a booster shot) elicits a secondary immune response, which causes the body to produce more antibodies and (20) _____ _____ to provide long-lasting protection.

## Complete the Table

Indicate with a check (√) the age(s) of vaccination.

*Age Vaccination Is Administered*

| Disease | (a) At birth | (b) At 2 mos. | (c) 2–4 mos. | (d) 4 mos. | (e) 6 mos. | (f) 6–18 mos. | (g) 12–15 mos. | (h) 12–18 mos. | (i) 1–6 mos. | (j) 11–12 mos. |
|---|---|---|---|---|---|---|---|---|---|---|
| 21. Diphtheria | | | | | | | | | | |
| 22. *Hemophilus influenzae* | | | | | | | | | | |
| 23. Hepatitis B | | | | | | | | | | |
| 24. Measles | | | | | | | | | | |
| 25. Mumps | | | | | | | | | | |
| 26. Polio | | | | | | | | | | |
| 27. Rubella | | | | | | | | | | |
| 28. Tetanus | | | | | | | | | | |
| 29. Whooping cough (Pertussis) | | | | | | | | | | |

## Fill-in-the-Blanks

(30) is an altered secondary response to a normally harmless substance that may actually cause injury to tissues. (31) _____ _____ is a disorder in which the body mobilizes its forces against certain of its own tissues. (32) _____ _____ is an example of this kind of disorder in which antibodies tag acetylcholine receptors on skeletal muscle cells and cause progressive weakness. (33) _____ _____ is a similar kind of disorder in which skeletal joints are chronically inflamed.

AIDS is a constellation of disorders that follow infection by the (34) _____ _____ _____ . In the United States, transmission has occurred most often among intravenous drug abusers who share needles and among (35) _____ _____ . HIV is a (36) _____; its genetic material is RNA rather than DNA, and it has several copies of an enzyme (37) _____ _____ , which uses the viral RNA as a template for making DNA, which is then inserted into a host chromosome.

HIV is transmitted when (38) _____ _____ of an infected person enter another person's tissues. The virus cripples the immune system by attacking (39) _____ _____ cells and (40) _____ .

More than (41) _____ (number) Americans are now infected with HIV. Worldwide, an estimated (42) _____ (number) are infected and (43) _____ (number) million are already dead. By the year 2000, the number of infected people may reach an estimated (44) _____ million.

# Self-Quiz

_____ 1. All the body's phagocytes are derived from stem cells in the _____ .
 a. spleen
 b. liver
 c. thymus
 d. bone marrow
 e. thyroid

_____ 2. The plasma proteins that are activated when they contact a bacterial cell are collectively known as the _____ system.
 a. shield
 b. complement
 c. Ig G
 d. MHC
 e. HIV

_____ 3. _____ are divided into two groups: T cells and B cells.
 a. Macrophages
 b. Lymphocytes
 c. Platelets
 d. Complement cells
 e. Cancer cells

_____ 4. _____ produce and secrete antibodies that set up bacterial invaders for subsequent destruction by macrophages.
 a. B cells
 b. Phagocytes
 c. T cells
 d. Bacteriophages
 e. Thymus cells

_____ 5. Antibodies are shaped like the letter _____ .
 a. Y
 b. W
 c. Z
 d. H
 e. E

_____ 6. The markers for every cell in the human body are referred to by the letters _____ .
 a. HIV
 b. MBC
 c. RNA
 d. DNA
 e. MHC

___ 7. Effector B cells _____ .
   a. fight against extracellular pathogens and toxins circulating in tissues
   b. develop from effector B cells
   c. manufacture and secrete antibodies
   d. do not divide and form clones
   e. all of the above

___ 8. Clones of B or T cells are _____ .
   a. being produced continually
   b. sometimes known as memory cells if they keep circulating in the bloodstream
   c. only produced when their surface proteins recognize other specific proteins previously encountered
   d. produced and mature in the bone marrow
   e. both (b) and (c)

___ 9. Whenever the body is reexposed to a specific sensitizing agent, IgE antibodies cause _____ .
   a. prostaglandins and histamine to be produced
   b. clonal cells to be produced
   c. histamine to be released
   d. the immune response to be suppressed
   e. none of the above

___10. The clonal selection hypothesis explains _____ .
   a. how self cells are distinguished from nonself
   b. how B cells differ from T cells
   c. how so many different kinds of antigen-specific receptors can be produced by lymphocytes
   d. how memory cells are set aside from effector cells
   e. how antigens differ from antibodies

## Matching

Choose the most appropriate description for each term.

11. ___allergy
12. ___antibody
13. ___antigen
14. ___macrophage
15. ___clone
16. ___complement
17. ___histamine
18. ___MHC marker
19. ___effector B cell
20. ___T cell

A. Begins its development in bone marrow, but matures in the thymus gland
B. Cells that have directly or indirectly descended from the same parent cell
C. A potent chemical that causes blood vessels to dilate and let protein pass through the vessel walls
D. Y-shaped immunoglobulin
E. A nonself marker
F. The progeny of turned-on B cells
G. A group of about fifteen proteins that participate in the inflammatory response
H. An altered secondary immune response to a substance that is normally harmless to other people
I. The basis for self-recognition at the cell surface
J. Principal perpetrator of phagocytosis

# Chapter Objectives/Review Questions

This section lists general and detailed chapter objectives that can be used as review questions. You can make maximum use of these items by writing answers on a separate sheet of paper. Fill in answers where blanks are provided. To check for accuracy, compare your answers with information given in the chapter or glossary.

*Page*          *Objectives/Questions*

(671–672;  1. Understand how vertebrates (especially mammals) recognize and discriminate between self
676–677)       and nonself tissues.
(672)      2. Describe typical external barriers that organisms present to invading organisms.
(672–673)  3. List and discuss four nonspecific defense responses that serve to exclude microbes from the
              body.
(673)      4. Explain how the complement system is related to an inflammatory response.
(674–675)  5. Describe the sequence of events that occur during inflammatory responses.
(676–677)  6. List the three general types of cells that form the basis of the vertebrate immune system.
(676–677)  7. Distinguish the roles of T cells from the roles of B cells.
(677)      8. Explain what is meant by *primary immune response* as contrasted with *secondary immune re-*
              *sponse*.
(678–681)  9. Distinguish between the antibody-mediated response pattern and the cell-mediated response
              pattern.
(681; 684) 10. Explain what monoclonal antibodies are and tell how they are currently being used in passive
              immunization and cancer treatment.
(682–683) 11. Describe how recognition proteins and antibodies are made. State how they are used in im-
              munity.
(682–683) 12. Describe the clonal selection hypothesis and tell what it helps to explain.
(684)     13. Describe two ways that people can be immunized against specific diseases.
(684–685) 14. Distinguish allergy from autoimmune disorder.
(685–687) 15. Describe some examples of immune failures and identify as specifically as you can which
              weapons in the immunity arsenal failed in each case.
(686–687) 16. Describe how AIDS specifically interferes with the human immune system.

# Integrating and Applying Key Concepts

Suppose you wanted to get rid of forty-seven warts that you have on your hands by treating them with monoclonal antibodies. Outline the steps you would have to take.

# 41

# RESPIRATION



## Interactive Exercises

### *Conquering Chomolungma* (pp. 690–691)

### 41.1. THE NATURE OF RESPIRATION (p. 692)

### 41.2. INVERTEBRATE RESPIRATION (p. 693)

### 41.3. VERTEBRATE RESPIRATION (p. 694)

***Selected Words:*** *hypoxia, external* gills, *internal* gills, *rete mirabile,* ventilation

***Boldfaced, Page-Referenced Terms***

(691) respiration ________________________________________

________________________________________

(691) respiratory systems ________________________________________

________________________________________

(692) pressure gradients ________________________________________

________________________________________

(692) partial pressure _____

_____

(692) respiratory surface _____

_____

(692) Fick's law _____

_____

(692) hemoglobin _____

_____

(693) integumentary exchange _____

_____

(693) gills _____

_____

(693) tracheal respiration _____

_____

(694) countercurrent flow _____

_____

(694) lungs _____

_____

(695) vocal cords _____

_____

(695) glottis _____

_____

## Fill-in-the-Blanks

A(n) (1) _____ is an outfolded, thin, moist membrane endowed with blood vessels; it may (as in fish) be protected by bony covering or (as in aquatic insects) it may be naked. Gas transfer is enhanced by (2) _____ _____ , in which water flows past the bloodstream in the opposite direction. Insects have (3) _____ (chitin-lined air tubes leading from the body surface to the interior). At sea level, atmospheric pressure is approximately 760 mm Hg, and oxygen represents about (4) _____ percent of the total volume.

The energy to drive animal activities comes mainly from (5) _____ _____ , which uses (6) _____ and produces (7) _____ _____ wastes. In a process called (8) _____ , animals move (6) into their internal environment and give up (7) to the external environment.

All respiratory systems make use of the tendency of any gas to diffuse down its (9) _____ _____ . Such a (9) exists between (6) in the atmosphere [(10) _____ pressure] and the metabolically active cells in body tissues [where (6) is used rapidly; pressure is (11) ☐ highest ☐ lowest here]. Another (9) exists between (7) in body tissues where (12) [choose one] □ high □ low pressure exists and the atmosphere, with its (13) □ higher □ lower amount of (7).

The more extensive the (14) _____ _____ of a respiratory surface membrane and the larger the differences in (15) _____ _____ across it, the faster a gas diffuses across the membrane.

(16) _____ is an important transport pigment, each molecule of which can bind loosely with as many as four $O_2$ molecules in the lungs.

A(n) (17) _____ is an internal respiratory surface in the shape of a cavity or sac. In all lungs, (18) _____ carry gas to and from one side of the respiratory surface, and (19) _____ in blood vessels carries gas to and from the other side.

## Labeling

Identify the numbered parts of the accompanying illustration, which shows the respiratory system of many fishes.

20. _____

21. _____
    _____

22. _____ -

    _____

    _____

23. _____ -

    _____

    _____

24. _____

25. _____

water in

**20** water out

gill arch

**21**

**22**

**23**

direction of **24** flow (gray arrow) and **25** flow (black arrow)

## Fill-in-the-Blanks

Oxygen is said to exert a (26) _____ _____ of 760/21 or 160 mm Hg. (27) _____ alone moves oxygen from the alveoli into the bloodstream, and it is enough to move (28) _____ _____ in the reverse direction. (29) _____ is the medical name for oxygen deficiency; it is characterized by faster breathing, faster heart rate, and anxiety at altitudes of 8,000 feet above sea level.

## 41.4. HUMAN RESPIRATORY SYSTEM (pp. 696–697)
## 41.5. BREATHING—CYCLIC REVERSALS IN AIR PRESSURE GRADIENTS (pp. 698–699)

**Selected Words:** *pleurisy, respiratory bronchioles, "bronchial tree", inhalation, exhalation, atmospheric pressure, intrapulmonary pressure, intrapleural pressure, laryngitis*

## Boldfaced, Page-Referenced Terms

(696) alveolus, alveoli _____

_____

(697) pharynx _____

_____

(697) larynx _____

_____

(697) epiglottis _____

_____

(697) trachea _____

_____

(697) bronchus, (pl., bronchi) _____

_____

(697) diaphragm _____

_____

(697) bronchioles _____

_____

(698) respiratory cycle _____

_____

(699) vital capacity _____

_____

(699) tidal volume _____

_____

## Fill-in-the-Blanks

During inhalation, the (1) _____ moves downward and flattens, and the (2) _____ _____

moves outward and upward; when these things happen, the chest cavity volume (3) [choose one]

□ increases, □ decreases, and the internal pressure (4) [choose one] □ rises, □ drops, □ stays the same.

Every time you take a breath, you are (5) _____ the respiratory surfaces of your lungs. The

(6) _____ _____ surrounds each lung. In succession, air passes through the nasal cavities,

pharynx, past the epiglottis, through the (7) _____ , (the space between the true vocal cords), and into

the (8) _____ , into the trachea, and then to the (9) _____ , (10) _____ , and alveolar ducts.

Exchange of gases occurs across the epithelium of the (11) _____ .

## Labeling

Identify each indicated part of the accompanying illustration.

12. _____ _____

13. _____

14. _____

15. _____

16. _____ _____

17. _____

18. _____

19. _____

20. _____ _____

21. _____ _____

22. _____ _____

23. _____

24. _____ (sing.),

_____ (plural)

25. _____

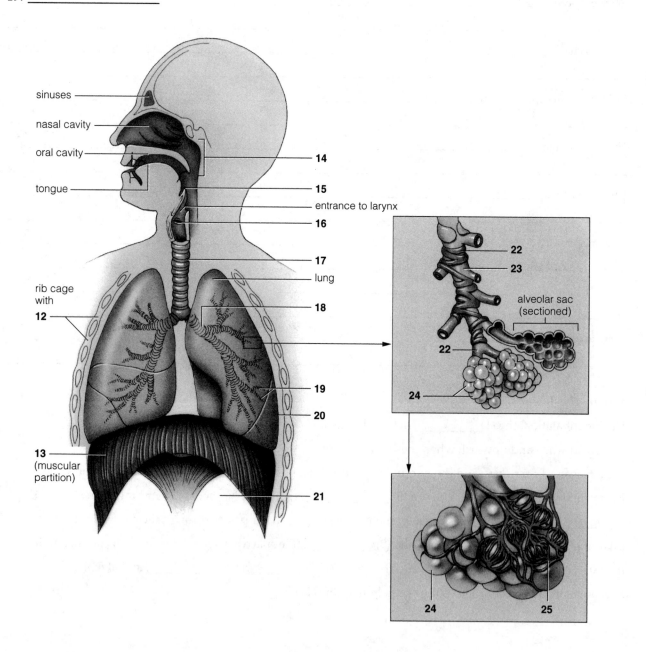

sinuses

nasal cavity

oral cavity

tongue

**14**

**15**

entrance to larynx

**16**

**17**

lung

rib cage with

**12**

**18**

**19**

**20**

**13**
(muscular partition)

**21**

**22**

**23**

alveolar sac (sectioned)

**22**

**24**

**24**          **25**

*Complete the Table*

26. Complete the following table with the structures that carry out the functions listed.

| Structure | Function |
|---|---|
| a. | Thin-walled sacs where $O_2$ diffuses into body fluids and $CO_2$ diffuses out |
| b. | Increasingly branched airways that connect the trachea and alveoli |
| c. | Muscle sheet that contracts during inhalation |
| d. | Airway where breathing is blocked while swallowing and where sound is produced |
| e. | Airway that enhances speech sounds; connects nasal cavity with larynx |

## 41.6. GAS EXCHANGE AND TRANSPORT (pp. 700–701)

## 41.7. *Focus on Health:* WHEN THE LUNGS BREAK DOWN (pp. 702–703)

## 41.8. RESPIRATION IN UNUSUAL ENVIRONMENTS (pp. 704–705)

## 41.9. *Focus on Science:* RESPIRATION IN LEATHERBACK SEA TURTLES (pp. 706–707)

**Selected Words:** bicarbonate $(HCO_3^-)$, breathing *rhythm*, breathing *magnitude*, *medulla oblongata*, partial pressure gradient, pons, aortic bodies, carotid bodies, *apnea*, *sudden infant death syndrome (SIDS)*, *bronchitis*, *emphysema*, *secondhand smoke*, *"smoker's cough"*, *pot*, *carbon monoxide poisoning*, *plexus*, *nitrogen narcosis*, *"the bends"*, *decompression sickness*

### *Boldfaced, Page-Referenced Terms*

(700) heme groups _____

_____

(700) oxyhemoglobin $(HbO_2)$ _____

_____

(700) carbaminohemoglobin $(HbCO_2)$ _____

_____

(700) carbonic anhydrase _____

_____

(704) acclimatization _____

_____

(704) erythropoietin _____

_____

(705) myoglobin _____

_____

## Fill-in-the-Blanks

An inward-directed gradient for (1) _____ is maintained because inhalations continually replenish (1) and cells continually use it during the last stage of the (2) _____ _____ process. An outward-directed gradient for (3) _____ _____ is maintained because cells continually produce (3) and exhalation removes it. Each (4) _____ molecule is constructed of four polypeptide chains compactly bound to four iron-containing, nitrogenous heme groups; the iron atom of each heme group binds reversibly with (1). 98.5 percent of all inhaled (1) is bound to the heme groups of (4).

About 70 percent of the carbon dioxide in the blood is transported as (5) _____ . Without (6) _____ , the plasma would be able to carry only about 2 percent of the oxygen that whole blood carries. When oxygen-rich blood reaches a(n) (7) _____ tissue capillary bed, oxygen diffuses outward, and carbon dioxide moves from tissues into the capillaries. When the (8) _____ _____ of carbon dioxide is lower in the alveoli than in the neighboring blood capillaries, carbonic acid dissociates to form water and carbon dioxide. Chemoreceptors in the brain monitor and coordinate signals coming in from arterial walls, from blood vessels, and from other brain regions. One respiratory center, the (9) _____ _____ , regulates contractions of the diaphragm and intercostal muscles associated with inhalation and exhalation.

(10) _____ is the distension of lungs and the loss of gas exchange efficiency such that running, walking, and even exhaling are painful experiences. At least 90 percent of all (11) _____ _____ deaths are the result of cigarette smoking; only about 10 percent of afflicted individuals will survive.

When a diver ascends, (12) _____ tends to move out of the tissues and into the bloodstream. If the ascent is too rapid, many bubbles of (12) collect at the (13) _____ , hence the common name, "the bends," for what is otherwise known as (14) _____ sickness.

---

## Self-Quiz

____ 1. Most forms of life depend on _____ to obtain oxygen and eliminate carbon dioxide.
   a. active transport
   b. bulk flow
   c. diffusion
   d. osmosis
   e. muscular contractions

____ 2. _____ is the most abundant gas in Earth's atmosphere.
   a. Water vapor
   b. Oxygen
   c. Carbon dioxide
   d. Hydrogen
   e. Nitrogen

____ 3. With respect to respiratory systems, counter-current flow is a mechanism that explains how _____ .
   a. oxygen uptake by blood capillaries in the lamellae of fish gills occurs
   b. ventilation occurs
   c. intrapleural pressure is established
   d. sounds originating in the vocal cords of the larynx are formed
   e. all of the above

____ 4. _____ have the most efficient respiratory system.
   a. Amphibians
   b. Reptiles
   c. Birds
   d. Mammals
   e. Humans

5. Immediately before reaching the alveoli, air passes through the _____ .
   a. bronchioles
   b. glottis
   c. larynx
   d. pharynx
   e. trachea

6. During inhalation, _____ .
   a. the pressure in the thoracic cavity is less than the pressure within the lungs
   b. the pressure in the chest cavity is greater than the pressure within the lungs
   c. the diaphragm moves upward and becomes more curved
   d. the thoracic cavity volume decreases
   e. all of the above

7. Hemoglobin _____ .
   a. releases oxygen more readily in tissues with high rates of cellular respiration
   b. tends to release oxygen in places where the temperature is lower
   c. tends to hold on to oxygen when the pH of the blood drops
   d. tends to give up oxygen in regions where partial pressure of oxygen exceeds that in the lungs
   e. all of the above

8. Oxygen moves from alveoli to the bloodstream _____ .
   a. whenever the concentration of oxygen is greater in alveoli than in the blood
   b. by means of active transport
   c. by using the assistance of carbaminohemoglobin
   d. principally due to the activity of carbonic anhydrase in the red blood cells
   e. by all of the above

9. Oxyhemoglobin releases $O_2$ when _____ .
   a. carbon dioxide concentrations are high
   b. body temperature is lowered
   c. pH values are high
   d. $CO_2$ concentrations are low
   e. all of the above occur

10. Nonsmokers live an average of _____ longer than people in their mid-twenties who smoke two packs of cigarettes each day.
    a. 6 months
    b. 1–2 years
    c. 3–5 years
    d. 7–9 years
    e. over 12 years

## Matching

11. ___bronchioles
12. ___bronchitis
13. ___carbonic anhydrase
14. ___emphysema
15. ___glottis
16. ___hypoxia
17. ___intercostal muscles
18. ___larynx
19. ___oxyhemoglobin
20. ___pharynx
21. ___pleurisy
22. ___rete mirabile
23. ___tidal volume
24. ___ventilation
25. ___vital capacity

A. membrane that encloses human lung becomes inflamed and swollen; painful breathing generally results
B. $HbO_2$
C. the amount of air inhaled and exhaled during normal breathing of a human at rest; generally about 500 ml.
D. throat passageway that connects to *both* the respiratory tract below *and* the digestive tract
E. greatly increases gas concentrations in a swim bladder
F. inflammation of the two principal passageways that lead air into the human lungs
G. contract when air is leaving the lungs; relax when lungs are filling with air
H. the opening into the "voicebox"
I. finer and finer branchings that lead to alveoli
J. maximum volume of air that can move out of your lungs after a single, maximal inhalation
K. an enzyme that increases the rate of production of $H_2CO_3$ from $CO_2$ and $H_2O$
L. lungs have become distended and inelastic so that walking, running and even exhaling are difficult
M. where sound is produced by vocal cords
N. movements that keep air or water moving across a respiratory surface
O. too little oxygen is being distributed in the body's tissues

# Chapter Objectives/Review Questions

This section lists general and detailed chapter objectives that can be used as review questions. You can make maximum use of these items by writing answers on a separate sheet of paper. Fill in answers where blanks are provided. To check for accuracy, compare your answers with information given in the chapter or glossary.

| Page | | Objectives/Questions |
|------|------|----------------------|
| (691, 700–701) | 1. | Understand how the human respiratory system is related to the circulatory system, to cellular respiration, and to the nervous system. |
| (692–697) | 2. | Understand the behavior of gases and the types of respiratory surfaces that participate in gas exchange. |
| (693, 696) | 3. | Describe how incoming oxygen is distributed to the tissues of insects and contrast this process with the process that occurs in mammals. |
| (694–695) | 4. | Define *countercurrent flow* and explain how it works. State where such a mechanism is found. |
| (696–699) | 5. | List all the principal parts of the human respiratory system and explain how each structure contributes to transporting oxygen from the external world to the bloodstream. |
| (698–699) | 6. | Describe the relationship of the human lung to the pleural sac and to the thoracic cavity. |
| (700) | 7. | Describe what happens to carbon dioxide when it dissolves in water under conditions normally present in the human body. |
| (700–701) | 8. | Explain why oxygen diffuses from the bloodstream into the tissues far from the lungs. Then explain why carbon dioxide diffuses into the bloodstream from the same tissues. |
| (700–701) | 9. | Explain why oxygen diffuses from alveolar air spaces, through interstitial fluid, and across capillary epithelium. Then explain why carbon dioxide diffuses in the reverse direction. |
| (701) | 10. | List the structures involved in detecting carbon dioxide levels in the blood and in regulating the rate of breathing. Name the location of each structure. |
| (702–703) | 11. | Distinguish bronchitis from emphysema. Then explain how lung cancer differs from emphysema. |
| (704–707) | 12. | List some of the ways that respiratory systems are adapted to unusual environments. |

# Integrating and Applying Key Concepts

Consider the amphibians—animals that generally have aquatic larval forms (tadpoles) and terrestrial adults. Outline the respiratory changes that you think might occur as an aquatic tadpole metamorphoses into a land-going juvenile.

# 42

# DIGESTION AND HUMAN NUTRITION



## Interactive Exercises

### *Lose It—And It Finds Its Way Back* (pp. 710–711)

### 42.1. THE NATURE OF DIGESTIVE SYSTEMS (pp. 712–713)

### 42.2. OVERVIEW OF THE HUMAN DIGESTIVE SYSTEM (p. 714)

### 42.3. INTO THE MOUTH, DOWN THE TUBE (p. 715)

***Selected Words:*** *anorexia nervosa, bulimia,* "oxlike appetite", *food-gathering* region, *food-processing* region, *crop, gizzard,* "chewing cud", *lumen, caries, gingivitis, periodontal disease*

***Boldfaced, Page-Referenced Terms***

(711) nutrition ____________________________________________

____________________________________________

(711) digestive system ____________________________________________

____________________________________________

(712) incomplete digestive system _____

_____

(712) complete digestive system _____

_____

(712) mechanical processing and motility _____

_____

(712) secretion _____

_____

(712) digestion _____

_____

(712) absorption _____

_____

(712) elimination _____

_____

(713) ruminants _____

_____

(714) gut _____

_____

(715) oral cavity _____

_____

(715) mouth _____

_____

(715) tooth _____

_____

(715) tongue _____

_____

(715) saliva _____

_____

(715) pharynx _____

_____

(715) esophagus _____

_____

(715) sphincter _____

_____

## Fill-in-the-Blanks

(1) _____ is a large concept that encompasses processes by which food is ingested, digested, absorbed and later converted to the body's own (2) _____ , lipids, proteins, and nucleic acids. A digestive system is some form of body cavity or tube in which food is reduced first to (3) _____ and then to small (4) _____ . Digested nutrients are then (5) _____ into the internal environment. A(n) (6) _____ digestive system has only one opening, two-way traffic, and a highly branched gut cavity that serves both digestive and (7) _____ functions. A(n) (8) _____ digestive system has a tube or cavity with regional specializations and a(n) (9) _____ at each end. (10) _____ involves the muscular movement of the gut wall, but (11) _____ is the release into the lumen of enzyme fluids and other substances required to carry out digestive functions.

The human digestive system is a tube, 21–30 feet long in an adult, that has regions specialized for different aspects of digestion and absorption; they are, in order, the mouth, pharynx, esophagus, (12) _____ , (13) _____ _____ , large intestine, rectum, and (14) _____ . Various (15) _____ structures secrete enzymes and other substances that are also essential to the breakdown and absorption of nutrients; these include the salivary glands, liver, gallbladder and (16) _____ . The (17) _____ system distributes nutrients to cells throughout the body. The (18) _____ system supplies oxygen to the cells so that they can oxidize the carbon atoms of food molecules, thereby changing them to the waste product, (19) _____ _____ which is eliminated by the same system. And if excess water, salts and wastes accumulate in the blood, the (20) _____ system and skin will maintain the volume and composition of blood and other body fluids.

Saliva contains an enzyme (21) _____ _____ that breaks down starch. Contractions force the larynx against a cartilaginous flap called the (22) _____ , which closes off the trachea. The (23) _____ is a muscular tube that propels food to the stomach. Any alternating progression of contracting and relaxing muscle movements along the length of a tube is known as (24) _____ . (25) _____ is an enzyme that works in the stomach to begin to break down proteins.

## Labeling

Identify each numbered structure in the accompanying illustration.

26. _____ _____
27. _____ _____
28. _____
29. _____
30. _____
31. _____ _____
32. _____ _____
33. _____
34. _____
35. _____
36. _____

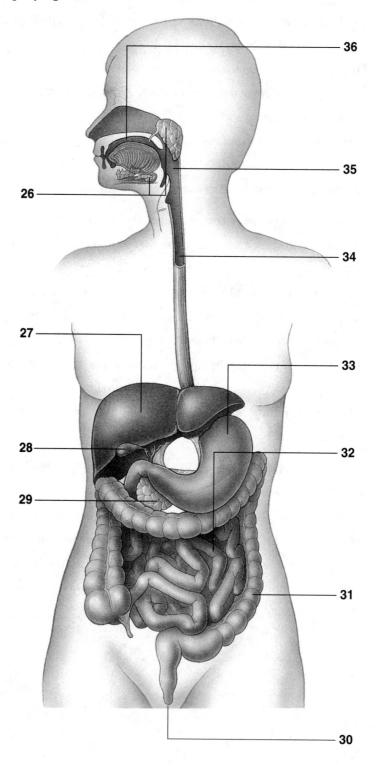

**42.4. DIGESTION IN THE STOMACH AND SMALL INTESTINE** (pp. 716–717)

**42.5. ABSORPTION IN THE SMALL INTESTINE** (pp. 718–719)

**42.6. DISPOSITION OF ABSORBED ORGANIC COMPOUNDS** (p. 720)

**42.7. THE LARGE INTESTINE** (p. 721)

*Selected Words:* *heartburn, peptic ulcer,* <u>Helicobacter</u> <u>pylori</u>, trypsin, chymotrypsin, "emulsion", CCK (cholecystokinin), GIP (glucose insulinotropic peptide), mucosa, gastrin, *constipation, appendicitis, colon cancer*

### Boldfaced, Page-Referenced Terms

(716) stomach _____

_____

(716) gastric fluid _____

_____

(716) chyme _____

_____

(716) pancreas _____

_____

(716) liver _____

_____

(716) gallbladder _____

_____

(717) bile _____

_____

(717) emulsification _____

_____

(718) villi (s., villus) _____

_____

(718) microvilli (s. microvillus) _____

_____

(718) segmentation _____

_____

(719) micelle formation _____

_____

(721) colon _____

_____

(721) bulk _____

_____

(721) appendix _____

_____

## Complete the Table

1. Complete the following table by naming the organs described.

| Organ | Main Functions |
| --- | --- |
| a. | Mechanically breaks down food, mixes it with saliva |
| b. | Moisten food; start polysaccharide breakdown; buffer acidic foods in mouth |
| c. | Stores, mixes, dissolves food; kills many microorganisms; starts protein breakdown; empties in a controlled way |
| d. | Digests and absorbs most nutrients |
| e. | Produces enzymes that break down all major food molecules; produces buffers against hydrochloric acid from stomach |
| f. | Secretes bile for fat emulsification; secretes bicarbonate, which buffers hydrochloric acid from stomach |
| g. | Stores, concentrates bile from liver |
| h. | Stores, concentrates undigested matter by absorbing water and salts |
| i. | Controls elimination of undigested and unabsorbed residues |

## Fill-in-the-Blanks

Carbohydrates include sugars and (2) _____ , the name commonly given to polysaccharides. Rice, cereal, pasta, bread and white potatoes are composed of many polysaccharide molecules that are too large to be absorbed into the internal environment. If these foods are chewed thoroughly, (3) _____ _____ in the mouth digests them to the (4) _____ (double sugar) level. Since no carbohydrate digestion occurs in the (5) _____ , if you gulped down your food, starch digestion would again begin in the (6) _____ _____ where (7) _____ produced by the pancreas would do what should have been done in the mouth. Digestion of disaccharides to monosaccharides (simple sugars) also occurs in the (8) _____ _____ . The enzymes responsible are (9) _____ with names such as sucrase, lactase and maltase.

Proteins are digested to protein fragments, beginning in the (10) _____ by (11) _____ secreted by the lining of the stomach. Protein fragments are subsequently digested to smaller protein fragments in the (12) _____ _____ by enzymes known as trypsin and chymotrypsin produced by the (13) _____ . Eventually the smaller protein fragments are digested to (14) _____ _____ by means of carboxypeptidase produced by the pancreas and by aminopeptidase produced by glands in the intestinal lining.

Another name for fat is triglycerides. (15) _____ , produced by the pancreas but acting in the (16) _____ _____ breaks down one triglyceride molecule into three (17) _____ _____ molecules and one glycerol molecule. (18) _____ , which was made by the liver, stored in the (19) _____ and which does not contain digestive enzymes emulsifies the fat droplet (converts it into small droplets coated with bile salts) thereby increasing the surface area of the substrate upon which (20) _____ can act.

Nutrients are also mostly digested and absorbed in the (21) _____ _____ . (18) is made by the liver, is stored in the gallbladder, and works in the (22) _____ _____ . (23) _____ - _____ is an example of an enzyme that is made by the pancreas but works in the small intestine to convert protein fragments to amino acids. (24) _____ _____ are made in the pancreas, but convert DNA and RNA into nucleotides in the small intestine. Any alternating progression of contracting and relaxing muscle movements along the length of a tube is known as (25) _____ .

### True-False

If the statement in true, write a T in the blank. If false, make it true by changing the underlined word.

_____26. Amylase digests starch, lipase digests lipids, and proteases break peptide bonds.

_____27. ATP is the end product of digestion.

_____28. The appendix has no known digestive functions.

_____29. Water and sodium ions are absorbed into the bloodstream from the lumen of the large intestine.

_____30. Fatty acids and monoglycerides recombine into fats inside epithelial cells lining the colon.

### Short Answer

To answer the following questions, consult Figure 42.11 and pp. 720; 722–723.

31. What is the pool of amino acids used for in the human body? _____

_____

32. Which breakdown products result from carbohydrate and fat digestion? _____

_____

33. Monosaccharides, free fatty acids, and monoglycerides all have three uses; identify them. _____

_____

## 42.8. HUMAN NUTRITIONAL REQUIREMENTS (pp. 722–723)

## 42.9. VITAMINS AND MINERALS (pp. 724–725)

## 42.10. *Focus on Science:* TANTALIZING ANSWERS TO WEIGHTY QUESTIONS
(pp. 726–727)

**Selected Words:** *the zone diet, sucrose polyester, complete* proteins, *incomplete* proteins, vitamin K, vitamins C and E, beta-carotene, vitamin A, free radicals, *caloric intake, energy output,* "set point"

## Boldfaced, Page-Referenced Terms

(722) kilocalories _____

_____

(722) food pyramids _____

_____

(723) essential fatty acids _____

_____

(723) essential amino acids _____

_____

(723) net protein utilization (NPU) _____

_____

(724) vitamins _____

_____

(724) minerals _____

_____

(726) obesity _____

_____

(727) *ob* gene _____

_____

(727) leptin _____

_____

## Complete the Table

1. Complete the following table by determining how many kilocalories the people described should take in daily, given the stated exercise level, in order to *maintain* their weight. Consult page 726 of the text.

| Height | Age | Sex | Level of Physical Activity | Present Weight (lbs.) | Number of Kilocalories/Day |
|--------|-----|-----|----------------------------|-----------------------|----------------------------|
| a. 5'6" | 25 | Female | Moderately active | 138 | a. |
| b. 5'10" | 18 | Male | Very active | 145 | b. |
| c. 5'8" | 53 | Female | Not very active | 143 | c. |

## Fill-in-the-Blanks

People who have gained 11 to (2) _____ pounds of excess fat in adult life are overweight individuals who have increased their risk of heart attack significantly. (3) _____ _____ are the body's main sources of energy; they should make up (4) _____ to _____ percent of the human daily caloric intake. (5) _____ and cholesterol are components of animal cell membranes.

Fat deposits are used primarily as (6) _____ _____ , but they also cushion many organs and provide insulation. Lipids should constitute less than (7) _____ percent of the human diet. One teaspoon a day of polyunsaturated oil supplies all (8) _____ _____ _____ that the body cannot synthesize. (9) _____ are digested to twenty common amino acids, of which eight are (10) _____ , cannot be synthesized, and must be supplied by the diet. Animal proteins such as (11) _____ and (12) _____ contain high amounts of essential amino acids (that is, they are complete). (13) _____ are organic substances needed in small amounts in order to build enzymes or help them catalyze metabolic reactions. (14) _____ are inorganic substances needed for a variety of uses.

## Review Problems

You are a 19-year-old male, very sedentary (TV, sleep, and computers), 6'0", medium frame, and you weigh 195 lbs.

15. Use Fig. 42.15, page 726, to calculate the number of calories required to sustain your desired weight. Use the average of that range. Are you underweight, overweight, or just right?
16. How many calories are you allowed to ingest every day?

## Complete the Table

Use the new, improved food pyramid (p. 722) to construct a one-day diet that would eventually allow you to reach that weight if you ate a similar diet every day. Place your choices in the table below as a diet for the person described above.

| *How many servings from each group below are you allowed to have daily?* | *What, specifically, could you choose to eat?* |
|---|---|
| 17. complex carbohydrates | 17b. |
| 18. fruits | 18b. |
| 19. vegetables | 19b. |
| 20. dairy group | 20b. |
| 21. assorted proteins | 21b. |
| 22. The "sin" group at the top | 22b. |

*Fill-in-the-Blanks*

Use the (23) _____ _____ diagram at the left, as revised in 1992, to devise a well-balanced diet for yourself. Group 1, the trapezoidal base, represents the group of complex (24) _____ , which includes rice, pasta, cereal and (25) _____ . From this group, (26) (choose 1) □ 0, □ 2–3, □ 2–4, □ 3–5, □ 6–11 servings from this group every day are needed to supply energy and fiber. Group 2 represents the (27) _____ group. Use the choices in (26) to indicate the number of servings (28) _____ that are needed from this group each day. Group 3 is the (29) _____ group, from which (30) _____ servings are needed each day. Choices include mango, oranges and (31) _____ , cantaloupe, pineapple or 1 cup of fresh (32) _____ . Group 4 includes foods that are a source of nitrogen: nuts, poultry, fish, legumes and (33) _____ . From this group, the body's (34) _____ and nucleic acids are constructed. (35) _____ servings are required every day because the human body cannot synthesize eight of the 20 essential (36) _____ _____ that are used to construct proteins, and must get them in their food supplies. The foods in Group 5, the (37) _____ , yogurt and cheese group, supply calcium, vitamins A and D, $B_2$, $B_{12}$. You need (38) _____ servings every day. The foods in Group 6 provide extra calories but few vitamins and minerals; (39) _____ servings are needed every day.

# Self-Quiz

___ 1. The process that moves nutrients into the blood or lymph is _____ .
   a. ingestion
   b. absorption
   c. assimilation
   d. digestion
   e. none of the above

___ 2. The enzymatic digestion of proteins begins in the _____ .
   a. mouth
   b. stomach
   c. liver
   d. pancreas
   e. small intestine

___ 3. The enzymatic digestion of starches begins in the _____ .
   a. mouth
   b. stomach
   c. liver
   d. pancreas
   e. small intestine

___ 4. The greatest amount of absorption of digested nutrients occurs in the _____ .
   a. stomach
   b. pancreas
   c. liver
   d. colon
   e. small intestine

___ 5. Glucose moves through the membranes of the small intestine mainly by _____ .
a. peristalsis
b. osmosis
c. diffusion
d. active transport
e. bulk flow

___ 6. Which of the following is *not* found in bile?
a. lecithin
b. salts
c. digestive enzymes
d. cholesterol
e. pigments

___ 7. The average American consumes approximately _____ pounds of sugar per year.
a. 25
b. 50
c. 75
d. 100
e. 125

___ 8. Of the following, _____ has (have) the highest net protein utilization.
a. beans
b. eggs
c. fish
d. meat
e. bread

___ 9. One hour after a meal, the blood richest in nutrients would be in the _____ .
a. abdominal aorta
b. hepatic portal vein
c. hepatic vein
d. pulmonary artery
e. vena cava

___10. The element needed by humans for blood clotting, nerve impulse transmission, and bone and tooth formation is _____ .
a. magnesium
b. iron
c. calcium
d iodine
e. zinc

## Matching

Match the best lettered item with its correct numbered item at the left.

11. ___anorexia nervosa

12. ___vitamins C and E

13. ___bulimia

14. ___complex carbohydrates

15. ___essential amino acids

16. ___essential fatty acids

17. ___mineral

18. ___rickets

19. ___scurvy

20. ___vitamin

A. Linoleic acid is one example
B. Phenylalanine, lysine, and methionine are 3 of 8
C. Combine with free radicals; counteract their destructive effects on DNA and cell membranes
D. Obsessive dieting + skewed perception of body weight
E. Vitamin C deficiency
F. Vitamin D deficiency in young children
G. Organic substances that help enzymes to do their jobs; required in small amounts for good health.
H. Feasting followed by vomiting or taking laxatives
I. Inorganic substances required for good health
J. Long chains of simple sugars; in pasta and white potatoes

# Chapter Objectives/Review Questions

This section lists general and detailed chapter objectives that can be used as review questions. You can make maximum use of these items by writing answers on a separate sheet of paper. Fill in answers where blanks are provided. To check for accuracy, compare your answers with information given in the chapter or glossary.

*Page*      *Objectives/Questions*

(712)     1.  Distinguish between incomplete and complete digestive systems and tell which is characterized by (a) specialized regions, (b) two-way traffic, and (c) discontinuous feeding.

(712)     2.  Define and distinguish among motility, secretion, digestion, and absorption.

(714)     3.  List all parts (in order) of the human digestive system through which food actually passes. Then list the auxiliary organs that contribute one or more substances to the digestive process.

(715)     4.  Explain how, during digestion, food is mechanically broken down. Then explain how it is chemically broken down.

(717)     5.  Tell which foods undergo digestion in each of the following parts of the human digestive system and state what the food is broken into: mouth, stomach, small intestine, large intestine.

(717)     6.  List the enzyme(s) that act in (a) the oral cavity, (b) the stomach, and (c) the small intestine. Then tell where each enzyme was originally made.

(717;     7.  Describe how the digestion and absorption of fats differ from the digestion and absorption of
719–720)      carbohydrates and proteins.

(717, 720)  8.  Explain how the human body manages to meet the energy and nutritional needs of the various body parts even though the person may be feasting sometimes and fasting at other times.

(718–720)  9.  Describe the cross-sectional structure of the small intestine and explain how its structure is related to its function.

(719)    10.  List the items that leave the digestive system and enter the circulatory system during the process of absorption.

(721)    11.  State which processes occur in the colon (large intestine).

(722)    12.  Reproduce from memory the Food Pyramid Diagram as revised in 1992. Identify each of the six components, list the numerical range of servings permitted from each group and also list some of the choices available.

(722)    13.  Construct an ideal diet for yourself for one 24-hour period. Calculate the number of calories necessary to maintain your weight (see p. 726) and then use the Food Pyramid (p. 722) to choose exactly what to eat and how much.

(722–725) 14.  Compare the contributions of carbohydrates, proteins, and fats to human nutrition with the contributions of vitamins and minerals.

(722–726) 15.  Summarize current ideas for promoting health by eating properly.

(722–726) 16.  Summarize the daily nutritional requirements of a 25-year-old man who weighs 135 pounds, works at a desk job and exercises very little. State what he needs in energy, carbohydrates, proteins, and lipids and name at least six vitamins and six minerals that he needs to include in his diet every day.

(724–725) 17.  Distinguish vitamins from minerals, and state what is meant by net protein utilization.

(725)    18.  Name five minerals that are important in human nutrition and state the specific role of each.

# Integrating and Applying Key Concepts

Suppose you could not eat solid food for two weeks and you had only water to drink. List in correct sequential order the measures your body would take to try to preserve your life. Mention the command signals that are given as one after another critical point is reached, and tell which parts of the body are the first and the last to make up for the deficit.

# 43

# THE INTERNAL ENVIRONMENT

## Interactive Exercises

*Tale of the Desert Rat* (pp. 730–731)

## 43.1. URINARY SYSTEM OF MAMMALS (pp. 732–733)

*Selected Words:* *solutes, glomerular* capillaries, *peritubular* capillaries

*Boldfaced, Page-Referenced Terms*

(731) interstitial fluid _____

_____

(731) blood _____

_____

(731) extracellular fluid _____

_____

(732) urinary excretion _____

_____

(732) urea _____

_____

(733) urinary system _____

_____

(733) kidneys _____

_____

(733) urine _____

_____

(733) ureter _____

_____

(733) urinary bladder _____

_____

(733) urethra _____

_____

(733) nephrons _____

_____

(733) Bowman's capsule _____

_____

(733) glomerulus _____

_____

(733) proximal tubule _____

_____

(733) loop of Henle _____

_____

(733) distal tubule _____

_____

(733) collecting duct _____

_____

## Fill-in-the-Blanks

The body gains water by absorbing water from the slurry in the lumen of the small intestine and from
(1) _____ during condensation reactions. The mammalian body loses water mostly by excretion of
(2) _____ , evaporation through the skin, and (3) _____ _____ , elimination of feces from the
gut, and (4) _____ as the body is cooled. (5) _____ behavior, in which the brain compels the
individual to seek liquids, influences the gain of water.

The body gains solutes by absorption of substances from the gut, by the secretion of hormones and other
substances, and by (6) _____ , which produces $CO_2$ and other waste products of degradative reactions.
Besides $CO_2$, there are several major metabolic wastes that must be eliminated: (7) _____ , formed by
amino groups being split from amino acids; (8) _____ , which is produced in the liver during reactions
that link two ammonia molecules to $CO_2$ and release a molecule of water; and (9) _____ _____ ,
which is formed in reactions that break down nucleic acids.

## Labeling

Identify each indicated part of the accompanying illustrations.

10. _____

11. _____

12. _____  _____

13. _____

14. _____

15. _____

16. _____

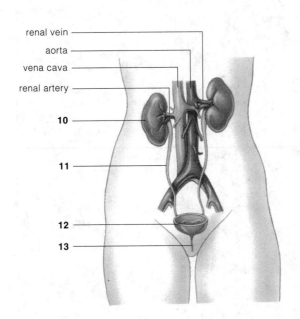

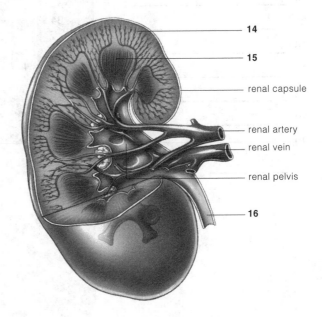

## Fill-in-the-Blanks

In mammals, urine formation occurs in a pair of (17) _____ . Each contains about a million tubelike

blood-filtering units called (18) _____ . The function of (17) depends on intimate links between the

(18) and the (19) _____ . Blood containing excess water and solutes enters each kidney by means of a

renal artery, which subdivides into smaller blood vessels called (20) _____ . Each (20) leads into a set of

capillaries inside a (21) _____ _____ , the wall of which forms a cup around a cluster

(22) _____ capillaries. Most of the water and solutes are driven by (23) _____ _____ from the

blood-filtering unit, the (24) _____ , across membranous barriers into the lumen of the (21) and on into

the (25) _____ _____ ; the remainder of the blood, now containing relatively little water, few

solutes and many large proteins, continues onward, back out of the (21) and into a second set of

(26) _____ capillaries which surround the slender urine-forming tubule and reabsorb most of the water

and needed solutes.  From the (25), that which was filtered out of the blood (the filtrate) continues around a

hairpin-shaped (27) _____ _____ _____ and up into the (28) _____ _____ . It ends

up in a (29) _____ _____ that leads to the tube (called the (30) _____ ) that conducts urine

from the kidney to the urinary bladder for storage.

Every vertebrate has a (31) _____ system that filters water and (32) _____ from the blood. Then it reclaims both in amounts necessary to maintain (33) _____ fluid, and it eliminates the rest in the form of urine.

### Labeling

Identify each indicated part of the accompanying illustration.

34. _____ _____

35. _____ _____

36. _____ _____

37. _____ _____

38. _____ _____

39. _____ _____

40. _____ _____

    _____

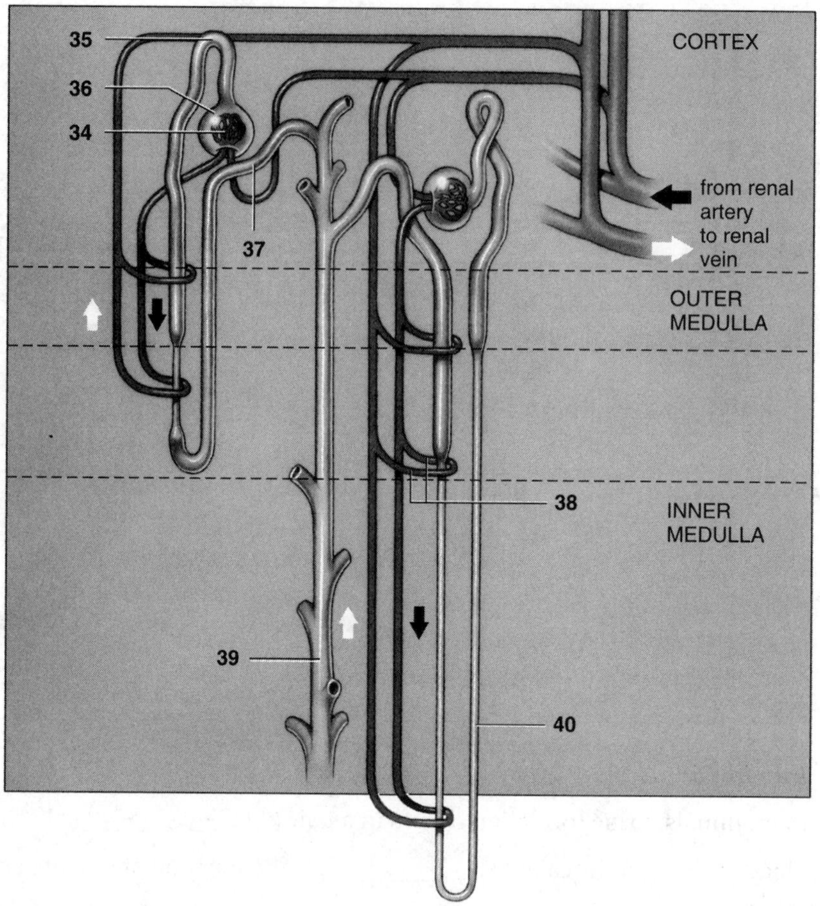

**43.2. URINE FORMATION** (pp. 734–735)

**43.3.** *Focus on Health:* **WHEN KIDNEYS BREAK DOWN** (p. 736)

**43.4. THE ACID-BASE BALANCE** (p. 736)

**43.5. ON FISH, FROGS, AND KANGAROO RATS** (p. 737)

*Selected Words:* hypothalamus, renin, adrenal cortex, *uremic toxicity, glomerulonephritis,* Streptococcus, *kidney stones, kidney dialysis machine, "dialysis", hemodialysis, peritoneal dialysis, metabolic acidosis, bicarbonate-carbon dioxide,* buffer system

### Boldfaced, Page-Referenced Terms

(734) filtration _____

_____

(734) tubular reabsorption _____

_____

(734) tubular secretion _____

_____

(735) ADH, antidiuretic hormone _____

_____

(735) aldosterone _____

_____

(735) angiotensin II _____

_____

(735) thirst center _____

_____

(736) renal failure _____

_____

(736) acid-base balance _____

_____

## Fill-in-the-Blanks

Three processes ((1) _____ , (2) _____ _____ , and (3) _____ _____ ,) cooperate to produce urine in nephrons. (1) occurs by blood pressure forcing most water and solutes (except proteins) out through the walls of the (4) _____ capillaries. (2) occurs along the (5) _____'s tubular regions as appropriate amounts of water and solutes move into neighboring (6) _____ capillaries. (3) occurs at the tubule wall but in the opposite direction of (2); ATP is required to pump any substance against its natural gradient in an active transport process.

Approximately two-thirds of the filtrate's water moves into the peritubular capillaries from the (7) _____ _____ , as do some of the sodium and other ions. As the filtrate moves along the loop of Henle, water from the filtrate is attracted into (6) by (8) _____ before the turn; after the turn, the wall of the loop is impermeable to water, but active transport pumps (9) _____ out. Filtrate arriving at the distal tubule is (10) _____ and is ready for hormone-induced adjustments.

Two hormones, ADH and (11) _____ , adjust the reabsorption of water and (12) _____ along the distal tubules and collecting ducts. An increase in the secretion of aldosterone causes (13) [choose one] □ more, □ less sodium to be excreted in the urine. When the body cannot rid itself of excess sodium, it inevitably retains excess water, and this leads to a rise in (14) _____ _____ . Abnormally high blood pressure is called (15) _____ ; it can damage the kidneys, vascular system, and brain. One way to control hypertension is to restrict the intake of (16) _____ _____ . Increased secretion of (17) _____ enhances water reabsorption at distal tubules and collecting ducts when the body must conserve water. When excess water must be excreted, ADH secretion is (18) [choose one] □ stimulated, □ inhibited.

*True-False*

If false, explain why.

_____19. When the body rids itself of excess water, urine becomes more dilute.

_____20. Water reabsorption into capillaries is achieved by diffusion and active transport.

*Fill-in-the-Blanks*

The (21) _____ control the acid-base balance of body fluids by controlling the levels of dissolved ions, especially (22) _____ ions. The extracellular pH of humans must be maintained between 7.37 and (23) _____ . (24) [choose one] □ Acids □ Bases lower the pH and (25) [choose one] □ acids □ bases raise it. If you were to drink a gallon of orange juice, the pH would be (26) [choose one] □ raised □ lowered, but the effect is minimized when excess (27) _____ ions are neutralized by (28) _____ ions in the bicarbonate-carbon dioxide buffer system. Only the (29) _____ system eliminates excess $H^+$ and restores buffers. Desert-dwelling kangaroo rats have very long (30) _____ _____ _____ so that nearly all (31) _____ that reaches their very long collecting ducts is reabsorbed. In freshwater, bony fishes and amphibians tend to gain (32) _____ and lose (33) _____ ; they produce (34) [choose one] □ very dilute □ very concentrated urine.

## 43.6. MAINTAINING THE BODY'S CORE TEMPERATURE (pp. 738–739)
## 43.7. TEMPERATURE REGULATION IN MAMMALS (pp. 740–741)

*Selected Words:* brown adipose tissue, hypothermia, frostbite, "panting", hyperthermia, fever

*Boldfaced, Page-Referenced Terms*

(738) core temperature _____

_____

(738) radiation _____

_____

(738) conduction _____

_____

(738) convection _____

_____

(739) evaporation _____

_____

(739) ectotherms _____

_____

(739) behavioral temperature regulation _____

_____

(739) endotherms _____

_____

(739) heterotherms _____

_____

(740) peripheral vasoconstriction _____

_____

(740) pilomotor response _____

_____

(740) shivering response _____

_____

(740) nonshivering heat production _____

_____

(741) peripheral vasodilation _____

_____

(741) evaporative heat loss _____

_____

## Fill-in-the-Blanks

In the brain of mammals, the (1) _____ is the seat of temperature control. Thermoreceptors located

deep in the body are called (2) _____ thermoreceptors. The (3) _____ _____ contains smooth

muscles that erect hairs or feathers and create an insulative layer of still air that helps prevent heat loss.

(4) _____ _____ is a response to cold stress in which the bloodstream's convective delivery of heat

to the body's surface is reduced. A drop in core body temperature below tolerance levels is referred to as

(5) _____ .

## True-False

If false, explain why.

_____6. Jackrabbits are endotherms.

_____7. When the core temperature of the human body is about 86°F, consciousness is lost and
heart muscle action becomes irregular.

_____8. When the core temperature of the human body reaches 77°F, ventricular fibrillation sets in
and death soon follows.

## Matching

Choose the most appropriate answer to match with the following terms.

9. ___conduction
10. ___convection
11. ___ectotherm
12. ___endotherm
13. ___evaporation
14. ___heterotherm
15. ___radiation

A. Body temperature determined more by heat exchange with the environment than by metabolic heat
B. Heat transfer by air or water heat-bearing currents moving away from or toward a body
C. Body temperature determined largely by metabolic activity and by precise controls over heat produced and heat lost
D. Direct transfer of heat energy between two objects in direct contact with each other
E. The emission of energy in the form of infrared or other wavelengths that are converted to heat by the absorbing body
F. Body temperature fluctuating at some times and heat balance controlled at other times
G. In changing from the liquid state to the gaseous state, the energy required is supplied by the heat content of the liquid

# Self-Quiz

___ 1. The most toxic waste product of metabolism is _____ . Animals that live in fresh water excrete this in their urine.
   a. water
   b. uric acid
   c. urea
   d. ammonia
   e. carbon dioxide

___ 2. An entire subunit of a kidney that purifies blood and restores solute and water balance is called a _____ .
   a. glomerulus
   b. loop of Henle
   c. nephron
   d. ureter
   e. none of the above

___ 3. In humans, the thirst center is located in the _____ .
   a. adrenal cortex
   b. thymus
   c. heart
   d. adrenal medulla
   e. hypothalamus

___ 4. The longer the _____ , the greater an animal's capacity to conserve water and to concentrate solutes to be excreted in the urine.
   a. loop of Henle
   b. proximal tubule
   c. ureter

   d. Bowman's capsule
   e. collecting tubule

___ 5. During reabsorption, sodium ions cross the proximal tubule walls into the interstitial fluid principally by means of _____ .
   a. phagocytosis
   b. countercurrent multiplication
   c. bulk flow
   d. active transport
   e. all of the above

___ 6. Filtration of the blood in the kidney takes place in the _____ .
   a. loop of Henle
   b. proximal tubule
   c. distal tubule
   d. Bowman's capsule
   e. all of the above

___ 7. _____ primarily controls the concentration of solutes in urine.
   a. Insulin
   b. Glucagon
   c. Antidiuretic hormone
   d. Aldosterone
   e. Epinephrine

___ 8. Hormonal control over excretion primarily affects _____ .
   a. Bowman's capsules
   b. distal tubules and collecting ducts
   c. proximal tubules
   d. the urinary bladder
   e. loops of Henle

___ 9. The last portion of the excretory system passed by urine before it is eliminated from the body is the _____ .
a. renal pelvis
b. bladder
c. ureter
d. collecting ducts
e. urethra

___ 10. Normally, the extracellular pH of the human body must be maintained between _____ and _____ ; only the urinary system eliminates excess _____ and restores _____ .
a. 6.45–7.30; $NH_4^+$; urea
b. 7.37–7.43; $H^+$; buffers
c. 7.50–7.85; $H^+$; glucose
d. 7.90–8.30; $NH_4^+$; urea
e. 8.15–8.35; $OH^-$; glucose

## Chapter Objectives/Review Questions

This section lists general and detailed chapter objectives that can be used as review questions. You can make maximum use of these items by writing answers on a separate sheet of paper. Fill in answers where blanks are provided. To check for accuracy, compare your answers with information given in the chapter or glossary.

| Page | | Objectives/Questions |
|---|---|---|
| (730–733) | 1. | List some of the factors that can change the composition and volume of body fluids. |
| (732) | 2. | List three soluble by-products of animal metabolism that various kinds of animals excrete in their urine. |
| (732–733) | 3. | List successively the parts of the human urinary system that constitute the path of urine formation and excretion. |
| (734–735) | 4. | Locate the processes of filtration, reabsorption, and tubular secretion along a nephron and tell what makes each process happen. |
| (735) | 5. | State explicitly how the hypothalamus, adrenal cortex, and distal tubules and collecting ducts of the nephrons are interrelated in regulating water and solute levels in body fluids. |
| (736) | 6. | List two kidney disorders and explain what can be done if kidneys become too diseased to work properly. |
| (736) | 7. | Describe the role of the kidney in maintaining the pH of the extracellular fluids between 7.37 and 7.43. |
| (738–739) | 8. | List the ways in which ectotherms are disadvantaged by not being able to maintain a particular body temperature. Describe the things ectotherms can do to lessen their vulnerability. |
| (739) | 9. | Distinguish between ectotherms and endotherms and give two examples of each group. |
| (739–741) | 10. | Explain how endotherms maintain their body temperature when environmental temperatures fall. |
| (740–741) | 11. | Define *hypothermia* and state the situations in which a human might experience the disorder. |
| (741) | 12. | Explain how endotherms maintain their body temperature when environmental temperatures rise 3 to 4 degrees Fahrenheit above standard body temperature. |

## Integrating and Applying Key Concepts

The hemodialysis machine used in hospitals is expensive and time-consuming. So far, artificial kidneys capable of allowing people who have nonfunctional kidneys to purify their blood by themselves, without having to go to a hospital or clinic, have not been developed. Which aspects of the hemodialysis procedure do you think have presented the most problems in development of a method of home self-care? If you had an unlimited budget and were appointed head of a team to develop such a procedure and its instrumentation, what strategy would you pursue?

## ACROSS

1. Connects urinary bladder to exterior
4. _____ enhances sodium reabsorption, and is produced by the adrenal cortex.
6. Extracellular _____ bathes the body's cells.
7. _____ acid is the least toxic nitrogenous waste and is constructed from nucleic acid breakdown.
8. Bowman's _____ encloses the glomerulus.
10. _____ is an enzyme secreted by kidney cells that detaches part of a protein circulating in the blood so that the new protein can be made into a hormone that acts on the adrenal cortex.
11. A cluster of capillaries enclosed by a Bowman's capsule that filters blood
13. Connects a kidney to the urinary bladder
14. An organ that adjusts the volume and composition of blood and helps maintain the composition of the extracellular fluid
16. A small tube
17. Relating to kidney function
19. _____ of Henle
20. Most toxic nitrogenous waste
21. Nitrogenous waste formed in the liver and is relatively harmless
22. Water and solutes move out of the nephron tubule, then into adjacent capillaries.
24. The urinary _____ includes three organs and three larger waste-containing tubes.
25. More than a million of these units are packed inside each fist-sized kidney
26. The elimination of fluid wastes
27. The _____ tubule is the part of the nephron that is farthest from the Bowman's capsule.

## DOWN

1. Product of kidneys
2. _____ fluid bathes the body's cells
3. Loop of _____
4. _____ glands are perched on top of the kidneys
5. _____ moves excess $H^+$ and a few other substances by active transport from the capillaries into the cells of the nephron wall and then into the urine.
6. Water and small-molecule solutes are forced from the blood into the Bowman's capsule.
9. Urinary _____ stores urine.
12. The middle region of the kidney
13. _____ excretion is a process that dumps excess mineral ions and metabolic wastes.
15. The _____ tubule is nearest to its Bowman's capsule.
18. The outer region of kidney, adrenal gland or brain
23. The "water conservation" hormone produced by the posterior pituitary

# 44

# PRINCIPLES OF REPRODUCTION AND DEVELOPMENT

*From Frog to Frog and Other Mysteries*

THE BEGINNING: REPRODUCTIVE MODES
    Sexual Versus Asexual Reproduction
    Costs and Benefits of Sexual Reproduction

STAGES OF DEVELOPMENT—AN OVERVIEW

EARLY MARCHING ORDERS
    Information in the Egg Cytoplasm
    Cleavage—The Start of Multicellularity
    Cleavage Patterns

HOW SPECIALIZED TISSUES AND ORGANS FORM
    Cell Differentiation
    Morphogenesis

*Focus on Science:* TO KNOW A FLY

PATTERN FORMATION
    A Theory of Pattern Formation
    Embryonic Induction

FROM THE EMBRYO ONWARD
    Post-Embryonic Development
    Aging and Death

*Commentary:* DEATH IN THE OPEN

---

## Interactive Exercises

---

*From Frog to Frog and Other Mysteries* (pp. 744–745)

## 44.1. THE BEGINNING: REPRODUCTIVE MODES (pp. 746–747)

## 44.2. STAGES OF DEVELOPMENT—AN OVERVIEW (pp. 748–749)

*Selected Words:* reproductive timing, oviparous, viviparous, ovoviviparous

*Boldfaced, Page-Referenced Terms*

(744) zygote _____

_____

(746) sexual reproduction _____

_____

(746) asexual reproduction _____

_____

(746) internal fertilization _____

_____

(747) yolk _____

_____

(748) embryo _____

_____

(748) gamete formation _____

_____

(748) fertilization _____

_____

(748) cleavage _____

_____

(748) blastomeres _____

_____

(748) gastrulation _____

_____

(748) ectoderm _____

_____

(748) endoderm _____

_____

(748) mesoderm _____

_____

(748) organ formation _____

_____

(748) growth and tissue specialization _____

_____

## Fill-in-the-Blanks

New sponges budding from parent sponges and a flatworm dividing into two flatworms represent examples of (1) _____ reproduction. This type of reproduction is useful when gene-encoded traits are strongly adapted to a limited set of (2) _____ conditions. Separation into male and female sexes requires special reproductive structures, control mechanisms, and behaviors; this cost is offset by a selective advantage: (3) _____ in traits among the offspring. Males and females of the same species must synchronize their (4) _____ _____ to ensure that their gametes mature and find each other at the correct time for fertilization to occur.

(5) _____ _____ is considered the first stage of animal development. Rich stores of substances such as yolk become assembled in localized regions of the (6) _____ cytoplasm. When sperm and egg unite and their DNA mingles and is reorganized, the process is referred to as (7) _____ . At the end of fertilization, a(n) (8) _____ is formed. (9) _____ includes the repeated mitotic divisions of a zygote

that segregate the egg cytoplasm into a cluster of cells known as (10) _____ ; the entire cluster is known as a blastula. (11) _____ is the process that arranges cells into three germ layers. Ectoderm eventually will give rise to skin epidermis and the (12) _____ system; endoderm forms the inner lining of the (13) _____ and associated digestive glands. Mesoderm forms the circulatory system, the (14) _____ , the muscles and connective tissues. (15) _____ animals are nourished by maternal tissues, not yolk only, until the time of birth.

Copperheads are (16) ____: fertilization is internal; the fertilized eggs develop inside the mother's body without additional nourishment, and the young are born live. Birds are (17) _____ : eggs with large yolk reserves are released from and develop outside the mother's body.

Each stage of (18) _____ development builds on structures that were formed during the stage preceding it. (19) _____ cannot proceed properly unless each stage is successfully completed before the next begins.

## True-False

If false, explain why.

_____20. Yolk is a substance rich in carbohydrates that nourishes embryonic stages.

_____21. Sperm penetration into the cytoplasm of the egg brings about specific structural changes and chemical reactions.

_____22. Most animals reproduce sexually.

_____23. Sexual reproduction is less advantageous in predictable environments; asexual reproduction is more advantageous in predictable environments.

_____24. The eggs of leopard frogs (Rana pipiens) are fertilized externally in the water.

_____25. Gastrulation precedes organ formation.

## Sequence

Arrange the following events in correct chronological sequence. Write the letter of the first step next to 26, the letter of the second step next to 27, and so on.

26. ___      A. Gastrulation

27. ___      B. Fertilization

28. ___      C. Cleavage

29. ___      D. Growth, tissue specialization

30. ___      E. Organ formation

31. ___      F. Gamete formation

## Complete the Table

32. Complete the table below by entering the correct germ layer (ectoderm, mesoderm, or endoderm) that forms the tissues and organs listed.

| Tissues/Organs | Germ Layer |
|---|---|
| Muscle, circulatory organs | a. |
| Nervous tissues | b. |
| Inner lining of the gut | c. |
| Circulatory organs (blood vessels, heart) | d. |
| Outer layer of the integument | e. |
| Reproductive and excretory organs | f. |
| Organs derived from the gut | g. |
| Most of the skeleton | h. |
| Connective tissues of the gut and integument | i. |

## 44.3. EARLY MARCHING ORDERS (pp. 750–751)

## 44.4. HOW SPECIALIZED TISSUES AND ORGANS FORM (pp. 752–753)

## 44.5. *Focus on Science:* TO KNOW A FLY (pp. 754–755)

*Selected Words:* *"maternal messages"*, *animal pole, vegetal pole, "inner cell mass", gastrula*

### Boldfaced, Page-Referenced Terms

(750) oocyte _____

_____

(750) sperm _____

_____

(750) gray crescent _____

_____

(750) cleavage furrow _____

_____

(750) cytoplasmic localization _____

_____

(751) blastula _____

_____

(751) blastocoel _____

_____

(751) morula _____

_____

(752) neural tube _____

_____

(752) cell differentiation _____

_____

(753) identical twins _____

_____

(753) morphogenesis _____

_____

(753) apoptosis _____

_____

(755) fate map _____

_____

## True-False

If false, explain why.

_____1. During gastrulation, <u>maternal</u> controls over gene activity are activated and begin the process of differentiation in each cell's nucleus.

_____2. In complex eukaryotes, development until gastrulation is governed by <u>DNA in the nucleus of the zygote</u>.

_____3. In a developing chick embryo, the heart begins to beat at some time between <u>30 and 36 hours</u> after fertilization.

## Fill-in-the-Blanks

The third stage of animal development, (4) _____ , is characterized by the subdividing and compartmentalizing of the zygote; no growth occurs at this stage, and usually a hollow ball of cells, the (5) _____ , is formed. Much of the information that determines how structures will be spatially organized in the embryo begins with the distribution of (6) _____ _____ in the oocyte. These consist of mRNAs, regionally distributed enzymes and other proteins, yolk, and other factors. As cleavage membranes divide up the cytoplasm, various cytoplasmic determinants become localized in different daughter cells; this process, called (7) _____ _____ , helps seal the developmental fate of the descendants of those cells. In amphibian eggs, sperm penetration on one side of an egg causes pigment granules on the opposite side of the egg to flow toward the (8) _____ _____ . A lightly pigmented area called the (9) _____ _____ results. It is a visible marker of the site where the (10) _____ _____ will be established and where gastrulation will begin.

The fourth stage, (11) _____ , is concerned with the formation of ectoderm, mesoderm, and endoderm, the (12) _____ layers of the embryo; at the end of this stage, the (13) _____ is formed, and resembles an early embryo. Cleavage of a frog's zygote is complete, because there is so little yolk that cleavage membranes can subdivide the entire cytoplasmic mass; in the chick, however, there is so much yolk that cleavage membranes cannot subdivide the entire mass. Cleavage is therefore said to be incomplete, and the chick grows from a primitive streak on the surface of the (14) _____ mass into a chick embryo complete with wing and leg buds and beating heart during the first (15) _____ days.

In every vertebrate, an imaginary straight line connecting the head to the tail end defines where a (16) _____ _____ , the forerunner of the brain and spinal cord, will form. All organs of the adult begin formation by two principal processes: (17) _____ _____ , in which a cell selectively activates specific genes (and not others) and synthesizes some proteins not found in other cell types, lays the ground work for (18) _____ , which is a program of orderly changes in an embryo's size, shape and proportions that results in tissues becoming specialized to function in a specific way and form the early stages of organs.

During (19) _____ _____ _____ , cells send out pseudopods and use them to move them along prescribed routes; forerunners of neurons interconnect this way as a nervous system is forming. (20) _____ cues tell the cells when to stop migration. As (21) _____ lengthen and rings of (22) _____ in cells constrict, sheets of cells expand and fold inward and outward, as in neural tube formation. Sometimes, as in the formation of the human hand, specific cells in the early form of the hand die on schedule according to cues in their genetic programs; this form of programmed cell death, called (23) _____ , rids the developing embryo of cells not needed in the next developmental phases.

## Matching

In *Drosophila*, specific types of genes act as master organizers that establish the location, shapes, and sizes of various body parts in the animal during pattern formation. Match the correct gene with its function.

24. ___ gap genes

25. ___ homeotic genes

26. ___ maternal effect genes

27. ___ pair-rule genes

28. ___ segment polarity genes

A. Divide the embryo into segment-size units

B. Produces substances that accumulate in bands that each correspond to two body segments

C. Map out broad regions of the body; different concentrations of these switch on pair-rule genes

D. Directly and collectively govern the developmental fate of each body segment

E. Transcribe specific mRNAs and indirectly cause specific regulatory proteins to be translated; these products become localized in different parts of the egg cytoplasm and are activated in the zygote, where they activate or suppress gap genes.

## 44.6. PATTERN FORMATION (pp. 756–757)
## 44.7. FROM THE EMBRYO ONWARD (p. 758)
## 44.8. *Commentary:* DEATH IN THE OPEN (p. 759)

**Selected Words:** *nonidentical twins, active cell migration, incomplete metamorphosis, complete metamorphosis, limited division potential*

## Boldfaced, Page-Referenced Terms

(756) pattern formation _____

_____

(756) homeotic genes _____

_____

(756) embryonic induction _____

_____

(757) morphogens _____

_____

(758) juvenile _____

_____

(758) metamorphosis _____

_____

(758) molting _____

_____

(758) aging _____

_____

## Fill-in-the-Blanks

Through (1) _____ _____ , a single fertilized egg gives rise to an assortment of different types of specialized cells; these differentiated cells have the same number and same kinds of (2) _____ because they are all descended by mitosis from the same zygote. Through gene controls, however, (3) _____ are placed on which genes may be expressed (translated) in a given cell. (4) _____ (from root words that mean "giving rise to shape") is a program of orderly changes in an early embryo's size, shape and proportions that creates the specialized tissues and early organs characteristic of that species. (5) _____ _____ consists of several processes that transform a gastrula (with its three germ layers) into an embryo with established developmental axes and organ rudiments (undeveloped lumps of tissue) in place and recognizable.

As the embryo develops, one group of cells may produce a substance (say, a growth factor) that diffuses to another group of cells and turns on protein synthesis in those cells. Such interaction among embryonic cells is called (6) _____ _____ . Sometimes entire organs (such as testes in human males) change position in the developing organism, but the inward or outward folding of (7) _____ _____ is seen more often. Spemann demonstrated that the process known as (8) _____ _____ occurs in salamander embryos, where one body part differentiates because of signals it receives from an adjacent body part. (9) _____ are slowly degradable proteins that form concentration gradients as they diffuse from an inducing tissue into adjoining tissues.

In many animals, the embryo develops into a motile, independent (10) _____ , which extends the food supply and range of the population. A larva necessarily must undergo (11) _____ in order to become a juvenile. Fruit flies show (12) _____ metamorphosis. Normal cell types have a (13) _____ _____ _____ , and mitosis is scheduled to quit after so many cell divisions.

## True-False

If false, explain why.

_____14. The aging and death of a cell may be coded in large part in its DNA; <u>external</u> signals activate those DNA messages and tell the cell that it is time to die.

_____15. A process of predictable cellular deterioration is built into the life cycle of all organisms that consist of <u>differentiated cells that show considerable specialization</u>.

_____16. Body parts become folded, tubes become hollowed out, and eyelids, lips, noses, and ears all become slit or perforated by <u>apoptosis</u>.

# Self-Quiz

___ 1. Animals such as birds lay eggs with large amounts of yolk; embryonic development happens within the egg covering outside the mother's body. This developmental strategy is called _____ .
a. ovoviviparity
b. viviparity
c. oviparity
d. parthenogenesis
e. none of the above

___ 2. The process of cleavage most commonly produces a(n) _____ .
a. zygote
b. blastula
c. gastrula
d. third germ layer
e. organ

___ 3. Incomplete cleavage of cells at the yolk periphery is characteristic of the cleavage pattern of _____ .
a. frogs
b. fruit flies
c. chickens
d. sea urchins
e. humans

___ 4. The formation of three germ (embryonic) tissue layers occurs during _____ .
a. gastrulation
b. cleavage

c. pattern formation
d. morphogenesis
e. neural plate formation

___ 5. The differentiation of a body part in response to signals from an adjacent body part is _____ .
a. contact inhibition
b. ooplasmic localization
c. embryonic induction
d. pattern formation
e. none of the above

___ 6. A homeotic mutation _____ .
a. may cause a leg to develop on the head where an antenna should grow
b. may affect pattern formation
c. affects morphogenesis
d. may alter the path of development
e. all of the above

___ 7. Shortly after fertilization, the zygote is subdivided into a multicelled embryo during a process known as _____ .
a. meiosis
b. parthenogenesis
c. embryonic induction
d. cleavage
e. invagination

___ 8. Muscles differentiate from _____ tissue.
   a. ectoderm
   b. mesoderm
   c. endoderm
   d. parthenogenetic
   e. yolky

___ 9. The gray crescent is _____ .
   a. formed where the sperm penetrates the egg
   b. part of only one blastomere after the first cleavage

   c. the yolky region of the egg
   d. where the first mitotic division begins
   e. formed opposite from where the sperm enters the egg

___ 10. The nervous system differentiates from _____ tissue.
   a. ectoderm
   b. mesoderm
   c. endoderm
   d. yolky
   e. homeotic

## Chapter Objectives/Review Questions

This section lists general and detailed chapter objectives that can be used as review questions. You can make maximum use of these items by writing answers on a separate sheet of paper. Fill in answers where blanks are provided. To check for accuracy, compare your answers with information given in the chapter or glossary.

*Page*      *Objectives/Questions*

(746)      1. Understand how asexual reproduction differs from sexual reproduction. Know the advantages and problems associated with having separate sexes.

(746)      2. Define *oviparous, viviparous,* and *ovoviviparous.* For each of the three developmental strategies, cite an example of an animal that goes through it.

(746–747) 3. Explain why evolutionary trends in many groups of organisms tend toward developing more complex, sexual strategies rather than retaining simpler, asexual strategies.

(747, 751–753) 4. Explain how the amount of yolk in an ovum can influence an animal's cleavage pattern.

(748)      5. Name each of the three embryonic tissue layers and the organs formed from each.

(748)      6. Describe early embryonic development and distinguish among the following: gamete formation, fertilization, cleavage, gastrulation, organ formation, and tissue specialization.

(749–753) 7. Compare the early stages of frog and chick development (see Figures 44.4, 44.8, and 44.9) with respect to egg size and type of cleavage pattern (incomplete or complete).

(750–751) 8. Explain what causes polarity to occur during oocyte maturation in the mother and state how polarity influences later development.

(752)      9. Define *gastrulation* and state what process begins at this stage that did not happen during cleavage.

(752)      10. Define *differentiation* and give two examples of cells in a multicellular organism that have undergone differentiation.

(753, 756–757) 11. Explain why the differentiation of cells in a multicellular organism goes hand in hand with the division of labor and the integration of life processes.

(753, 758) 12. Define what is meant by *larva.* Distinguish metamorphosis from morphogenesis.

(758)      13. Explain how a spherical zygote becomes a multicellular adult with arms and legs.

(758)      14. Distinguish complete from incomplete metamorphosis.

## Integrating and Applying Key Concepts

If embryonic induction did not occur in a human embryo, how would the eye region appear? What would happen to the forebrain and epidermis? If controlled cell death did not happen in a human embryo, how would its hands appear? its face?

# 45

# HUMAN REPRODUCTION AND DEVELOPMENT

## Interactive Exercises

**Selected Words:** *testicular cancer, prostate cancer, seminal vesicles, vas deferens, bulbourethral glands, epididymis, urethra, ejaculatory ducts, scrotum*

## Boldfaced, Page-Referenced Terms

(763) testes (testis, sing.) _____

_____

(763) ovaries (ovary, sing.) _____

_____

(763) secondary sexual traits _____

_____

(764) seminiferous tubules _____

_____

(764) semen _____

_____

(766) Sertoli cells _____

_____

(766) Leydig cells _____

_____

(766) testosterone _____

_____

(767) LH, luteinizing hormone _____

_____

(767) FSH, follicle-stimulating hormone _____

_____

(767) GnRH _____

_____

## Fill-in-the-Blanks

The numbered items on the illustrations that follow represent missing information; complete the numbered blanks in the narrative below to supply the missing information on the illustration. Some illustrated structures are numbered more than once to aid identification.

Within each testis and following repeated (1) _____ divisions of undifferentiated diploid cells just inside the (2) _____ tubule walls, (3) _____ occurs to form haploid, mature (4) _____ . Males produce sperm continuously from puberty onward.  Sperm leaving a testis enter a long coiled duct, the (5) _____ ; the sperm are stored in the last portion of this organ.  When a male is sexually aroused, muscle contractions quickly propel the sperm through a thick-walled tube, the (6) _____ _____ , then to ejaculatory ducts and finally the (7) _____ , which opens at the tip of the penis.  During the trip to the urethra, glandular secretions become mixed with the sperm to form semen.  (8) _____ _____ secrete fructose to nourish the sperm and prostaglandins to induce contractions in the female reproductive tract.  (9) _____ _____ secretions help neutralize vaginal acids.  (10) _____ glands secrete mucus to lubricate the penis, aid vaginal penetration, and improve sperm motility.

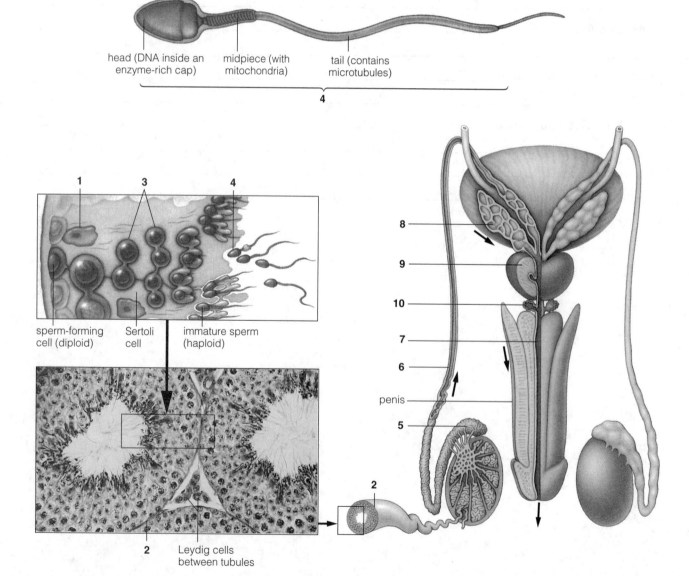

head (DNA inside an enzyme-rich cap)    midpiece (with mitochondria)    tail (contains microtubules)

4

1    3    4

sperm-forming cell (diploid)    Sertoli cell    immature sperm (haploid)

2    Leydig cells between tubules

8
9
10
7
6
penis
5

2

## Dichotomous Choice

Circle one of two possible answers given between parentheses in each statement.

11. Testosterone is secreted by (Leydig/hypothalamus) cells.
12. (Testosterone/FSH) governs the growth, form, and functions of the male reproductive tract.
13. Sexual behavior, aggressive behavior, and secondary sexual traits are associated with (LH/testosterone).
14. LH and FSH are secreted by the (anterior/posterior) lobe of the pituitary gland.
15. The (testes/hypothalamus) governs sperm production by controlling interactions among testosterone, LH, and FSH.
16. When blood levels of testosterone (increase/decrease), the hypothalamus stimulates the pituitary to release LH and FSH, which travel the bloodstream to the testes.
17. Within the testes, (LH/FSH) acts on Leydig cells; they secrete testosterone, which enters the sperm-forming tubes.
18. FSH enters the sperm-forming tubes and diffuses into (Sertoli/Leydig) cells to improve testosterone uptake.
19. When blood testosterone levels (increase/decrease) past a set point, negative feedback loops to the hypothalamus slow down testosterone secretion.

## Labeling

20. _____

21. _____ _____

22. _____ _____

23. _____ _____

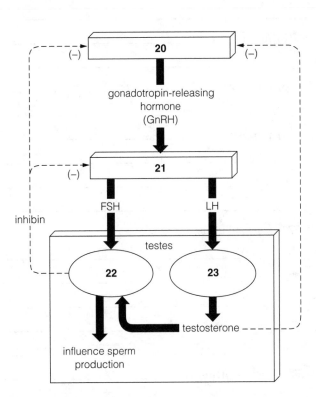

## 45.3. REPRODUCTIVE SYSTEM OF HUMAN FEMALES (pp. 768–769)

## 45.4. FEMALE REPRODUCTIVE FUNCTION (pp. 770–771)

## 45.5. VISUAL SUMMARY OF THE MENSTRUAL CYCLE (p. 772)

*Selected Words:* *endometriosis, primary oocyte, oviducts, vagina, clitoris, labium major, labium minor, cervix, myometrium.*

## Boldfaced, Page-Referenced Terms

(768) oocyte _____

_____

(768) uterus _____

_____

(768) endometrium _____

_____

(768) menstrual cycle _____

_____

(768) follicular phase _____

_____

(768) ovulation _____

_____

(768) luteal phase _____

_____

(768) estrogens _____

_____

(768) progesterone _____

_____

(770) follicle _____

_____

(770) zona pellucida _____

_____

(770) secondary oocyte _____

_____

(770) polar bodies _____

_____

(771) corpus luteum _____

_____

## Fill-in-the-Blanks

The numbered items on the illustrations that follow represent missing information; complete the numbered blanks in the narrative below to supply the missing information on the illustration. Some illustrated structures are numbered more than once to aid identification.

An immature egg (oocyte) is released from one (1) _____ of a pair. From each ovary, an

(2) _____ forms a channel for transport of the immature egg to the (3) _____ , a hollow, pear-shaped organ where the embryo grows and develops. The lower narrowed part of the uterus is the

(4) _____ . The uterus has a thick layer of smooth muscle, the (5) _____ , lined inside with

connective tissue, glands, and blood vessels; this lining is called the (6) _____ . The (7) _____ ,

a muscular tube, extends from the cervix to the body surface; this tube receives sperm and functions as part

of the birth canal. At the body surface are external genitals (vulva) that include organs for sexual stimulation.

Outermost is a pair of fat-padded skin folds, the (8) _____ _____ . Those folds enclose a

smaller pair of skin folds, the (9) _____ _____ . The smaller folds partly enclose the (10)

_____ , an organ sensitive to stimulation.

The location of the (11) _____ is about midway between the clitoris and the vaginal opening. (12) _____ occurs in the ovaries so that a normal female infant has about 2 million primary oocytes with the division process halted in the (13) ☐ I ☐ II stage. By age seven, only about (14) _____ remain. A primary oocyte surrounded by a nourishing layer of granulosa cells is called a(n)

(15) _____ .

When a female enters puberty, the (16) _____ secretes a hormone (GnRH) that makes the (17) _____ _____ secrete follicle-stimulating hormone (FSH) and luteinizing hormone (LH). These hormones are carried by the blood to all part of the body, but each month, one (or more) follicles respond by growing and secreting the steroid hormones called (18) _____ . The primary oocyte completes the (12) (13) division 8–10 hours before (19) _____ occurs in response to a surge of (20) _____ being produced by the (17) on day 12 or 13 of a 28-day cycle.

The first 5 days of the cycle are occupied by (21) _____ , the deterioration and expulsion of (22) _____ tissues that line the uterus. During the next week, (18) _____ stimulate new (22) _____ tissues to be constructed.

Following ovulation, the granulosa cells of the follicle are transformed into a (23) _____ _____ by the midcycle surge of LH. (23) secretes its key hormone (24) _____ and some estrogen. (24) prepares the reproductive tract for the arrival of the (25) _____ , which develops after an egg is fertilized. (24) also maintains the (26) _____ during pregnancy. If a (25) does not burrow into the (26), the (23) self-destructs after approximately twelve days by secreting prostaglandins that poison its own functioning.

After this, (24) and estrogen levels in the blood decline rapidly, and the (26) tissues die and disintegrate; together with blood escaping from the (26) tissues, the (27) _____ _____ begins and continues for three to six days.

## 45.6. PREGNANCY HAPPENS (p. 773)
## 45.7. FORMATION OF THE EARLY EMBRYO (pp. 774–775)
## 45.8. EMERGENCE OF THE VERTEBRATE BODY PLAN (p. 776)
## 45.9. ON THE IMPORTANCE OF THE PLACENTA (p. 777)
## 45.10. EMERGENCE OF DISTINCTLY HUMAN FEATURES (pp. 778–779)
## 45.11. *Focus on Health:* MOTHER AS PROVIDER, PROTECTOR, POTENTIAL THREAT (pp. 780–781)

*Selected Words:* erection, ejaculation, embryonic period, fetal period, first trimester, second trimester, third trimester, morula, blastocoel, amniotic cavity, pregnancy tests, "primitive streak", teratogens, thalidomide, anti-acne drugs, fetal alcohol syndrome, second-hand smoke

### Boldfaced, Page-Referenced Terms

(773) coitus _____

_____

(773) orgasm _____

_____

(773) ovum (plural, ova) _____

_____

(774) fetus _____

_____

(774) blastocyst _____

_____

(774) implantation _____

_____

(775) amnion _____

_____

(775) yolk sac _____

_____

(775) chorion _____

_____

(775) allantois _____

_____

(775) HCG (Human Chorionic Gonadotropin) _____

_____

(776) gastrulation _____

_____

(776) somites _____

_____

(777) placenta _____

_____

## Fill-in-the-Blanks

Fertilization generally takes place in the (1) _____ ; six or seven days after conception, (2) _____ begins as the blastocyst sinks into the endometrium. Extensions from the chorion fuse with the endometrium of the uterus to form a (3) _____ , the organ of interchange between mother and fetus. By the end of the (4) _____ week, all major organs have formed; the offspring is now referred to as a(n) (5) _____ .

## Labeling

Identify each indicated part of the accompanying illustrations.

6. _____  _____

7. _____  _____

8. _____  _____

9. _____  _____

10. _____  _____

11. _____

12. _____

13. _____

14. _____  _____

15. _____  _____

16. _____

17. _____  _____

18. _____

19. _____

20. _____

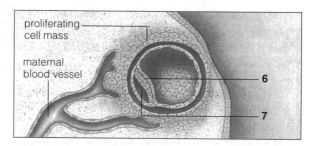

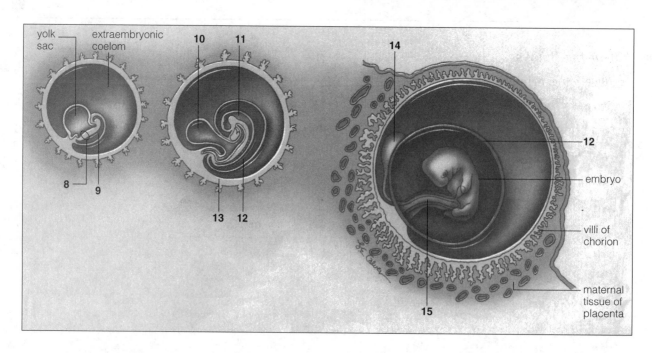

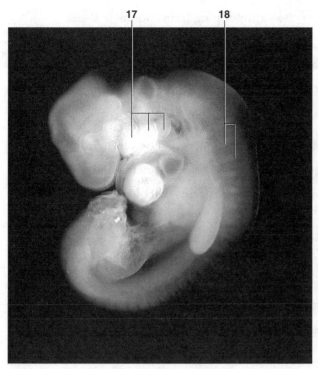

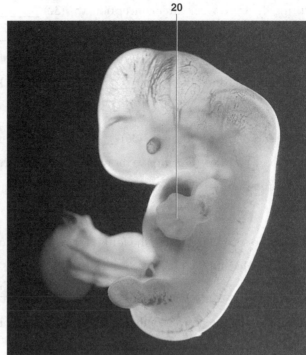

A human embryo at (**16**) weeks after conception.

A human embryo at (**19**) weeks after conception.

## Matching

Link each of the four extra embryonic membranes with its function.

21. ___allantois

22. ___amnion

23. ___chorion

24. ___yolk sac

A. Outermost membrane; will become part of the spongy blood-engorged placenta
B. Directly encloses the embryo and cradles it in a buoyant protective fluid
C. In humans, some of this membrane becomes a site for blood cell formation, some will become the forerunner of gametes
D. In humans, this membrane serves in early blood formation and formation of the urinary bladder

## Fill-in-the-Blanks

At-home human pregnancy tests use a treated "dip-stick" that changes color when (25) _____ _____ _____ secreted by the blastocyst is present in the mother's urine. (25) also stimulates the (26) _____ _____ to continue secreting progesterone and estrogen.

By the time a woman has missed her first menstrual period, generally during the third week after conception, cleavage is completed and (27) _____ is well under way as the three primary tissue layers (ectoderm, mesoderm, and endoderm) are forming. On day 15, a faint band, the (28) _____ _____ , appears around a depression along the axis of the embryonic disk; it marks the beginning of (27) in vertebrate embryos. During days 18–23, morphogenetic processes cause neural folds to arch upward and merge to form the (29) _____ _____ . (30) _____ appear that will later form most of the axial skeleton, skeletal muscles, and much of the dermal layer of the skin. By days 24–25 (during the fourth week), (31) _____ _____ are visible that will contribute in forming the face, neck, mouth, larynx, pharynx, and nasal cavities.

During the third week after conception, the (32) _____ was also forming from endometrial and extraembryonic membrane tissues; a(an) (33) _____ _____ connects the developing embryo to it. Oxygen and (34) _____ exit from maternal blood vessels into the embryo. Coming from the embryo are carbon dioxide and other (35) _____ that will quickly be disposed of by the mother's lungs and kidneys.

By the end of the fourth week (36) _____ form from limb buds and fingers and toes form from embryonic paddles. During the months that follow, organs become recognizable from the undeveloped lumps and bumps that preceded them. Growth of the (37) _____ surpasses that of any other region of the body.

During the second trimester, the fetus is moving most of its (38) _____ and is covered with hair. The eyes are sealed shut.

During the last trimester, the fetus gains weight, develops its lungs, immune system and temperature regulation. By the ninth month, the rate of surviving birth has increased to (39) _____ percent.

## 45.12. FROM BIRTH ONWARD (pp. 782–783)
## 45.13. CONTROL OF HUMAN FERTILITY (pp. 784–785)
## 45.14. *Focus on Health:* SEXUALLY TRANSMITTED DISEASES (pp. 786–787)
## 45.15. *Focus on Science:* TO SEEK OR END PREGNANCY (p. 788)

*Selected Words:* breech birth, "afterbirth", *afterbaby blues, breast cancer, prenatal, postnatal, aging, gamete, zygote, abstinence, rhythm method, withdrawal, douching, vasectomy, tubal ligation, spermicidal foam, spermicidal jelly, diaphragm, condoms, birth control pill, Depo-Provera, Norplant, morning-after pill,* Herpes, *AIDS, behavioral controls,* Neisseria gonorrhoeae, *gonorrhea, syphilis,* Treponema pallidum, *pelvic inflammatory disease (PID), genital herpes, Type II* Herpes simplex, *genital warts,* Chlamydia trachomatis, *NGU (chlamydial nongonococcal urethritis)*

### Boldfaced, Page-Referenced Terms

(782) labor _____

_____

(782) oxytocin _____

_____

(782) lactation _____

_____

(783) prolactin _____

_____

(786) sexually transmitted diseases (STDs) _____

_____

(788) in vitro fertilization _____

_____

(788) abortion _____

_____

## Fill-in-the-Blanks

On Earth each week, (1) _____ more babies are born than people die.  Each year in the United States, we still have about (2) _____ teenage pregnancies and (3) _____ abortions.  The most effective method of preventing conception is complete (4) _____ .  (5) _____ are about 85–93 percent reliable and help prevent venereal disease.  A (6) _____ is a flexible, dome-shaped disk, used with a spermicidal foam or jelly, that is placed over the cervix.  In the United States, the most widely used contraceptive is the Pill—an oral contraceptive of synthetic (7) _____ and (8) _____ that suppress the release of (9) _____ from the pituitary and thereby prevent the cyclic maturation and release of eggs.  Two forms of surgical sterilization are vasectomy and (10) _____ _____ .

## Matching

Match each of the following with *all* applicable diseases.

11. ___can damage the brain and spinal cord in ways leading to various forms of insanity and paralysis

12. ___has no cure

13. ___can cause violent cramps, fever, vomiting, and sterility due to scarring and blocking of the oviducts

14. ___caused by a motile, corkscrew-shaped bacterium, *Treponema pallidum*

15. ___infected women typically have miscarriages, stillbirths, or sickly infants

16. ___caused by a bacterium with pili, *Neisseria gonorrhoeae*

17. ___caused by direct contact with the viral agent; about 25 million people in the United States infected by it

18. ___produces a chancre (localized ulcer) 1–8 weeks following infection

19. ___chronic infections by this can lead to cervical cancer

20. ___can lead to lesions in the eyes that cause blindness in babies born to mothers with this disease

21. ___acyclovir decreases the healing time and may also decrease the pain and viral shedding from the blisters

22. ___8 weeks following infection, a flattened, painless chancre harbors many motile bacteria

23. ___may be cured by antibiotics but can be infected again

24. ___can be treated with tetracycline and sulfonamides

25. ___generally preventable by correct condom usage

A. AIDS
B. Chlamydial infection
C. Genital herpes
D. Gonorrhea
E. Pelvic inflammatory disease
F. Syphilis

# Self-Quiz

For questions 1–5, choose from the following answers:

    a. AIDS
    b. Chlamydial infection
    c. Genital herpes
    d. Gonorrhea
    e. Syphilis

___ 1. _____ is a disease caused by a spherical bacterium (*Neisseria*) with pili; it is curable by prompt diagnosis and treatment.

___ 2. _____ is a disease caused by a spiral bacterium (*Treponema*) that produces a localized ulcer (a chancre).

___ 3. _____ is an incurable disease caused by a retrovirus (an RNA-based virus).

___ 4. _____ is a disease caused by an intracellular parasite that lives in the genital and urinary tracts; also causes NGU.

___ 5. _____ is an extremely contagious viral infection (DNA-based) that causes sores on the facial area and reproductive tract; it is also incurable.

For questions 6–8, choose from the following answers:

    a. blastocyst
    b. allantois
    c. yolk sac
    d. oviduct
    e. cervix

___ 6. The _____ lies between the uterus and the vagina.

___ 7. The _____ is a pathway from the ovary to the uterus.

___ 8. The _____ results from the process known as cleavage.

For questions 9–12, choose from the following answers:

    a. Leydig cells
    b. seminiferous tubules
    c. vas deferens
    d. epididymis
    e. prostate

___ 9. The _____ connects a structure on the surface of the testis with the ejaculatory duct.

___10. Testosterone is produced by the _____ .

___11. Meiosis occurs in the _____ .

___12. Sperm mature and become motile in the _____ .

---

# Chapter Objectives/Review Questions

This section lists general and detailed chapter objectives that can be used as review questions. You can make maximum use of these items by writing answers on a separate sheet of paper. Fill in answers where blanks are provided. To check for accuracy, compare your answers with information given in the chapter or glossary.

| Page | Objectives/Questions |
|---|---|
| (763) | 1. Distinguish between primary and secondary sexual traits and between gonads and accessory reproductive organs. |
| (764) | 2. Follow the path of a mature sperm from the seminiferous tubules to the urethral exit. List every structure encountered along the path and state the contribution to the nurture of the sperm. |
| (765) | 3. Describe how a man examines himself for testicular cancer. |
| (766–767) | 4. Diagram the structure of a sperm, label its components, and state the function of each. |
| (766–767) | 5. List in order the stages that compose spermatogenesis. |

(766–767)  6. Name the four hormones that directly or indirectly control male reproductive function. Diagram the negative feedback mechanisms that link the hypothalamus, anterior pituitary, and testes in controlling gonadal function.

(766;  7. Compare the function of the Leydig cells of the testis with the function of the ovarian follicle
770–771)  and the corpus luteum.

(768)  8. Distinguish the follicular phase of the menstrual cycle from the luteal phase and explain how the two cycles are synchronized by hormones from the anterior pituitary, hypothalamus, and ovaries.

(768; 773)  9. Trace the path of a sperm from the urethral exit to the place where fertilization normally occurs. Mention in correct sequence all major structures of the female reproductive tract that are passed along the way and state the principal function of each structure.

(770–772) 10. State which hormonal event brings about ovulation and which other hormonal events bring about the onset and finish of menstruation.

(773)  11. List the physiological factors that bring about erection of the penis during sexual stimulation and the factors that bring about ejaculation.

(773)  12. List the similar events that occur in both male and female orgasm.

(773–778) 13. Describe the events that occur during the first month of human development. State how much time cleavage and gastrulation require, when organogenesis begins, and what is involved in implantation and placenta formation.

(778)  14. State when the embryo begins to be referred to as a fetus and during which month fetuses born prematurely survive.

(780–781) 15. Explain why the mother must be particularly careful of her diet, health habits, and life-style during the first trimester after fertilization (especially during the first six weeks).

(784)  16. Describe two different types of sterilization.

(784–785) 17. Identify the factors that encourage and discourage methods of human birth control.

(784–785) 18. State which birth control methods help prevent venereal disease.

(785)  19. Identify the three most effective birth control methods used in the United States and the four least effective birth control methods.

(786–787) 20. For each STD described in the Focus, know the causative organism and the symptoms of the disease.

(788)  21. State the physiological circumstances that would prompt a couple to try in vitro fertilization.

---

## Integrating and Applying Key Concepts

What rewards do you think a society should give a woman who has at most two children during her lifetime? In the absence of rewards or punishments, how can a society encourage women not to have abortions and yet ensure that the human birth rate does not continue to increase?

# 46

# POPULATION ECOLOGY

## Interactive Exercises

*Tales of Nightmare Numbers* (pp. 792–793)

## 46.1. CHARACTERISTICS OF POPULATIONS (pp. 794–795)

## 46.2. POPULATION SIZE AND EXPONENTIAL GROWTH (pp. 796–797)

*Selected Words* In addition to the boldfaced terms, the text features other important terms essential to understanding the assigned material. "Selected Words" is a list of these terms, which appear in the text in italics, in quotation marks, and occasionally in roman type. Latin binomials found in this section are underlined and in roman type to distinguish them from other italicized words.

*habitat, pre-reproductive, reproductive, and post-reproductive ages, crude density,* Larrea

## Boldfaced, Page-Referenced Terms

The page-referenced terms are important; they were in boldface type in the chapter. Write a definition for each term in your own words without looking at the text. Next, compare your definition with that given in the chapter or in the text glossary. If your definition seems inaccurate, allow some time to pass and repeat this procedure until you can define each term rather quickly (how fast you answer is a gauge of your learning effectiveness).

(794) demographics _____

_____

(794) population size _____

_____

(794) population density _____

_____

(794) population distribution _____

_____

(794) age structure _____

_____

(794) reproductive base _____

_____

(796) immigration _____

_____

(796) emigration _____

_____

(796) migrations _____

_____

(796) zero population growth _____

_____

(796) per capita _____

_____

(796) net reproduction per individual per unit time or $r$ _____

_____

(797) exponential growth _____

_____

(797) doubling time _____

_____

(797) biotic potential _____

_____

## Matching

Choose the most appropriate answer for each.

1. ___demographics
2. ___population size
3. ___population density
4. ___habitat
5. ___population distribution
6. ___age structure
7. ___reproductive base
8. ___crude density
9. ___pre-reproductive, reproductive, and post-reproductive
10. ___clumped dispersion
11. ___nearly uniform dispersion
12. ___random dispersion

A. Includes pre-reproductive and reproductive age categories
B. The general pattern in which the individuals of the population are dispersed through a specified area
C. When individuals of a population are more evenly spaced than they would be by chance alone
D. The number of individuals in a given area or volume of a habitat
E. Occurs only when individuals of a population neither attract nor avoid one another when conditions are fairly uniform through the habitat, and when resources are available all the time
F. The number of individuals in each of several to many age categories
G. The measured number of individuals in a specified area
H. The number of individuals that share the population's gene pool
I. The type of place where a species normally lives
J. Categories of a population's age structure
K. The vital statistics of a population
L. Individuals of a population form aggregations at specific habitat sites; most common dispersion pattern

## Problems

For exercises 13–17, consider the equation $G = rN$, where $G$ = the population growth rate, $r$ = the N reproduction per individual per unit of time, and $N$ = the number of individuals in the population.

13. Assume that $r$ remains constant at 0.2.
    a. As the value of $G$ increases, what happens to the value of $N$?

    _____

    b. If the value of $G$ decreases, what happens to the value of $N$?

    _____

    c. If the net reproduction per individual stays the same and the population grows faster, then what must happen to the number of individuals in the population?

    _____

14. If a society decides it is necessary to lower its value of $N$ through reproductive means because supportive resources are dwindling, it must lower either its net reproduction per individual per unit or its

    _____

15. The equation $G = rN$ expresses a direct relationship between $G$ and $r \times N$. If $G$ remains constant and $N$ increases, what must the value of $r$ do? (In this situation, $r$ varies inversely with $N$.)

    _____

16. Look at line (a) in the graph on the next page. After seven hours have elapsed, approximately how many individuals are in the population?

    _____

17. Look at line (b) in the same graph.
    a. After 24 hours have elapsed, approximately how many individuals are in the population? _____
    b. After 28 hours have elapsed, approximately how many individuals are in the population? _____

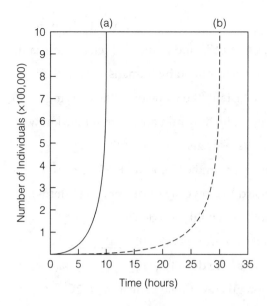

## 46.3. LIMITS ON THE GROWTH OF POPULATIONS (pp. 798–799)

## 46.4. LIFE HISTORY PATTERNS (pp. 800–801)

## 46.5. *Focus on Science:* NATURAL SELECTION AND THE GUPPIES OF TRINIDAD
(pp. 802–803)

*Selected Words:* *sustainable* supply of resources, *bubonic plague, pneumonic plague,* Yersinia pestis, *Type I curves, Type II curves, Type III curves,* Poecilia reticulata, Rivulus hartii, Crenicichla alta

### Boldfaced, Page-Referenced Terms

(798) limiting factor _____

_____

(798) carrying capacity _____

_____

(798) logistic growth _____

_____

(800) cohort _____

_____

(800) life table _____

_____

(801) survivorship curves _____

_____

### Fill-in-the-Blanks

If (1) _____ factors (essential resources in short supply) act on a population, population growth tapers off. (2) _____ _____ refers to the maximum number of individuals of a population that can be sustained indefinitely by the environment. S-shaped growth curves are characteristic of (3) _____ population growth. The plot of (3) growth levels off once the (4) _____ _____ is reached. In the equation $G = r_{max}N[(K - N)/K]$, as the value of $N$ approaches the value of $K$, and $K$ and $r_{max}$ remain constant, the value of $G$ (5) [choose one] ☐ increases ☐ decreases ☐ cannot be determined by humans, even if they know algebra. As the value of $r_{max}$ increases and $G$ and $N$ remain constant, the value of $K$ (the carrying capacity) (6) [choose one] ☐ increases ☐ decreases ☐ cannot be determined by humans, even if they know algebra. In an overcrowded population, predators, parasites, and disease agents serve as (7) _____ - _____ controls. When an event such as a freak summer snowstorm in the Colorado Rockies causes more deaths or fewer births in a butterfly population (with no regard to crowding or dispersion patterns), the controls are said to be (8) _____ - _____ factors. Food availability is a density-(9.) _____ factor that works to cut back population size when it approaches the environment's (10) _____ _____ . Environmental disruptions such as forest fires and floods are density-(11) _____ factors that may push a population above or below its tolerance range for a given variable. The term, (12) _____ _____ pattern, refers to the individuals of a species in a population that exhibit particular morphological, physiological, and behavioral traits that are adaptive to different conditions at different times in the life cycle. (13) Life and health _____ companies use information summarized from life tables, which show trends in mortality and life expectancy. A (14) _____ is a group of individuals that is tracked from the time of birth until the last one dies. A (15) _____ table lists the completed data on a population's age-specific death schedule. Such tables can be converted to more cheery (16) "_____" schedules, which lists the number of individuals that reach some specified age ($x$). (17) _____ curves are the graph lines that emerge when ecologists plot the age-specific survival of a cohort in a particular habitat. Type (18) _____ survivorship curves illustrate populations having low survivorship early in life. Organisms whose life cycles reflect high survivorship until fairly late in life with a later large increase in deaths have a type (19) _____ survivorship curve. Organisms like lizards, small mammals, and some songbirds are just as likely to be killed or die of disease at any age and reflect the type (20) _____ survivorship curve. Provided people have access to good health care, human populations typically show a type (21) _____ survivorship curve.

## Matching

Choose the most appropriate answer for each.

22. ___cohort
23. ___type III survivorship curves
24. ___life tables
25. ___type I survivorship curves
26. ___type II survivorship curves

A. Survivorship curves that reflect a fairly constant death rate at all ages; typical of some song birds, lizards, and small mammals
B. Survivorship curves that reflect high survivorship until fairly late in life; produce a few large offspring provided with extended parental care; examples are elephants and humans
C. A group tracked by researchers from birth until the last survivor dies
D. Survivorship curves that reflect a high death rate early in life; typical of sea stars and other invertebrate animals, insects, many fishes, plants, fungi
E. Summaries of age-specific patterns of birth and death

## Matching

Choose the most appropriate answer(s) for each. A letter may be used more than once, and a blank may contain more than one letter.

27. ___cohort example
28. ___example(s) of density-independent controls
29. ___example(s) of density-dependent controls
30. ___selective effects of predation by killifish and pike cichlids

A. Drought, floods, earthquakes
B. Bubonic plague and pneumonic plague
C. Food availability
D. Adrenal enlargement in wild rabbits in response to crowding
E. Heavy applications of pesticides in your backyard
F. All of the 1987 human babies of New York City
G. Natural selection in the life history patterns of Trinidadian guppies
H. Freak snowstorms in the Rocky Mountains

## 46.6. HUMAN POPULATION GROWTH (pp. 804–805)

## 46.7. CONTROL THROUGH FAMILY PLANNING (pp. 806–807)

## 46.8. POPULATION GROWTH AND ECONOMIC DEVELOPMENT (pp. 808–809)

## 46.9. SOCIAL IMPACT OF NO GROWTH (p. 809)

*Selected Words:* <u>Vibrio</u> <u>cholerae</u>, *baby boomers, preindustrial stage, transitional stage, industrial stage, postindustrial stage, postpone*

### Boldfaced, Page-Referenced Terms

(806) family planning programs _____

_____

(806) total fertility rate _____

_____

(808) demographic transition model _____

_____

## Graph Construction

1. Construct a graph of the following data in the space provided on page 543.

| Year | Estimated World Population |
|------|---------------------------|
| 1650 | 500,000,000 |
| 1850 | 1,000,000,000 |
| 1930 | 2,000,000,000 |
| 1975 | 4,000,000,000 |
| 1986 | 5,000,000,000 |
| 1993 | 5,500,000,000 |
| 1995 | 5,700,000,000 |
| 1997 | 5,800,000,000 |

a. Estimate the year that the world contained 3 billion humans. _____

b. Estimate the year that Earth will house 8 billion humans. _____

c. Do you expect Earth to house 8 billion humans within your lifetime? _____

## True/False

If the statement is true, write a T in the blank. If the statement is false, make it correct by changing the underlined word(s) and writing the correct word(s) in the answer blank.

_____ 2. During 1997, the human population surpassed 8.2 billion.

_____ 3. Even if we could double our present food supply, death from starvation could still reach 20 million to 40 million people a year.

_____ 4. Compared with the geographic spread of other organisms, it has taken the human population an extremely long period of time to expand into diverse environments.

_____ 5. Managing food supplies through agriculture has had the effect of increasing the carrying capacity for the human population.

_____ 6. By bringing many disease agents under control and by tapping into concentrated existing forms of energy, humans utilized factors that had previously limited their population growth.

_____ 7. The stupendously accelerated growth of the human population continues and can be sustained indefinitely.

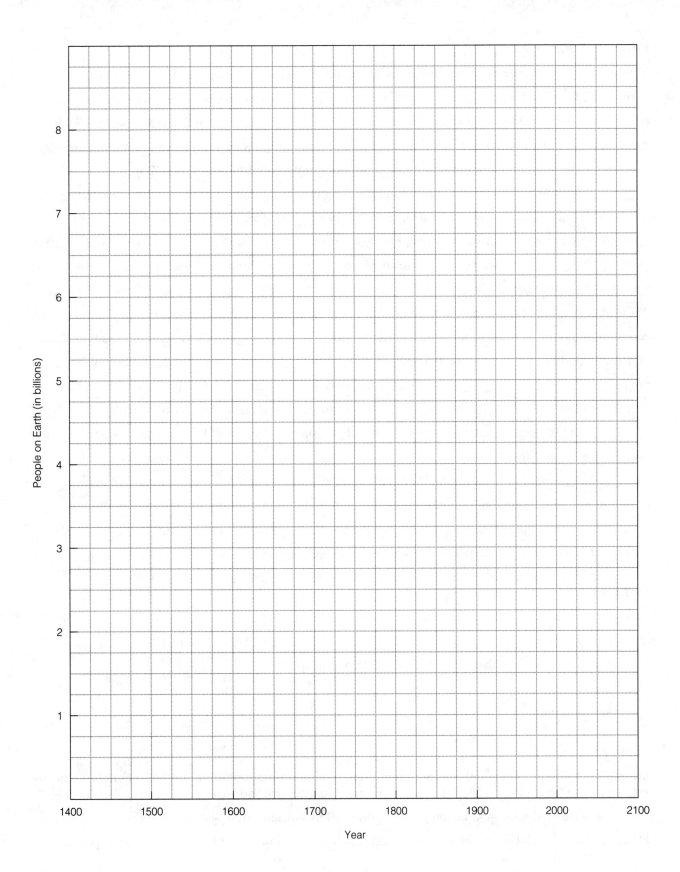

## Fill-in-the-Blanks

If the annual current average population growth rate of 1.55 percent is maintained for the human population, there may be (8) _____ billion to (9) _____ billion people by the year 2050. Large efforts will have to be made to increase the amount of (10) _____ to sustain that many people. Such gross manipulation of resources is likely to intensify (11) _____ , which almost certainly will adversely affect our water supplies, the atmosphere, and productivity on land and in the seas. (12) _____ _____ programs educate individuals about choosing how many children they will have, and when. Carefully developed and administered programs may bring about a long-term decline in (13) _____ rates. To arrive at zero population growth, the average "replacement rate" would have to be slightly higher than (14) _____ children per couple, for some (15) _____ children die before reaching reproductive age. A more useful measure of global population trends is the total (16) _____ rate. This is defined as the average number of children born to women during their reproductive years, as estimated on the basis of current age-specific birth rates. In 1996, the average rate was (17) _____ children per woman, an impressive decline from 1950, when the rate was (18) _____ . In the United States, a cohort of 78 million (19) _____ - _____ is being tracked since it formed in 1946 when soldiers returned home from World War II. Although the United States has a narrow reproductive base, more than (20) _____ - _____ of the world population falls in the broad pre-reproductive base. One way to slow growth is to bear children during the early (21) _____ , rather than in the mid-teens or early twenties. (22) _____ is the country described to have the world's most extensive family planning program. Family planning programs on a global scale are designed to help (23) _____ the size of the human population. Even if the human population reaches a level of zero growth, it will continue to grow for sixty years, for its (24) _____ base already consists of a staggering number of individuals.

## Matching

Choose the most appropriate examples to match with each population age structure (text, p. 807, Figure 46.15).

25. ___ zero growth
26. ___ rapid growth (more than a third of the world population)
27. ___ negative growth
28. ___ slow growth

A. United States, Canada
B. Germany, Hungary
C. Kenya, Nigeria, Saudi Arabia
D. Denmark, Italy

## Sequence

Arrange the following stages of the demographic transition model in correct chronological sequence. Write the letter of the first step next to 7, the letter of the second step next to 8, and so on.

29. ___ A. Industrial stage: population growth slows and industrialization is in full swing

30. ___ B. Preindustrial stage: harsh living conditions, high birth rates and low death rates, slow population growth

31. ___ C. Postindustrial stage: zero population growth is reached; birth rate falls below death rate, and population size slowly decreases

32. ___ D. Transitional stage: industrialization begins, food production rises, and health care improves; death rates drop, birth rates remain high to give rapid population growth

*Fill-in-the-Blanks*

Today, the United States, Canada, Australia, Japan, nations of the former Soviet Union, and most countries of western Europe are in the (33) _____ stage. Their growth rate is slowly (34) [choose one] □ increasing □ decreasing. In Germany, Bulgaria, Hungary, and some other countries, the death rates exceed the birth rates, and the populations are getting (35) _____ . Mexico and other less developed countries are in the (36) _____ stage. In many countries, population growth outpaces economic growth, death rates will increase, and they may be stalled in the (37) _____ stage. Enormous disparities in (38) _____ development are driving forces for immigration and emigration. Claiming that population growth affects economic health, many governments restrict (39) _____ .

India has a whopping (40) _____ percent of the human population. By comparison, the United States has merely (41) _____ percent. The highly industrialized United States produces (42) _____ percent of all goods and services. It uses (43) _____ percent of the world's processed minerals and available, nonrenewable sources of energy. Its people consume (44) _____ times as much as the average person in India. The United States generates at least (45) _____ percent of the global pollution and trash. By contrast, India produces about (46) _____ percent of all goods and services and uses (47) _____ percent of the available minerals and nonrenewable energy resources. India generates only about (48) _____ percent of the pollution and trash. Tyler Miller, Jr., estimated that it would take (49) _____ billion impoverished individuals in India to have as much environmental impact as (50) _____ million Americans.

For all species on our planet, the biological implications of extremely rapid (51) _____ are truly staggering. Yet so are the (52) _____ implications of what will happen when and if the human population declines to the point of zero population growth—and stays there. If the population ever does reach and maintain zero growth over time, a larger proportion of the individuals will end up in the (53) _____ age brackets. To care for older, nonproductive citizens, fewer productive ones must carry more and more of the (54) _____ burden. Through our special abilities to undergo rapid cultural evolution, we have (55) _____ the action of most of the factors that limit growth. No amount of (56) _____ intervention can repeal the ultimate laws governing population growth, as imposed by the (57) _____ _____ of the environment.

# Self-Quiz

_____ 1. The number of individuals that share the population's gene pool is _____ .
   a. the population density
   b. the population growth
   c. the population birth rate
   d. the population size

_____ 2. The number of individuals in a given area or volume of a habitat is _____ .
   a. the population density
   b. the population growth
   c. the population birth rate
   d. the population size

___ 3. A population that is growing exponentially in the absence of limiting factors can be illustrated accurately by a(n) _____ .
a. S-shaped curve
b. J-shaped curve
c. curve that terminates in a plateau phase
d. tolerance curve

___ 4. How are the individuals in a population most often dispersed?
a clumped
b. very uniform
c. nearly uniform
d. random

___ 5. Assuming immigration is balancing emigration over time, _____ may be defined as an interval in which the number of births is balanced by the number of deaths.
a. the lack of a limiting factor
b. exponential growth
c. saturation
d. zero population growth

___ 6. Assuming the birth and death rate remain constant, both can be combined into a single variable, $r$, or _____ .
a. the per capita rate
b. the minus migration factor
c. exponential growth
d. the net reproduction per individual per unit time

___ 7. _____ is a way to express the growth rate of a given population.
a. Doubling time
b. Population density
c. Population size
d. Carrying capacity

___ 8. Any population that is not restricted in some way will grow exponentially _____ .
a. except in the case of bacteria
b. irrespective of doubling time
c. if the death rate is even slightly greater than the birth rate
d. all of the above

___ 9. The maximum number of individuals of a population (or species) that a given environment can sustain indefinitely defines _____ .
a. the carrying capacity of the environment
b. exponential growth
c. the doubling time of a population
d. density-independent factors

___10. In natural communities, some feedback mechanisms operate whenever populations change in size; they are _____ .
a. density-dependent factors
b. density-independent factors
c. always intrinsic to the individuals of the community
d. always extrinsic to the individuals of the community

___11. _____ are any essential resources that are in short supply for a population.
a. Density-independent factors
b. Extrinsic factors
c. Limiting factors
d. Intrinsic factors

___12. Which of the following is *not* characteristic of logistic growth?
a. S-shaped curve
b. leveling off of growth as carrying capacity is reached
c. unrestricted growth
d. slow growth of a low-density population followed by rapid growth

___13. The population growth rate ($G$) is equal to the _____ net population growth rate per individual ($r$) and number of individuals ($N$).
a. sum of
b. product of
c. doubling of
d. difference between

___14. $G = r_{max} N (K - N/K)$ represents _____ .
a. exponential growth
b. population density
c. population size
d. logistic growth

___15. The beginning of industrialization, a rise in food production, improvement of health care, rising birth rates, and declining death rates describes the _____ stage of the demographic transition model.
a. preindustrial
b. transitional
c. industrial
d. postindustrial

___16. A group of individuals that is typically tracked from the time of birth until the last species dies is a _____ .
  a. cohort
  b. species
  c. type I curve group
  d. density-independent group

___17. The survivorship curve typical of industrialized human populations is type _____ .
  a. I
  b. II
  c. III
  d. none of the above types

## Dichotomous Choice

Circle one of two possible answers given between parentheses in each statement.

18. Population (size/density) is the number of individuals per unit of area or volume.
19. Population (size/density) refers to the number of members that make up the gene pool.
20. The general pattern of dispersal of members of a population through the habitat, e.g., clumped or random, is its (density/dispersion).
21. Dividing a population into pre-reproductive, reproductive, and post-reproductive categories characterizes its (age/distribution) structure.
22. The "reproductive base" of a population refers to the number of individuals in the (pre-reproductive and reproductive/reproductive and post-reproductive) age structure categories.
23. When the number of population members stabilizes due to a balance in births/immigrations and deaths/emigrations, a (biotic potential/zero population growth) is demonstrated.
24. A plot of population size against time has a characteristic (S-shaped/J-shaped) curve if the population is growing exponentially.
25. The maximum rate of increase per individual under ideal conditions in a population is (logistic growth/biotic potential).
26. As long as the per capita birth rate remains even slightly above the per capita death rate, a population will grow (exponentially/logistically).
27. When limiting factors act on a population, population growth (increases/decreases).

---

# Chapter Objectives/Review Questions

This section lists general and detailed chapter objectives that can be used as review questions. You can make maximum use of these items by writing answers on a separate sheet of paper. Fill in answers where blanks are provided. To check for accuracy, compare your answers with information given in the chapter or glossary.

| Page | Objectives/Questions |
|---|---|
| (794) | 1. Be able to define: *demographics, population, habitat, population size, population density, population distribution, age structure,* and *reproductive base.* |
| (794) | 2. In a population, the pre-reproductive ages and the reproductive ages together are counted as the _____ _____ . |
| (794) | 3. The _____ _____ is no more than the measured number of individuals in a specified area or volume of a habitat. |
| (794–795) | 4. List and describe the three patterns of dispersion illustrated by populations in a habitat. |
| (796) | 5. Distinguish immigration from emigration; define *migration.* |
| (796) | 6. Define *zero population growth* and describe how achieving it would affect population size. |
| (796–797) | 7. In the equation $G = rN$, as long as $r$ holds constant, any population will show ____ growth. |
| (796–797) | 8. Calculate a population growth rate ($G$); use values for birth, death, and number of individuals ($N$) that seem appropriate. |
| (797) | 9. State how increasing the death rate of a population affects its doubling time. |
| (797) | 10. _____ _____ is the maximum rate of increase per individual under ideal conditions. |
| (797) | 11. The _____ rate of increase in population growth depends on the age at which each individual reproduces, and how many offspring are produced. |

(798)   12. List several examples of limiting factors and explain how they influence population curves.

(798)   13. Explain the phrase "exceeding the carrying capacity," as applied to a population of organisms in an environment.

(798)   14. Understand the meaning of the logistic growth equation and know how to calculate values of $G$ by using the logistic growth equation. Understand the meaning of $r_{max}$ and $K$.

(796–798) 15. Contrast the conditions that promote J-shaped growth curves with those that promote S-shaped curves in populations.

(799)   16. Define density-dependent growth of populations; cite one example.

(799)   17. Define *density-independent factors* and list two examples; indicate how such factors affect populations.

(800)   18. Most species have a _____ _____ pattern in that its individuals exhibit particular morphological, physiological, and behavioral traits that are adaptive to different conditions at different times in the life cycle.

(800)   19. Life insurance companies and ecologists track a _____ , a group of individuals from the time of birth until the last one dies.

(800)   20. Understand the significance and use of life tables; be able to list and interpret the three survivorship curves.

(800–801) 21. Explain how the construction of life tables and survivorship curves can be useful to humans in managing the distribution of scarce resources.

(802–803) 22. Guppy populations targeted by killifish tend to be larger, less streamlined, and more brightly colored, and guppy populations targeted by pike-cichlids tend to be smaller, more streamlined, and duller in color patterning. Other life history pattern differences exist between the two groups. After consideration of the research results obtained by Reznick and Endler, provide an explanation for these differences.

(804)   23. Be able to list three possible reasons that growth of the human population is out of control.

(806)   24. Most governments are trying to lower birth rates by _____ _____ programs.

(806)   25. Define *total fertility rate*.

(807)   26. What is the significance of the cohort of 78 million baby boomers?

(808)   27. List and describe the four stages of the demographic transition model.

(809)   28. Differences in population growth among countries correlate with levels of _____ development.

(809)   29. Be able to generally compare the implications of the resource consumptions of India and the United States.

(809)   30. In the final analysis, no amount of _____ can repeal the ultimate laws governing population growth, as imposed by the _____ _____ of the environment.

---

# Integrating and Applying Key Concepts

Assume that the world has reached zero population growth. The year is 2110, and there are 10.5 billion individuals of *Homo pollutans* on Earth. You have seen stories on the community television screen about how people used to live 120 years ago. List the ways that life has changed and comment on the events that no longer happen because of the enormous human population.

# 47

# COMMUNITY INTERACTIONS

## Interactive Exercises

*No Pigeon Is An Island* (pp. 812–813)

### 47.1. FACTORS THAT SHAPE COMMUNITY STRUCTURE (p. 814)

### 47.2. MUTUALISM (p. 815)

*Selected Words:* potential niche, realized niche, obligatory

*Boldfaced, Page-Referenced Terms*

(814) habitat _____

_____

(814) community _____

_____

(814) niche _____

_____

(814) commensalism _____

_____

(814) mutualism _____

_____

(814) interspecific competition _____

_____

(814) predation _____

_____

(814) parasitism _____

_____

(814) symbiosis _____

_____

## Fill-in-the-Blanks

The type of place where you will normally find an organism is its (1) _____ . The populations of all species in a habitat associate with one another as a (2) _____ . The (3) _____ of an organism is defined by its "profession" in a community, in the sum of activities and relationships in which it engages to secure and use the resources necessary for its survival and reproduction. The (4) _____ niche is the one that could prevail in the absence of competition and other factors that could constrain its acquisition and use of resources. The (5) _____ niche is one that is more constrained and that shifts in large and small ways over time, as individuals of the species respond to a mosaic of changes.

A given species has a (6) _____ relationship with most species in its habitat, that is, they do not affect each other (7) _____ . Tree-roosting birds are (8) _____ with trees; the trees get nothing but are not harmed. The flowering plants and their pollinators form (9) _____ relationships from which both participating populations benefit (two-way exploitation). In (10) _____ competition, disadvantages flow both ways between species. An extreme interaction that benefits one species, the predator, is known as (11) _____ ; the other organism in this relationship is hurt and is known as the (12) _____ . Another extreme interaction where one species, the parasite, directly benefits is called (13) _____ ; the other organism in this relationship is hurt and is known as the (14) _____ . Commensalism, mutualism, and parasitism are all forms of (15) _____ ; it means "living together."

## Complete the Table

16. Complete the following table to describe how each of the organisms listed is intimately dependent on the other for survival and reproduction in an obligatory mutualistic symbiotic interaction.

| Organism | Dependency |
|----------|------------|
| a. Yucca moth | |
| b. Yucca plant | |

## 47.3. COMPETITIVE INTERACTIONS (pp. 816–817)

## 47.4. PREDATION (pp. 818–819)

## 47.5. *Focus on the Environment:* THE COEVOLUTIONARY ARMS RACE (pp. 820–821)

**Selected Words:** *intraspecific* competition, *interspecific* competition, Paramecium, Plethodon glutinosus, P. jordani, *carrying capacity*, *three-level* interaction, Lithops, Dendrobates, Ranunculus

## Boldfaced, Page-Referenced Terms

(817) competitive exclusion _____

_____

(817) resource partitioning _____

_____

(818) predators _____

_____

(818) prey _____

_____

(818) parasites _____

_____

(818) host _____

_____

(818) coevolution _____

_____

(820) camouflage _____

_____

(820) warning coloration _____

_____

(820) mimicry _____

_____

## Fill-in-the-Blanks

(1) _____ competition means individuals of the same species compete with one another. (2) _____ competition occurs between populations of different species. (3) _____ competition is usually the least intense of the two types. According to the concept of (4) _____ _____ , two species that require identical resources cannot coexist indefinitely. In (5) _____ _____ , competing species coexist by subdividing a category of similar resources. (6) _____ are animals that feed on other living organisms—their (7) _____—but do not take up residence on or in them. (8) _____ take up residence in or on other living organisms—their (9) _____—and feed on specific (10) _____ tissues for part of their live cycle. The (11) _____ organisms may or may not die as a result of the interaction. (12) _____ refers to the joint evolution of two (or more) species that exert selection pressure on each other through their close ecological interaction. Organisms that exhibit (13) _____ have adaptations in form, patterning, color, and behavior that help them blend in with their surroundings and escape detection. Toxic types of prey often have (14) _____ _____ , or conspicuous patterns and colors that predators learn to recognize as "avoid me" signals. (15) _____ is a term applied to prey organisms that bear close resemblance to dangerous, unpalatable, or hard-to-catch species. (16) _____ -of- _____ defenses refers to last-ditch tricks used by prey species for their survival as they are cornered or under attack. Predators may counter prey defenses with their own marvelous (17) _____ responses.

## Matching

Match each of the following with the most appropriate answer. The same letter may be used more than once. Use only one letter per blank.

18. ___Cornered earwigs, skunks, and stink beetles producing awful odors

19. ___Different chipmunk species are forced to use different mountain habitats

20. ___Different species of New Guinea pigeons coexisting in the same environment by eating fruits of different sizes

21. ___Tapeworms and humans

22. ___Body coloration of a caterpillar resembles bird droppings and enables it to hide in the open from birds that prey on it

23. ___Each species of the genus *Yucca* is pollinated exclusively by one species of yucca moth

24. ___Endosymbiotic origins of eukaryotes

25. ___Canadian lynx and snowshoe hare

26. ___Trees provide New Guinea pigeons with food and the pigeons help disperse seeds from the trees to new germination sites

27. ___Grasshopper mice plunging the noxious chemical-spraying tail end of their beetle prey into the ground to feast on the head end

A. parasitism
B. mutualism
C. predator-prey interactions
D. camouflage
E. warning coloration
F. mimicry
G. moment-of-truth defense
H. competitive exclusion
I. resource partitioning
J. Adaptive responses to prey defenses
K. Effects of competitive interactions

28. ___Resemblance of *Lithops*, a desert plant, to a small rock

29. ___Reef corals that poison and grow on top of other species of reef corals

30. ___Nibbling on the seemingly luscious yellow petals of a buttercup inflicts a chemical burn upon the lining of the mouth

31. ___An inedible butterfly is a model for the edible *Dismorphia*

32. ___In the same area, drought-tolerant foxtail grasses have a shallow, fibrous root system, mallow plants have a deeper taproot system, and smartweed has a taproot system that branches in topsoil and in soil below the roots of other species

33. ___Aggressively stinging yellow jackets are the likely model for nonstinging edible wasps

34. ___Mycorrhizae

35. ___Migrating rufous hummingbirds that are more aggressive than native broadtailed hummingbirds evict them from broadtailed territories along their migration route

36. ___Two species of *Paramecium* requiring identical resources cannot coexist indefinitely

37. ___A slowed growth rate in populations of *Paramecium* that did have much overlap in their requirements and continued to coexist

38. ___Different species of salamanders belonging to the genus *Plethodon* compete, yet they coexist at suppressed population sizes

39. ___Most flowering plants and the insects and birds, bats, and other animals that pollinate them

40. ___Baboon on the run turns to give canine tooth display to a pursuing leopard

41. ___An aggressive yellow jacket is the probable model for similar-looking but edible flies

42. ___Least bittern with coloration similar to surrounding withered reeds

## 47.6. PARASITIC INTERACTIONS (pp. 822–823)

*Selected Words: ecto*parasites, *endo*parasites, *holo*parasitic, *hemi*parasitic, *biological controls*

### Boldfaced, Page-Referenced Terms

(822) microparasites _____

_____

(822) macroparasites _____

_____

(822) social parasites _____

_____

(823) parasitoids _____

_____

## Choice

For questions 1–10, choose from the following:

a. ectoparasites  b. endoparasites  c. microparasites  d. macroparasites
e. holoparasitic plants  f. hemiparasitic plants  g. social parasites  h. parasitoids

1. ____Adult female cuckoos lay eggs in the nests of other species

2. ____Parasitic species which live on the body surface of a host

3. ____Insect larvae that always kill what they eat

4. ____Members of the broomrape family parasitizing the roots of oak and birch trees

5. ____The North American brown-headed cowbird removes an egg from the nest of another bird and lays one as a "replacement"

6. ____Mistletoe is an example; its extensions invade the sapwood of host trees

7. ____Parasitic species which live within the host's body

8. ____Includes a great number of bacteria, viruses, and protistans

9. ____Wasps that lay eggs on the cocoons of a sawfly species

10. ____Includes a variety of invertebrates such as flatworms, roundworms, and many arthropods

## Short Answer

11. Explain why parasite and host associations tend to favor long-term relationships rather than death.

_____

_____

_____

_____

_____

_____

12. Discuss the reasons that the use of parasites as biological control agents is risky.

_____

_____

_____

_____

_____

_____

_____

**47.7. FORCES CONTRIBUTING TO COMMUNITY STABILITY** (pp. 824–825)

**47.8. COMMUNITY INSTABILITY** (pp. 826–827)

**47.9.** *Focus on the Environment:* **NILE PERCH AND RABBITS AND KUDZU, OH. MY!**
(pp. 828–829)

*Selected Words:* *natural* and *active* restoration of the climax community, <u>Pisaster</u>, <u>Mytilus</u>, <u>Littorina</u>, <u>Enteromorpha</u>, <u>Chondrus</u>, <u>Cryphonectria</u>, <u>Solenopsis</u>, <u>Oryctolagus</u>, *myxamatosis*, <u>Pueraria</u>

*Boldfaced, Page-Referenced Terms*

(824) ecological succession _____

_____

(824) pioneer species _____

_____

(824) primary succession _____

_____

(824) secondary succession _____

_____

(824) climax-pattern model _____

_____

(826) keystone species _____

_____

(827) geographic dispersal _____

_____

(827) jump dispersal _____

_____

(828) exotic species _____

_____

*Fill-in-the-Blanks*

When a community develops in a predictable sequence from pioneers to a stable end array of species over some region, this is known as ecological (1) _____ . (2) _____ species are opportunistic colonizers of vacant habitats that are noted for their high dispersal rates and rapid growth. As time passes, the (3) _____ are replaced by species that are more competitive. These species are in turn replaced in an orderly progression of species change until a stable, self-perpetuating (4) _____ community results. When pioneer species begin to colonize a barren habitat such as a volcanic island, it is known as (5) _____ succession. Once established, the pioneers improve conditions for other species and often set the stage for their own (6) _____ . When an area within a community is disturbed and then recovers to move again toward the stable, self-perpetuating climax community, it is known as (7) _____ succession.

Some scientists hypothesize that the colonizers (8) _____ their own replacement while other scientists subscribe to the idea that the sequence of (9) _____ depends on what species arrive first to compete successfully with species that could replace them. According to the (10) _____ - _____ model, a community is adapted to a total pattern of environmental factors—including climate, soil, topography, wind, species interactions, and recurring disturbances such as fires and chance events. Small-scale changes contribute to the internal dynamics of the (11) _____ as a whole. For example, a tree falling in a tropical forest opens a gap in the forest canopy that allows more light to reach that gap. Here the growth of previously suppressed small trees, germination of (12) _____ or shade-intolerant species is encouraged. Giant sequoia trees in climax communities of the California Sierra Nevada are best maintained by modest (13) _____ that eliminate trees and shrubs that compete with young sequoias but do not damage the older sequoias. Thus, recurring, small-scale changes are built into the overall workings of many communities. The example of secondary succession regrowth following Mount St. Helen's violent eruption in 1980 is classified as (14) _____ restoration of the climax community. Attempts to reestablish biodiversity in abandoned farmlands as well as other large areas disturbed by agriculture, mining, and other human activities, is known as (15) _____ restoration.

## Choice

For questions 16–25, choose from the following:

> a. primary succession    b. secondary succession

16.____Following a disturbance, a patch of habitat or a community moves once again toward the climax state.

17.____Successional changes begin when a pioneer population colonizes a barren habitat.

18.____A disturbed area within a community recovers and moves again toward the climax state.

19.____Many plants in this succession arise from seeds or seedlings that are already present when the process begins.

20.____Many types are mutualists with nitrogen-fixing bacteria and initially outcompete other plants in nitrogen-poor habitats.

21.____Involves typically small plants, with brief life cycles, adapted to grow in exposed areas with intense sunlight, swings in temperature, and nutrient-deficient soil.

22.____A successional pattern that occurs in abandoned fields, burned forests, and storm-battered intertidal zones.

23.____Might occur on a new volcanic island or on land exposed by the retreat of a glacier.

24.____Once established, the pioneers improve conditions for other species and often set the stage for their own replacement.

## Matching

Choose the most appropriate example for each.

25. ___community stability
26. ___keystone species
27. ___geographic dispersal
28. ___jump dispersal
29. ___exotic species

A. A rapid geographic dispersal mechanism as when an insect might travel in a ship's hold from an island to the mainland

B. An outcome of forces that have come into an uneasy balance

C. Resident of an established community that has moved from its home range and successfully taken up residence elsewhere; Nile perch, European rabbits, and kudzu are examples

D. A dominant species that can dictate community structure; sea stars and periwinkles are examples

E. Residents of established communities move out from their home range and successfully take up residence elsewhere; the process may be slow or rapid by expansion of home range, jump dispersal, or extremely slow as in continental drift

## 47.10. PATTERNS OF BIODIVERSITY (pp. 830–831)

### Boldfaced, Page-Referenced Terms

(831) distance effect _____

_____

(831) area effect _____

_____

### Short Answer

1. List three factors responsible for creating the higher species diversity values as related to the distance of land and sea from the equator. _____

a. _____

_____

_____

b. _____

_____

_____

_____

c. _____

_____

_____

_____

*Fill-in-the-Blanks*

The number of coexisting species is highest in the (2) _____ , and it systematically declines toward the poles. In 1965, a volcanic eruption formed a new island southwest of (3) _____ which was named Surtsey. The new island served as a laboratory for studying (4) _____ . Within six months, bacteria, fungi, seeds, flies and some seabirds became established on the island. After two years one vascular plant appeared and two years after that, a moss plant was observed. These organisms were all colonists from (5) _____ .

Islands at great distances from potential colonists receive few colonists; the few that arrive are adapted for long distance (6) _____ . This is called the (7) _____ effect. Larger islands tend to support more species than small islands at equivalent distances from source areas. This is the (8) _____ effect. (9) _____ islands tend to have more and varied habitats, more complex topography, and extend farther above sea level. Thus, they favor species (10) _____ . Also, being bigger (11) _____ , they may intercept more colonists.

Most important, extinctions suppress (12) _____ on (13) _____ islands. There, the (14) _____ populations are far more vulnerable to storms, volcanic eruptions, diseases, and random shifts in birth and death rates. As for any island, the number of species reflects a balance between (15) _____ rates for new species and (16) _____ rates for established ones. Small islands that are distant from a source of colonists have low (17) _____ and high (18) _____ rates, so they support few species once the balance has been struck for their populations.

*Problem*

19. After consideration of the distance effect, the area effect, and species diversity patterns as related to the equator, answer the following question. There are two islands (B and C) of the same size and topography that are equidistant from the African coast (A), as shown in the illustration below. Which will have the higher species diversity values?

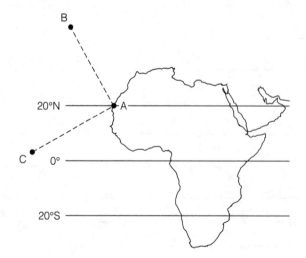

# Self-Quiz

___ 1. All the populations of different species that occupy and are adapted to a given habitat are referred to as a(n) _____ .
   a. biosphere
   b. community
   c. ecosystem
   d. niche

___ 2. The range of all factors that influence whether a species can obtain resources essential for survival and reproduction is called the _____ of a species.
   a. habitat
   b. niche
   c. carrying capacity
   d. ecosystem

___ 3. A one-way relationship in which one species benefits and the other is directly harmed is called _____ .
   a. commensalism
   b. competitive exclusion
   c. parasitism
   d. mutualism

___ 4. A lopsided interaction that directly benefits one species but does not harm or help the other much, if at all, is _____ .
   a. commensalism
   b. competitive exclusion
   c. predation
   d. mutualism

___ 5. An interaction in which both species benefit is best described as _____ .
   a. commensalism
   b. mutualism
   c. predation
   d. parasitism

___ 6. A striped skunk being pursued by a predator suddenly turns and releases its foul-smelling odor. This is an example of _____ .
   a. warning coloration
   b. mimicry
   c. camouflage
   d. moment-of-truth defense

___ 7. When an inexperienced predator attacks a yellow-banded wasp, the predator receives the pain of a stinger and will not attack again. This is an example of _____ .
   a. mimicry
   b. camouflage
   c. a prey defense
   d. warning coloration
   e. both c and d

___ 8. _____ is represented by foxtail grass, mallow plants, and smartweed because their root systems exploit different areas of the soil in a field.
   a. Succession
   b. Resource partitioning
   c. A climax community
   d. A disturbance

___ 9. During the process of community succession, _____ .
   a. pioneer populations adapt to growing in habitats that cannot support most species
   b. pioneers set the stage for their own replacement
   d. species composition eventually is stable in the form of the climax community
   e. all of the above

___10. G. Gause utilized two species of *Paramecium* in a study that described _____ .
   a. interspecific competition and competitive exclusion
   b. resource partitioning
   c. the establishment of territories
   d. coevolved mutualism

___11. The most striking patterns of species diversity on land and in the seas relate to _____ .
   a. distance effect
   b. area effect
   c. immigration rate for new species
   d. distance from the equator

___12. In a community, friction between two competing populations might be minimized by _____ .
   a. partitioning the niches in time
   b. partitioning the niches spatially
   c. fights to the death
   d. both a and b

___13. Robins probably have a(n) _____ relationship with humans.
a. parasitic
b. mutualistic
c. obligate
d. commensal

___14. The relationship between an insect and the
plants it pollinates (for example, apple blos-
soms, dandelions, and honeysuckle) is best
described as _____ .
a. mutualism
b. competitive exclusion
c. parasitism
d. commensalism

___15. The relationship between the yucca plant
and the yucca moth that pollinates it is best
described as _____ .
a. camouflage
b. commensalism
c. competitive exclusion
d. obligatory mutualism

## Chapter Objectives/Review Questions

This section lists general and detailed chapter objectives that can be used as review questions. You can make
maximum use of these items by writing answers on a separate sheet of paper. Fill in answers where blanks are
provided. To check for accuracy, compare your answers with information given in the chapter or glossary.

| Page | | Objectives/Questions |
|---|---|---|
| (814) | 1. | The type of place where you normally find a maple is its _____ . |
| (814) | 2. | List five factors that shape the structure of a biological community. |
| (814) | 3. | An organism's _____ is the sum of activities and relationships in which it engages to se-cure and use the resources necessary for its survival and reproduction. |
| (814) | 4. | Contrast the terms, potential niche and realized niche. |
| (814) | 5. | The interaction of a bird's nest and a tree is known as _____ . |
| (814) | 6. | In forms of _____ , each of the participating species reaps benefits from the interaction. |
| (814) | 7. | In _____ competition, disadvantages flow both ways between species. |
| (814) | 8. | _____ and _____ are interactions that directly benefit one species and directly hurt the other. |
| (814) | 9. | Commensalism, mutualism, and parasitism are all forms of _____ , which means "living together." |
| (815) | 10. | Define and cite an example of *obligatory mutualism*. |
| (816) | 11. | Define intraspecific competition and interspecific competition. |
| (816–817) | 12. | Describe a study that demonstrates laboratory evidence supporting the competitive exclusion concept. |
| (817) | 13. | Cite one example of resource partitioning. |
| (818) | 14. | A predator gets food from other living organisms, its _____ . |
| (818) | 15. | _____ take up residence in or on other living organisms—their hosts—and feed on spe-cific host tissues for part of the life cycle. |
| (818) | 16. | Many of the adaptations of predators and their victims arose through _____ . |
| (820–821) | 17. | Be able to completely define and give examples of the following prey defenses: warning col-oration, mimicry, moment-of-truth defenses, and camouflage. |
| (821) | 18. | Describe two examples of how predators counter prey defenses with their own marvelous adaptations. |
| (822) | 19. | What are the general body locations of ectoparasites and endoparasites? |
| (822) | 20. | List the types of organisms classified as microparasites and those classified as macroparasites. |
| (822) | 21. | Describe the distinctions between holoparasitic and hemiparasitic plants. |
| (822) | 22. | Cuckoos and the North American brown-headed cowbird are classified as _____ para-sites. |

(823) 23. In an evolutionary sense, _____ tend to interact with their hosts in ways that produce less-than-fatal effects.

(823) 24. Describe risks associated with the use of parasites as biological control agents.

(824) 25. Define *ecological succession*.

(824) 26. A _____ community is a stable, self-perpetuating array of species in equilibrium with one another and their habitat.

(824) 27. Distinguish between primary and secondary succession.

(824) 28. By a _____ - _____ model, a community is adapted to a total pattern of environmental factors.

(825) 29. Describe how fire disturbances positively affect a community of giant sequoias.

(825) 30. Distinguish between natural and active restoration.

(826) 31. Community _____ is an outcome of forces that have come into uneasy balance.

(826) 32. Explain how the presence of a keystone species can increase biodiversity in a community; relate an example.

(827–829) 33. Explain how the introduction of nonnative species can be disastrous. List five specific examples of species introductions into the United States that have had adverse results (see Table 47-2).

(830) 34. List three factors that underlie the existing patterns of biodiversity.

(831) 35. Describe the distance effect and the area effect.

(831) 36. Estimate qualitatively the differences in species diversity and abundance of organisms likely to exist on two islands with the following characteristics: Island A has an area of 6,000 square miles, and Island B has an area of 60 square miles; both islands lie at 10° N latitude and are equidistant from the same source area of colonizers.

---

# Integrating and Applying Key Concepts

If you were Ruler of All People on Earth, how would you organize industry and human populations in an effort to solve our most pressing pollution problems?

Is there a *fundamental niche* that is occupied by humans? If you think so, describe the minimal abiotic and biotic conditions required by populations of humans in order to live and reproduce. (Note that "*thrive* and *be happy*" are not criteria.) If you do not think so, state why.

These minimal niche conditions can be viewed as resource categories that must be protected by populations if they are to survive. Do you believe that the cold war between the United States and the Soviet Union primarily involved protection of minimal niche conditions, or do you believe that the cold war was based on other, more (or less) important factors?

a. If the former, how do you think *minimal* niche conditions might have been guaranteed for all humans willing and able to accept certain responsibilities as their contribution toward enabling this guarantee to be met?

b. If the latter, identify what you think those factors are and explain why you consider them more (or less) important than minimal niche conditions.

# 48

# ECOSYSTEMS

## Interactive Exercises

*Crepes for Breakfast, Pancake Ice for Dessert* (pp. 834–835)

### 48.1. THE NATURE OF ECOSYSTEMS (pp. 836–837)

### 48.2. ENERGY FLOW THROUGH ECOSYSTEMS (pp. 838–839)

### 48.3. *Focus on Science:* ENERGY FLOW AT SILVER SPRINGS, FLORIDA (p. 840)

*Selected Words:* herbivores, carnivores, omnivores, parasites, energy input, nutrient inputs, troph, cross-connecting

*Boldfaced, Page-Referenced Terms*

(836) primary producers _____

_____

(836) consumers _____

_____

(836) decomposers _____

_____

(836) detritivores _____

_____

(836) ecosystem _____

_____

(836) trophic levels _____

_____

(837) food chain _____

_____

(837) food webs _____

_____

(838) primary productivity _____

_____

(838) grazing food webs _____

_____

(838) detrital food webs _____

_____

(838) ecological pyramid _____

_____

(839) energy pyramid _____

_____

## Fill-in-the-Blanks

In an ecosystem, autotrophs known as primary (1) _____ are able to secure energy from the
environment for the entire ecosystem of which they are a part. All other organisms in the system are not
self-feeders and, in general, are known as (2) _____ . There are several types of these that extract
energy from compounds put together by the primary producers: (3) _____ eat plants; (4) _____ eat
animals; (5) _____ eat both plants and animals; (6) _____ extract energy from living hosts that they
live in or on. Still other heterotrophs, known as (7) _____ , include fungi and bacteria that obtain
energy by breaking down the remains or products of organisms. Other heterotrophs, the (8) _____ ,
obtain nutrients from decomposing particles of organic matter. Ecosystems are (9) _____ systems, so
they cannot be self-sustaining. They require (10) _____ input (the sun) and (11) _____ inputs
(minerals); they also have outputs of the same components. However, (12) _____ cannot be recycled
and much is lost to the environment; (13) _____ do get cycled, but some are still lost. Within virtually
any ecosystem, the members fit into a hierarchy of energy relationships called (14) _____ levels. A
straight-line sequence of "who eats whom" in a ecosystem (for example, cow eats grass, man eats cow) is
called a (15) _____ _____ . In reality, the same food source is most likely part of more than one
chain, especially at low trophic levels. It is more accurate to think of different food chains as cross-
connecting with one another—that is, as a (16) _____ _____ . The rate at which an ecosystem's
primary producers capture and store a given amount of energy in a specified time interval is the primary

(17) _____ . Energy from a primary source flows in one direction through two categories of food webs. In (18) _____ food webs, the energy flows from plants to herbivores, and then through an assortment of carnivores. In (19) _____ food webs, it flows mainly from photosynthetic organisms through detritivores and decomposers. These two kinds of food webs often cross-connect. Ecologists often represent the trophic structure of an ecosystem in the form of an ecological (20) _____ . In such forms, the primary (21) _____ form a broad base for successive tiers of consumers above them. Some pyramids are based on (22) _____ or the weight of all the members at each trophic level. Sometimes, a pyramid of (23) _____ can be "upside down," with the (24) [choose one] ☐ smallest ☐ largest tier on the bottom. A more useful way to depict an ecosystem's trophic structure is with an (25) _____ pyramid. These pyramids show energy losses at each transfer to a different (26) _____ level in the ecosystem. They have a (27) [choose one] ☐ small ☐ large energy base at the bottom and are always "right-side up." The ecological study of energy flow at Silver Springs, Florida found that about (28) [choose one] ☐ 1 ☐ 5 percent of all incoming solar energy was captured by producers prior to transfer to the next trophic level. It was also reported that only about (29) [choose one] ☐ 6 ☐ 8 to (30) [choose one] ☐ 10 ☐ 16 percent of the energy entering one trophic level becomes available for organisms at the next level. In general, the study showed that the efficiency of the energy transfers is so (31) [choose one] ☐ high ☐ low, ecosystems have usually no more than (32) [choose one] ☐ 4 ☐ 6 consumer trophic levels.

*Choice*

For questions 33–47, choose from the following:

a. primary producer    b. consumer    c. detritivore    d. decomposer

33. ____ Mule deer

34. ____ Earthworm

35. ____ Parasites

36. ____ The only category lacking heterotrophs

37. ____ Omnivores

38. ____ Tapeworm

39. ____ Herbivores

40. ____ Wolf

41. ____ Fungi and bacteria

42. ____ Carnivores

43. ____ Green plants

44. ____ Homo sapiens

45. ____ Crabs

46. ____ Autotrophs

47. ____ Grasshopper

## Dichotomous Choice

Circle one of two possible answers given between parentheses in each statement.

48. Ecosystems are (open/closed) systems, and so are not self-sustaining.
49. Minerals carried by erosion into a lake represent nutrient (input/output).
50. Energy (can/cannot) be recycled.
51. Nutrients typically (can/cannot) be cycled.
52. Ecosystems have energy and nutrient inputs as well as nutrient output and (have/lack) energy output.

## Matching

Match each organism in the Antarctic food chain given below with the principal trophic level it occupies. Some letters may not be needed. (See Figure 48.3 in the text.)

53. ___emperor penguins    A. Chemosynthetic autotrophs
54. ___krill                         B. Herbivores
                               C. Photosynthetic autotrophs
55. ___blue whale           D. Secondary consumers
                               E. Tertiary consumers
56. ___diatoms
57. ___leopard seal
58. ___fishes, small squids
59. ___killer whale

## 48.4. BIOGEOCHEMICAL CYCLES—AN OVERVIEW (p. 841)

## 48.5. HYDROLOGIC CYCLE (pp. 842–843)

## 48.6. CARBON CYCLE (pp. 844–845)

## 48.7. *Focus on the Environment:* FROM GREENHOUSE GASES TO A WARMER. PLANET? (pp. 846–847)

## 48.8. NITROGEN CYCLE (pp. 848–849)

*Selected Words:* availability of *nutrients*, *hydrologic* cycle, *atmospheric* cycles, *sedimentary* cycles, Anabaena, Nostoc, Rhizobium, Azotobacter

## Boldfaced, Page-Referenced Terms

(841) biogeochemical cycle _____

_____

(842) hydrologic cycle _____

_____

(842) watershed _____

_____

(844) carbon cycle _____

_____

(846) greenhouse effect _____

_____

(846) global warming _____

_____

(848) nitrogen cycle _____

_____

(848) nitrogen fixation _____

_____

(848) decomposition _____

_____

(848) ammonification _____

_____

(848) nitrification _____

_____

(849) denitrification _____

_____

## Short Answer

1. In what form(s) are elements used as nutrients usually available to producers? _____

_____

_____

2. How is the ecosystem's reserve of nutrients maintained? _____

_____

_____

3. How does the amount of a nutrient being cycled through most major ecosystems compare with the amount entering or leaving in a given year? _____

_____

_____

4. What are the common input sources for an ecosystem's nutrient reserves? _____

_____

_____

5. What are the output sources of nutrient loss for land ecosystems? _____

_____

_____

## Complete the Table

6. Complete the following table to summarize the functions of the three types of biogeochemical cycles.

| Biogeochemical Cycle | General Function(s) |
|---|---|
| a. Hydrologic cycle | |
| b. Atmospheric cycles | |
| c. Sedimentary cycles | |

## Matching

Choose the most appropriate answer for questions 7–10; questions 11–14 may have more than one answer.

7. ___ solar energy

8. ___ source of most water

9. ___ watershed

10. ___ water and plants taking up water

11. ___ forms of precipitation falling to land

12. ___ deforestation

13. ___ have important roles in the global hydrologic cycle

14. ___ forms of atmospheric water

A. Mostly rain and snow
B. Important in moving nutrients in biochemical cycles
C. Water vapor, clouds, and ice crystals
D. Where precipitation of a specified region becomes funneled into a single stream or river
E. May have long-term disruptive effects on nutrient availability for an entire ecosystem
F. Slowly drives water through the atmosphere, on or through land mass surface layers, to oceans, and back again
G. Ocean currents and wind patterns
H. Evaporation from the oceans

## Fill-in-the-Blanks

A watershed may be any (15) _____ that is of interest to an investigator. Most of the water that enters a watershed seeps into the (16) _____ or becomes surface runoff that moves into (17) _____ . Plants take up water and dissolved minerals from the soil, then they lose water by (18) _____ .

Watershed studies revealed the importance of (19) _____ cover in the movements of (20) _____ through the ecosystem phase of biogeochemical cycles.

In studies of young, undisturbed forests in the Hubbard Brook watersheds, each hectare lost only about 8 kilograms of (21) _____ . Rainfall and the weathering of rocks brought in replacements of this element. Tree roots were also "mining" the soil, so (22) _____ was being stored in a growing (23) _____ of tree tissues.

In some experimental watersheds, researchers stripped the vegetation cover but did not disturb the (24) _____ . Then they applied (25) _____ to the soil for three years to prevent regrowth. After that time, stream outflow from the experimental watersheds carried more than (26) _____ times as much calcium as the stream outflow from undisturbed ones.

Because calcium and other nutrients move so slowly through geochemical cycles, (27) _____ may disrupt nutrient availability for an entire ecosystem.

In ecosystems on land, plants (28) _____ the soil and absorb dissolved (29) _____ . By doing so, they minimize the loss of soil (30) _____ in runoff from land.

## Matching

Choose the one most appropriate answer for each.

31. ___greenhouse gases

32. ___carbon dioxide fixation

33. ___carbon cycle

34. ___ways carbon enters the atmosphere

35. ___greenhouse effect

36. ___carbon dioxide ($CO_2$)

37. ___oceans and plant biomass accumulation

   A. Form of most of the atmospheric carbon

   B. Aerobic respiration, fossil fuel burning, and volcanic eruptions

   C. $CO_2$, CFCs, $CH_4$, and $N_2O$

   D. Photosynthesizers incorporate carbon atoms into organic compounds

   E. Annual "holding stations" for about half of all atmospheric carbon

   F. Carbon reservoirs → atmosphere and oceans → through organisms → carbon reservoirs

   G. Warming of Earth's lower atmosphere due to accumulation of certain gases

## Fill-in-the-Blanks

Atmospheric concentrations of (38) _____ molecules play a profound role in shaping the average temperature near the surface of the Earth. That temperature has enormous effect on the global (39) _____ . In the (40) _____ _____ , Earth's surface is warmed by sunlight and radiates (41) _____ (infrared wavelengths) to the atmosphere and space. Greenhouse gases such as (42) _____ , which are used as refrigerants, solvents and plastic foams, and (43) _____ _____ , which is released from burning fossil fuels, deforestation, car exhaust, and factory emissions, are together responsible for the alarming global (44) _____ trend. Collectively, the greenhouse gases act somewhat like a pane of glass in a greenhouse. The gaseous molecules can absorb the infrared wavelengths and then (45) _____ much of the absorbed (46) _____ back toward the Earth. This constant reradiation of heat energy from greenhouse gases and absorption of wavelengths from the sun results in a heat build-up in the lower (47) _____ . The (48) _____ effect is the name for this warming action. Greenhouse gases may be contributing to long-term higher temperatures at the Earth's surface, an effect called (49) _____ warming. The effects of a warmer planet are alarming. Even a slight temperature increase might cause (50) _____ levels to rise as well as faster melting of glaciers and polar ice sheets. Such increases in water volume would (51) _____ coastal regions and bring disturbing changes in world (52) _____ . In general, scientists have determined that atmospheric levels of the greenhouse gases are (53) [choose one] □ decreasing □ increasing.

*Choice*

For questions 54–58, choose from the following:

   a. nitrogen cycle description    b. nitrogen fixation    c. decomposition and ammonification
                    d. nitrification    e. denitrification

54.____Ammonia or ammonium in soil are stripped of electrons, and nitrite ($NO_2^-$) is the result; other bacteria convert nitrite to nitrate ($NO_3^-$)

55.____Bacteria convert nitrate or nitrite to $N_2$ and a bit of nitrous oxide ($N_2O$).

56.____Bacteria and fungi break down nitrogen-containing wastes and plant and animal remains; released amino acids and proteins are used for growth, with the excess given up as ammonia or ammonium ions that plants can use.

57.____Occurs in the atmosphere (largest reservoir); only certain bacteria, volcanic action, and lightning can convert $N_2$ into forms that can enter food webs.

58.____A few kinds of bacteria convert $N_2$ to ammonia ($NH_3$), which dissolves quickly in water to form ammonium ($NH_4^+$).

*Short Answer*

59. List reasons that an insufficient soil nitrogen supply is a problem for land plants. _____

_____

_____

_____

_____

_____

## 48.9. SEDIMENTARY CYCLES (p. 850)

## 48.10. PREDICTING THE IMPACT OF CHANGE IN ECOSYSTEMS (p. 851)

*Selected Words:* Plasmodium, Rickettsia

*Boldfaced, Page-Referenced Terms*

(850) phosphorus cycle _____

_____

(850) eutrophication _____

_____

(851) ecosystem modeling _____

_____

(851) biological magnification _____

_____

## Fill-in-the-Blanks

The Earth's crust is the largest (1) _____ for phosphorus and other minerals that move through ecosystems as part of (2) _____ cycles. In rock formations on land, phosphorus is typically in the form of (3) _____ . By natural weathering and erosion of rock formations, phosphates enter rivers and streams that transport them to the (4) _____ . Phosphorus slowly accumulates mainly on (5) _____ shelves. Following long time periods, crustal plates can uplift part of the seafloor and expose the phosphate on drained land surfaces. Over time, weathering releases the phosphates from the exposed rocks, and the cycle's (6) _____ phase begins again. The (7) _____ phase of the cycle is more rapid than the long-term geochemical phase. All (8) _____ require phosphorus for synthesizing phospholipids, NADPH, ATP, nucleic acids, and other compounds. Plants take up dissolved, (9) _____ forms of phosphate very rapidly. Herbivores obtain phosphorus by eating (10) _____ ; carnivores get it by eating (11) _____ . These organisms excrete phosphates in urine and feces. (12) _____ also releases phosphates to the soil and again. Plants can take it in and help to rapidly recycle it within the (13) _____ .

(14) _____ use phosphates as key ingredients. Where these compounds are heavily applied, phosphorus that becomes concentrated in runoff from fields alters living conditions in (15) _____ and other aquatic ecosystems. Nitrogen, potassium, and phosphorus are three important nutrients that algae and aquatic plants require for their (16) _____ . Bacteria fix enough (17) _____ . Freshwater ecosystems usually have excess amounts of (18) _____ . Most of the phosphorus is locked in (19) _____ , so it tends to be a limiting factor on plant growth. If the water is enriched with (20) _____ -containing fertilizers, the outcome will be dense (21) _____ blooms. Activities that increase the concentrations of dissolved nutrients can lead to (22) _____ , a term referring to nutrient enrichment of any aquatic ecosystem.

(23) _____ _____ is a method of identifying and combining crucial bits of information about an ecosystem through computer programs and models in order to predict the outcome of the next disturbance. During World War II, DDT was sprayed in the tropical Pacific to control (24) _____ responsible for transmitting the organisms that cause a dangerous disease, (25) _____ . Because of its stability, DDT is a prime candidate for (26) _____ _____—the increasing concentration of a nondegradable substance as it moves up through trophic levels. Following WW II, DDT began to spread through the global environment, infiltrate (27) _____ _____ , and affect organisms in ways that no one had predicted. DDT was indiscriminately killing off natural (28) _____ that had been keeping pest populations in check. Most devastated were those organisms at the (29) _____ of the food webs. Then side effects of (30) _____ _____ began to show up in places that were far removed from areas of DDT application—and much later in time. (31) _____ to one aspect of an ecosystem typically have unexpected effects on other, seemingly unrelated parts.

# Self-Quiz

___ 1. A network of interactions that involve the cycling of materials and the flow of energy between a community and its physical environment is a(n) _____ .
  a. population
  b. community
  c. ecosystem
  d. biosphere

___ 2. _____ consume dead or decomposing particles of organic matter.
  a. Herbivores
  b. Parasites
  c. Detritivores
  d. Carnivores

___ 3. The members of feeding relationships are structured in a hierarchy, the steps of which are called _____ .
  a. organism levels
  b. energy source levels
  c. eating levels
  d. trophic levels

___ 4. In the Antarctic, blue whales feed mainly on _____ .
  a. petrels
  b. krill
  c. seals
  d. fish and small squids

___ 5. Which of the following is a primary consumer?
  a. cow
  b. dog
  c. hawk
  d. all of the above

___ 6. In a natural community, the primary consumers are _____ .
  a. herbivores
  b. carnivores
  c. scavengers
  d. decomposers

___ 7. A straight-line sequence of who eats whom in an ecosystem is sometimes called a(n) _____ .
  a. trophic level
  b. food chain
  c. ecological pyramid
  d. food web

___ 8. Of the 1,700,000 kilocalories of solar energy that entered an aquatic ecosystem in Silver Springs, Florida, investigators determined that about _____ percent of incoming solar energy was trapped by photosynthetic autotrophs.
  a. 1
  b. 10
  c. 25
  d. 74

___ 9. A biogeochemical cycle that deals with phosphorus and other nutrients that do not have gaseous forms is the _____ type.
  a. sedimentary
  b. hydrologic
  c. nutrient
  d. atmospheric

___ 10. _____ is a process in which nitrogenous waste products or organic remains of organisms are decomposed by soil bacteria and fungi that use the amino acids being released for their own growth and release the excess as ammonia or ammonium, which plants take up.
  a. Nitrification
  b. Ammonification
  c. Denitrification
  d. Nitrogen fixation

___ 11. In the carbon cycle, carbon enters the atmosphere through _____ .
  a. carbon dioxide fixation
  b. respiration, burning, and volcanic eruptions
  c. oceans and accumulation of plant biomass
  d. release of greenhouse gases

___ 12. _____ refers to an increase in concentration of a nondegradable (or slowly degradable) substance in organisms as it is passed along food chains.
  a. Ecosystem modeling
  b. Nutrient input
  c. Biogeochemical cycle
  d. Biological magnification

# Chapter Objectives/Review Questions

This section lists general and detailed chapter objectives that can be used as review questions. You can make maximum use of these items by writing answers on a separate sheet of paper. Fill in answers where blanks are provided. To check for accuracy, compare your answers with information given in the chapter or glossary.

*Page*      *Objectives/Questions*

(836)      1.   List the principal trophic levels in an ecosystem of your choice; state the source of energy for each trophic level and give one or two examples of organisms associated with each trophic level.

(836)      2.   Explain why nutrients can be completely recycled but energy cannot.

(836)      3.   An _____ is a complex of organisms and their physical environment, all interacting through a flow of energy and a cycling of materials.

(836)      4.   Members of an ecosystem fit somewhere in a hierarchy of energy transfers (feeding relationships) called _____ levels.

(837)      5.   Distinguish between food chains and food webs.

(838)      6.   Define *primary productivity*.

(838)      7.   Compare grazing food webs with detrital food webs. Present an example of each.

(838–839)  8.   Understand how materials and energy enter, pass through, and exit an ecosystem.

(838–839)  9.   Ecological pyramids that are based on _____ are determined by the weight of all the members of each trophic level; _____ pyramids reflect the energy losses at each transfer to a different trophic level.

(841)      10.  In _____ cycles, the nutrient is transferred from the environment to organisms, then back to the environment—which serves as a large reservoir for it.

(842)      11.  Be able to discuss water movements through the hydrologic cycle.

(843)      12.  Explain what studies in the Hubbard Brook watershed have taught us about the movement of substances (water, for example) through a forest ecosystem.

(844)      13.  The carbon cycle traces carbon movement from reservoirs in the _____ and oceans, through organisms, then back to reservoirs.

(846)      14.  Certain gases cause heat to build up in the lower atmosphere, a warming action known as the _____ effect.

(848)      15.  A major element found in all proteins and nucleic acids moves in an atmospheric cycle called the _____ cycle.

(848)      16.  Define the chemical events that occur during nitrogen fixation, decomposition and ammonification, and nitrification.

(850)      17.  Describe the geochemical and ecosystem phase of the phosphorus cycle.

(850)      18.  Define *eutrophication* and be able to discuss its causes.

(850)      19.  Explain why agricultural methods in the United States tend to put more energy into mechanized agriculture in the form of fertilizers, pesticides, food processing, storage, and transport than is obtained from the soil in the form of energy stored in foods.

(851)      20.  Through _____ modeling, crucial bits of information about different ecosystem components are identified and used to build computer models for predicting outcomes of ecosystem disturbances.

(851)      21.  Describe how DDT damages ecosystems; discuss biological magnification.

# Integrating and Applying Key Concepts

In 1971, *Diet for a Small Planet* was published. Frances Moore Lappé, the author, felt that people in the United States of America wasted protein and ate too much meat. She said, "We have created a national consumption pattern in which the majority, who can pay, overconsume the most inefficient livestock products [cattle] well beyond their biological needs (even to the point of jeopardizing their health), while the minority, who cannot pay, are inadequately fed, even to the point of malnutrition." Cases of marasmus (a nutritional disease caused by prolonged lack of food calories) and kwashiorkor (caused by severe, long-term protein deficiency) have been found in Nashville, Tennessee, and on an Indian reservation in Arizona, respectively. Lappé's partial solution to the problem was to encourage people to get as much of their protein as possible directly from plants and to supplement that with less meat from the more efficient converters of grain to protein (chickens, turkeys, and hogs) and with seafood and dairy products. Most of us realize that feeding the hungry people of the world is not just a matter of distributing the abundance that exists—that it is being prevented in part by political, economic, and cultural factors. Devise two full days of breakfasts, lunches, and dinners that would enable you to exploit the lowest acceptable trophic levels to sustain yourself healthfully.

# 49

# THE BIOSPHERE

## Interactive Exercises

**Does a Cactus Grow in Brooklyn?** (pp. 854–855)

## 49.1. AIR CIRCULATION PATTERNS AND REGIONAL CLIMATES (pp. 856–857)
## 49.2. THE OCEAN, LAND FORMS, AND REGIONAL CLIMATES (pp. 858–859)

*Selected Words:* leeward, windward

*Boldfaced, Page-Referenced Terms*

(855) biosphere _____

_____

(855) hydrosphere _____

_____

(855) lithosphere _____

_____

(855) atmosphere _____

_____

(855) climate _____

_____

(856) ozone layer _____

_____

(856) temperature zones _____

_____

(858) ocean _____

_____

(859) rain shadow _____

_____

(859) monsoons _____

_____

## Matching

Choose the one most appropriate answer for each.

1. ___biosphere
2. ___hydrosphere
3. ___lithosphere
4. ___atmosphere
5. ___climate
6. ___ozone layer
7. ___temperature zones
8. ___ocean
9. ___rain shadow
10. ___leeward
11. ___windward
12. ___monsoons
13. ___"mini-monsoons"

A. One continuous body of water that covers 71 percent of the Earth's surface
B. Made up of gases and airborne particles that envelop the Earth
C. Of the world, defined by differences in solar heating at different latitudes and the modified air circulation patterns
D. The sum total of all places in which organisms live
E. Refers to the direction not facing a wind
F. An atmospheric region between 17 and 25 kilometers above sea level
G. Recurring sea breezes along coastlines; warmed air above land rises and cooler marine air moves in
H. The waters of the Earth, including the ocean, polar ice caps, and other forms of liquid and frozen water
I. Refers to the direction from which the wind is blowing
J. Average weather conditions, such as temperature, humidity, wind speed, cloud cover, and rainfall, over time
K. The outer, rocky layer of the Earth
L. An arid or semiarid region of sparse rainfall on the leeward side of high mountains
M. Patterns of air circulation that affect conditions on the continents lying poleward of warm oceans; intensely heated land draws in moisture-laden air that forms above ocean water

## Fill-in-the-Blanks

The numbered items on the accompanying illustrations represent missing information. Complete the matching numbered blanks in the narrative to supply the missing information.

The sun's rays are more direct in (14) _____ regions than in polar regions. The global pattern of air circulation begins as (15) _____ equatorial air (16) _____ and spreads northward and southward giving up much (17) _____ as precipitation (warm air can hold more moisture than cold air) that supports luxuriant tropical forests. The cooled, drier air then (18) _____ at latitudes of about 30°, becomes warmer and drier in areas where deserts form. Air even farther from the equator picks up some (19) _____ , and (20) _____ to higher altitudes, cools, and then gives up (21) _____ at latitudes of about 60° to create another moist belt. The cooled, dry air then (22) _____ at the polar regions, where the low temperatures and almost nonexistent precipitation give rise to the cold, dry, polar deserts. The Earth's rotation then modifies the air circulation by deflecting it into worldwide belts of prevailing (23) _____ and (24) _____ winds. The world's temperature zones, beginning at the equator and moving toward the poles, are the (25) _____ , the (26) _____ temperate, the (27) _____ temperate, and the (28) _____ . Finally, the amount of the (29) _____ radiation reaching the surface varies annually, owing to the Earth's (30) _____ around the sun. This leads to seasonal changes in daylength, prevailing wind directions, and temperature. These factors influence the locations of different ecosystems.

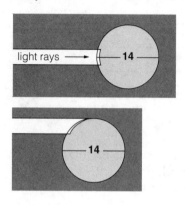

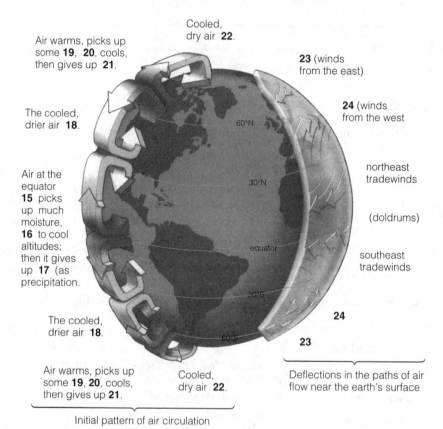

Air warms, picks up some **19**, **20**, cools, then gives up **21**.

Cooled, dry air **22**.

The cooled, drier air **18**.

**23** (winds from the east)

**24** (winds from the west

Air at the equator **15** picks up much moisture, **16** to cool altitudes; then it gives up **17** (as precipitation.

60°N

30°N

equator

30°S

60°S

northeast tradewinds

(doldrums)

southeast tradewinds

The cooled, drier air **18**.

**24**

Air warms, picks up some **19**, **20**, cools, then gives up **21**.

Cooled, dry air **22**.

**23**

Deflections in the paths of air flow near the earth's surface

Initial pattern of air circulation

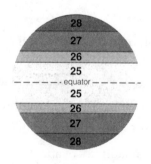

28
27
26
25
- - - - - equator - - - - -
25
26
27
28

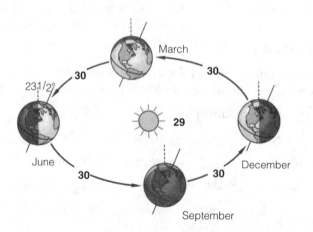

March

30

30

23,1/2°

29

June

30

30

December

September

## Identification Matching

Notice the numbered currents shown in the accompanying map. Match their numbers with the lettered choices below.

31. ___
32. ___
33. ___
34. ___
35. ___
36. ___
37. ___

A. Benguela current
B. California current
C. Canary current
D. Gulf Stream
E. Humboldt current
F. Japan current
G. Labrador current
H. North Atlantic drift

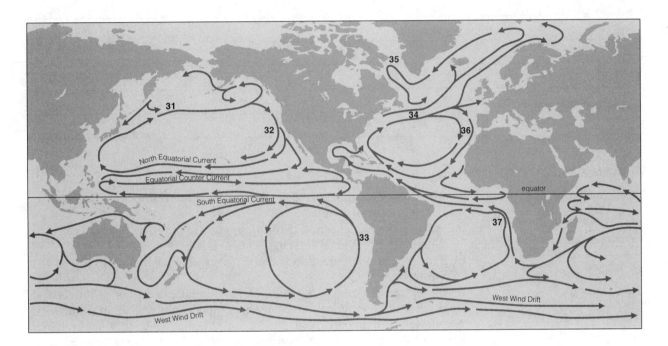

## 49.3. THE WORLD'S BIOMES (pp. 860–861)

## 49.4. SOILS OF MAJOR BIOMES (p. 862)

## 49.5. DESERTS (p. 863)

## 49.6. DRY SHRUBLANDS, DRY WOODLANDS, AND GRASSLANDS (pp. 864–865)

*Selected Words:* soil *profile, shortgrass* prairie, *tallgrass* prairie, *monsoon* grasslands

*Boldfaced, Page-Referenced Terms*

(860) biogeography _____

_____

(860) biogeographic realms _____

_____

(861) biome _____

_____

(862) soils _____

_____

(863) deserts _____

_____

(863) desertification _____

_____

(864) dry shrublands _____

_____

(864) dry woodlands _____

_____

(864) grasslands _____

_____

(864) savannas _____

_____

## Matching

1. ___biogeography
2. ___biogeographic realms
3. ___biome
4. ___soil
5. ___deserts
6. ___desertification
7. ___dry shrublands
8. ___soil profile
9. ___dry woodlands
10. ___grasslands
11. ___savannas
12. ___monsoon grasslands

A. Mixtures of mineral particles and variable amounts of decomposing organic material (humus)
B. Areas receiving less than 25 to 60 centimeters of rain per year; local names include fynbos and chaparral
C. The conversion of grasslands and other productive biomes to dry wastelands
D. Sweep across much of the interior of continents, in the zones between deserts and temperate forests; warm temperatures prevail in summer, winters are extremely cold
E. Areas that dominate when annual rainfall is about 40 to 100 centimeters; dominant trees can be tall but do not form a dense, continuous canopy
F. The study of the global distribution of species, each of which is adapted to regional conditions
G. Subdivision of biogeographic realms; a large region of land characterized by the climax vegetation of the ecosystems within its boundaries
H. Formed in land regions where the potential for evaporation exceeds sparse rainfall
I. Six vast land areas on the Earth, each with distinguishing plants and animals
J. Broad belts of grasslands with a smattering of shrubs or trees; rainfall averages 90 to 150 centimeters a year, with prolonged seasonal droughts common
K. The layered structure of soils
L. Form in southern Asia where heavy rains alternate with a dry season; dense stands of tall, coarse grasses form, then die back and often burn in the dry season

## Identification

Identify the numbered regions in the world map by matching their numbers with the lettered choices below.

13. ___
14. ___
15. ___
16. ___
17. ___
18. ___
19. ___
20. ___

A. desert
B. dry woodlands and shrublands (e.g., chaparral)
C. evergreen broadleaf forest (e.g., tropical rain forest)
D. evergreen coniferous forest (e.g., boreal forest, montane coniferous forest)
E. temperate deciduous forest
F. temperate grassland
G. tropical deciduous forest
H. tundra

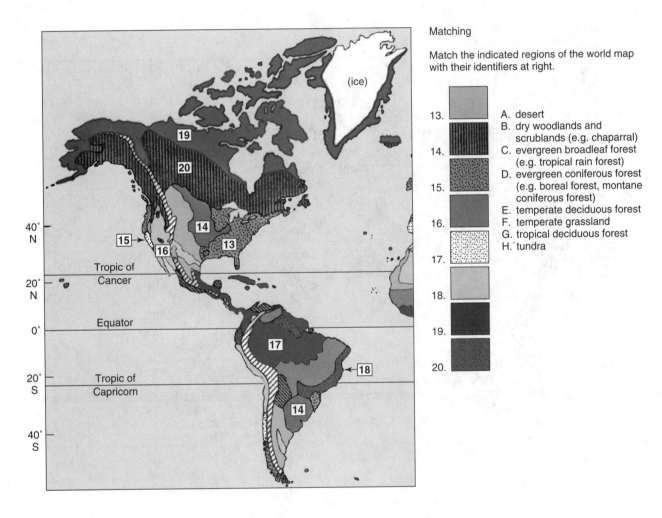

### Matching

Match the indicated regions of the world map with their identifiers at right.

13.
14.
15.
16.
17.
18.
19.
20.

A. desert
B. dry woodlands and scrublands (e.g. chaparral)
C. evergreen broadleaf forest (e.g. tropical rain forest)
D. evergreen coniferous forest (e.g. boreal forest, montane coniferous forest)
E. temperate deciduous forest
F. temperate grassland
G. tropical deciduous forest
H. tundra

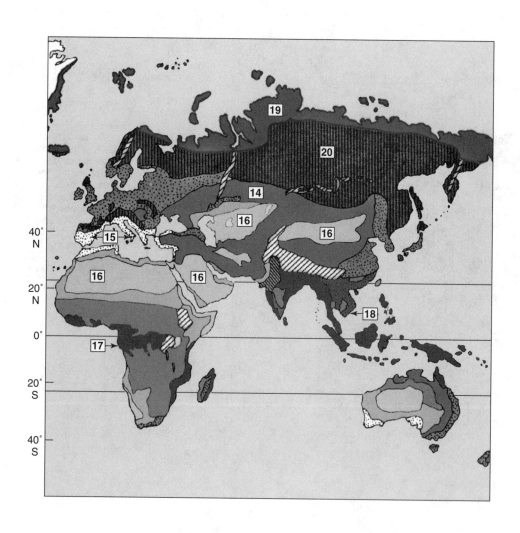

*Choice*

For questions 21–30, choose from the following:

a. deserts          b. dry shrublands          c. dry woodlands          d. grasslands          e. savannas

21.____Monsoon type that form dense stands of tall, coarse plants in parts of southern Asia where heavy rains alternate with a dry season

22.____Biome where the potential for evaporation greatly exceeds rainfall

23.____Steinbeck's *Grapes of Wrath* speaks eloquently of the disruption of this biome

24.____Local names for this biome include fynbos and chaparral; plants are woody, multibranched, and only a few meters tall

25.____A biome in which dominant trees can be tall but do not form a dense canopy; includes Eucalyptus woodlands of southwestern Australia and oak woodlands of California and Oregon

26.____Home to deep-rooted evergreen shrubs, fleshy-stemmed, shallow-rooted cacti, saguaro, short prickly pear, and ocotillo

27.____Broad belts of grasslands with a smattering of shrubs and trees; prolonged seasonal droughts are common

28.____The dominant animals are grazing and burrowing types; grazing and periodic fires maintain the fringes of this biome

29.____Within this biome, dominate fast-growing grasses dominate where rainfall is low but acacia and other shrubs grow where there is slightly more moisture

30.____In summer, lightning-sparked, wind-driven firestorms can sweep quickly through these biomes and burn shrubs with highly flammable leaves

## 49.7. TROPICAL RAIN FORESTS AND OTHER BROADLEAF FORESTS (pp. 866–867)

## 49.8. CONIFEROUS FORESTS (p. 868)

## 49.9. TUNDRA (p. 869)

*Selected Words:* *tropical* deciduous forests, *temperate* deciduous forests, *taiga*, *tuntura*, *alpine* tundra

### Boldfaced, Page-Referenced Terms

(866) evergreen broadleaf forests _____

_____

(866) tropical rain forest _____

_____

(867) deciduous broadleaf forests _____

_____

(868) coniferous forests _____

_____

(868) boreal forests _____

_____

(868) pine barrens _____

_____

(869) tundra _____

_____

(869) permafrost _____

_____
•

## Choice

For questions 1–10, choose from the following:

a. deciduous broadleaf forests     b. coniferous forests     c. evergreen broadleaf forests     d. tundra

1. ____Nearly continuous sunlight in summer; short plants grow and flower profusely with rapidly ripening seeds

2. ____Near the equator, highly productive, rainfall regular and heavy, tropical rain forests

3. ____Highly productive forest; decomposition and mineral cycling are rapid in the hot, humid climate

4. ____Within this biome, complex forests of ash, beech, chestnut, elm, and deciduous oaks once stretched across northeastern North America

5. ____Boreal forests or taiga; conifers are the primary producers in this biome

6. ____Spruce and balsam fir dominate this North American biome

7. ____Biome of the temperate zone; cold winter temperatures; many trees drop all their leaves in winter

8. ____Lies between the polar ice cap and belts of boreal forests in North America, Europe, and Asia

9. ____Soils are highly weathered, humus-poor, and not good nutrient reservoirs

10. ____Sandy, nutrient-poor soil of New Jersey's coastal plain supports scrub forests known as the pine barrens

## Fill-in the-Blanks

Warm temperatures combined with uniform, abundant rainfall encourage the growth of the highly productive (11) _____ _____ forests. This type of biome exists where the annual mean temperature is about (12) _____ °C and a (13) _____ of 80 percent or more; it produces more litter than any other forest biome. However, (14) _____ and mineral cycling are rapid in the hot, humid climate, and its soils are not a good reservoir of nutrients. In (15) _____ _____ forests, most plants lose some or all of their leaves during a pronounced dry season. In (16) _____ _____ forests, decomposition is not as rapid as in the humid tropics, and many nutrients are conserved in accumulated litter on the forest floor. Conifers are the primary producers of (17) _____ forests. Most have (18) _____ -shaped leaves with a thick cuticle and recessed (19) _____—(20) _____ that help the trees conserve water through dry times. (21) _____ forests stretch across northern Europe, Asia, and North America. These "swamp forests" are also known by the name (22) _____ . In the Northern Hemisphere, (23) _____ coniferous forests extend southward through the great mountain ranges. (24) _____ and balsam fir dominate in the north and at higher elevations. They give way to fir and (25) _____ in the south and at lower elevations. New Jersey's coastal plain is composed of sandy, nutrient-poor soil that supports (26) _____ _____ or scrub forests in which grasses and low shrubs grow beneath the open stands of pitch pine and oak trees. Just beneath the surface of the (27) _____ (a Finnish word for treeless plain) is the (28) _____ , a permanently frozen layer. Here anaerobic conditions and low temperatures limit nutrient cycling. A similar biome at high elevations in mountains throughout the world is the (29) _____ tundra. However, here the permanently frozen layer, the (30) _____ is lacking. Dominant plants form low cushions or mats that withstand strong winds.

## 49.10. FRESHWATER PROVINCES (pp. 870–871)
## 49.11. THE OCEAN PROVINCES (pp. 872–873)

*Selected Words:* *phyto*plankton, *zoo*plankton, *oligotrophic* lakes, *eutrophic* lakes, *benthic* province, *pelagic* province

### Boldfaced, Page-Referenced Terms

(870) lake _____

_____

(870) spring overturn _____

_____

(870) fall overturn _____

_____

(871) eutrophication _____

_____

(871) streams _____

_____

(872) ultraplankton _____

_____

(872) hydrothermal vents _____

_____

### Fill-in-the-Blanks

A (1) _____ is a standing body of freshwater with three zones. The shallow, usually well-lit (2) _____ zone extends around the shore to the depth at which rooted aquatic plants stop growing. The diversity of organisms is greatest here. The (3) _____ zone is the open, sunlit water past the littoral and extends to a depth where photosynthesis is insignificant. Aquatic communities of (4) _____ abound here. The (5) _____ zone includes all open water below the depth at which wavelengths suitable for photosynthesis can penetrate. Bacterial decomposers in bottom sediments of this zone release nutrients into the water. Water is densest at 4°C; at this temperature, it sinks to the bottom of its basin, displacing the nutrient-rich bottom water upward and giving rise to spring and fall (6) _____ . In spring, ice melts, daylength increases, and the surface waters of a lake slowly warm to (7) _____ °C. Surface winds cause a (8) _____ overturn in which strong vertical water movements of water carry dissolved oxygen from a lake's surface layer to its depths, and nutrients released by decomposition are brought from the bottom sediments to the surface layer. The surface layer warms above 4°C by midsummer, becomes less dense, and the lake has developed a middle layer, the (9) _____ that prevents vertical mixing. When autumn comes, the upper layer (10) _____ , becomes denser, then sinks, and the thermocline vanishes. This is termed the (11) _____ overturn. Water then mixes vertically allowing dissolved oxygen to move (12) [choose one] □ up □ down and nutrients to move (13) [choose one] □ up □ down. Primary productivity of a lake corresponds with the seasons. Following a spring overturn, longer daylengths and cycled nutrients support (14) [choose one] □ lower □ higher rates of photosynthesis. By late summer, nutrient shortages are limiting photosynthesis. After the fall overturn, nutrient cycling drives a (15) [choose one] □ short □ long burst of primary activity. Lakes have a trophic nature. (16) _____ lakes are deep, nutrient-poor, and low in primary productivity. Lakes that are (17) _____ are often shallow, nutrient-enriched, and high in primary productivity. Human activities can determine the trophic condition of lakes. The term (18) _____ refers to nutrient enrichment of a lake resulting in reduced water transparency and a community rich in phytoplankton. (19) _____ are flowing-water ecosystems that begin as freshwater springs or seeps. Three habitat types are found between headwaters and the river's end: riffles, pools, and (20) _____ .

## Label-Match

Label each numbered item on the accompanying illustration. Complete the exercise by matching and entering the letter of the proper description in the parentheses following each label.

21. _____ province (   )
22. _____ province (   )
23. _____ zone (   )
24. _____ zone (   )

A. The entire volume of ocean water
B. All the water above the continental shelves
C. Includes all sediments and rocks of the ocean bottom; begins with continental shelves and extends to deep-sea trenches
D. Water of the ocean basin

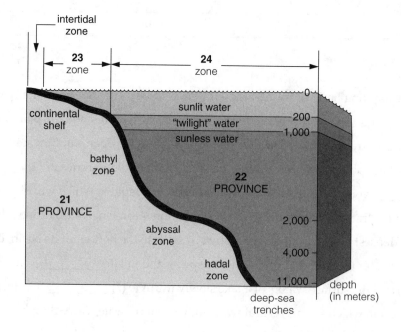

*Choice*

For questions 25–44, choose from the following:

a. stream ecosystems     b. lake ecosystems     c. ocean

25.\_\_\_\_Riffles, pools, and runs

26.\_\_\_\_Hydrothermal vent communities of chemosynthetic bacteria in the abyssal zone; possible sites of life's origin

27.\_\_\_\_70 percent of the primary productivity is contributed by ultraplankton

28.\_\_\_\_Can be oligotrophic or eutrophic

29.\_\_\_\_Submerged mountains, valleys, and plains

30.\_\_\_\_Begin as freshwater springs or seeps

31.\_\_\_\_Sewage and logging on adjacent lands can contribute to eutrophication of this water

32.\_\_\_\_Average flow volume and temperature depend on rainfall, snowmelt, geography, altitude, and even shade cast by plants

33.\_\_\_\_Spring and fall overturns

34.\_\_\_\_Has benthic and pelagic provinces

35.\_\_\_\_Especially in forests, these waters import most of the organic matter that supports food webs

36.\_\_\_\_Created by geologic processes such as retreating glaciers

37.\_\_\_\_They grow and merge as they flow downslope and then often combine

38.\_\_\_\_Seasonal variations in primary production correspond to latitude; phytoplankton blooms are brought about by increases in daylight

39.\_\_\_\_The final successional stage is a filled-in basin

40.\_\_\_\_Vast "pastures" of phytoplankton and zooplankton become the basis of detrital food webs

41.\_\_\_\_Since cities formed, these waters have been sewers for industrial and municipal wastes

42.\_\_\_\_Has a thermocline by midsummer

43.\_\_\_\_Littoral, limnetic, and profundal zones

44.\_\_\_\_bathyal, abyssal, and hadal zones

## 49.12. CORAL REEFS AND BANKS
(pp. 874–875)

## 49.13. LIFE ALONG THE COASTS
(pp. 876–877)

## 49.14. *Focus on Science:* EL NIÑO AND SEESAWS IN THE WORLD'S CLIMATES
(pp. 878–879)

*Selected Words:* <u>Corallina</u>, <u>Lophelia</u>, *sandy* shores, *muddy* shores, *El Niño Southern Oscillation* (ENSO)

*Boldfaced, Page-Referenced Terms*

(874) coral reef _____

_____

(874) coral banks _____

_____

(876) estuary _____

_____

(876) intertidal zone _____

_____

(877) upwelling _____

_____

(877) El Niño _____

_____

## Complete the Table

1. Complete the following table, which describes three types of coral reefs.

| Reef Type | Description |
|---|---|
| a. Atolls | |
| b. Fringing reefs | |
| c. Barrier reefs | |

## Choice

For questions 2–21, choose from the following:

a. coral reefs and banks b. estuary     c. intertidal zone       d. coastal upwelling    e. ENSO

2. ____The fogbanks that form along the California coast are one outcome

3. ____*Lophelia* has constructed large ones in the cold waters of Norway's fjords

4. ____Waves batter its resident organisms

5. ____Pulses of heat from the Earth may be the trigger for warming the southern surface waters; more warm air arises here than anyplace else

6. ____Organisms living there must contend with the tides

7. ____Constructed mainly from the remains of countless organisms

8. ____Primary producers here include phytoplankton, salt-tolerant plants, and some algae that grow in mud and on plant surfaces

9. ____In general, an upward movement of cold, deep, often nutrient-rich ocean waters occurring in equatorial currents as well as along the coasts of both continents

10. ____A partly enclosed coastal region where seawater swirls and mixes with nutrient-rich freshwater from rivers, streams, and land runoff

11. ____The larger ones are found in clear, warm waters between latitudes 25° north and south

12. ____Salt marshes are common

13.____El Niño

14.____Wind friction causes surface waters to begin moving and under the force of the Earth's rotation, moving water is deflected west, away from a coast

15.____Many of these structures are being destroyed by human activities

16.____Commercial fishing industries of Peru and Chile depend on it

17.____A recurring seesaw in atmospheric pressure in the western equatorial Pacific

18.____Home to exotic and rare fish sold in pet stores and eaten in Japanese restaurants

19.____Sandy and muddy shores

20.____Reverses the usual westward flow of air and water displaces the cold, deep Humboldt current; this stops nutrient upwelling along the western coast of South America

21.____Found along rocky and sandy coastlines

22.____All show spectacular biodiversity, and vulnerability

# Self-Quiz

____ 1. The distribution of different types of ecosystems is influenced by _____ .
   a. air currents
   b. variation in the amount of solar radiation reaching the earth through the year
   c. ocean currents
   d. all of the above

____ 2. In a(n) _____ , water draining from the land mixes with seawater carried in on tides.
   a. pelagic province
   b. rift zone
   c. upwelling
   d. estuary

____ 3. A biome with grasses as primary producers and scattered trees adapted to prolonged dry spells is known as a _____ .
   a. warm desert
   b. savanna
   c. tundra
   d. taiga

____ 4. Located at latitudes of about 30° north and south, limited vegetation, and rapid surface cooling at night describe a _____ biome.
   a. shrubland
   b. savanna
   c. taiga
   d. desert

____ 5. In tropical rain forests, _____ .
   a. competition for available sunlight is intense
   b. diversity is limited because the tall forest canopy shuts out most of the incoming light
   c. conditions are extremely favorable for growing luxuriant crops
   d. all of the above

____ 6. In a lake, the open sunlit water with its suspended phytoplankton is referred to as its _____ zone.
   a. epipelagic
   b. limnetic
   c. littoral
   d. profundal

____ 7. The lake's upper layer cools, the thermocline vanishes, lake water mixes vertically, and once again dissolved oxygen moves down and nutrients move up. This describes the _____ .
   a. spring overturn
   b. summer overturn
   c. fall overturn
   d. winter overturn

___ 8. The _____ is a permanently frozen, water-impermeable layer just beneath the surface of the _____ biome.
a. permafrost; alpine tundra
b. hydrosphere; alpine tundra
c. permafrost; arctic tundra
d. taiga; arctic tundra

___ 9. _____ refers to a season of heavy rain that corresponds to a shift in prevailing winds over the Indian Ocean.
a. Geothermal ecosystem
b. Upwelling
c. Taiga
d. Monsoon

___10. All of the water above the continental shelves is in the _____ .
a. neritic zone of the benthic province
b. oceanic zone of the pelagic province
c. neritic zone of the pelagic province
d. oceanic zone of the benthic province

___11. Complex forests of ash, beech, birch, chestnut, elm, and oaks are found in the _____ .
a. tropical deciduous forest
b. monsoon forest
c. temperate deciduous forest
d. evergreen broadleaf forest

___12. Permafrost is characteristic of the _____ forest.
a. boreal
b. tundra
c. deciduous broadleaf
d. evergreen broadleaf

## Chapter Objectives/Review Questions

This section lists general and detailed chapter objectives that can be used as review questions. You can make maximum use of these items by writing answers on a separate sheet of paper. Fill in answers where blanks are provided. To check for accuracy, compare your answers with information given in the chapter or glossary.

| Page | | Objectives/Questions |
|---|---|---|
| (855) | 1. | The _____ is the entire realm in which organisms live. |
| (855) | 2. | _____ refers to average weather conditions, such as temperature, humidity, wind speed, cloud cover, and rainfall. |
| (856) | 3. | State the reason that most forms of life depend on the ozone layer. |
| (856) | 4. | _____ energy drives Earth's weather systems. |
| (856–857) | 5. | Be able to describe the causes of global air circulation patterns. |
| (856–857) | 6. | Describe how the tilt of the earth's axis affects annual variation in the amount of incoming solar radiation. |
| (858–859) | 7. | Atmospheric _____ patterns and oceanic _____ influence the distribution of different types of ecosystems. |
| (858) | 8. | Mountains, valleys, and other land formations influence _____ climates. |
| (859) | 9. | Describe the cause of the rain shadow effect. |
| (860) | 10. | Broadly, there are six distinct land realms, the _____ realms that were named by W. Sclater and Alfred Wallace. |
| (861) | 11. | Realms are divided into _____ . |
| (862) | 12. | _____ is a mixture of rock, mineral ions, and organic matter in some state of physical or chemical breakdown. |
| (862–869) | 13. | Be able to list the major biomes and briefly characterize them in terms of climate, topography, and organisms. |
| (863) | 14. | The wholesale conversion of grasslands and other productive biomes to dry wastelands is known as _____ . |
| (870) | 15. | A _____ is a standing body of freshwater with littoral, limnetic, and profundal zones. |
| (870) | 16. | Define plankton, phytoplankton, and zooplankton. |

(870–871) 17. Describe the spring and fall overturn in a lake in terms of causal conditions and physical outcomes.

(871) 18. _____ refers to nutrient enrichment of a lake or some other body of water

(871) 19. _____ lakes are often deep, poor in nutrients, and low in primary productivity; _____ lakes are often shallow, rich in nutrients, and high in primary productivity.

(871) 20. Describe a stream ecosystem.

(872) 21. Be able to fully describe the benthic and pelagic provinces of the ocean.

(872) 22. Within the pelagic province, all the water above the continental shelves is the _____ zone; the _____ zone is the water of the ocean basin.

(872) 23. As much as 70 percent of the ocean's primary productivity may be the contribution of _____ .

(872–873) 24. Describe the unusual hydrothermal vent ecosystems.

(874–875) 25. List the three types of coral reefs and be able to describe their formation.

(874–875) 26. List reasons that reefs are being ecologically threatened.

(876) 27. Be able to descriptively distinguish between an estuary and an intertidal zone.

(877) 28. State the significance of ocean upwelling.

(877–879) 29. Describe conditions of ENSO occurrence and its significance to the ocean.

---

## Integrating and Applying Key Concepts

One species, *Homo sapiens*, uses about 40 percent of all of Earth's productivity, and its representatives have invaded every biome, either by living there or by dumping waste products there. Many of Earth's residents are being denied the minimal resources they need to survive, while human populations continue to increase exponentially. Can you suggest a better way of keeping Earth's biomes healthy while providing for at least the minimal needs of all Earth's residents (not just humans)? If so, outline the requirements of such a system and devise a way in which it could be established.

# 50

# HUMAN IMPACT ON THE BIOSPHERE

*An Indifference of Mythic Proportions*

**AIR POLLUTION—PRIME EXAMPLES**
    Smog
    Acid Deposition

**OZONE THINNING—GLOBAL LEGACY OF AIR POLLUTION**

**WHERE TO PUT SOLID WASTES, WHERE TO PRODUCE FOOD**
    Oh Bury Me Not, In Our Own Garbage
    Converting Marginal Lands for Agriculture

**DEFORESTATION—CONCERTED ASSAULTS ON FINITE RESOURCES**

*Focus on Bioethics:* **YOU AND THE TROPICAL RAIN FOREST**

**TRADING GRASSLANDS FOR DESERTS**

**A GLOBAL WATER CRISIS**
    Consequences of Heavy Irrigation
    Water Pollution
    The Coming Water Wars

**A QUESTION OF ENERGY INPUTS**
    Fossil Fuels
    Nuclear Energy

**ALTERNATIVE ENERGY SOURCES**
    Solar-Hydrogen Energy
    Wind Energy
    Fusion Power

*Focus on Bioethics:* **BIOLOGICAL PRINCIPLES AND THE HUMAN IMPERATIVE**

---

## Interactive Exercises

---

*An Indifference of Mythic Proportions* (pp. 882–883)

## 50.1. AIR POLLUTION—PRIME EXAMPLES (pp. 884–885)

## 50.2. OZONE THINNING—GLOBAL LEGACY OF AIR POLLUTION (p. 886)

*Selected Words:* methyl bromide

*Boldfaced, Page-Referenced Terms*

(884) pollutants _____

_____

(884) thermal inversion _____

_____

(884) industrial smog _____

_____

(884) photochemical smog _____

_____

(884) PANs _____

_____

(884) dry acid deposition _____

_____

(884) acid rain _____

_____

(886) ozone thinning _____

_____

(886) chlorofluorocarbons _____

_____

## Fill-in-the-Blanks

(1) _____ are substances with which ecosystems have had no prior evolutionary experience. Adaptive mechanisms are not in place to deal with them. When a layer of dense, cool air gets trapped beneath a layer of warm air, the situation is known as a(n) (2) _____ _____ ; this has been a key factor in some of the worst air pollution disasters. Where (3) _____ are cold and wet, (4) _____ _____ develops as a gray haze over industrialized cities that burn coal and other fossil fuels for manufacturing, heating and generating electric power. In warm climates, (5) _____ _____ develops as a brown haze over large cities located in natural basins. The key culprit is nitric oxide. After release from vehicles, it reacts with (6) _____ in the air to form (7) _____ _____ . When (7) is exposed to sunlight, it reacts with hydrocarbons, and (8) _____ oxidants result. Most hydrocarbons come from spilled or partially burned (9) _____ . The main oxidants are ozone and (10) _____ (peroxyacyl nitrates). Even traces can sting eyes, irritate lungs, and damage crops. Oxides of (11) _____ and (12) _____ are among the worst pollutants. Coal-burning power plants, metal smelters, and factories emit most (13) _____ dioxides. Motor vehicles, gas- and oil-burning power plants, and (14) _____ -rich fertilizers produce (15 ) _____ oxides.

During dry weather, fine particles of oxides may be briefly airborne and then fall to Earth as dry (16) _____ _____ . When the oxides dissolve in atmospheric water, they form weak solutions of (17) _____ and (18) _____ acids. Strong winds may distribute them over great distances; when they fall to Earth in rain and snow, it is called wet acid deposition, or (19) _____ _____ . (19) can be 10 to 100 times more acidic than normal rainwater that has a pH of about (20) _____ . The deposited acids eat away at marble buildings, metals, rubber, plastics, and even nylon stockings. They also have the potential to disrupt the physiology of organisms and the chemistry of ecosystems. (21) _____ (CFCs) are compounds of chlorine, fluorine and carbon that are odorless and invisible. They are major factors in reduction of the ozone layer in the atmosphere.

*Choice*

For questions 22–43, choose from the following aspects of atmospheric pollution.

a. thermal inversion    b. industrial smog    c. photochemical smog    d. acid deposition
            e. chlorofluorocarbons    f. ozone layer

22.____Develops as a brown, smelly haze over large cities.

23.____Includes the "dry" and "wet" types.

24.____Contributes to ozone reduction more than any other factor.

25.____Where winters are cold and wet, this develops as a gray haze over industrialized cities that burn coal and other fossil fuels.

26.____Weather conditions trap a layer of cool, dense air under a layer of warm air.

27.____The cause of London's 1952 air pollution disaster, in which 4,000 died.

28.____Each year, from September through mid-October, it thins down at higher altitudes.

29.____Methyl bromide, a fungicide, will account for about 15 percent of its thinning if production does not stop.

30.____Intensifies a phenomenon called smog.

31.____Today most of this forms in cities of China, India, and other developing countries, as well as in Hungary, Poland, and other countries of eastern Europe.

32.____Contains airborne pollutants, including dust, smoke, soot, ashes, asbestos, oil, bits of lead and other heavy metals, and sulfur oxides.

33.____Depending on soils and vegetation cover, some regions are more sensitive than others to this.

34.____Have contributed to some of the worst local air pollution disasters.

35.____Chemically attack marble buildings, metals, mortar, rubber, plastic, and even nylon stockings.

36.____Its reduction allows more ultraviolet radiation to reach the earth's surface.

37.____Reaches harmful levels where the surrounding land forms a natural basin, as it does around Los Angeles and Mexico City.

38.____Tall smokestacks were added to power plants and smelters in an unsuccessful attempt to solve this problem.

39.____Those already in the air will be there for one or two centuries before they are neutralized by natural processes.

40.____A dramatic rise in skin cancers, eye cataracts, immune system weakening, and harm to photosynthesizers is related to its reduction.

41.____Widely used as propellants in aerosol spray cans, refrigerator coolants, air conditioners, industrial solvents, and plastic foams; enter the atmosphere slowly and resist breakdown.

42.____The main culprit is nitric oxide, produced mainly by cars and other vehicles with internal combustion engines.

43.____Oxides of sulfur and nitrogen dissolve in water to form weak solutions of sulfuric acid and nitric acid that may fall with rain or snow.

## 50.3. WHERE TO PUT SOLID WASTES, WHERE TO PRODUCE FOOD (p. 887)

## 50.4. DEFORESTATION—CONCERTED ASSAULTS ON FINITE RESOURCES (pp. 888–889)

## 50.5. *Focus on Bioethics:* YOU AND THE TROPICAL RAIN FOREST (p. 890)

*Selected Words:* subsistence agriculture, animal-assisted agriculture, mechanized agriculture

### Boldfaced, Page-Referenced Terms

(887) green revolution _____

_____

(888) deforestation _____

_____

(888) shifting cultivation _____

_____

### Matching

Choose the most appropriate answer for each.

1. ___throwaway mentality
2. ___green revolution
3. ___shifting cultivation
4. ___animal-assisted agriculture
5. ___deforestation
6. ___recycling
7. ___watersheds of forested regions
8. ___subsistence agriculture
9. ___new genetic resources
10. ___mechanized agriculture

A. Agriculture in developing countries runs on energy inputs from sunlight and human labor
B. An affordable, technologically feasible alternative to "throwaway technology"
C. Act like giant sponges that absorb, hold, and gradually release water
D. Potential benefits to be obtained by genetic engineering and tissue culture methods in the rainforests
E. Runs on energy inputs from oxen and other draft animals
F. An attitude prevailing in the United States and other developed countries that greatly adds to solid waste accumulation
G. Requires massive inputs of fertilizers, pesticides, fossil fuel energy, and ample irrigation to sustain high-yield crops
H. Research directed toward improving crop plants for higher yields and exporting modern agricultural practices and equipment to developing countries
I. Trees are cut and burned, then ashes tilled into the soil; crops are grown for one to several seasons on quickly leached soils that become infertile
J. Removal of all trees from large land tracts; leads to loss of fragile soils and disrupts watersheds; greatest today in Brazil, Indonesia, Columbia, and Mexico

## 50.6. TRADING GRASSLANDS FOR DESERTS (p. 891)

## 50.7. A GLOBAL WATER CRISIS (pp. 892–893)

*Selected Words:* primary, secondary, and tertiary wastewater treatment

## Boldfaced, Page-Referenced Terms

(891) desertification _____

_____

(892) desalination _____

_____

(892) salination _____

_____

(892) water table _____

_____

(893) wastewater treatment _____

_____


## Dichotomous Choice

Circle one of two possible answers given between parentheses in each statement.

1. Conversion of large tracts of grasslands, rain-fed or irrigated croplands to a more barren state is known as (subsistence agriculture/desertification).
2. Presently, (too many cattle in the wrong places/overgrazing on marginal lands) is the main cause of large-scale desertification.
3. In Africa, (domestic cattle/native wild herbivores) trample grasses and compact the soil surfaces as they wander about looking for water.
4. A 1978 study by biologist David Holpcraft demonstrated that African range conditions improved in land areas where (domestic cattle/native wild herbivores) were ranched.
5. Without irrigation and conservation practices, grasslands that were converted for agriculture often end up as (deserts/forested watersheds).


## Fill-in-the-Blanks

The supply of Earth's seawater is essentially unlimited. The removal of salt from seawater is called (6) _____ . For most countries this process is not practical due to the costly fuel (7) _____ necessary to drive it. Large-scale (8) _____ accounts for nearly two-thirds of the human population's use of freshwater. Irrigation of otherwise useless soil can cause a salt buildup, or (9) _____ due to evaporation in areas of poor soil drainage. Land that drains poorly also becomes waterlogged and raises the (10) _____ _____ . When the (10) is too close to the ground's surface, soil becomes saturated with (11) _____ water, which can damage plant roots. A large problem is the fact that water tables are subsiding. For example, overdrafts have depleted half of the Ogallala aquifer that supplies irrigation water for (12) [choose one] ☐ 40 ☐ 20 percent of the croplands in the United States. Human sewage, animal wastes, toxic chemicals, agricultural runoff, sediments, pesticides, and plant nutrients are all sources of water (13) _____ that amplifies the problem of water scarcity. There are three levels of (14) _____ treatment. They are primary, secondary, and (15) _____ treatments. Most of the (14) is not treated adequately. If the current rates of population growth and water depletion hold, the amount of freshwater available for each person on the planet will be (16) [choose one] ☐ 45–56 ☐ 55–66 ☐ 65–76 percent less than it was in 1976. In the future, water, not (17) _____ , may become the most important fluid.  Unless long overdue planning for the future occurs, wars over water rights are likely to occur.

## Complete the Table

18. Complete the following table, which summarizes three levels of treatment methods for maintaining the water quality of polluted wastewater.

| Treatment Method | Description |
|---|---|
| a. | Primary treatment |
| b. | Secondary treatment |
| c. | Tertiary treatment |

## 50.8. A QUESTION OF ENERGY INPUTS (pp. 894–895)

## 50.9. ALTERNATIVE ENERGY SOURCES (p. 896)

## 50.10. *Focus on Bioethics:* BIOLOGICAL PRINCIPLES AND THE HUMAN IMPERATIVE (p. 897)

*Selected Words: total* energy, *net* energy, supertanker *Valdez*

## Boldfaced, Page-Referenced Terms

(894)  fossil fuels _____

_____

(894)  meltdown _____

_____

(896)  solar-hydrogen energy _____

_____

(896)  fusion power _____

_____

## Dichotomous Choice

Circle one of two possible answers given between parentheses in each statement.

1. Paralleling the (S-shaped/J-shaped) curve of human population growth is a steep rise in total and per capita energy consumption.
2. The increase in per capita energy consumption is due to (increased numbers of energy users and to extravagant consumption and waste/energy used to locate, extract, transport, store, and deliver energy to consumers).
3. (Total energy/Net energy) is that left over after subtracting the energy used to locate, extract, transport, store, and deliver energy to consumers.

4. Fossil fuels are the carbon-containing remains of (plants/plants and animals) that lived hundreds of millions of years ago.
5. Even with strict conservation efforts, known petroleum and natural gas reserves may be used up during the (current/next) century.
6. The net energy (decreases/increases) as costs of extraction and transportation to and from remote areas increase.
7. World coal reserves can meet human energy needs for several centuries, but burning releases sulfur dioxides into the atmosphere and adds to the global problem of (photochemical smog/acid deposition).
8. By 1990 in the United States, it cost slightly more to generate electricity by nuclear energy than by using coal but today, it costs (more/less).
9. The danger in the use of radioactivity as an energy supply during normal operation is with potential (radioactivity escape/meltdown).
10. After nearly fifty years of research, scientists (have/have not) agreed on the best way to store high-level radioactive wastes.
11. When electrodes in photovoltaic cells exposed to sunlight produce an electric current to split water molecules into oxygen and hydrogen gas (potential fuels), it is known as (fusion power/solar-hydrogen energy).
12. California gets 1 percent of its electricity from (fusion power/wind farms).
13. (Solar-hydrogen energy/Fusion power) involves a mimic of a process occurring in the sun's environment and may provide a good energy source in about fifty years.

## Sequence-Classify

Arrange the consumption of world resources in correct hierarchical order. Enter the letter of the energy source of highest consumption next to 14, the letter of the next highest next to 15, and so on. Enter an (N) in the parentheses following the letter of the resource if it is nonrenewable and an (R) if the resource is renewable.

14. ___ (  )   A. Hydropower, geothermal, solar

15. ___ (  )   B. Natural gas

16. ___ (  )   C. Oil

17. ___ (  )   D. Nuclear power

18. ___ (  )   E. Biomass

19. ___ (  )   F. Coal

---

# Self-Quiz

___ 1. Which of the following processes is not generally considered a component of secondary wastewater treatment.
   a. screens and settling tanks remove sludge
   b. microbial populations are used to break down organic matter
   c. removal of all nitrogen, phosphorus, and toxic substances
   d. chlorine is often used to kill pathogens in the water

2. When fossil-fuel burning gives dust, smoke, soot, ashes, asbestos, oil, bits of lead, other heavy metals, and sulfur oxides, it forms _____ .
   a. photochemical smog
   b. industrial smog
   c. a thermal inversion
   d. both a and c

___ 3. _____ result(s) when nitrogen dioxide and hydrocarbons react in the presence of sunlight.
   a. Photochemical smog
   b. Industrial smog
   c. A thermal inversion
   d. Both a and c

___ 4. When weather conditions trap a layer of cool, dense air under a layer of warm air, _____ occurs.
   a. photochemical smog
   b. a thermal inversion
   c. industrial smog
   d. acid deposition

___ 5. The most abundant fossil fuel in the United States is _____ .
   a. carbon monoxide
   b. oil
   c. natural gas
   d. coal

___ 6. Sulfur and nitrogen dioxides dissolve in atmospheric water to form a weak solution of sulfuric acid and nitric acid; this describes _____ .
   a. photochemical smog
   b. industrial smog
   c. ozone and PANs
   d. acid rain

___ 7. Which of the following statements is false?
   a. Ozone reduction allows more ultraviolet radiation to reach the earth's surface.
   b. CFCs enter the atmosphere and resist breakdown.
   c. Salination of soils aids plant growth and increases yields.
   d. CFCs already in the air will be there for over a century.

___ 8. "Adequately reduces pollution but is largely experimental and expensive," describes _____ wastewater treatment.
   a. quaternary
   b. secondary
   c. tertiary
   d. primary

___ 9. "Photovoltaic cells exposed to sunlight produce an electric current that splits water molecules into oxygen and hydrogen gas" refers to _____ .
   a. fusion power
   b. wind energy
   c. water power
   d. solar-hydrogen energy

___10. Energy inputs from sunlight and human labor best describes _____ .
   a. animal-assisted agriculture
   b. subsistence agriculture
   c. the green revolution
   d. mechanized agriculture

## Chapter Objectives/Review Questions

| Page | | Objectives/Questions |
|---|---|---|
| (884) | 1. | Identify the principal air pollutants, their sources, their effects, and the possible methods for controlling each pollutant. |
| (884) | 2. | During a _____ _____ , weather conditions trap a layer of cool, dense air under a layer of warm air; trapped pollutants may reach dangerous levels. |
| (884) | 3. | Distinguish photochemical smog from industrial smog. |
| (884) | 4. | Explain what acid rain does to an ecosystem. Contrast those effects with the action of CFCs. |
| (885) | 5. | Be able to discuss the significance of the effects of the thinning of the ozone layer to life on Earth. |
| (885) | 6. | List the key sources of air pollutants. |
| (887) | 7. | Under the banner of the _____ _____ research has been directed toward improving the genetics of crop plants for higher yields and exporting. |
| (887) | 8. | _____ agriculture runs on energy inputs from sunlight and human labor; _____ - _____ agriculture runs on energy inputs from oxen and other draft animals; _____ agriculture requires massive inputs of fertilizers, pesticides, and ample irrigation to sustain high-yield crops. |

(888)    9. Explain the repercussions of deforestation that are evident in soils, water quality, and genetic diversity in general.

(888–889) 10. _____ _____ involves cutting and burning trees, tilling ashes with soil, plant crops from one to several seasons, and then abandon the clear plots.

(890)   11. Examine the effects that modern agriculture has wrought on desert and grassland ecosystems.

(891)   12. The name for the conversion of large tracts of natural grasslands to a more desertlike state is _____ .

(892)   13. Explain why desalination is not a practical solution for the shortage of freshwater.

(893)   14. Define *primary*, *secondary*, and *tertiary* wastewater treatment and list some of the methods used in each of the three types of treatment.

(893)   15. Explain the meaning of, "the coming water wars."

(894)   16. _____ energy refers to the amount left over after subtracting the energy that is used to locate, extract, transport, store, and deliver energy to consumers.

(894)   17. Be able to describe the dangers accompanying a meltdown.

(894–895) 18. Describe how our use of fossil fuels, solar energy, and nuclear energy affects ecosystems.

(897)   19. Characterize the magnitude of pollution problems in the United States and the entire Earth.

(897)   20. List five ways in which you could become personally involved in ensuring that institutions serve the public interest in a long-term, ecologically sound way.

## Integrating and Applying Key Concepts

If you were Ruler of All People on Earth, how would you encourage people to depopulate the cities and adopt a way of life by which they could supply their own resources from the land and dispose of their own waste products safely on their own land?

Explain why some biologists believe that the endangered species list now includes all species.

# 51

# AN EVOLUTIONARY VIEW OF BEHAVIOR

## Interactive Exercises

*Deck the Nest With Sprigs of Green Stuff* (pp. 900–901)

### 51.1. THE HERITABLE BASIS OF BEHAVIOR (pp. 902–903)

### 51.2. LEARNED BEHAVIOR (p. 904)

*Selected Words:* <u>Sturnus</u> <u>vulgaris</u>, *intermediate* response

*Boldfaced, Page-Referenced Terms*

(903) song system _____

_____

(903) instinctive behavior _____

_____

(903) sign stimuli _____

_____

(903) fixed action pattern _____

_____

(904) learned behavior _____

_____

(904) imprinting _____

_____

## Matching

Choose the most appropriate answer for each.

1. ___intermediate response
2. ___song system
3. ___instinctive behavior
4. ___sign stimuli
5. ___fixed action pattern
6. ___learned behavior
7. ___imprinting

A. Animals process and integrate information gained from specific experiences
B. Well-defined environmental cues that trigger suitable responses
C. Term applied to genetically-based behavioral reactions of hybrid offspring
D. Time-dependent form of learning; triggered by exposure to sign stimuli and usually occurring during sensitive periods of young animals
E. A behavior performed without having been learned by actual environmental experience
F. Consists of several parts of the brain that will govern activities of the muscles of a local organ; these brain regions differ noticeably in size and structure between male and female birds
G. A program of coordinated muscle activity that runs to completion independently of feedback from the environment

## Dichotomous Choice

Circle one of two possible answers given between parentheses in each statement.

8. For garter snake populations living along the California coast, the food of choice is (the banana slug/tadpoles and small fishes).
9. In Stevan Arnold's experiments, newborn garter snakes that were offspring of coastal parents usually (ate/ignored) a chunk of slug as the first meal.
10. Newborn garter snake offspring of (coastal/inland) parents ignored cotton swabs drenched in essence of slug and only rarely ate the slug meat.
11. The differences in the behavioral eating responses of coastal and inland snakes (were/were not) learned.
12. Hybrid garter snakes with coastal and inland parents exhibited a feeding response that indicated a(n) (environmental/genetic) basis for this behavior.
13. In zebra finches and some other songbirds, singing behavior is an outcome of seasonal differences in the secretion of melatonin, a hormone secreted by the (gonads/pineal gland).
14. In spring, melatonin secretion is suppressed and gonads grow; their secretion of estrogen and testosterone is increased to (indirectly/directly) influence singing behavior.
15. Even before a male bird hatches, a high (estrogen/testosterone) level triggers development of a masculinized brain.
16. Later, at the start of the breeding season, a male's enlarged gonads secrete even more (estrogen/testosterone) which acts on cells in the sound system to prepare the bird to sing when properly stimulated.
17. (Hormones/Genes) influence the organization and activation of mechanisms required for particular forms of behavior.

## Complete the Table

18. Complete the following table to consider examples of instinctive and learned behavior.

| Category | Examples |
|---|---|
| a. Instinctive behavior | |
| b. Learned behavior | |

## Matching

Choose the most appropriate answer for each category of learned behavior.

19. ___imprinting

20. ___classical conditioning

21. ___operant conditioning

22. ___habituation

23. ___spatial or latent learning

24. ___insight learning

A. Birds living in cities learn not to flee from humans or cars, which pose no threat to them.

B. Chimpanzees abruptly stack several boxes and use a stick to reach suspended bananas out of reach.

C. In response to a bell, dogs salivate even in the absence of food.

D. Bluejays store information about dozens or hundreds of places where they have stashed food.

E. Baby geese formed an attachment to Konrad Lorenz if separated from the mother shortly after hatching.

F. A toad learns to avoid stinging or bad-tasting insects after attempts to eat them.

## 51.3. THE ADAPTIVE VALUE OF BEHAVIOR (p. 905)

### Boldfaced, Page-Referenced Terms

(905) natural selection _____

_____

(905) reproductive success _____

_____

(905) adaptive behavior _____

_____

(905) social behavior _____

_____

(905) selfish behavior _____

_____

(905) altruistic behavior _____

_____

(905) territory _____

_____

## Fill-in-the-Blanks

(1) _____ _____ is the result of differences in survival and reproduction among individuals of a population that differ from one another in heritable traits. Some versions of a trait are better than others at helping the individual survive and reproduce. The frequency of the helpful (2) _____ increase in a population while the others do not. Using the theory of (3) _____ by natural selection as a starting point, one should be able to identify (4) _____ forms of behavior—and to discern how they bestow (5) _____ benefits that offset reproductive costs or disadvantages that might be associated with them. Solitary animals or groups of animals may be studied. If the behavior is adaptive, it must promote the survival and production of offspring of the (6) _____ .

## Complete the Table

7. Complete the following table of terms describing forms of individual adaptive behavior.

| Behavior | Description |
| --- | --- |
| a. Reproductive success | |
| b. Adaptive behavior | |
| c. Social behavior | |
| d. Selfish behavior | |
| e. Altruistic behavior | |

## 51.4. COMMUNICATION SIGNALS (pp. 906–907)

**Selected Words:** *signaling* pheromones, *priming* pheromones

### Boldfaced, Page-Referenced Terms

(906) communication signals _____

_____

(906) signaler _____

_____

(906) signal receivers _____

_____

(906) pheromones _____

_____

(906) composite signal _____

_____

(906) communication display _____

_____

(906) threat display _____

_____

(907) courtship displays _____

_____

(907) tactile displays _____

_____

(907) illegitimate receiver _____

_____

(907) illegitimate signalers _____

_____

## Choice

For questions 1–10, choose from the following:

a. chemical signal, signaling pheromones    b. chemical signal, priming pheromones    c. acoustical signal
d. composite signal    e. communication display, social signal    f. communication display, threat display
g. communication display, courtship    h. communication display, tactile signal    i. illegitimate signaler
j. illegitimate receiver

1. ____ Assassin bugs hook dead termite bodies on their dorsal surfaces and acquire termite scent; this deception allows assassin bugs to hunt termite victims more easily.

2. ____ Male songbirds sing to secure territory and attract a female

3. ____ The play bow of dogs and wolves.

4. ____ Bombykol molecules released by female silk moths serve as sex attractants.

5. ____ A male bird might emit calls while bowing low, as if to peck the ground for food.

6. ____ Termites act defensively when detecting scents from invading ants whose scent signals are meant to elicit cooperation from other ants.

7. ____ A dominant male baboon's "yawn" that exposes large canines.

8. ____ A volatile odor in the urine of certain male mice triggers and enhances estrus in female mice.

9. ____ After finding a source of pollen or nectar, a foraging honeybee returns to its colony, a hive and performs a complex dance

10. ____ Tungara frogs issue nighttime calls to females and rival males; a "whine" followed by a "chuck."

11. ____ A male firefly has a light-generating organ that emits a bright, flashing signal.

12. ____ Ears laid back against the head of a zebra convey hostility, but ears pointing up convey its absence; a zebra with laid-back ears isn't too riled up when its mouth is open just a bit, but when the mouth is gaping, watch out.

## 51.5. MATING, PARENTING, AND INDIVIDUAL REPRODUCTIVE SUCCESS
(pp. 908–909)

*Selected Words:* Harpobittacus, Centrocercus

*Boldfaced, Page-Referenced Terms*

(908) sexual selection _____

(909) lek _____

### Complete the Table

1. Complete the following table to supply the common names of the animals that fit the text examples of sexual selection.

| Animals | Descriptions of Sexual Selection |
|---|---|
| a. | Females select the males that offer them superior material goods; females permit mating only after they have eaten the "nuptial gift" for about five minutes. |
| b. | Males congregate in a lek or communal display ground; each male stakes out a few square meters as his territory; females are attracted to the lek to observe male displays and usually select and mate with only one male. |
| c. | Females of a species cluster in defendable groups at a time they are sexually receptive; males compete for access to the clusters; combative males are favored. |
| d. | Extended parental care improves the likelihood that the current generation of offspring will survive; this behavior comes at a reproductive cost to the adults. |

## 51.6. BENEFITS OF LIVING IN SOCIAL GROUPS (pp. 910–911)

## 51.7. COSTS OF LIVING IN SOCIAL GROUPS (p. 912)

## 51.8. EVOLUTION OF ALTRUISM (p. 913)

## 51.9. *Focus on the Social Environment:* ABOUT THOSE SELF-SACRIFICING INSECTS
(pp. 914–915)

## 51.10. *Focus on Science:* ABOUT THOSE NAKED MOLE RATS (p. 916)

*Selected Words:* Parus, caring for relatives, *indirect* genetic contribution, *DNA fingerprinting*

*Boldfaced, Page-Referenced Terms*

(910) cost-benefit approach _____

_____

(910) selfish herd _____

_____

(911) dominance hierarchy _____

_____

(913) theory of indirect selection _____

_____

## Matching

Choose the most appropriate answer for each.

1. ___cost-benefit approach
2. ___disadvantages to sociality
3. ___dominance hierarchy
4. ___cooperative predator avoidance
5. ___the selfish herd

A. Competition for resources, rapid depletion of food resources, cannibalism, and greater vulnerability to disease
B. A simple society brought together by reproductive self-interest; larger, more powerful male bluegills tend to claim the central locations
C. A consideration of social life in terms of reproductive success of the individual
D. Adult musk oxen form a circle around their young while they face outward, and the "ring of horns" successfully deters the wolves; writhing, regurgitating reaction of Australian sawfly caterpillars to a disturbance
E. Some individuals of a baboon adopt a subordinate status with respect to the others

## Complete the Table

6. Complete the following table to supply examples of each listed category of the costs of living in social groups.

| Costs of Living in Social Groups | Examples |
| --- | --- |
| a. An increase in the competition for food | |
| b. Encourages the spread of contagious diseases and parasites | |
| c. Risks of being killed or exploited by others in the group | |

## Fill-in-the-Blanks

A (7) _____ animal that gives way to a dominant one is acting in its own self-interest. Such (8) _____ animals are found among many vertebrate groups. A female wolf's (9) _____ success increases when she monopolizes the benefits that males offer. Within a wolf pack, there is usually a (10) _____ breeding female and male. The other wolves are (11) _____ sisters, aunts, brothers and uncles. Their (12) _____ behavior is to hunt and bring back food to members that remain in the den and guard the pups. Altruistic behavior is most extreme in certain (13) _____ societies. When a (14) _____ bee plunges its stinger into an invader of the hive, it commits suicide. Altruistic individuals of a social group do not contribute their (15) _____ to the next generation but yet seem to perpetuate the genetic basis for their altruistic behavior over evolutionary time. According to William Hamilton's theory of (16) _____ _____ , those genes associated with caring for relatives—not one's direct descendants—tend to be favored in certain situations. This form of altruism can be thought of as an extension of (17) _____ . For example, if an uncle helps his niece survive long enough to reproduce, he has made an (18) _____ genetic contribution to the next generation, as measured in terms of the genes that he and his relatives share. Altruism costs him; he may lose his own opportunities to (19) _____ . Similarly, nonbreeding workers in insect societies indirectly promote their "self-sacrifice" genes through (20) _____ behavior directed toward relatives. Thus, when a guard bee drives her stinger into a raccoon, she inevitably dies—but her siblings in the hive will perpetuate some of her (21) _____ . Sterility and extreme self-sacrifice are rare among social groups of (22) _____ . Unlike any other known vertebrate, the highly social naked mole-rat individuals live out their lives as (23) _____ helpers in their social group. In each mole-rat clan, there is a single reproducing female, and she mates with one to three males. All other members of the clan care for the "queen" and "king" (or kings) and their offspring. A self-sacrificing naked mole-rat helps to (24) _____ a very high proportion of the forms of genes that it carries. As it turns out, the (25) _____ of helpers and the helped might be as much as 90 percent identical.

## 51.11. AN EVOLUTIONARY VIEW OF HUMAN SOCIAL BEHAVIOR (p. 917)

*Selected Words: redirected* adaptive behaviors

*Boldfaced, Page-Referenced Terms*

(917) adoption _____

_____

### *Dichotomous Choice*

Circle one of two possible answers given between parentheses in each statement.

1. Many people seem to believe that attempts to identify the adaptive value of a particular (animal/human) trait is an attempt to define its moral or social advantage.
2. "Adaptive" refers to (a trait with moral value/a trait valuable in gene transmission).
3. In many species, adults that have lost their offspring will, if presented with a substitute, (adopt/reject) it.
4. John Alcock suggests that husbands and wives who have lost an only child or who fail to produce children themselves should be especially prone to adopt (strangers/relatives).
5. The human adoption process (can/cannot) be considered adaptive when indirect selection favors adults who direct parenting assistance to relatives.
6. Joan Silk showed that in some traditional societies, children (are not/are) adopted overwhelmingly by relatives.
7. In large, industrialized societies in which agencies and other means of adoption exist, adoption of relatives (is/is not) predominant.
8. It (is/is not) possible to test evolutionary hypotheses about the adaptive value of human behaviors.
9. *Adaptive* behavior and *socially desirable* behavior (are not/are) separate issues.

10. Strong parenting mechanisms evolved in the past, and it may be that their redirection toward a (relative/nonrelative) says more about human evolutionary history than it does about the transmission of one's genes.

# Self-Quiz

___ 1. The observable, coordinated responses that animals make to stimuli are what we call _____ .
   a. imprinting
   b. instinct
   c. behavior
   d. learning

___ 2. In _____ , components of the nervous system allow an animal to carry out complex, stereotyped responses to certain environmental cues, which are often simple.
   a. natural selection
   b. altruistic behavior
   c. sexual selection
   d. instinctive behavior

___ 3. Newly hatched goslings follow any large moving objects to which they are exposed shortly after hatching; this is an example of _____ .
   a. homing behavior
   b. imprinting
   c. piloting
   d. migration

___ 4. A young toad flips its sticky-tipped tongue and captures a bumblebee that stings its tongue; in the future, the toad leaves bumblebees alone. This is _____ .
   a. instinctive behavior
   b. a fixed reaction pattern
   c. altruistic
   d. learned behavior

___ 5. Pavlov's dog experiments represent an example of _____ .
   a. classical conditioning
   b. latent learning
   c. selfish behavior
   d. habituation

___ 6. _____ provides an example of an illegitimate signaler.
   a. A soldier termite killing an ant on cue
   b. An assassin bug with acquired termite odor
   c. A termite pheromone alarm signal
   d. The "yawn" of a dominant male baboon

___ 7. The claiming of the more protected central locations of the bluegill colony by the largest, most powerful males suggests _____ .
   a. cooperative predator avoidance
   b. the selfish herd
   c. a huge parent cost
   d. self-sacrificing behavior

___ 8. A chemical odor in the urine of male mice triggers and enhances estrus in female mice. The source of stimulus for this response is a _____ .
   a. generic mouse pheromone
   b. signaling pheromone
   c. priming pheromone
   d. cue from male mice

___ 9. Female insects often attract mates by releasing sex pheromones. This is an example of a(n) _____ signal.
   a. chemical
   b. visual
   c. acoustical
   d. tactile

___10. Male birds sing to stake out territories, attract females, and discourage males. This is an example of a _____ signal.
   a. chemical
   b. visual
   c. acoustical
   d. tactile

___11. An example of dominance hierarchy and self-sacrificing behavior is _____ .
   a. cannibalistic behavior of a breeding pair of herring gulls in a huge nesting colony
   b. members of wolf packs helping others by sharing food or fending off predators even though they do not breed
   c. clumps of regurgitating Australian sawfly caterpillars
   d. a huge colony of prairie dogs being ravaged by a parasite

___12. When musk oxen form a "ring of horns" against predators, it is _____ .
   a. a selfish herd
   b. cooperative predator avoidance
   c. self-sacrificing behavior
   d. dominance hierarchy

___13. Caring for nondescendant relatives favors the genes associated with helpful behavior and is classified as _____ .
   a. dominance hierarchy
   b. indirect selection
   c. altruism
   d. both b and c

___14. "_____" means only that a given trait has proved beneficial in the transmission of individual's genes.
   a. Dominance hierarchy
   b. Indirect selection
   c. Adaptive
   d. Altruism

___15. _____ also favors adults who direct parenting behavior toward relatives and so indirectly perpetuate their shared genes.
   a. Indirect selection
   b. Moral selection
   c. Redirected selection
   d. Perpetuated selection

# Chapter Objectives/Review Questions

This section lists general and detailed chapter objectives that can be used as review questions. You can make maximum use of these items by writing answers on a separate sheet of paper. Fill in answers where blanks are provided. To check for accuracy, compare your answers with information given in the chapter or glossary.

| Page | | Objectives/Review Questions |
|---|---|---|
| (902) | 1. | What explains the fact that coastal and inland garter snakes of the same species have different food preferences? |
| (902) | 2. | Describe the intermediate response obtained in Arnold's experiment with coastal and inland garter snakes. |
| (903) | 3. | Describe the origin and formation of a song system. |
| (903) | 4. | Tongue-flicking, body orientation, and strikes at prey by newborn garter snakes are good examples of _____ behavior. |
| (903) | 5. | Define *sign stimuli*. |
| (903) | 6. | Describe and cite an example of a fixed action pattern. |
| (904) | 7. | When animals incorporate and process information gained from specific experiences and then use the information to vary or change responses to stimuli, it is _____ behavior. |
| (904) | 8. | Define each of the following categories of learned behavior and give one example of each: imprinting, classical conditioning, operant conditioning, habituation, spatial or latent learning, and insight learning. |
| (905) | 9. | What is meant by reproductive success? |
| (905) | 10. | _____ behavior is any behavior by which an individual protects or increases its own chance of producing offspring, regardless of the consequences for the group to which it belongs. |
| (905) | 11. | _____ behavior refers to the cooperative, interdependent relationships among individuals of the species. |

| | | |
|---|---|---|
| (905) | 12. | Distinguish between selfish behavior and altruistic behavior. |
| (905) | 13. | A _____ is an area that one or more individuals defend against competitors. |
| (906) | 14. | Examples of _____ signals are chemical, visual, acoustical, and tactile. |
| (906) | 15. | Define the roles of signalers and signal receivers. |
| (906) | 16. | Distinguish between signaling and priming pheromones and cite an example of each. |
| (906) | 17. | A _____ signal is illustrated by a zebra with laid-backears and a gaping mouth. |
| (906) | 18. | Describe one example of a threat display. |
| (907) | 19. | Ritualization is often developed to an amazing degree in _____ displays between potential mates. |
| (907) | 20. | An example of a _____ signal is the physical contact of bees in a hive maintaining physical contact during the dance of a returning foraging bee. |
| (907) | 21. | When soldier termites detect ant scents meant for other ants and kill ants on cue, the termites are said to be _____ _____ of a signal meant for individuals of a different species. |
| (907) | 22. | Assassin bugs covered with termite scent are able to use deception to hunt termite victims more easily and as such are acting as _____ signalers. |
| (907) | 23. | Natural selection tends to favor communication signals that promote _____ success. |
| (908) | 24. | Competition among members of one sex for access to mates and selection of mates are the result of a microevolutionary process called _____ _____ . |
| (908–909) | 25. | Be able to discuss mate selection processes by female hangingflies and the female sage grouse. |
| (909) | 26. | List the costs and benefits of parenting in the example of adult Caspian terns. |
| (910) | 27. | Explain the "cost-benefit approach" that evolutionary biologists utilize to find answers to the questions about social life. |
| (910) | 28. | Studies of Australian sawfly caterpillars indicate _____ predator avoidance. |
| (910–911) | 29. | Define selfish herd; cite an example. |
| (911) | 30. | Members of baboon troops recognize a _____ _____ in which some individuals have adopted a subordinate status with respect to the others. |
| (912) | 31. | List disadvantages to sociality. |
| (913) | 32. | Hamilton's theory of _____ selection relates to caring for nondescendant relatives and how this favors genes associated with helpful behavior. |
| (913) | 33. | With _____ behavior, an individual behaves in a self-sacrificing way that helps others but decreases its own chance of reproductive success. |
| (914–915) | 34. | Be able to discuss the self-sacrificing behavior of termites in a eucalyptus forest in Australia and honeybees. |
| (916) | 35. | Explain how DNA fingerprinting was used to establish that self-sacrificing mole-rats help to perpetuate a very high proportion of genes (alleles) that it carries—even though they are not the reproducing mole-rats. |
| (917) | 36. | It is possible to test evolutionary hypotheses about the _____ value of human behaviors; discuss human adoption practices in this context. |
| (917) | 37. | "_____" means only that a given trait has proved beneficial in the transmission of an individual's genes. |

## Integrating and Applying Key Concepts

Think about communication signals that humans use and list them. Do you believe a dominance hierarchy exists in human society? Think of examples.

# ANSWERS

## Chapter 1  Methods and Concepts in Biology

**Biology Revisited** (pp. 2–3)
**1.1. DNA, Energy, and Life** (pp. 4–5)
1. cells;  2. DNA;  3. proteins;  4. amino;  5. Enzymes;
6. RNAs;  7. RNAs;  8. proteins;  9. reproduction;
10. Energy;  11. transfer;  12. Metabolism;  13. photosynthesis;  14. ATP;  15. respiration;  16. receptors;
17. internal  18. homeostatis.

**1.2. Energy and Life's Organization** (pp. 6–7)
1. G;  2. F;  3. I;  4. L;  5. H;  6. K;  7. C;  8. N;  9. B;
10. O;  11. D;  12. E;  13. M;  14. A;  15. J;  16. D;  17. G;
18. B;  19. J;  20. L;  21. E;  22. M;  23. A;  24. I;  25. F;
26. C;  27. H;  28. K;  29. producers;  30. consumers;
31. Decomposers;  32. energy;  33. cycling.

**1.3. So Much Unity, Yet so Many Species** (pp. 8–9)
1. species;  2. genus;  3. genus;  4. species;  5. family;
6. order;  7. class;  8. phylum;  9. prokaryotic;  10. eukaryotic;  11. a. Plantae;  b. Eubacteria;  c. Fungi;
d. Archaebacteria;  e. Protista;  f. Animalia;  12. D;
13. F;  14. A;  15. E;  16. B;  17. C;  18. G.

**1.4. An Evolutionary View of Life's Diversity** (pp. 10–11)
1. a;  2. b;  3. b;  4. a;  5. b;  6. b;  7. a;  8. a;  9. b;  10. b.

**1.5. The Nature of Biological Inquiry** (pp. 12–13)
**1.6.** *Focus on Science:* **The Power and Pitfalls of Experimental Tests** (pp. 14–15)
**1.7. The Limits of Science** (p. 16)
1. G;  2. A;  3. D;  4. B;  5. E;  6. C;  7. F;  8. O;  9. O;
10. C;  11. O;  12. C;  13. a. Hypothesis;  b. Prediction;
c. Theory;  d. Scientific experiment;  e. Control group;
f. Variable;  g. Inductive logic;  h. Deductive logic;
14. experiment;  15. control;  16. hypothesis;  17. prediction;  18. belief;  19. inductive logic;  20. deductive logic;
21. variable;  22. test predictions;  23. cause and effect;
24. sampling error;  25. quantitative;  26. subjective;
27. supernatural;  28. conviction.

**Self-Quiz**
1. d;  2. a;  3. b;  4. c;  5. d;  6. b;  7. d;  8. a;  9. c;  10. e.

## Chapter 2  Chemical Foundations for Cells

**Leafy Clean-Up Crews** (pp. 20–21)
**2.1. Regarding the Atoms** (pp. 22–23)
**2.2.** *Focus on Science:* **Using Radioisotopes to Date the Past, Track Chemicals, and Save Lives** (pp. 24–25)
1. F;  2. O;  3. D;  4. K;  5. P;  6. C;  7. H;  8. I;  9. M;
10. N;  11. G;  12. J;  13. B;  14. E;  15. A;  16. L;  17.
equation;  18. formula;  19. yields;  20. reactants;  21.
products;  22. 12;  23. 98 grams.

**2.3. The Nature of Chemical Bonds** (pp. 26–27)
1. C;  2. D;  3. H;  4. G;  5. A;  6. E;  7. B;  8. F;  9. a.
Calcium, Ca;  b. Carbon, C;  c. Chlorine, Cl;  d. Hydrogen, H;  e. Sodium, Na;  f. Nitrogen, N;  g. Oxygen, O;

10.

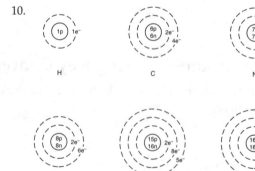

## 2.4. Important Bonds in Biological Molecules
(pp. 28–29)

1.

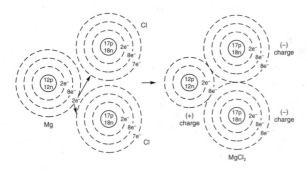

MgCl₂

2.

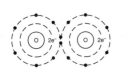 

3. In a covalent bond, atoms share electrons to fill their outermost shells. In a nonpolar covalent bond, atoms attract shared electrons equally. An example is the $H_2$ molecule. In a polar covalent bond, atoms do not share electrons equally, and the bond is positive at one end, negative at the other (for example, the water molecule).
4. In the linear DNA molecule, the two nucleotide chains are held together by hydrogen bonds.

## 2.5. Properties of Water (pp. 30–31)
1. polarity; 2. hydrophilic; 3. hydrophobic; 4. Temperature; 5. hydrogen; 6. evaporation; 7. ice; 8. cohesion; 9. solvent; 10. solutes; 11. dissolved; 12. hydration.

## 2.6. Acids, Bases, and Buffers (pp. 32–33)
1. M; 2. K; 3. F; 4. J; 5. H; 6. D; 7. I; 8. A; 9. B; 10. C; 11. N; 12. L; 13. E; 14. G; 15. a. 7.3–7.5, slightly basic; b. 6.2–7.4, variable or slightly basic or slightly acid; c. 5.0–7.0, slightly acid to neutral; d. 1.0–3.0, acid; e. 7.8–8.3, basic; f. 3.0, acid.

## Self-Quiz
1. c; 2. a; 3. d; 4. d; 5. c; 6. c; 7. a; 8. c; 9. a; 10. d; 11. e.

---

# Chapter 3   Carbon Compounds in Cells

**Carbon, Carbon, In the Sky—Are You Swinging Low and High?** (pp. 36–37)
**3.1. Properties of Organic Compounds** (pp. 38–39)
**3.2. How Cells Use Organic Compounds** (p. 40)
**3.3.** *Focus on the Environment:* **Food Production and a Chemical Arms Race** (p. 41)
1. methyl; 2. hydroxyl; 3. ketone; 4. amino; 5. phosphate; 6. carboxyl; 7. aldehyde; 8. Enzymes represent a special class of proteins that speed up specific metabolic reactions. Enzymes mediate five categories of reactions by which most of the biological molecules are assembled, rearranged, and broken apart; 9. a. A juggling of internal bonds converts one type of organic compound into another; b. A molecule splits into two smaller ones; c. Through covalent bonding, two molecules combine to form a larger molecule; d. One or more electrons stripped from one molecule are donated to another molecule; e. One molecule gives up a functional group, which another molecule accepts; 10.

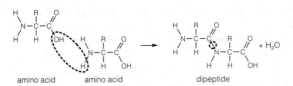

amino acid     amino acid          dipeptide

11. Hydrolysis reactions reverse the chemistry of condensation reactions; in the presence of water, large molecules. Both condensation and hydrolysis require the presence of enzymes specific to the particular molecules involved.

## 3.4. Carbohydrates (pp. 42–43)
1.

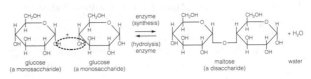

glucose          glucose          maltose          water
(a monosaccharide)  (a monosaccharide)   (a disaccharide)

2. a. Sucrose, Oligosaccharide (disaccharide); b. Deoxyribose, Monosaccharide; c. Glucose, Monosaccharide; d. Cellulose, Polysaccharide; e. Ribose, Monosaccharide; f. Lactose, Oligosaccharide (disaccharide); g. Amylopectin, Polysaccharide; h. Chitin, Polysaccharide; i. Glycogen, Polysaccharide; j. Amylose, Polysaccharide.

## 3.5. Lipids (pp. 44–45)
1. a. unsaturated; b. saturated
2.

glycerol     three fatty     triglyceride
             acids           (a complete fat
                              molecule)

3. Phospholipids have two fatty acid tails attached to a glycerol backbone; they have hydrophilic heads that dissolve in water. Phospholipids are the main structural materials of cell membranes; 4. B; 5. B; 6. D; 7. B; 8. A; 9. B; 10. E; 11. C; 12. E; 13. D; 14. B; 15. E; 16. A; 17. B; 18. B; 19. D.

**3.6. Amino Acids and the Primary Structure of Proteins** (pp. 46–47)
**3.7. Emergence of the Three-Dimensional Structure of Proteins** (pp. 48–49)
1. A; 2. C; 3. B;
4.

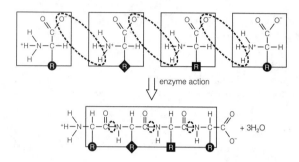

5. K; 6. F; 7. B; 8. J; 9. L; 10. A; 11. D; 12. I; 13. G; 14. E; 15. C; 16. H.

**3.8. Nucleotides and the Nucleic Acids** (pp. 50–51)
1. B; 2. A; 3. C; 4. Three as shown:
5. B; 6. A; 7. E; 8. D;
9. C; 10. a. Lipids;
b. Proteins; c. Proteins;
d. Nucleic acids;
e. Carbohydrates; f. Lipids;
g. Lipids; h. Nucleic acids;
i. Lipids; j. Carbohydrates

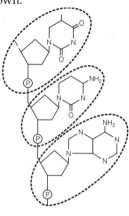

**Self-Quiz**
1. d; 2. a; 3. c; 4. c; 5. c; 6. b; 7. d; 8. c; 9. b; 10. b.

---

# Chapter 4   Cell Structure and Function

*Animalcules and Cells Fill'd With Juices* (pp. 54–55)
**4.1. Basic Aspects of Cell Structure and Function** (pp. 56–57)
**4.2.** *Focus on Science:* **Microscopes—Gateways to the Cell** (pp. 58–59)
1. plasma membrane (B); 2. cytoplasm (A); 3. DNA-containing region (C); 4. Cell size is constrained by the surface-to-volume ratio. Past a certain point of growth, the surface area will not be sufficient to admit enough nutrients and to allow enough wastes to exit the cell. As a cell grows, the surface area increases with the square and the volume increases with the cube.
5. F; 6. E; 7. G; 8. C; 9. B; 10. A; 11. D.

**4.3. The Defining Features of Eukaryotic Cells** (pp. 60–63)
1. a. Nucleus; b. Ribosomes; c. Endoplasmic reticulum; d. Golgi body; e. Various vesicles; f. mitochondria; g. cytoskeleton.

**4.4. The Nucleus** (pp. 64–65)
1. First, the nucleus physically separates the cell's DNA from the metabolic machinery of the cytoplasm; this permits easier copying of DNA and the distribution of new DNA molecules into new cells following division. Second, the nuclear envelope helps control the exchange of substances and signals between the nucleus and the cytoplasm.

2. a. Nucleolus; b. Nuclear envelope; c. Chromatin; d. Nucleoplasm; e. Chromosome; 3. Enzymes; 4. cytoplasm; 5. cytomembrane; 6. organelles; 7. polypeptides; 8. vesicles; 9. cytomembrane; 10. lipids; 11. membranes; 12. plasma.

**4.5. The Cytomembrane System** (pp. 66–67)
1. I; 2. A; 3. B; 4. H; 5. C; 6. C; 7. F; 8. L; 9. K; 10. E; 11. D; 12. G; 13. J.

**4.6. Mitochondria** (p. 68)
**4.7. Specialized Plant Organelles** (p. 69)
1. b; 2. a; 3. c; 4. a; 5. e; 6. a; 7. d; 8. b; 9. e; 10. a; 11. d; 12. b; 13. b; 14. a; 15. c; 16. a; 17. b; 18. d; 19. a; 20. b.

**4.8. Components of the Cytoskeleton** (pp. 70–71)
**4.9. The Structural Basis of Cell Motility** (pp. 72–73)
1. eukaryotic; 2. Protein; 3. movements; 4. tubulin; 5. plus; 6. minus; 7. microtubule organizing center; 8. centrioles, centrosomes, and kinetochores; 9. inhibits assembly and promotes disassembly of microtubules; 10. stabilizes existing microtubules and prevents formation of new ones; 11. Taxol; 12. microfilaments; 13. actin; 14. at cell surfaces; 15. accessory proteins; 16. intermediate; 17. strength and support; 18. assembly; 19. lengthen slide and shorten; 20. cytoplasmic streaming; 21. Flagella; 22. Cilia; 23. cilia; 24. cilia; 25. centriole.

**4.10. Cell Surface Specializations** (pp. 74–75)
1. C;  2. E;  3. B;  4. A;  5. F;  6. D.

**4.11. Prokaryotic Cells—The Bacteria** (pp. 76–77)
1. bacteria;  2. nucleus;  3. wall;  4. cytoplasm;  5. flagella;  6. ribosomes;  7. nucleoid.

**Self-Quiz**
1. Golgi body (J);  2. vesicle (G);  3. microfilaments (L); 4. mitochondrion (O);  5. chloroplast (M);  6. microtubules (H);  7. central vacuole (C);  8. rough endoplasmic reticulum (P);  9. ribosomes (R);  10. smooth endoplasmic reticulum (D);  11. DNA + nucleoplasm (Q);  12. nucleolus (K);  13. nuclear envelope (A); 14. nucleus (I);  15. plasma membrane (N);  16. cell wall (B);  17. microfilaments (L);  18. microtubules (H); 19. plasma membrane (N);  20. mitochondrion (O); 21. nuclear envelope (A);  22. nucleolus (K);  23. DNA + nucleoplasm (Q);  24. nucleus (I);  25. vesicle (G); 26. lysosome (E);  27. rough endoplasmic reticulum (P); 28. ribosomes (R);  29. smooth endoplasmic reticulum (D);  30. vesicle (G);  31. Golgi body (J);  32. centrioles (F);  33. d;  34. c;  35. d;  36. d;  37. b;  38. a;  39. c; 40. a;  41. b;  42. c;  43. b;  44. d;  45. c;  46. b, c, d, e; 47. a, b, c, d;  48. b, c, d, e;  49. b, c, d, e;  50. c, d;  51. a; 52. a, b, d;  53. b, d;  54. b, c, d, e;  55. b, c, d, e;  56. b, c, d, e;  57. b, c, d, e.

# Chapter 5  A Closer Look at Cell Membranes

*It Isn't Easy Being Single* (pp. 80–81)
**5.1. Membrane Structure and Function** (pp. 82–83)
**5.2.** *Focus on Science:* **Testing Ideas About Cell Membranes** (pp. 84–85)
1. cytoskeletal;  2. adhesion;  3. open channel;  4. gated channel;  5. gated channel;  6. active transport; 7. transport;  8. receptor;  9. recognition;  10. oligosaccharide;  11. phospholipid;  12. cholesterol;  13. J;  14. E; 15. H;  16. A;  17. I;  18. F;  19. B;  20. G;  21. D;  22. C.

**5.3. How Substances Cross Cell Membranes** (pp. 86–87)
**5.4. The Directional Movement of Water Across Membranes** (pp. 88–89)
1. permeability;  2. Concentration;  3. concentration; 4. diffusion;  5. faster;  6. slower;  7. equilibrium; 8. faster;  9. electric;  10. attraction;  11. pressure;  12. C; 13. F;  14. A;  15. H;  16. E;  17. B;  18. I;  19. D;  20. G;

21. a. osmosis;  b. diffusion;  c. active transport; d. facilitated diffusion;  e. diffusion;  f. active transport; g. facilitated diffusion;  h. active transport;  i. facilitated diffusion;  j. diffusion;  22. T;  23. T;  24. tonicity; 25. hypotonic;  26. T;  27. isotonic;  28. less;  29. bulk flow;  30. T.

**5.5. Protein-Mediated Transport** (pp. 90–91)
1. b;  2. c;  3. a;  4. c;  5. a;  6. b;  7. c;  8. b;  9. c;  10. b; 11. b;  12. a;  13. a;  14. c.

**5.6. Exocytosis and Endocytosis** (pp. 92–93)
1. D;  2. E;  3. C;  4. A;  5. B.

**Self-Quiz**
1. d;  2. c;  3. a;  4. d;  5. c;  6. d;  7. d;  8. c;  9. d;  10. b.

# Chapter 6  Ground Rules of Metabolism

*Growing Old with Molecular Mayhem* (pp. 96–97)
**6.1. Energy and the Underlying Organization of Life** (pp. 98–99)
1. second;  2. T;  3. T;  4. T;  5. increasing;  6. II;  7. II; 8. I;  9. I;  10. I;  11. II;  12. I;  13. I;  14. I;  15. II;  16. b; 17. e;  18. d;  19. a;  20. c.

**6.2. The Directional Nature of Metabolism** (pp. 100–101)
**6.3. Energy Transfers and Cellular Work** (pp. 102–103)
1. products;  2. substrates (reactants);  3. reversible; 4. equilibrium;  5. rate;  6. exergonic;  7. endergonic; 8. exergonic;  9. exergonic;  10. exergonic;  11. B;  12. A; 13. A;  14. B;  15. F;  16. G(A);  17. I;  18. D;  19. J; 20. A;  21. E;  22. C;  23. B(A);  24. H;  25. C;  26. B; 27. D;  28. F;  29. E;  30. A;  31. ATP;  32. adenosine

triphosphate;  33. adenine;  34. ribose;  35. phosphate; 36. covalent;  37. hydrolysis;  38. energy;  39. energy; 40. ATP;  41. phosphorylation;  42. triphosphate; 43. ribose sugar;  44. adenine (a nitrogen-containing molecule);  45. adenosine triphosphate;  46. linear; 47. cyclic;  48. branched.

**6.4. Enzyme Structure and Function** (pp. 104–105)
**6.5. Factors Influencing Enzyme Activity** (pp. 106–107)
1. Enzymes;  2. equilibrium;  3. substrate;  4. active site; 5. induced-fit;  6. activation energy;  7. collision; 8. hydrogen;  9. hydrolysis;  10. rate;  11. activation energy without enzyme;  12. activation energy with enzyme;  13. uncatalyzed;  14. catalyzed;  15. energy released by the reaction;  16. D;  17. F;  18. C;  19. B; 20. A;  21. E;  22. 40°C;  23. 60°C;  24. C;  25. B;  26. A;

27. Temperature (pH, saltiness); 28. pH or saltiness (temperature); 29. metabolism; 30. bonds; 31. 7; 32. tryptophan; 33. amino acids; 34. falls; 35. feedback inhibition; 36. enzyme; 37. allosteric; 38. control; 39. rises; 40. hormones; 41. D; 42. E; 43. A; 44. B; 45. C.

## 6.6. Electron Transfers Through Transport Systems (p. 108)
## 6.7. *Focus on Science:* You Light Up My Life—Visible Signs of Metabolic Activity (p. 109)

1. intermediate; 2. enzyme; 3. electron transport; 4. cofactors; 5. electron; 6. oxidized; 7. reduced; 8. energy; 9. most; 10. energy; 11. work; 12. ATP; 13. E; 14. B; 15. C; 16. A; 17. D; 18. bioluminescence; 19. luciferin; 20. ATP; 21. $O_2$; 22. luciferase; 23. electrons.

**Self-Quiz**
1. d; 2. c; 3. e; 4. d; 5. a; 6. c; 7. d; 8. c; 9. d; 10. a.

---

# Chapter 7   Energy-Acquiring Pathways

*Sun, Rain, and Survival* (pp. 112–113)
## 7.1. Photosynthesis: An Overview (pp. 114–115)
1. Autotrophs; 2. carbon dioxide; 3. Photosynthetic; 4. Chemosynthetic; 5. electrons; 6. Heterotrophs; 7. animals; 8. aerobic respiration; 9. $12 H_2O + 6CO_2 \rightarrow 6O_2 + C_6H_{12}O_6 + 6H_2O$; 10. (a) Twelve (b) carbon dioxide (c) oxygen (d) glucose (e) six; 11. light-dependent (light-independent); 12. light-independent (light-dependent); 13. Carbon dioxide; 14. water; 15. glucose; 16. thylakoid; 17. grana; 18. hydrogen; 19. stroma; 20. b; 21. a; 22. a; 23. b; 24. a; 25. a; 26. b; 27. a; 28. a; 29. a; 30. b; 31. a; 32. $O_2$; 33. ATP; 34. NADPH; 35. $CO_2$; 36. chloroplasts; 37. thylakoid membrane system; 38. stroma.

## 7.2. Sunlight as an Energy Source (pp. 116–117)
## 7.3. The Rainbow Catchers (pp. 118–119)
1. thylakoid; 2. photon; 3. pigments; 4. chlorophylls; 5. red (blue); 6. blue (red); 7. Carotenoids; 8. photosystem; 9. light (photon); 10. electron; 11. acceptor; 12. Phosphorylation; 13. P700; 14. channel proteins; 15. ADP; 16. chemiosmotic; 17. H; 18. G; 19. E; 20. F; 21. B; 22. A; 23. C; 24. I; 25. J; 26. D.

## 7.4. The Light-Dependent Reactions (pp. 120–121)
## 7.5. A Closer Look at ATP Formation in Chloroplasts (p. 122)
1. a. A pigment cluster dominated by P700; b. Electrons representing energy are ejected from P700 to an electron acceptor but move over the electron transport system, where some of the energy is used to produce ATP; c. A special chlorophyll molecule that absorbs wavelengths of 700 nanometers and then ejects electrons; d. A molecule that accepts electrons ejected from chlorophyll P700 and then passes electrons down the electron transport system; e. Electrons flow through this system which is composed of a series of molecules bound in the thylakoid membrane that drive photophosphorylation; f. ADP undergoes photophosphorylation in cyclic photophosphorylation to become ATP; 2. P680; 3. electron transport system; 4. photosystem II; 5. photosystem I; 6. photolysis; 7. NADPH; 8. ATP; 9. electrons; 10. thylakoid; 11. gradients; 12. water ($H_2O$); 13. ATP synthases; 14. ATP; 15. chemiosmotic; 16. NADPH; 17. electrons; (18–53). The following numbers should have a check mark ($\sqrt{}$): 20, 22, 25, 28, 30, 32, 33, 34, 35, 37, 40, 41, 42, 45, 50, and 51. All others should be blank.

## 7.6. Light-Independent Reactions (p. 123)
## 7.7. Fixing Carbon—So Near, Yet So Far (pp. 124–125)
## 7.8. *Focus on the Environment:* Autotrophs, Humans, and the Biosphere (p. 126)
1. carbon dioxide (D); 2. carbon dioxide fixation (E); 3. phosphoglycerate (F); 4. adenosine triphosphate (H); 5. NADPH (G); 6. phosphoglyceraldehyde (A); 7. sugar phosphates (B); 8. Calvin-Benson cycle (I); 9. ribulose bisphosphate (C); 10. ATP (NADPH); 11. NADPH (ATP); 12. carbon dioxide; 13. ribulose bisphosphate; 14. PGA; 15. fixation; 16. PGA; 17. PGAL; 18. six; 19. RuBP; 20. carbon dioxide; 21. PGALs; 22. sugar phosphate; 23. fixation; 24. light-dependent; 25. ATP (NADPH); 26. NADPH (ATP); 27. Sugar phosphate; 28. photorespiration; 29. $C_3$; 30. food; 31. oxygen; 32. oxaloacetate; 33. chemosynthetic; 34. organic (food); 35. protons (electrons); 36. electrons (protons); 37. nitrate (nitrite); 38. nitrite (nitrate); 39. fertility; (40 - 83). The following numbers should have a check mark ( $\sqrt{}$ ): 40, 45, 50, 52, 53, 54, 55, 57, 59, 63, 64, 69, 70, 71, 80, and 81. All others should be blank.

**Self-Quiz**
1. a; 2. b; 3. c; 4. a; 5. c; 6. d; 7. a; 8. b; 9. c; 10. d.

# Chapter 8    Energy-Releasing Pathways

*The Killers Are Coming!* (pp. 130–131)
**8.1. How Cells Make ATP** (pp. 132–133)
1. Adenosine triphosphate (ATP);  2. Oxygen withdraws electrons from the electron transport system and joins with $H^+$ to form water;  3. Glycolysis followed by some end reactions (called fermentation), and anaerobic electron transport. Some organisms (including humans) use fermentation pathways when oxygen supplies are low; many microbes rely exclusively on anaerobic pathways;  4. ATP;  5. photosynthesis;  6. aerobic respiration;  7. glycolysis;  8. pyruvate;  9. Krebs;  10. water;  11. ATP;  12. electrons;  13. transport;  14. phosphorylation;  15. ATP;  16. Oxygen;  17. anaerobic;  18. Fermentation;  19. electron transport;  20. $C_6H_{12}O_6 + 6O_2 \rightarrow 6CO_2 + 6H_2O$;  21. One molecule of glucose plus six molecules of oxygen (in the presence of appropriate enzymes) yield six molecules of carbon dioxide plus six molecules of water;  22. G;  23. I;  24. H;  25. D;  26. F;  27. J;  28. A;  29. B;  30. C;  31. E.

**8.2. Glycolysis: First Stage of the Energy-Releasing Pathways** (pp. 134–135)
1. Autotrophic;  2. Glucose;  3. pyruvate;  4. ATP (NADH);  5. NADH (ATP);  6. glucose;  7. ATP;  8. PGAL;  9. phosphate;  10. hydrogen;  11. ATP;  12. water;  13. phosphate;  14. ATP;  15. substrate-level;  16. glucose;  17. pyruvate;  18. three;  19. D;  20. F;  21. B;  22. H;  23. G;  24. A;  25. E;  26. C.

**8.3. Second Stage of the Aerobic Pathway** (pp. 136–137)
1. acetyl-CoA;  2. Krebs;  3. electron transport;  4. thirty-four;  5. ATP;  6. carbon dioxide;  7. electrons;  8. $NAD^+$ (FAD);  9. FAD ($NAD^+$);  10. inner compartment;  11. inner membrane;  12. outer compartment;  13. outer membrane;  14. cytoplasm;  15. ATP;  16. oxygen ($O_2$);  17. $FADH_2$;  18. NADH;  19. electron transport system.

**8.4. Third Stage of the Aerobic Pathway** (pp. 138–139)
1. three;  2. two;  3. transport;  4. chemiosmotic;  5. synthases;  6. ATP;  7. oxygen;  8. mitochondria;  9. thirty-six (thirty-eight);  10. thirty-eight (thirty-six);  11. c;  12. a, b;  13. c;  14. a;  15. b;  16. a;  17. b;  18. b;  19. a, b, c;  20. b;  21. c;  22. a, b;  23. b;  24. c;  25. a;  26. c;  27. c;  28. a;  29. b;  30. a;  31. c;  32. c.

**8.5. Anaerobic Routes of ATP Formation** (pp. 140–141)
1. oxygen ($O_2$);  2. fermentation;  3. lactate;  4. ethanol;  5. carbon dioxide;  6. Anaerobic;  7. electron;  8. Glycolysis;  9. pyruvate;  10. NADH;  11. pyruvate;  12. lactate;  13. acetaldehyde;  14. carbon dioxide;  15. ethanol;  16. glycolysis;  17. $NAD^+$;  18. bacteria;  19. ATP;  20. sulfate;  21. nitrite;  22. nitrogen;  (23–74) With a check ($\sqrt{}$): 23, 25, 31, 32, 34, 35, 37, 42, 47, 50, 54, 56, 59, 61, 63, 65, 66, 71;  all others lack a check.

**8.6. Alternative Energy Sources in the Human Body** (pp. 142–143)

**8.7.** *Commentary:* **Perspective On Life** (p. 144)
1. Figure 8.1 shows how any complex carbohydrate or fat can be broken down and at least part of those molecules can be fed into the glycolytic pathway;  2. T;  3. T;  4. Page 110 (last paragraph) tells us that energy flows through time in one direction—from organized to less organized forms;  thus energy cannot be completely recycled;  5. Page 110 tells us that Earth's first organisms were anaerobic fermenters;  6. fatty acids;  7. glycerol;  8. glycolysis;  9. amino acids;  10. Krebs cycle;  11. acetyl-CoA;  12. pyruvate;  (13–102) With a check ($\sqrt{}$): 13, 15, 16, 17, 19, 22, 23, 25, 29, 31, 32, 39, 46, 47, 48, 49, 53, 54, 65, 66, 70, 71, 72, 80, 84, 87, 95, 102;  all others lack a check;  103. e;  104. b;  105. d;  106. a;  107. c;  108. h;  109. e, j;  110. b;  111. b;  112. h;  113. i;  114. e;  115. c;  116. h;  117 g.

**Self-Quiz**
1. c;  2. c;  3. d;  4. b;  5. d;  6. d;  7. a;  8. d;  9. d;  10. d;  11. C;  12. A, B, D;  13. B, (C), D;  14. C, E;  15. B, C;  16. A, D;  17. A;  18. B, D;  19. E;  20. A, E.

# Chapter 9    Cell Division and Mitosis

*Silver In the Stream of Time* (pp. 148–149)
**9.1. Dividing Cells: The Bridge Between Generations** (p. 150)
**9.2. The Cell Cycle** (p. 151)
1. G;  2. J;  3. C;  4. E;  5. B;  6. A;  7. I;  8. H;  9. F;  10. D;  11. interphase;  12. mitosis;  13. G1;  14. S;  15. G2;  16. prophase;  17. metaphase;  18. anaphase;  19. telophase;  20. cytoplasm divided or cytokinesis;  21. 15;  22. 12;  23. 14;  24. 13;  25. 11;  26. 20;  27. 11.

**9.3. The Stages of Mitosis—An Overview** (pp. 152–153)
1. interphase-daughter cells (F);  2. anaphase (A);  3. late prophase (G);  4. metaphase (D);  5. interphase-parent cell (E);  6. early prophase (C);  7. transition to metaphase (B);  8. telophase (H).

**9.4. A Closer Look at the Cell Cycle** (pp. 154–155)
1. F;  2. I;  3. H;  4. A;  5. D;  6. J;  7. B;  8. G;  9. E;  10. C.

**9.5. Division of the Cytoplasm** (pp. 156–157)
**9.6.** *Focus on Science:* **Henrietta's Immortal Cells** (p. 158)
1. a;  2. b;  3. a;  4. a;  5. b;  6. a;  7. b;  8. b;  9. a;  10. a;
11. HeLa cells are tumor cells taken from and named for a cancer patient, Henrietta Lacks, in 1951. Henrietta Lacks died (at age thirty-one) two months following her diagnosis of cancer. HeLa cells have continued to divide in culture and are used for cancer research in laborato-ries all over the world. The legacy of Henrietta Lacks continues to benefit humans everywhere.

**Self-Quiz**
1. a;  2. c;  3. d;  4. a;  5. d;  6. c;  7. e;  8. c;  9. c;  10. d;
11. d.

---

# Chapter 10    Meiosis

*Octopus Sex and Other Stories* (pp. 160–161)
**10.1. Comparison of Asexual and Sexual Reproduction**
 (p. 162)
**10.2. How Meiosis Halves the Chromosome Number**
 (pp. 162–163)
1. b;  2. a;  3. b;  4. b;  5. a;  6. b;  7. a;  8. b;  9. a;  10. b;
11. allele;  12. Meiosis;  13. gamete;  14. Diploid;
15. Diploid;  16. Meiosis;  17. sister chromatids;
18. sister chromatids;  19. one;  20. four;  21. two;
22. interphase preceding meiosis I;  23. chromosome;
24. haploid;  25. meiosis II;  26. twenty-three

**10.3. A Visual Tour of the Stages of Meiosis**
 (pp. 164–165)
**10.4. Key Events of Meiosis I** (pp. 166–167)
1. E $(2n = 2)$;  2. D $(2n = 2)$;  3. B $(2n = 2)$;  4. A $(n = 1)$;
5. C $(n = 1)$;  6. anaphase II (H);  7. metaphase II (F);
8. metaphase I (A);  9. prophase II (B);  10. telophase II
(C);  11. telophase I (G);  12. prophase I (E);
13. anaphase I (D);  14. F;  15. I;  16. G;  17. D;  18. B;
19. J;  20. A;  21. H;  22. C;  23. E

**10.5. From Gametes to Offspring** (pp. 168–169)
1. b;  2. c;  3. a;  4. c;  5. b;  6. b;  7. b;  8. b;  9. c;  10. b;
11. 2 $(2n)$;  12. 5 $(n)$;  13. 4 $(n)$;  14. 1 $(2n)$;  15. 3 $(n)$;
16. A;  17. B;  18. E;  19. D;  20. C;  21. During prophase I of meiosis, crossing over and genetic recombination occur. During metaphase I of meiosis, the two members of each homologous chromosome assort independently of the other pairs. Fertilization is a chance mix of different combinations of alleles from two different gametes

**10.6. Meiosis and Mitosis Compared** (pp. 170–171)
1. a. Mitosis;  b. Mitosis;  c. Meiosis;  d. Meiosis;
e. Meiosis;  f. Meiosis;  g. Meiosis;  h. Mitosis;  i. Meiosis;
2. C;  3. F;  4. D;  5. A;  6. B;  7. E;  8. 4;  9. 8;  10. 4;
11. 8;  12. 2

**Self-Quiz**
1. a;  2. a;  3. d;  4. b;  5. c;  6. b;  7. d;  8. a;  9. b;  10. c

---

# Chapter 11    Observable Patterns of Inheritance

*A Smorgasbord of Ears and Other Traits* (pp. 174–175)
**11.1. Mendel's Insights into the Patterns of Inheritance**
 (pp. 176–177)
**11.2. Mendel's Theory of Segregation** (pp. 178–179)
**11.3. Independent Assortment** (pp. 180–181)
1. I;  2. B;  3. D;  4. J;  5. H;  6. M;  7. O;  8. E;  9. F;
10. K;  11. A;  12. G;  13. N;  14. C;  15. L;
16. monohybrid;  17. Probability;  18. fertilization;
19. segregation;  20. testcross;  21. 1:1;  22. dihybrid;
23. independent assortment;  24. genotype: 1/2 *Tt*;  1/2
*tt*, phenotype: 1/2 tall;  1/2 short;  25. a. 1 tall : 1 short;
1 heterozygous tall : 1 homozygous short;  b. all tall;  1
homozygous tall : 1 heterozygous tall;  c. all short;  all
homozygous short;  d. 3 tall : 1 short;  1 homozygous
tall : 2 heterozygous tall : 1 homozygous short;  e. 1 tall :
1 short;  1 homozygous short : 1 heterozygous tall;  f. all
tall;  all heterozygous tall;  g. all tall;  all homozygous
tall;  h. all tall;  1 heterozygous tall : 1 homozygous tall;
26. a. 9/16 pigmented eyes, right-handed;  b. 3/16 pig-mented eyes, left-handed;  c. 3/16 blue-eyed, right-handed;  d. 1/16 blue-eyed, left-handed.
27. Albino = *aa*, normal pigmentation = *AA* or *Aa*. The woman of normal pigmentation with an albino mother is genotype *Aa*; the woman received her recessive gene (*a*) from her mother and her dominant gene (*A*) from her father. It is likely that half of the couple's children will be albinos (*aa*) and half will have normal pigmentation but be heterozygous (*Aa*). 28. a. $F_1$: black trotter; $F_2$: nine black trotters, three black pacers, three chestnut trotters, one chestnut pacer; b. black pacer; c. *BbTt*; d. *bbtt*, chestnut pacers and *BBTT*, black trotters.

**11.4. Dominance Relations** (p. 182)
**11.5. Multiple Effects of Single Genes** (p. 183)
**11.6. Interactions Between Gene Pairs** (pp. 184–185)
1. a. Incomplete dominance;  b. Codominance;
c. Multiple alleles;  d. Epistasis;  e. Pleiotropy.

2.a. phenotype: all pink, genotype: all RR';
b. phenotype: all white, genotype: all R'R';
c. phenotype: 1/2 red; 1/2 pink, genotype: 1/2 RR;
1/2 RR'; d. phenotype: all red, genotype: all RR.
3. The man must have sickle-cell trait with the genotype
$Hb^A Hb^S$, and the woman he married would have a normal genotype, $Hb^A Hb^A$. The couple could be told that the
probability is 1/2 that any child would have sickle-cell
trait and 1/2 that any child would have the normal genotype.
4. Both the man and the woman have the genotype $Hb^S Hb^S$. The probability of children from this marriage is:
1/4 normal, $Hb^A Hb^A$; 1/2 sickle-cell trait, $Hb^A Hb^S$; 1/4
sickle-cell anemia, $Hb^S Hb^S$.
5. Genotypes: 1/4 $I^A I^A$; 1/4 $I^A I^B$; 1/4 $I^A i$; 1/4 $I^B i$, phenotypes: 1/2 A; 1/4 AB; 1/4 B.
6. Genotypes: 1/4 $I^A I^B$; 1/4 $I^B i$; 1/4 $I^A i$; 1/4 $ii$, phenotypes: 1/4 AB; 1/4 B; 1/4 A; 1/4 O.
7. Genotypes: all $I^A i$, phenotypes: all A.
8. Genotypes: all $ii$, phenotypes: all O.
9. Genotypes: 1/4 $I^A I^A$; 1/2 $I^A I^B$; 1/4 $I^B I^B$, phenotypes:
1/4 A; 1/2 AB; 1/4 B.
10. 1/4 color; 3/4 white. 11. 3/4 color; 1/4 white. 12.
1/4 color; 3/4 white.

13. 3/8 black; 1/2 yellow; 1/8 brown.
14. The genotype of the male parent is RrPp and the
genotype of the female parent is rrpp. The offspring are
1/4 walnut comb, RrPp; 1/4 rose comb, Rrpp; 1/4 pea
comb, rrPp; 1/4 single comb, rrpp.
15. The genotype of the walnut-combed male is RRpp
and the genotype of the single-combed female is rrpp. All
offspring are rose comb with the genotype Rrpp.

**11.7. Less Predictable Variations in Traits** (pp. 186–187)
**11.8. Examples of Environmental Effects on Phenotype**
(p. 188)
1. b; 2. b; 3. a; 4. b; 5. a.

**Self-Quiz**
1. d; 2. b; 3. a; 4. c; 5. d; 6. b; 7. e; 8. a; 9. a; 10. c;
11. d.

**Integrating and Applying Key Concepts**
*DdPp × Ddpp*

---

# Chapter 12  Chromosomes and Human Genetics

*Too Young To Be Old* (pp. 192–193)
**12.1. The Chromosomal Basis of Inheritance—An
Overview** (p. 194)
**12.2.** *Focus on Science:* **Preparing a Karyotype Diagram**
(p. 195)
1. genes; 2. homologous; 3. Alleles; 4. wild; 5. crossing over; 6. recombination; 7. sex; 8. autosomes;
9. karyotype; 10. in vitro; 11. colchicine; 12. metaphase; 13. centrifugation.

**12.3. Sex Determination in Humans** (pp. 196–197)
**12.4. Early Questions About Gene Locations**
(pp. 198–199)
1. F; 2. C; 3. A; 4. E; 5. B; 6. D; 7. Two blocks of the
Punnett square should be XX and two blocks should be
XY; 8. sons; 9. mothers; 10. daughters; 11.a. $F_1$ flies
all have red eyes: 1/2 heterozygous red females : 1/2
red-eyed males; b. F2 Phenotypes: females all have red
eyes; males 1/2 red eyes, 1/2 white eyes; Genotypes:
females are $X^W X^W$; 12. a.

**12.5. Recombination Patterns and Chromosome
Mapping** (pp. 200–201)
**12.6. Human Genetic Analysis** (pp. 202–203)
1. Crossing over would be expected to occur twice as
often between genes A and B as it would between genes
C and D; 2. a; 3. Linked genes of the children: daughter
1 is *On/An*, daughter 2 is *ON/On*, and the son is *On/On*;
4. The son has genotype *On/On* because a crossover must

have occurred during meiosis in his mother, resulting in
the recombination *On*; 5. A; 6. E; 7. F; 8. B; 9. C;
10. G; 11. D; 12. A genetic abnormality is nothing more
than a rare, uncommon version of a trait that society may
judge as abnormal or merely interesting; a genetic disorder is an inherited condition that sooner or later causes
mild to severe medical problems; an illness is classified
as a genetic disease when a person's genes increase sus-
ceptibility to infection or weaken the response to it;
13.a. Autosomal recessive; b. Autosomal dominant;
c. X-linked recessive; d. X-linked recessive; e. Autosomal dominant; f. X-linked recessive; g. Changes in
chromosome number; h. X-linked dominant;
i. Changes in chromosome structure; j. Changes in chromosome number; k. X-linked recessive; l. Changes in
chromosome structure; m. Autosomal dominant;
n. Changes in chromosome number; o. Autosomal dominant; p. Autosomal recessive.

**12.7. Patterns of Autosomal Inheritance** (pp. 204–205)
**12.8. Patterns of X-Linked Inheritance** (pp. 206–207)
1. b and d; 2. a and c; 3. b; 4. c; 5. a; 6. b; 7. b; 8. b
and d; 9. a; 10. c; 11. c; 12. b; 13. b; 14. a; 15. b;
16. c; 17. a; 18. d; 19. The woman's mother is heterozygous normal, *Aa*, the woman is also heterozygous normal, *Aa*. The albino man, *aa*, has two heterozygous
normal parents, *Aa*. The two normal children are heterozygous normal, *Aa*; the albino child is *aa*; 20. Assuming the father is heterozygous with Huntington's

disorder and the mother normal, the chances are 1/2 that the son will develop the disease; 21. If only male offspring are considered, the probability is 1/2 that the couple will have a color-blind son; 22. The probability is that 1/2 of the sons will have hemophilia; the probability is 0 that a daughter will express hemophilia; the probability is that 1/2 of the daughters will be carriers; 23. If the woman marries a normal male, the chance that her son would be color-blind is 1/2. If she marries a color-blind male, the chance that her son would be color blind is also 1/2.

**12.9. Changes in Chromosome Number** (pp. 208–209)
**12.10. Changes in Chromosome Structure** (pp. 210–211)
1.a. With aneuploidy, individuals have one extra or one less chromosome; a major cause of human reproductive failure; b. With polyploidy, individuals have three or more each type of chromosome; common in flowering plants and some animals but lethal in humans; c. Nondisjunction is due to a failure of one or more pairs of chromosomes to separate in mitosis or meiosis; some

or all forthcoming cells will have too many or too few chromosomes; 2. All gametes will be abnormal; 3. One-half of the gametes will be abnormal; 4. About half of all flowering plant species are polyploids but this condition is lethal for humans; 5. A tetraploid cell has four of each type of chromosome; a trisomic (2n + 1) individual will have three of one type of chromosome and two of every other type; a monosomic (2n – 1) individual will have only one of one type of chromosome but two of every other type; 6. c; 7. b; 8. c; 9. d; 10. a; 11. b; 12. d; 13. c; 14. a; 15. d; 16. duplication (B); 17. inversion (C); 18. deletion (A); 19. translocation (D).

**12.11.** *Focus on Science:* **Prospects in Human Genetics** (pp. 212–213)
1.a. Prenatal diagnosis; b. Genetic counseling; c. Phenotypic treatments; d. Genetic screening.

**Self-Quiz**
1. d; 2. d; 3. b; 4. c; 5. b; 6. a; 7. c; 8. b; 9. b; 10. c.

---

# Chapter 13  DNA Structure and Function

*Cardboard Atoms and Bent-Wire Bonds* (pp. 216–217)
**13.1. Discovery of DNA Function** (pp. 218–219)
1.a. Miescher: identified "nuclein" from nuclei of pus cells and fish sperm; discovered DNA; b. Griffith: discovered the transforming principle DNA; b. Griffith: discovered the transforming principle in *Streptococcus pneumoniae*; live, harmless R cells were mixed with dead S cells; R cells became S cells; c. Avery: reported that the transforming substance in Griffith's bacteria experiments was probably DNA, the substance of heredity; d. Delbrück, Hershey, and Luria: studied the infectious cycle of bacteriophages; e. Hershey and Chase: worked with radioactive sulfur (protein) and phosphorus (DNA) labels; T4 bacteriophage and *E. coli* demonstrated that labeled phosphorus was in bacteriophage DNA and contained hereditary instructions for new bacteriophages.

**13.2. DNA Structure** (pp. 220–221)
1. A five-carbon sugar called deoxyribose, a phosphate group, and one of the four nitrogen-containing bases; 2. guanine; 3. cytosine; 4. adenine; 5. thymine; 6. T; 7. T; 8. F, sugar; 9. T; 10. T; 11. pairing; 12. constant; 13. sequence; 14. different; 15. deoxyribose (B); 16. phosphate group (G); 17. purine (C); 18. pyrimidine (A); 19. purine (E); 20. pyrimidine (D); 21. nucleotide (F); 22. Living organisms have so many diverse body structures and behave in different ways because the many different habitats of Earth have selected those genotypes most able to survive in those habitats. The remaining genotypes have perished. The directions that code for the building of those body structures and which

enable the specific successful behaviors reside in DNA, or in a few cases, RNA. All living organisms follow the same rules for base pairing between the two nucleotide strands in DNA; adenine always pairs with thymine in undamaged DNA and cytosine always pairs with guanine. All living organisms must extract energy from food molecules and the reactions of glycolysis occur in virtually all of Earth's species. That means that similar enzyme sequences enable similar metabolic pathways to occur. While virtually all living organisms on Earth use the same code and the same enzymes during replication, transcription and translation, the particular array of proteins being formed differs from individual to individual even of the same species according to the sequences of nitrogenous bases that make up an individual's chromosome(s), and therein lies the key to the enormous diversity of life on Earth: no two individuals have the exact same array of proteins in their phenotypes.

**13.3. DNA Replication and Repair** (pp. 222–223)
**13.4.** *Focus on Health:* **When DNA Can't Be Fixed** (p. 224)

1.

| T– | A | T | –A |
|----|---|---|----|
| G– | C | G | –C |
| A– | T | A | –T |
| C– | G | C | –G |
| C– | G | C | –G |
| C– | G | C | –G |
| old | new | new | old |

2. F, adenine bonds to thymine (during replication) or uracil (during transcription); 3. F, it is a conserving process because each "new" DNA molecule contains one

"old" strand from the parent cell attached to a strand of "new" complementary nucleotides that were assembled from stockpiles in the cell;  4. T;  5. T;  6. C;  7. E;  8. D; 9. A;  10. F;  11. B.

## Crossword Puzzle: DNA Structure and Function

```
 1D     2C    3T  4H  Y  M  I  N  E              5N  U  C  L  E  O  T  I  D  E        6D  E
 I      H         E                                                            N          7P
 8F  R  A  N  K   L  I  N             9E  A  S  E           10R  E  P  A  I  R
 F      S         I                           R                     N                     I
 R      E     11X  R  A  Y        12C  A  R  C  I  N  O  M  A                              M
 A               T                            T                          13G             E
14C  R  I  C  K          15P  O  L  Y  M  E  R  A  S  E                   U               R
 T                            R                                          A
 I           16C  Y  T  O  S  I  N  E                         O          N
 O   17W                                        O        18L          19P
 N    A             20R  E  P  L  I  C  A  T  I  O  N                 I
 T                            H                          G       E    R
21S  E  M  I  C  O  N  S  E  R  V  A  T  I  V  E          A       S    I
 O                            G                                  A    N
 N   22P  Y  R  I  M  I  D  I  N  E         23A  D  E  N  I  N  E
```

# Chapter 14  From DNA to Proteins

*Beyond Byssus* (pp. 226–227)

**14.1.** *Focus on Science:* **Discovering The Connection Between Genes And Proteins** (pp. 228–229)

**14.2. Transcription of DNA into RNA** (pp. 230–231)

1. sequence;  2. gene;  3. transcription (translation); 4. translation (transcription);  5. transcription;  6. translation;  7. protein;  8. folded;  9. structural (functional); 10. functional (structural);  11. a. ribosomal RNA; rRNA;  RNA molecule that associates with certain proteins to form the ribosome, the "workbench" on which polypeptide chains are assembled;  b. messenger RNA; mRNA;  RNA molecule that moves to the cytoplasm, complexes with the ribosome where translation will result in polypeptide chains;  c. transfer RNA;  tRNA; RNA molecule that moves into the cytoplasm, picks up a specific amino acid, and moves it to the ribosome where tRNA pairs with a specific mRNA code word for that amino acid;  12. RNA molecules are single-stranded, while DNA has two strands;  uracil substitutes in RNA molecules for thymine in DNA molecules;  ribose sugar is found in RNA, while DNA has deoxyribose sugar;

13. Both DNA replication and transcription follow base-pairing rules;  nucleotides are added to a growing RNA strand one at a time as in DNA replication;  14. Only one region of a DNA strand serves as a template for transcription;  transcription requires different enzymes (three types of RNA polymerase);  the results of transcription are single-stranded RNA molecules, but replication results in DNA, a double-stranded molecule;  15. C; 16. B;  17. E;  18. A;  19. D;  20. A-U-G-U-U-C-U-A-U-U-G-U-A-A-U-A-A-A-G-G-A-U-G-G-C-A-G-U-A-G; 21. DNA(E);  22. introns (B);  23. cap (F);  24. exons (A); 25. tail (D);  26. mature mRNA transcript (C).

**14.3. Deciphering the mRNA Transcripts** (pp. 232–233)

**14.4. Stages of Translation** (pp. 234–235)

1. F;  2. B;  3. G;  4. H;  5. C;  6. A;  7. E;  8. D;  9. a. initiation;  b. chain elongation;  c. chain termination; 10. mRNA transcript: AUG UUC UAU UGU AAU AAA GGA UGG CAG UAG;  11. tRNA anticodons: UAC AAG

AUA ACA UUA UUU CCU ACC GUC AUC;  12. amino acids: (start) met phe tyr cys asn lys gly try gln (stop);  13. amino acids;  14. three;  15. one;  16. mRNA;  17. codon;  18. mRNA;  19. assembly (synthesis);  20. Transfer;  21. amino acid;  22. protein (polypeptide);  23. codon;  24. anticodon.

## 14.5. How Mutations Affect Protein Synthesis (pp. 236–237)

**Summary of Protein Synthesis (p. 238)**

1. Mutagens are environmental agents that attack a DNA molecule and modify its structure. Viruses, ultraviolet radiation, and certain chemicals are examples;  2. mutagens;  3. substitution;  4. amino acid;  5. hemoglobin;  6. frameshift;  7. transposable;  8. DNA;  9. gene mutations;  10. nucleotide (base);  11. mutation rate;  12. ionizing;  13. DNA (H);  14. new mRNA transcript (J);  15. intron (E);  16. exon (A);  17. mature mRNA transcript (L);  18. tRNAs (C);  19. rRNA subunits (G);  20. mRNA (B);  21. anticodon (K);  22. amino acids (D);  23. tRNA (F);  24. ribosome-mRNA complex (I);  25. polypeptide (M);  26. amino acids;  27. three;  28. one;  29. mRNA;  30. codon;  31. mRNA;  32. assembly (synthesis);  33. Transfer;  34. amino acid;  35. protein (polypeptide);  36. codon;  37. anticodon.

## Self-Quiz

1. c;  2. b;  3. c;  4. a;  5. c;  6. a;  7. a;  8. d;  9. d;  10. d;  11. b;  12. d;  13. e;  14. a;  15. c.

## Crossword Answers

```
 1:TRANSFER        2:TRANSPOSABLE   3:across/down marker   4:EXON (down)
 5:ALKYLATING      6:POLYSOME       7:SEQUENCE             8:ANTICODON
 9:CODON           10:FRAMESHIFT    11:MUTAGEN / MESSENGER 12:INSERTION
 13:RIBOSOMAL      14:TRANSCRIPTION 15:I                   16:/17:ELECTROPHORESIS
 18:INTRON         19:DELETION      20:PROMETER            21:URACIL
 Down also: 1:TRANSLATION

  T R A N S F E R . T R A N S P O S A B L E
  R . . . . . . . . . . . . . . . E . . . X
  A L K Y L A T I N G . . P O L Y S O M E O
  N . . . . . . . . . . . . . . . E . . . N
  S E Q U E N C E . A N T I C O D O N . . .
  L . . . . . . . . . . . . A . . . . . . F
  A M U T A G E N . I N S E R T I O N . R R
  T E . . . . . . . . . . . C . . . . . I A
  I S T R A N S C R I P T I O N . . . . B M
  O S . . . . . . . . . . . N . . . . . O E
  N E . . I E L E C T R O P H O R E S I S S
  I N T R O N . . O . . . . . G . . . . O H
  . G . . . . . . D E L E T I O N . . . M I
  . E . . . . . . O . . . . . N . . . . A F
  P R O M E T E R N . . . . . U R A C I L T
```

# Chapter 15   Controls over Genes

*Here's To Suicidal Cells!* (pp. 242–243)
**15.1. Overview of Gene Control** (p. 244)
**15.2. Controls in Bacterial Cells** (pp. 244–245)
1. a. Prevents transcription enzymes (RNA polymerases) from binding to DNA; this is negative transcription control; b. Activator protein; c. Specific base sequences on DNA that serve as binding sites for control agents; before RNA assembly can occur on DNA, the enzymes must bind with the promoter site; d. Operators; 2. In any organism, gene controls operate in response to chemical changes within the cell or its surroundings; 3. transcription controls; 4. regulator; 5. promoter; 6. negative control; 7. low; 8. RNA polymerase (mRNA transcription); 9. blocks; 10. repressor protein; 11. operator; 12. needed (required); 13. regulator gene (K); 14. enyzme genes (G); 15. repressor protein (E); 16. promoter (J); 17. operator (B); 18. lactose operon (A); 19. repressor-operator complex (I); 20. RNA polymerase (D); 21. repressor-lactose complex (F); 22. lactose (C); 23. mRNA transcript (L); 24. lactose enzymes (H).

**15.3. Controls in Eukaryotic Cells** (pp. 246–247)
**15.4 Evidence of Gene Control** (pp. 248–249)
1. All cells in the body descend from the same zygote; as cells divide to form the body, they become specialized in composition, structure, and function—they differentiate through selective gene expression; 2. replicational control (D); 3. transcriptional control (E); 4. transcript processing control (B); 5. translational control (A); 6. post-translational control (C); 7. DNA (genes); 8. cell differentiation; 9. selective; 10. controls; 11. regulatory; 12. activators; 13. Transcription; 14. Barr; 15. mosaic; 16. anhidrotic ectodermal dysplasia; 17. selective; 18. F; 19. T.

**15.5. Examples of Signaling Mechanisms** (pp. 250–251)
**15.6.** *Focus On Science:* **Cancer and the Cells That Forget to Die** (pp. 252–253)
1. signals (molecules); 2. hormones; 3. receptors; 4. enhancer; 5. ecdysone; 6. amplification; 7. polytene; 8. prolactin; 9. receptors; 10. phytochrome; 11. cancer; 12. tumor; 13. metastasis; 14. T; 15. increase, high, start; 16. T; 17. T; 18. F, bring about; 19. T; 20. F, abnormal.

**Self Quiz**
1. d; 2. b; 3. d; 4. b; 5. b; 6. c; 7. c; 8. a; 9. b; 10. c.

# Chapter 16   Recombinant DNA and Genetic Engineering

*Mom, Dad, and Clogged Arteries* (pp. 256–257)
**16.1. Recombination in Nature—And in the Laboratory** (pp. 258–259)
1. The bacterial chromosome, a circular DNA molecule, contains all the genes necessary for normal growth and development. Plasmids, small, circular molecules of "extra" DNA, carry only a few genes and are self-replicating.
2. F, small circles of DNA in bacteria; 3. T; 4. mutations; 5. recombinant DNA; 6. species; 7. amplify; 8. protein; 9. research; 10. Genetic engineering; 11. plasmid; 12. viruses; 13. B; 14. F; 15. E; 16. C; 17. A; 18. D; 19. T; 20. F; 21. F; 22. T; 23. F; 24. "Good" alleles that code for producing normal LDL receptors on cell surfaces exist in DNA libraries. Through PCR, they can be copied over and over again. Through recombinant DNA techniques, the good alleles can be spliced into harmless viruses that are allowed to infect liver cells surgically removed from the patient and cultured in the laboratory. Later about a billion of the modified cells can be infused into the patient's hepatic portal vein, which leads directly to the liver. Some modified cells with the substituted "good" alleles may take up residence, produce the receptors and begin removing cholesterol from the bloodstream.

**16.2. Working with DNA Fragments** (pp. 260–261)
**16.3.** *Focus on Science:* **Rife-Lips and DNA Fingerprints** (p. 262)
1. a. bacterial enzymes that cut apart DNA molecules injected into the cell by viruses; several hundred have been identified;
b. a replication enzyme that joins the short fragments of DNA;
c. a collection of DNA fragments produced by restriction enzymes and incorporated into plasmids;
d. An electrical field forces molecules (generally DNA or proteins) to move through a viscous medium and separate from each other according to their different physical and chemical properties;
e. A method for amplifying DNA fragments in a test tube (see Fig. 16.5);
f. Several procedures allow researchers to determine the nucleotide sequence of a DNA fragment (see Fig. 16.6);
g. Variations in the banding patterns of DNA fragments from different individuals; the variations occur because no two individuals (except identical twins) have identical base sequences in their entire DNA set. Restriction enzymes cut different DNA molecules at different sites and into different numbers of DNA fragments;
h. A unique array of RFLPs;

2. B;  3. D;  4. G;  5. C;  6. H;  7. I;  8. E;  9. F;  10. A;
11. fingerprint;  12. sequencing;  13. TACAAGATAA CATTAGTCATC

**16.4. Modified Host Cells** (p. 263)
**16.5. Bacteria, Plants, and the New Technology**
  (pp. 264–265)
**16.6. Genetic Engineering of Animals** (p. 266)
**16.7.** *Focus on Bioethics:* **Regarding Gene Therapy**
  (p. 267)
1. a. a nucleic acid hybridization technique used to identify bacterial colonies harboring the DNA (gene) of interest; a short nucleotide sequence is assembled from radioactively labeled subunits (part of the sequence must be complementary to that of the desired gene);
b. base-pairing between nucleotide sequences from different sources can indicate that the two different sources carry the same gene(s);
c. any DNA molecule "copied" from mRNA;
d. a process that uses a viral enzyme to transcribe mRNA into DNA which can then be inserted into a plasmid for amplification;  the same cDNA can be inserted into bacteria, which they will command to make a specific protein;  2. probes (hybridization);  3. expression (translation);  4. engineered;  5. introns;  6. transcriptase;
7. mRNA;  8. cDNA;  9. Enzyme;  10. DNA;  11. cDNA;
12. transcript;  13. Researchers are working to sequence the estimated 3 billion nucleotides present in human chromosomes. 14. If human collagen can be produced, it may be used to correct various skin, cartilage, and bone

disorders. 15. The plasmid of *Agrobacterium* can be used as a vector to introduce desired genes into cultured plant cells;  *Agrobacterium* was used to deliver a firefly gene into cultured tobacco plant cells.
16. Certain cotton plants have been genetically engineered for resistance to worm attacks.
17. In separate experiments the rat and human somatotropin genes became integrated into the mouse DNA. The mice grew much larger than their normal littermates.
18. Bacteria which have specific proteins on their cell surfaces facilitate ice crystal formation on whatever substrate the bacteria are located; bacteria without the ability to synthesize those proteins ("ice-minus") have been genetically engineered and were sprayed upon strawberry plants. Nothing bad happened.
19. In 1996, researchers produced a genetic duplicate of an adult ewe by fusing a reprogrammed mammary gland cell with an egg from which the nucleus was removed. When the fused cell was implanted into a surrogate mother, Dolly was born. Her existence means that genetically engineered clones of domestic animals may supply reliable quantities of specific proteins for research and for medical uses;  20. ice-minus;  21. genome (DNA);  22. gene therapy;  23. eugenic engineering.

**Self-Quiz**
1. a;  2. a;  3. b;  4. c;  5. d;  6. a;  7. c;  8. b;  9. a;  10. d.

---

# Chapter 17    Emergence of Evolutionary Thought

*Fire, Brimstone, and Human History* (pp. 270–271)
**17.1. Early Beliefs, Confounding Discoveries**
  (pp. 272–273)
1. G;  2. H;  3. J;  4. I;  5. D;  6. E;  7. F;  8. C;  9. B;  10. A.

**17.2. A Flurry of New Theories** (pp. 274–275)
**17.3. Darwin's Theory Takes Form** (pp. 276–277)
**17.4.** *Focus On Science:* **Regarding the First of the
  "Missing Links"** (p. 278)

1. b;  2. a;  3. b;  4. a;  5. a;  6. b;  7. a;  8. b;  9. a;  10. b;
11. b;  12. b;  13. a. John Henslow;  b. Cambridge University;  c. H.M.S. *Beagle;*  d. Charles Lyell;  e. Thomas Malthus;  f. Gal(pagos Islands;  g. Natural selection;  h. Alfred Wallace;  i. *Archaeopteryx;*  14. F(D);  15. D(F);
16. B;  17. E;  18. A;  19. C;  20. G.

**Self-Quiz**
1. c;  2. d;  3. d;  4. b;  5. d;  6. c;  7. b;  8. a;  9. c;  10. e.

---

# Chapter 18    Microevolution

*Designer Dogs* (pp. 280–281)
**18.1. Individuals Don't Evolve—Populations Do**
  (pp. 282–283)
**18.2.** *Focus On Science:* **When Is a Population <u>Not</u>
  Evolving?** (pp. 284–285)
1. M;  2. P;  3. M;  4. B;  5. P;  6. M;  7. B;  8. P;  9. B;
10. M;  11. D;  12. C;  13. E;  14. B;  15. A;

16. Phenotypic variation is due to the effects of genes and the environment. The environment can mediate how the genes governing traits are expressed in an individual. For example, ivy plants grown from cuttings of the same parent all have the same genes, but the amount of sunlight affects the genes governing leaf growth. Ivy plants growing in full sun have smaller leaves than those growing in full shade. Offspring inherit genes, not phenotype;

17. There are five conditions: no genes are undergoing mutation; a very, very large population; the population is isolated from other populations of the species; all members of the population survive and reproduce; and mating is random;

18. a. 0.64 *BB*, 0.16 *Bb*, 0.16 *Bb*, and 0.04 *bb*;  b. genotypes: 0.64 *BB*, 0.32 *Bb*, and 0.04 *bb*;  phenotypes: 96% black, 4% gray;  c.

| Parents | B sperm | b sperm |
|---------|---------|---------|
| 0.64 *BB* | 0.64 | 0 |
| 0.32 *Bb* | 0.16 | 0.16 |
| 0.04 *bb* | 0 | 0.04 |
| Totals = | 0.80 | 0.20 |

19. Find (b) first, then (c), and finally (a). a. $2pq = 2$ X (0.9) X (0.1) = 2 X (0.09) = 0.18 = 18%, which is the percentage of heterozygotes;  b. $p^2 = 0.81$, $p = \sqrt{0.81} = 0.9 =$ the frequency of the dominant allele;  c. $p + q = 1$, $q = 1.00 - 0.9 = 0.1 =$ the frequency of the recessive allele;

20. a. homozygous dominant $= p^2$ X 200 $= (0.8)^2$ X 200 = 0.64 X 200 = 128 individuals;  b. q = (1.00 – p) = 0.20; homozygous recessive $= q^2$ X 200 $= (0.2)^2$ X 200 = (0.04) X (200) = 8 individuals;  c. heterozygotes $= 2pq$ X 200 = 2 X 0.8 X 0.2 X 200 = 0.32 X 200 = 64 individuals. Check: 128 + 8 + 64 = 200;

21. If $p = 0.70$, since $p + q = 1$, 0.70 + q = 1;  then q = 0.30, or 30 percent;

22. If $p = 0.60$, since $p + q = 1$, 0.60 + q = 1;  then q = 0.40; thus, $2pq = 0.48$, or 48 percent;

23. D;  24. H;  25. I;  26. J;  27. G;  28. C;  29. A;  30. F; 31. B;  32. E;  33. population;  34. Hardy-Weinberg; 35. gene frequency;  36. genetic equilibrium;  37. mutations;  38. genetic drift;  39. Gene;  40. Natural selection; 41. Natural selection.

**18.3. Natural Selection Revisited** (p. 285)

**18.4. Directional Change in the Range of Variation** (pp. 286–287)

**18.5. Selection Against or in Favor of Extreme Phenotypes** (pp. 288–289)

**18.6. Special Types of Selection** (pp. 290–291)

1. a. the most common phenotypes are favored;  average human birth weight of 7 pounds;  b. allele frequencies shift in a steady, consistent direction in response to a new environment or a directional change in an old one; light to dark forms of peppered moths;  c. forms at both ends of the phenotypic range are favored, and intermediate forms are selected against;  finches on the Galapagos Islands;  2. a. directional;  b. disruptive;  c. stabilizing; 3. natural;  4. Stabilizing;  5. Directional;  6. Disruptive; 7. Horsetails;  8. Sickle-cell anemia;  9. stabilizing; 10. allele;  11. balanced polymorphism;  12. sexual dimorphism;  13. selection;  14. Sexual;  15. sexual; 16. fitness;  17. c, d;  18. e;  19. c;  20. b;  21. d;  22. b; 23. a;  24. c, d;  25. b;  26. b;  27. b;  28. e;  29. a;  30. e.

**18.7. Gene Flow** (p. 291)

**18.8. Genetic Drift** (pp. 292–293)

1. a;  2. b;  3. a;  4. b;  5. a;  6. a;  7. a;  8. b;  9. a;  10. b; 11. In the founder effect, a few individuals leave a population and establish a new one;  by chance, allele frequencies will differ from the original population. In bottlenecks, disease, starvation, or some other stressful situation nearly eliminates a population;  relative allele frequencies are randomly changed.

**Self-Quiz**

1. b ;  2. a;  3. d;  4. c;  5. a;  6. b;  7. c;  8. b;  9. d;  10. c; 11. a;  12. b.

# Chapter 19   Speciation

*The Case of the Road-Killed Snails* (pp. 296–297)

**19.1. On the Road to Speciation** (pp. 298–299)

**19.2. Reproductive Isolating Mechanisms** (pp. 300–301)

1. c;  2. a;  3. b;  4. a;  5. b;  6. a;  7. c;  8. c;  9. a;  10. c; 11. C (pre);  12. A (pre);  13. B (post);  14. G (pre);  15. H (post);  16. E (pre);  17. D (pre);  18. F (post);  19. f; 20. c;  21. a;  22. e;  23. d;  24. b;  25. f;  26. a. one; b. two;  c. barely between A and B, but considerable divergence begins between B and C;  d. D;  e. B and C.

**19.3. Speciation in Geographically Isolated Populations** (pp. 302–303)

**19.4. Models for Other Speciation Routes** (pp. 304–305)

1. a. Sympatric;  b. Allopatric;  c. Parapatric;  2. sympatric;  3. parapatric;  4. allopatric;  5. allopatric;  6. allopatric;  7. a. Seven pairs;  b. AA;  c. Seven pairs; d. BB;  e. The chromosomes fail to pair in meiosis; f. Fertilization of nonreduced gametes produced a fertile tetraploid;  g. Sterile hybrid, ABD;  h. Fertilization of nonreduced gametes resulted in a fertile hexaploid; i. *Triticum monococcum*;  the unknown wild wheat, *T. tauschii*.

**19.5. Patterns of Speciation** (pp. 306–307)

1. a. Horizontal branching;  b. Many branchings of the same lineage at or near the same point in geologic time; c. A branch that ends before the present;  d. Softly angled branching;  e. Vertical continuation of a branch; f. A dashed line;  2. I;  3. H;  4. E;  5. C;  6. J;  7. D; 8. A;  9. F;  10. B;  11. G;  12. Speciation;  13. species; 14. Divergence;  15. reproductive isolating;  16. geographic;  17. polyploidy (hybridization);  18. hybridization (polyploidy).

**Self-Quiz**

1. c;  2. d;  3. d;  4. a;  5. c;  6. c;  7. b;  8. a;  9. c;  10. d; 11. e;  12. c;  13. c;  14. b.

# Chapter 20   The Macroevolutionary Puzzle

*Of Floods and Fossils* (pp. 310–311)
**20.1. Fossils—Evidence of Ancient Life** (pp. 312–313)
**20.2. Evidence of a Changing Earth** (pp. 314–315)
1. F;  2. J;  3. I;  4. A;  5. E;  6. H;  7. C;  8. D;  9. B;
10. G. 11. D;  12. C;  13. E;  14. A;  15. B;  16. E;  17. F;
18. I;  19. B;  20. G;  21. H;  22. C;  23. A;  24. J;  25. D;
26. Cenozoic;  27. Mesozoic;  28. Paleozoic;  29. Proterozoic;  30. Archean;  31. Cretaceous;  32. Jurassic;
33. Permian;  34. Cambrian;  35. 65;  36. 240;  37. 550;
38. 2,500;  39. 4,600.

**20.3. Evidence from Comparative Embryology**
(pp. 316–317)
**20.4. Evidence of Morphological Divergence**
(pp. 318–319)
**20.5. Evidence from Comparative Biochemistry**
(pp. 320–321)
1. T;  2. F, low;  3. F, convergence;  4. T;  5. F, differences;  6. d;  7. c;  8. a;  9. d;  10. b;  11. M;  12. 1M;
13. M;  14. C;  15. C;  16. C;  17. M;  18. C;  19. M;
20. M;  21. Systematics relies on taxonomy, the naming
and identification of organisms. It also depends heavily
on phylogenetic reconstruction, the identification of
evolutionary patterns that unite different organisms.
Systematics also uses classification, the retrieval systems
that consist of many hierarchical levels, or ranks. The
work of classification feeds back into taxonomy, since
naming species or higher taxa should reflect their place
in the hierarchy.

**20.6. Identifying Species, Past and Present** (pp. 322–323)
**20.7. Finding Evolutionary Relationships Among**
**Species** (pp. 324–325)
**20.8.** *Focus on Science:* **Constructing a Cladogram** (pp.
326–327)
**20.9. How Many Kingdoms?** (pp. 328–329)
1. G;  2. C;  3. J;  4. H;  5. B;  6. A;  7. E;  8. D;  9. I;
10. F;  11. kingdom;  phylum (or division);  class;
order;  family;  genus;  species;  12. kingdom: Plantae;
division: Anthophyta;  class: Dicotyledonae;  order:
Asterales;  family: Asteraceae;  genus: *Archibaccharis*;
species: *lineariloba*;  13. Systematics relies on (a) taxonomy, the naming and identification of organisms;
(b) phylogenetic reconstruction, the identification of
evolutionary patterns that unite different organisms;
and (c) classification, the grouping of organisms using
systems that consist of many hierarchical levels, or ranks
which reflect evolutionary relationships;  14. a. Differences and similarities between organisms are compared,
in a relatively imprecise, subjective manner. Groups of
organisms are defined by traditionally accepted morphological features, such as pollen size and morphology, leaf
size, leaf shape, and flower structure. Such subjective
methods may result in several different interpretations of
the same evidence;  b. Using cladistics, organisms are
linked or grouped together that have shared characteristics (homologies);  the similarities are derived from a
common ancestor;  15. Monophyletic lineages are lineages that share a common evolutionary heritage;
16. The cladogram is a branching diagram (a type of
family tree) that represents the patterns of relationships
of organisms, based on their derived traits;  17. An out-
group is the cladogram taxon exhibiting the fewest derived characteristics;  18. A derived trait is one that is
shared by members of a lineage but not present in their
common ancestor;  19. Cladograms portray relative
relationships among organisms. Distribution patterns of
many homologous structures are examined. Taxa closer
together on a cladogram share a more recent common
ancestor than those that are farther apart. Cladograms do
not convey direct information about ancestors and descendants;
20. a. The numbered traits could be discrete morphological, physiological, and behavioral traits;  b. trait 8;
c. trait 4;  d. trait 5;  e. Taxa that are closer together
share a more recent ancestor than those that are farther
apart;  f. taxon A;  g. A monophyletic taxon is one in
which all members have the same derived traits and
share a single node on the cladogram;  h. The monophyletic taxa and their derived traits are BCDEFG (derived trait 2), CDEFG (derived trait 4), CD (derived trait
5), EFG (derived trait 8), EF (derived traits 9 and 10);
21. B;  22. E;  23. A;  24. D;  25. C.

**Self-Quiz**
1. c;  2. b;  3. d;  4. c;  5. d;  6. b;  7. a;  8. b;  9. a;  10. e;
11. d;  12. c;  13. a.

# Chapter 21  The Origin and Evolution of Life

*In the Beginning* ... (pp. 332–333)
**21.1. Conditions on the Early Earth (pp. 334–335)**
**21.2. Emergence of the First Living Cells (pp. 336–337)**
1. f;  2. e;  3. a;  4. c;  5. d;  6. b;  7. f;  8. c;  9. b;  10. a;
11. d;  12. f;  13. e;  14. c;  15. a;  16. b;  17. e;  18. c;
19. a;  20. b.

**21.3. Origin of Prokaryotic and Eukaryotic Cells (pp. 338–339)**
**21.4. *Focus on Science:* Where Did Organelles Come From? (pp. 340–341)**
1. b;  2. b;  3. a;  4. b;  5. a;  6. a;  7. b;  8. a;  9. a;  10. b;
11. a;  12. b;  13. a;  14. a;  15. b;  16. organelles;
17. gene;  18. plasma;  19. channel;  20. ER;  21. genes;
22. plasmids;  23. prokaryotic;  24. endosymbiosis;
25. eukaryotes;  26. Electron;  27. oxygen;  28. bacteria;
29. ATP;  30. host;  31. incapable;  32. mitochondria;
33. ATP;  34. DNA;  35. chloroplasts;  36. host;  37. eubacteria;  38. protistans;  39. Archaebacterial;  40. Eukaryotes;  41. Eubacterial;  42. Eukaryote;  43. Animals;
44. Thermophiles;  45. Animals;  46. Fungi;  47. Mitochondria;  48. Chloroplasts;  49. Plants;  50. Paleozoic;
51. Mesozoic;  52. Fungi;  53. Plants.

**21.5. Life in the Paleozoic Era (pp. 342–343)**
**21.6. Life in the Mesozoic Era (pp. 344–345)**
**21.7. *Focus on Science:* Horrendous End to Dominance (p. 346)**
**21.8. Life in the Cenozoic Era (p. 347)**
**21.9. Summary of Earth and Life History (pp. 348–349)**
1. G;  2. H;  3. E;  4. K;  5. D;  6. L;  7. F;  8. M;  9. A;
10. N;  11. C;  12. J;  13. B;  14. I;  15. G, H;  16. E;
17. D, F, K, L;  18. M;  19. A, B. C, I, J, N;  20. a. Mesozoic, 240–205;  b. Archean, 4,600–3,800;  c. Paleozoic, 435-360;  d. Paleozoic, 550-500;  e. Archean, 4,600-3,800;
f. Mesozoic, 135-65;  g. Paleozoic, 360-280;  h. Proterozoic, 2,500-570;  i. Archean, 3,800-2,500;  j. Paleozoic, 290-240;  k. Proterozoic, 700-500;  l. Mesozoic, 181-65;
m. Proterozoic, 2,500-570;  n. Cenozoic, 65-1.65;
o. Cenozoic, 1.65-present;  p. Paleozoic, 440-435;  q. Paleozoic, 435-360.

**Crossword Puzzle**

**Self-Quiz**
1. b;  2. c;  3. c;  4. b;  5. e;  6. d;  7. a;  8. d;  9. c;  10. b.

# Chapter 22  Bacteria and Viruses

*The Unseen Multitudes* (pp. 352–353)
**22.1. Characteristics of Bacteria** (pp. 354–355)
**22.2. Bacterial Growth and Reproduction** (pp. 356–357)
1. b;  2. d;  3. c;  4. e;  5. c;  6. prokaryotic;  7. plasmids;
8. wall;  9. capsule;  10. micrometers;  11. binary fission;
12. cocci;  13. bacilli;  14. spiral;  15. positive.


**22.3. Bacterial Classification** (p. 357)
**22.4. Eubacteria** (pp. 358–359)
**22.5. Archaebacteria** (p. 360)
**22.6. Summary of Major Bacterial Groups** (p. 361)
1. Archaebacteria;  2. peptidoglycan;  3. eubacteria;
4. cyanobacteria;  5. nitrogen-fixation;  6. phosphorus;
7. nitrogen;  8. *Rhizobium*;  9. endospores;  10. pressure;
11. *Clostridium botulinum* (*C. tetani*);  12. *Clostridium tetani*
(*C. botulinum*);  13. Lyme disease;  14. spirochete;
15. membrane receptors;  16. sunlight;
17. decomposers;  18. Actinomycetes;  19. *Streptomyces*;
20. K;  21. d, L;  22. c, C;  23. c, J;  24. a, G;  25. c, E;
26. a, A;  27. b, I;  28. c, H;  29. e, B;  30. c, K;  31. c, N;
32. c, D;  33. c, M;  34. a, F.


**22.7. The Viruses** (pp. 362–363)
**22.8. Viral Multiplication Cycles** (pp. 364–365)
**22.9. *Focus On Health*: The Nature of Infectious
  Diseases** (pp. 366–367)
1. a. Nonliving, infectious agents, smaller than the smallest cells; require living cells to act as hosts for their replication; not acted upon by antibiotics.
b. The core can be DNA or RNA; the capsid can be protein and/or lipid.
c. Bacteriophage viruses may use the lytic pathway, in which the virus quickly subdues the host cells and replicates itself and descendants are released as the cell undergoes lysis; or they may use a temperate pathway, in which viral genes remain inactive inside the host cell during a period of latency, which may be a long time, before activation and lysis.

2. There can be multiple answers (see Table 22.1 in text).
a. Possible answers include Herpes simplex (a Herpesvirus), Varicella-zoster (a Herpesvirus), Rhinovirus (a Picornavirus), Poliovirus (an enterovirus of the Picorna group), and HIV (a Retrovirus).
b. Herpes simplex: DNA virus. Initial infection is a lytic cycle that causes Herpes (sores) on mucous membranes on mouth or genitals. Recurrent infections are temperate. Most cells are in nerves and skin. No immunity. No cure. Varicella-zoster: DNA virus. Initial infection is a lytic cycle that causes sores on skin. Generally, immunity is conferred by one infection, but in some people subsequent infections follow temperate cycles and cause "shingles." Rhinovirus: RNA virus. Causes the *common cold*. Host cells are generally mucus-producing cells of respiratory tract. Poliovirus: RNA virus causes *polio*. Host cells are in motor nerves that lead to the diaphragm and other important muscles. Destruction of these nerve cells causes paralysis that may be temporary or permanent. Recurrences can occur. Immunize your children! HIV: RNA virus. Host cells are specific white blood cells. Temperate cycle has a latency period that may last longer than a year before host tests positive for HIV. As white blood cells are destroyed, the host's immune system is progressively destroyed (AIDS). No cure exists.
3. virus;  4. nucleic acid;  5. protein coat (viral capsid);
6. Viruses;  7. *Herpesvirus* (or *Varicella*);  8. viroids;
9. Retroviruses (HIV);  10. temperate;  11. Nanometers;
12. micrometers;  13. 86,000;  14. latency;  15. prions;
16. Rhinoviruses, RNA;  17. Retroviruses, RNA;
18. Herpesviruses, DNA;  19. C;  20. H;  21. E;  22. D;
23. K;  24. I;  25. G;  26. L;  27. F;  28. B;  29. A;  30 J.


**Self-Quiz**
1. d;  2. a;  3. A, C;  4. A;  5. A;  6. B, E;  7. B, F;  8. A, D;  9. A, D;  10. E;  11. F;  12. D;  13. B;  14. C;  15. A.

# Crossword Puzzle

**Across / filled answers:**

- 2. PHOTOAUTOTROPHS
- 3. FISSION
- 6. PATHOGENS
- 7. CONJUGATION
- 8. ANTIBIOTIC
- 10. FLAGELLUM
- 14. CYANOBACTERIA
- 16. GLYCOCALLY
- 17. PROKARYOTIC
- 19. PHOTOHETEROTROPHS
- 20. BACTERIOPHAGE
- 23. MICROORGANISM
- 24. PILUS
- 25. PANDEMIC
- 26. ENDOSHORES
- 27. EPIDEMIC
- 28. HALOPHILE
- 29. METHANOGENS
- 30. COCCUS

**Down / filled answers include:**

- 1. BACILLUS
- 5. GRAM
- 9. CHEMO...
- 11. LYTIC
- 22. PLASMID

## Integrating and Applying Key Concepts

The text identifies natural gas as a fossil fuel along with coal and petroleum because most supplies of it are produced as a result of drilling and mining rather than through methane generation by bacteria.

$$2\,CH_4 + 4\,O_2 \rightarrow 4H_2O + 2\,CO_2 + heat```$$
(for cooking and heating)

Plants then photosynthesize $CO_2$ to produce carbon-based organismal flesh. Methanogens generate methane ($CH_4$) from dead organisms or dung.

---

# Chapter 23  Protistans

*Kingdom at the Crossroads* (pp. 370–371)

**23.1. Parasitic and Predatory Molds** (pp. 372–373)

1. a. E;  b. E;  c. P;  d. E;  e. P;  f. E;  g. E;  2. a. A C, D, E;  b. F, G, J, K;  d. B;  e. H, I, (K), L;  f. O;  g. B;  h. H, I, L;  i. P;  j. A, E;  k. F, G, K;  l. M, N, Q;  3. Chytrids;  4. water molds (Oomycetes);  5. red;  6. slime molds;  7. fruiting bodies;  8. myxobacteria (monerans);  9. fungi;  10. animals;  11. chitin;  12. chytrids;  13. parasitic;  14. mycelium;  15. decomposers;  16. enzymes;  17. saprobes;  18. parasites;  19. spore.

**23.2. The Animal-Like Protistans** (p. 374)
**23.3. Amoeboid Protozoans** (pp. 374–375)
**23.4. Major Parasites—Animal-Like Flagellates and Sporozoans** (p. 376)
**23.5.** *Focus on Health:* **Malaria and the Night-Feeding Mosquitoes** (p. 377)
**23.6. A Sampling of Ciliated Protozoans** (pp. 378–379)
1. pseudopods; 2. Foraminiferans; 3. spines; 4. radiolarians; 5. trypanosomes; 6. freshwater; 7. contractile vacuoles; 8. gullet; 9. enzyme-filled vesicles; 10. *Plasmodium* ; 11. mosquito; 12. gametes; 13. C, E; 14. C, E, I; 15. C, E, K; 16. A; 17. D, H; 18. B, F; 19. B, F, G; 20. A; 21. D; 22. C; 23. B; 24. E.

**23.7. A Sampling of the (Mostly) Single-Celled Algae** (pp. 380–381)
**23.8. Red Algae** (p. 382)
**23.9. Brown Algae** (p. 383)
**23.10. Green Algae** (pp. 384–385)
1. chloroplasts ; 2. eyespot; 3. heterotrophic; 4. algae; 5. Chrysophytes; 6. diatoms; 7. fucoxanthin; 8. silica (glass); 9. diatom shells; 10. filtering; 11. phytoplankton; 12. Dinoflagellates (*Gymnodinium*); 13. agar; 14. marine; 15. walls; 16. brown; 17. algin; 18. photosynthetic; 19. cellulose; 20. starch; 21. a. chlorophyll $\underline{a}$, phycobilins; b. cyanobacteria;

c. agar, used as a moisture-preserving agent and culture medium; carrageenan is a stabilizer of emulsions; d. *Bonnemaisonia, Euchema, Porphyra*; e. chlorophyll $\underline{a}$, $c_1$, and $c_2$, funcoxanthin and other carotenoids; f. chrysophytes; g. algin, used as a thickener, emulsifier and stabilizer of foods, cosmetics, medicines, paper and floor polish; also are sources of mineral salts and fertilizer; h. *Postelsia* (sea palm), *Sargassum, Laminaria*; i. chlorophylls $\underline{a}$ and $\underline{b}$; j. a heterotrophic prokaryote fusing with endosymbiont autotrophic prokaryote; k. chlorophytes form much of the phytoplankton base of many food webs that support humans; l. *Volvox, Ulva, Spirogyra, Chlamydomonas*; 22. zygote, F; 23. resistant zygote, B; 24. meiosis and germination, H; 25. asexual reproduction, C; 26. gamete production, E; 27. gametes meet, G; 28. cytoplasmic fusion, A; 29. fertilization, D; 30. (+): a, c, d, e, g; (–) b, f, h, i, j.

**Self-Quiz**
1. b; 2. c; 3. b; 4. b; 5. a; 6. a; 7. b; 8. c; 9. a; 10. b; 11. d; 12. A, E; 13. A, I; 14. A, B; 15. A, E, J; 16. A, C, D; 17. A, H; 18. A, F, G; 19. A, C; 20. B; 21. D; 22. C; 23. A.

# Chapter 24   Fungi

*Dragon Run* (pp. 388–389)
**24.1. Characteristics of Fungi** (p. 390)
1. D; 2. E; 3. C; 4. H; 5. G; 6. B; 7. A; 8. F.

**24.2. Consider the Club Fungi** (pp. 390–391)
1. reproductive; 2. club; 3. Nuclear; 4. diploid; 5. Meiosis; 6. spores; 7. cytoplasmic; 8. dikaryotic; 9. C; 10; D; 11. B; 12. A; 13. E.

**24.3. Spores and More Spores** (pp. 392–393)
1. multicelled; 2. T; 3. zygosporangium; 4. *Penicillium*; 5. *Neurospora crassa*; 6. truffles; 7. asexual; 8. sexual; 9. T; 10. meiosis; 11. hyphae; 12. gametangia; 13. zygosporangium; 14. spores; 15. mycelium; 16. If the sexual phase is unknown for a particular fungus, it is referred to the informal group known as the "imperfect" fungi. If mycologists later detect a sexual phase in the life cycle of the fungus, it is then assigned to (usually) the sac fungi or the club fungi. An example of an imperfect fungus is the fungus that is predatory on roundworms, *Arthrobotrys dactyloides*.

**24.4. Beneficial Associations Between Fungi and Plants** (pp. 394–395)

**24.5.** *FOCUS ON SCIENCE:* **A Look at the Unloved Few** (p. 396)
1. Symbiosis; 2. mutualism; 3. lichen; 4. mycobiont; 5. photobiont; 6. fungus; 7. hostile; 8. photobiont; 9. photobiont; 10. lichen's; 11. mycobiont; 12. substrate; 13. hypha; 14. cytoplasmic; 15. mycobiont; 16. photobiont; 17. layers; 18. growth; 19. mineral ions; 20. nitrogen; 21. antibiotics; 22. toxins; 23. soil; 24. colonization; 25. nitrogen; 26. environment; 27. C; 28. E; 29. B; 30. F; 31. A; 32. D; 33.a. Basidiomycetes; b. Ascomycetes; c. Zygomycetes; d. Ascomycetes.

**Self-Quiz**
1. *Polyporus* (C); 2. *Pilobolus* (B); 3. fly agaric mushroom (C); 4. cup fungus (A); 5. algae and fungi—lichen (E); 6. algae and fungi—lichen (E); 7. mushroom (C) ; 8. *Penicillium* (A); 9. b; 10. d; 11. d; 12. c; 13. a; 14. a; 15. c; 16. c; 17. d; 18. a; 19. a and c; 20. b.

# Chapter 25  Plants

*Pioneers in a New World* (pp. 398–399)

**25.1. Evolutionary Trends Among Plants** (pp. 400–401)

1. C;  2. D;  3. E;  4. A;  5. B;  6.a. Well-developed root systems;  b. Well-developed shoot systems;  c. Lignin production;  d. Xylem and phloem;  e. Cuticle production;  f. Stomata;  g. Gametophytes;  h. Sporophytes;  i. Spores;  j. Heterospory;  k. Pollen grains;  l. Seeds.
7.

```
                [1]C
                   U
[2]G  A  M  E  T  O  P  H  Y  T  E  S
                   I
                   C        [3]B
[4]V [5]A  S  C  U  L  A  R
     N           E           Y
     G                       O
[6]L  I  G  N  I  N     [7]P  H  L  O  E  N
     O                      H
     S           [8]G  Y  M  N  O  [9]S  P  E  [10]R  M  [11]S
     P               T               H          O          T
     E        [12]X  Y  L  E  M               O          O
     R               S               O          T          M
     M                               T                     A
[13]S [14]P  O  R  O  P  H  Y  T  E                         T
     O                                                     A
     L
     L
     E
     N
```

**25.2. Bryophytes**
(pp. 402–403)

1. air;  2. T;  3. rhizoids;  4. T;  5. mosses;  6. sporophytes;  7. T;  8. Peat;  9. water;  10. liverwort;  11. gametophytes;  12. gametophytes;  13. sperms;  14. Fertilization;  15. zygote;  16. sporophyte;  17. Meiosis;  18. spores;  19. gametophytes.

**25.3. Existing Seedless Vascular Plants** (pp. 404–405)

**25.4.** *Focus on Science:* **Ancient Carbon Treasures**
(p. 406)

1. c;  2. b;  3. e;  4. b;  5. d;  6. d;  7. e;  8. c;  9. a;  10. d;  11. e;  12. d;  13. c;  14. c;  15. e;  16. b;  17. d;  18. d;  19. b;  20. e;  21. rhizome;  22. sorus;  23. Meiosis;  24. spores;  25. spores;  26. gametophyte;  27. gametophyte;  28. sperm;  29. egg;  30. fertilization;  31. zygote;  32. embryo.

**25.5. The Rise of the Seed-Bearing Plants** (p. 407)

1. D;  2. F;  3. B;  4. C;  5. E;  6. A.

**25.6. Gymnosperms—Plants with "Naked" Seeds**
(pp. 408–409)

**25.7. A Closer Look at the Conifers** (pp. 410–411)

1. b;  2. d;  3. a;  4. b;  5. c;  6. a;  7. e;  8. b;  9. e;  10. c;  11. d;  12. e;  13. sporophyte;  14. cones;  15. cones;  16. ovule;  17. meiosis;  18. gametophyte;  19. eggs;  20.

meiosis; 21. pollen; 22. Pollination; 23. tube; 24. Sperm; 25. Fertilization; 26. embryo.

## 25.8. Angiosperms—The Flowering, Seed-Bearing Plants (pp. 412–413)
1. B; 2. C; 3. E; 4. F; 5. D; 6. A; 7. H. 8. G; 9. a. gametophyte, none or simple vascular tissue, no; b. sporophyte, yes, no; c. sporophyte, yes, no; d. sporophyte, yes, no; e. sporophyte, yes, yes; f. sporophyte, yes, yes.

### Self-Quiz
1. b; 2. c; 3. a; 4. c; 5. d; 6. b; 7. a; 8. d; 9. c; 10. c.

---

# Chapter 26    Animals: The Invertebrates

*Madeleine's Limbs* (pp. 416–417)
### 26.1. Overview of the Animal Kingdom (pp. 418–419)
### 26.2. Puzzles About Origins (p. 420)
### 26.3. Sponges—Success in Simplicity (pp. 420–421)
1. O; 2. I; 3. H; 4. E; 5. P; 6. A; 7. F, 8. K; 9. C; 10. J; 11. D; 12. G; 13. N; 14. M; 15. L; 16. B; 17. a. Placozoan; b. Sponges; c. Cnidaria; d. Turbellarians, flukes, tapeworms; e. Nematoda; f. Rotifera; g. Snails, slugs, clams, squids, octopuses; h. Annelida; i. Crustaceans, spiders, insects; j. Echinodermata; k. Chordata; l. Vertebrates; 18. c; 19. a; 20. c; 21. d; 22. c; 23. b; 24. c; 25. a; 26. c; 27. b; 28. K; 29. H; 30. L; 31. D; 32. J. 33. B ; 34. C; 35. I; 36. E; 37. A; 38. G; 39. F.

### 26.4. Cnidarians—Tissues Emerge (pp. 422–423)
### 26.5. Variations on the Cnidarian Body Plan (pp. 424–425)
### 26.6. Comb Jellies (p. 425)
1. Cnidaria; 2. nematocysts; 3. medusa; 4. polyp; 5. gastrodermis; 6. epidermis; 7. epithelium; 8. Nerve; 9. receptor (contractile); 10. contractile (receptor); 11. mesoglea; 12. hydrostatic; 13. corals; 14. gonads; 15. planulas; 16. feeding polyp; 17. reproductive polyp; 18. female medusa; 19. planula larva; 20. multiple; 21. Ctenophora; 22. mesoderm; 23. mirror; 24. radial; 25. eight; 26. forward; 27. do not; 28. lips; 29. two.

### 26.7. Acoelomate Animals—And the Simplest Organ Systems (pp. 426–427)
1. b, c; 2. a; 3. c; 4. a; 5. c; 6. d; 7. c; 8. d; 9. a; 10. a; 11. b; 12. d; 13. a; 14. c; 15. d; 16. a; 17. c; 18. c; 19. d; 20. c; 21. b; 22. c; 23. d; 24. d; 25. b; 26. branching gut; 27. pharynx (protruding); 28. brain; 29. nerve cord; 30. ovary; 31. testis; 32. planarian (genus name = *Dugesia*); 33. no; 34. yes; 35. no coelom or acoelomate.

### 26.8. Roundworms (p. 428)
### 26.9. *Focus on Science:* A Rogue's Gallery of Worms (pp. 428–429)
### 26.10. Rotifers (p. 430)
1. roundworm; 2. no; 3. a "false" coelom (pseudocoelomate); 4. rotifer; 5. bilateral; 6. yes; 7. false coelom; 8. Rotifers have a crown of cilia at the head end that assists in swimming and in wafting food toward the mouth; its rhythmic motions reminded early microscopists of a turning wheel; 9. b; 10. a; 11. a; 12. a; 13. b; 14. a; 15. a; 16. b; 17. a; 18. a; 19. b; 20. b; 21. a; 22. b; 23. a; 24. b; 25. a; 26. a; 27. sexually; 28. larvae; 29. human; 30. larvae; 31. human; 32. Larvae; 33. intermediate; 34. human; 35. small intestine; 36. proglottids; 37. organs; 38. proglottids; 39. feces; 40. larval; 41. intermediate.

### 26.11. Two Major Divergences (p. 430)
### 26.12. A Sampling of Molluscan Diversity (p. 431)
### 26.13. Evolutionary Experiments with Molluscan Body Plans (pp. 432–433)
1. b; 2. b; 3. a; 4. b; 5. a; 6. b; 7. a; 8. a; 9. b; 10. a; 11. III; 12. I; 13. II; 14. mouth; 15. anus; 16. gill; 17. heart; 18. radula; 19. foot; 20. shell; 21. stomach; 22. mouth; 23. gill; 24. mantle; 25. muscle; 26. foot; 27. stomach; 28. internal shell; 29. mantle; 30. reproductive organ; 31. gill; 32. ink sac; 33. tentacle; 34. soft; 35. bilateral; 36. mantle; 37. shell; 38. gills; 39. foot; 40. head; 41. radula; 42. cephalopods; 43. gastropods; 44. torsion; 45. slugs; 46. bivalves; 47. Humans; 48. respiration; 49. mucus; 50. palps; 51. siphons; 52. wastes; 53. Cephalopods; 54. smartest; 55. tentacles; 56. jet; 57. oxygen; 58. closed; 59. hearts; 60. nervous; 61. brains; 62. humans; 63. invertebrates.

### 26.14. Annelids—Segments Galore (pp. 434–435)
1. F; 2. L; 3. B; 4. I; 5. D; 6. G; 7. N; 8. C; 9. K; 10. J; 11. A; 12. H; 13. E; 14. M; 15. brain; 16. pharynx; 17. nerve cord; 18. hearts; 19. blood vessel; 20. crop; 21. gizzard; 22. earthworm; 23. Annelida; 24. segmentation and a closed circulatory system; 25. protostome; 26. yes; 27. bilateral; 28. yes.

### 26.15. Arthropods—The Most Successful Organisms on Earth (p. 436)
1. e; 2. f; 3. a; 4. b; 5. f; 6. a; 7. f; 8. d; 9. b; 10. c; 11. e; 12. d; 13. b; 14. a; 15. a; 16. Arthropods, as a group, have the highest number of species, occupy the most habitats, and have very efficient defenses against predators and competitors, and the capacity to exploit the greatest amounts and kinds of foods.

### 26.16. A Look at Spiders and Their Kin (p. 437)

**26.17. A Look at the Crustaceans** (pp. 438–439)

**26.18. How Many Legs?** (p. 439)

1. E; 2. G; 3. F; 4. D; 5. C; 6. B; 7. A; 8. an exoskeleton; 9. lobsters and crabs; 10. similar; 11. Lobsters and crabs; 12. Barnacles; 13. Copepods; 14. barnacles; 15. barnacles; 16. molts; 17. millipedes; 18. centipedes; 19. Millipedes; 20. Centipedes; 21. cephalothorax; 22. abdomen; 23. swimmerets; 24. legs; 25. cheliped; 26. antennae; 27. lobster; 28. crustaceans; 29. exoskeleton and jointed legs; 30. protostome; 31. yes; 32. bilateral; 33. yes; 34. poison gland; 35. brain; 36. heart; 37. spinnerets; 38. book lung; 39. chelicerates.

**26.19. A Look at Insect Diversity** (pp. 440–441)

1. C; 2. E; 3. A; 4. D; 5. B.

**26.20. The Puzzling Echinoderms** (pp. 442–443)

1. deuterostomes; 2. echinoderms; 3. calcium carbonate; 4. skeleton; 5. radial; 6. bilateral; 7. brain; 8. nervous; 9. arm; 10. tube; 11. water; 12. ampulla; 13. muscle; 14. whole; 15. digesting; 16. anus; 17. lower stomach; 18. upper stomach; 19. anus; 20. gonad; 21. coelom; 22. digestive gland; 23. eyespot; 24. tube feet; 25. starfish; 26. yes; 27. deuterostome; 28. water vascular; 29.a. brittle star; b. sea urchin; c. feather star (crinoid); d. sea cucumber; 30. Echinodermata; 31. A water vascular system and a body wall with spines, spicules, or plates; 32. radial.

**Self-Quiz**

1. a; 2. c; 3. a; 4. d; 5. b; 6. c; 7. d; 8. c; 9. d; 10. a; 11. a; 12. b; 13. h, I; 14. b, FJL; 15. c, CDQR; 16. g, BG; 17. k, H; 18. e, M; 19. l, P; 20. j, A; 21. f, EO; 22. a, K; 23. d, Q; 24. i, N.

# Chapter 27    Animals: The Vertebrates

*Making Do (Rather Well) With What You've Got* (pp. 446–447)

**27.1. The Chordate Heritage** (p. 448)

**27.2. Invertebrate Chordates** (pp. 448–449)

**27.3. Evolutionary Trends Among the Vertebrates** (pp. 450–451)

**27.4. Existing Jawless Fishes** (p. 451)

**27.5. Existing Jawed Fishes** (pp. 452–453)

1. nerve cord; 2. pharynx; 3. notochord; 4. invertebrate; 5. vertebrates; 6. lancelets; 7. filter-feeding; 8. gill slits; 9. Tunicates (sea squirts); 10. tadpoles; 11. notochord; 12. spine (backbone); 13. lungs; 14. circulatory; 15. larva; 16. mutation; 17. sex organs; 18. predators; 19. cephalochordates (lancelets); 20. vertebrae; 21. predators; 22. jaws; 23. brain; 24. paired; 25. fleshy; 26. tunicate adult (sea squirt); 27. lancelet; 28. notochord; 29. pharynx with gill slits; 30. mouth; 31. sievelike pharynx; 32. stomach; 33. intestine; 34. anus; 35. atrial opening; 36. oral opening; 37. pharyngeal gill slits; 38. anus; 39. notochord; 40. dorsal, tubular nerve cord; 41. early jawless fish (agnathan); 42. supporting structure; 43. gill slit; 44. placoderm; 45. jaws; 46. acorn worms; 47. tunicates; 48. lancelets; 49. lampreys, hagfishes; 50. early jawed fishes; 51. cartilaginous fishes; 52. bony fishes; 53. Amphibia; 54. Reptilia; 55. Aves; 56. Mammalia; 57. sharks; 58. gill slits; 59. bony; 60. ray-finned; 61. C; 62. G; 63. J; 64. F; 65. M; 66. H; 67. N; 68. I; 69. B; 70. O; 71. D; 72. E; 73. E; 74. K; 75. L; 76. A; 77. all jawless; 78. filter- feeders; 79. jaws; 80. cartilage; 81. bone; 82. ray-finned fishes; 83. lobe-finned fishes; 84. branch 5; 85. branch 4; 86. Silurian; 87. about 375 million years ago.

**27.6. Amphibians** (pp. 454–455)

**27.7. The Rise of Reptiles** (pp. 456–457)

**27.8. Major Groups of Existing Reptiles** (pp. 458–459)

1. lungs; 2. fins; 3. skeletal elements; 4. brains; 5. balance; 6. circulatory; 7. blood; 8. insects; 9. salamanders; 10. water; 11. reproduce; 12. toxins; 13. insects; 14. reptilian; 15. limb; 16. shelled amniote; 17. turtles (lizards); 18. lizards (turtles); 19. internal; 20. Carboniferous; 21. turtles; 22. Carboniferous; 23. Triassic; 24. Jurassic; 25. Birds; 26. four; 27. synapsid; 28. Carboniferous; 29. G; 30. D; 31. E; 32. B; 33. C; 34. A; 35. F; 36. A; 37. A; 38. A; 39. C; 40. C; 41. D; 42. D; 43. a. dry, scaly skin; b. four-chambered heart; c. feather development; d. hair development; e. loss of limbs.

**27.9. Birds** (pp. 460–461)

**27.10. The Rise of Mammals** (pp. 462–463)

**27.11. A Portfolio of Existing Mammals** (pp. 464–465)

1. embryo (notochord); 2. albumin; 3. yolk sac; 4. reptiles; 5. feathers; 6. sternum (breastbone); 7. air cavities; 8. shelled amniote; 9. temperature; 10. teeth; 11. high; 12. oxygen, $O_2$; 13. 4; 14. hummingbird; 15. ostrich; 16. mammals; 17. hair; 18. platypus; 19. pouched (marsupials); 20. placental; 21. four; 22. hair; 23. dentition; 24. orders; 25. synapsids; 26. therians; 27. dinosaurs; 28. convergent evolution.

**Self-Quiz**

1. d; 2. a; 3. d; 4. c; 5. early amphibian, Amphibia; 6. Arctic fox, Mammalia; 7. soldier fish, Osteichthyes; 8. Ostracoderm, Agnatha; 9. owl, Aves; 10. sea turtle, Reptilia; 11. shark, Chondrichthyes; 12. coelacanth, Osteichthyes (lobe-finned fish); 13. reef ray, Chondrichthyes; 14. tunicate, Urochordata; 15. lancelet, Cephalochordata; 16. d, H; 17. b, B; 18. i, F; 19. g, D; 20. h, A, C; 21. c, E; 22. a, I; 23. e, G; 24. f, C.

# Chapter 28  Human Evolution: A Case Study

*The Cave at Lascaux and the Hands of Gargas*
(pp. 468–469)

### 28.1. Evolutionary Trends Among the Primates
(pp. 470–471)

### 28.2. From Primates to Hominids (pp. 472–473)
1. mammals;  2. hair (mammary glands);  3. Orders;
4. Primates;  5. depth;  6. prehensile;  7. grasping;
8. smell;  9. brains;  10. culture;  11. handbones;
12. teeth;  13. 60;  14. rodents;  15. insects;  16. Miocene;
17. drier;  18. grasslands;  19. dryopiths;  20. Asia;
21. 10;  22. 5;  23. L;  24. N;  25. A;  26. F;  27. E;  28. H;
29. M;  30. I;  31. J;  32. C;  33. K;  34. D;  35. B;  36. G;
37. O;  38. chimpanzee-like apes;  39. 10;  40. 5;
41. Four;  42. australopiths;  43. bipedal;  44. b;  45. c;
46. c;  47. c;  48. c;  49. b;  50. d;  51. a;  52. a;  53. c;
54. B, G, K;  55. A, G, K;  56. C, D, J;  57. B, G, H, K;
58. E, F, I.

### 28.3. Emergence of Early Humans (pp. 474–475)
### 28.4. *Focus on Science:* Out of Africa—Once, Twice, or . . . (p. 476)
1. Hominids;  2. five;  3. Australopiths;  4. brain;
5. manual dexterity;  6. *habilis*;  7. 2.5;  8. *Homo erectus*;
9. fire;  10. tool;  11. Neandertals;  12. *Homo erectus*;
13. nucleic acid hybridization;  14. immunological;
15. Africa;  16. 100,000;  17. 40,000;  18. *Homo sapiens*;
19. anthropoids;  20. hominoids;  21. hominids;  22. Gorilla;  23. Chimpanzee;  24. *Australopithecus afarensis*
(Lucy);  25. *Homo erectus*.

### Self-Quiz
1. c;  2. a;  3. b;  4. a;  5. a;  6. b;  7. d;  8. B;  9. H;
10. C;  11. D;  12. A;  13. E;  14. G;  15. F.

### Crossword Puzzle

| 1 P | R | I | M | 2 A | R | Y | | 3 H | 4 O | M | I | N | O | 5 I | D |
|---|---|---|---|---|---|---|---|---|---|---|---|---|---|---|---|
| | | | | U | | | | | A | | | | | N | |
| 6 H | | 7 B | U | S | H | Y | | 8 S | | M | | 9 B | | C | |
| O | | | | T | | | | A | | M | | 10 I | R | I | S |
| 11 M | O | L | A | R | | | | P | | A | | P | | S | |
| I | | | | A | | | | I | | L | | E | | O | |
| N | | 12 C | U | L | T | U | R | E | | | 13 A | D | O | R | E |
| I | | | | O | | | | N | | | A | | | | |
| 14 D | R | Y | O | P | I | T | H | S | | 15 E | | L | | 16 T | |
| | | | | I | | | | | 17 C | A | N | I | N | E | S |
| 18 E | R | E | C | T | U | S | | | R | | | S | | E | |
| | | | | H | | | | | L | | | M | | T | |
| | 19 T | A | R | S | I | E | R | S | Y | | | | | H | |

---

# Chapter 29  Plant Tissues

*Plants Versus the Volcano* (pp. 480–481)

### 29.1. Overview of the Plant Body (pp. 482–483)
1. ground tissue (C);  2. vascular tissues (E);  3. dermal
tissues (D);  4. shoot system (A);  5. root system (B);
6. E;  7. D;  8. B;  9. F;  10. A;  11. C;  12. radial;
13. tangential;  14. transverse or cross section;  15. shoot
apical meristem (D);  16. shoot transitional meristems
(B);  17. root transitional meristems (E);  18. root apical
meristems (A);  19. lateral meristems (C);  20.a. Epidermis;  primary;  b. Ground tissues;  primary;  c. Vascular tissues;  primary;  d. Vascular tissues;  secondary;
e. Periderm;  secondary.

### 29.2. Types of Plant Tissues (pp. 484–485)
1. sclerenchyma;  2. sclerenchyma;  3. collenchyma;
4. sclerenchyma;  5. parenchyma;  6. sclerenchyma;
7. b;  8. c;  9. a;  10. a;  11. b;  12. b;  13. a;  14. c;  15. a;
16. a;  17. c;  18. pits (xylem);  19. cytoplasm (xylem);

20. tracheids (xylem); 21. vessel (xylem); 22. vessel (xylem); 23. sieve (phloem); 24. companion (phloem); 25. sieve (phloem); 26. sieve (phloem); 27.a. Vessel members and tracheids; no; conduct water and dissolved minerals absorbed from soil, mechanical support. b. Sieve-tube members and companion cells; yes; transports sugar and other solutes. 28.a. Primary plant body; cutin in the cuticle layer over epidermal cells restricts water loss and resists microbial attack; openings (stomata) permit water vapor and gases to enter and leave the plant. b. Secondary plant body; replaces epidermis to cover roots and stems. 29.a. One; in threes or multiples thereof, usually parallel; one pore or furrow; distributed throughout ground stem tissue. b. Two, in fours or fives or multiples thereof; usually netlike; three pores or pores with furrows; positioned in a ring in the stem.

### 29.3. Primary Structure of Shoots (pp. 486–487)
1. primordium; 2. apical meristem; 3. bud; 4. shoot; 5. leaf; 6. procambium; 7. protoderm; 8. procambium; 9. ground meristem; 10. epidermis; 11. cortex; 12. procambium; 13. pith; 14. primary xylem; 15. primary phloem; 16. epidermis; 17. ground tissue; 18. vascular bundle; 19. sclerenchyma cells; 20. air space; 21. xylem vessel; 22. sieve-tube; 23. companion cell; 24. epidermis; 25. cortex; 26. vascular bundle; 27. pith; 28. vessel; 29. meristematic cells; 30. sieve tube; 31. fibers.

### 29.4. A Closer Look at Leaves (pp. 488–489)

### 29.5. Commentary: Uses and Abuses of Shoots (pp. 490–491)
1. blade; 2. petiole (leaf stalk); 3. axillary bud; 4. node; 5. stem; 6. sheath; 7. dicot; 8. monocot; 9. palisade mesophyll (D); 10. spongy mesophyll (B); 11. lower epidermis (A); 12. stoma (C); 13. leaf vein (E); 14. G; 15. K; 16. C; 17. L; 18. E; 19. I; 20. D; 21. J; 22. B; 23. A; 24. F; 25. M; 26. H; 27. N.

### 29.6. Primary Structure of Roots (pp. 492–493)
1. root hair (H); 2. endodermis (G); 3. pericycle (B); 4. epidermis (E); 5. cortex (D); 6. root apical meristem (F); 7. root cap (A); 8. endodermis (G); 9. pericycle (B); 10. primary phloem (C); 11. primary; 12. lateral; 13. taproot; 14. lateral; 15. fibrous; 16. hairs; 17. vascular; 18. pericycle; 19. cortex; 20. pith; 21. air; 22. oxygen; 23. endodermis; 24. control; 25. pericycle; 26. lateral; 27. lateral.

### 29.7. Accumulated Secondary Growth—The Woody Plants (pp. 494–495)
### 29.8. A Look at Wood and Bark (pp. 496–497)
1. C; 2. M; 3. Q; 4. N; 5. G; 6. F; 7. P; 8. R; 9. H; 10. O; 11. B; 12. A; 13. I; 14. L; 15. D; 16. J; 17. E; 18. K; 19. vessel (E); 20. early wood (F); 21. late wood (B); 22. bark (D); 23. one (C); 24. two, three (A); 25. vascular cambium (G).

### Self-Quiz
1. b; 2. a; 3. b; 4. c; 5. a; 6. d; 7. d; 8. d; 9. c; 10. c; 11. b; 12. b.

---

# Chapter 30    Plant Nutrition and Transport

*Flies for Dinner* (pp. 500–501)
### 30.1. Soil and Its Nutrients (pp. 502–503)
1. F; 2. C; 3. I; 4. B; 5. H; 6. A; 7. J; 8. E; 9. G; 10. D; 11. hydrogen; 12. thirteen; 13. mineral ions; 14. macronutrients; 15. micronutrients; 16.a. iron, micronutrient; b. potassium, macronutrient; c. magnesium, macronutrient; d. chlorine, micronutrient; e. manganese, micronutrient; f. molybdenum, micronutrient; g. nitrogen, macronutrient; h. sulfur, macronutrient; i. boron, micronutrient; j. zinc, micronutrient; k. calcium, macronutrient; l. copper, micronutrient; m. phosphorus, macronutrient.

### 30.2. Absorption of Water and Mineral Ions into Roots (pp. 504–505)
1. exodermis (B); 2. root hair (E); 3. epidermis (H); 4. vascular cylinder (J); 5. cortex (C); 6. endodermis (I); 7. cytoplasm (A); 8. water movement (G); 9. Casparian strip (D); 10. endodermal cell wall (F); 11. C; 12. D; 13. A; 14. B; 15. E; 16. mutualism; 17. Nitrogen "fixed" by bacteria; 18. root nodules; 19. scarce minerals that the fungus is better able to absorb; 20. obtaining sugars and nitrogen-containing compounds.

### 30.3. A Theory of Water Transport Through Plants (pp. 506–507)
1. water; 2. xylem; 3. tracheids; 4. vessel; 5. transpiration; 6. hydrogen; 7. cohesion; 8. tension; 9. roots; 10. xylem.

### 30.4. Conservation of Water in Stems and Leaves (pp. 508–509)
1. 90 percent; 2. T; 3. T; 4. stomata; 5. opens; 6. T; 7. T; 8. decrease; 9. opens; 10. T; 11. night; 12. day.

### 30.5. Distribution of Organic Compounds Through the Plant (pp. 510–511)
1. photosynthesis; 2. starch; 3. proteins; 4. seeds; 5. Starch; 6. Proteins; 7. Storage; 8. solutes; 9. sucrose; 10. sucrose; 11. D; 12. F; 13. B; 14. G; 15. A; 16. C; 17. E.

### Self-Quiz
1. c; 2. e; 3. c; 4. d; 5. b; 6. d; 7. b; 8. a; 9. b; 10. c.

# Chapter 31   Plant Reproduction

*A Coevolutionary Tale* (pp. 514–515)

**31.1. Reproductive Structures of Flowering Plants** (pp. 516–517)

**31.2.** *Focus on the Environment:* **Pollen Sets Me Sneezing** (p. 517)

1. sporophyte (C);  2. flower (B);  3. meiosis (D); 4. gametophyte (F);  5. gametophyte (E);  6. fertilization (A);  7. sepal;  8. petal;  9. stamen;  10. filament; 11. anther;  12. carpel;  13. stigma;  14. style;  15. ovary; 16. ovule;  17. receptacle;  18. I;  19. C;  20. G;  21. F; 22. B;  23. E;  24. H;  25. A;  26. D.

**31.3. A New Generation Begins** (pp. 518–519)

1. anther;  2. pollen sac;  3. microspore mother;  4. meiosis;  5. microspores;  6. pollen tube;  7. sperm-producing;  8. pollen;  9. stigma;  10. male gametophyte; 11. ovule;  12. integuments;  13. meiosis;  14. megaspores;  15. megaspore;  16. megaspore;  17. mitosis; 18. eight;  19. embryo sac;  20. female gametophyte; 21. two;  22. endosperm;  23. egg;  24. double fertilization;  25. endosperm;  26. embryo;  seed;  27. seed; 28. coat;  29. Chemical and molecular cues guide a pollen tube's growth through tissues of the style and the ovary, toward the egg chamber and sexual destiny; 30. The embryo sac is the site of fertilization;  31.a. Fusion of one egg nucleus (*n*) with one sperm nucleus (*n*); the plant embryo (2*n*);  eventually develops into a new sporophyte plant. b. Fusion of one sperm nucleus (*n*) with the endosperm mother cell (2*n*);  endosperm tissues (3*n*);  nourishes the embryo within the seed.

**31.4. From Zygote to Seeds and Fruits** (pp. 520–521)

1.a. "Seed leaves" that develop from two lobes of meristematic tissue of the embryo;  b. Seeds are mature ovules;  c. Integuments of the ovule harden into the seed coat;  d. A fruit is a mature ovary;  2. nucleus;  3. vacuole;  4. zygote;  5. embryo;  6. embryo;  7. seed coat; 8. shoot tip;  9. cotyledons;  10. sporophyte;  11. endosperm;  12. root tip;  13. mature;  14. fruit;  15. mitotic;  16. sporophyte;  17. ovule;  18. fruit; 19. cotyledons;  20. two;  21. one;  22. endosperm; 23. germinates;  24. thin;  25. enzymes;  26. seedling; 27. ovule;  28. endosperm;  29. ovary;  30. coat; 31. seed;  32. ovule;  33. fruits;  34. dry;  35. ovaries; 36. receptacle;  37. fleshy;  38. endocarp;  39. mesocarp; 40. exocarp;  41. pericarp;  42. f;  43. a;  44. g;  45. b; 46. e;  47;  c;  48. d.

**31.5. Dispersal of Fruits and Seeds** (p. 522)

**31.6.** *Focus on Science:* **Why So Many Flowers And So Few Fruits?** (p. 523)

1. c;  2. c;  3. b;  4. a;  5. a;  6. c;  7. b;  8. b;  9. b;  10. a; 11. b.

**31.7. Asexual Reproduction of Flowering Plants** (pp. 524–525)

1. E;  2. G;  3. I;  4. F;  5. H;  6. D;  7. B;  8. C;  9. A; 10. D;  11. E;  12. A;  13. B;  14. C.

**Self-Quiz**

1. b;  2. c;  3. d;  4. c;  5. c;  6. d;  7. b;  8. c, 9. c;  10. d.

---

# Chapter 32   Plant Growth and Development

*Foolish Seedlings and Gorgeous Grapes* (pp. 528–529)

**32.1. Patterns of Early Growth and Development—An Overview** (pp. 530–531)

1. embryo;  2. germination;  3. environmental;  4. imbibition;  5. ruptures;  6. aerobic;  7. meristematic; 8. root;  9. primary root;  10. germination;  11. heritable (genetic);  12. genes;  13. zygote;  14. genetic;  15. cytoplasmic;  16. metabolism;  17. selective;  18. hormones; 19. Interactions;  20. cotyledons;  21. hypocotyl; 22. branch;  23. primary;  24. primary;  25. cotyledons; 26. foliage;  27. primary;  28. cotyledons;  29. coleoptile; 30. branch;  31. primary;  32. foliage;  33. stem; 34. adventitious;  35. branch;  36. primary;  37. prop; 38. foliage;  39. coleoptile.

**32.2. Hormonal Effects on Plant Growth and Development** (pp. 532–533)

1. e;  2. c;  3. a;  4. d;  5. a;  6. e;  7. a;  8. b;  9. a;  10. c; 11. e;  12. c;  13. e;  14. c;  15. F;  16. H;  17. E;  18. J; 19. G;  20. C;  21. I;  22. B;  23. A;  24. K;  25. D.

**32.3. Adjustments in the Rate and Direction of Growth** (pp. 534–535)

1. a;  2. c;  3. d;  4. b;  5. d;  6. b;  7. a;  8. a;  9. c;  10. b.

**32.4. Biological Clocks and Their Effects** (pp. 536–537)

1. D;  2. I;  3. H;  4. J;  5. B;  6. A;  7. G;  8. C;  9. F; 10. E;  11. Pr;  12. Pfr;  13. Pfr;  14. Pr;  15. response; 16. flowering;  17. longer;  18. shorter;  19. mature; 20. long-day;  21. short-day;  22. Day-neutral.

**32.5. Life Cycles End, and Turn Again** (pp. 538–539)

**32.6.** *Focus on Science:* **The Rise and Fall of a Giant**
(p. 540)
1. b;  2. e;  3. d;  4. b;  5. a;  6. a;  7. d;  8. c;  9. a;  10. a;
11. d;  12. a;  13. a.

**Self-Quiz**
1. b;  2. d;  3. c;  4. a;  5. d;  6. a;  7. d;  8. d;  9. d;  10. c.

# Chapter 33  Tissues, Organ Systems, and Homeostasis

*Meerkats, Humans, It's All the Same* (pp. 544–545)
**33.1. Epithelial Tissue** (pp. 546–547)
1. C;  2. F;  3. B;  4. H;  5. G;  6. D;  7. A;  8. E;  9. epithelial;  10. Simple epithelium;  11. Stratified epithelium;  12. Tight;  13. Adhering;  14. Gap;  15. Tight;  16. peptic;  17. Exocrine;  18. epithelial;  19. Endocrine;  20. hormones;  21. bloodstream.

**33.2. Connective Tissue** (pp. 548–549)
**33.3. Muscle Tissue** (p. 550)
**33.4. Nervous Tissue** (p. 551)
**33.5.** *Focus on Science:* **Frontiers in Tissue Research**
(p. 551)
1. b;  2. c;  3. a;  4. c;  5. b;  6. a;  7. c;  8. a;  9. c;  10. b;
11.a. Adipose;  b. Cartilage;  c. Blood;  d. Bone;
12. smooth;  13. cardiac;  14. smooth;  15. skeletal;
16. Skeletal;  17. smooth;  18. striped;  19. Skeletal;
20. move internal organs;  21. cardiac;  22. Nervous;
23. neurons;  24. Neuroglia;  25. neuron;  26. neurons;
27. a designer organ;  28. connective, D, 9, 11;  29. epithelial, G, 1, 6, 12;  30. muscle, I, 8, (11);  31. muscle, J, 5,
(11);  32. connective, E, 7, 11, (14);  33. gametes, 2;
34. connective, B , 10;  35. epithelial, H, 1, 6, 12;
36. connective, 1, 6, 15;  37. nervous, 3;  38. muscle, C,
13;  39. epithelial, F, 1, 6, 12;  40. connective, A, 4, 14.

**33.6. Organ Systems** (pp. 552–553)
1. cranial;  2. spinal;  3. thoracic;  4. abdominal;
5. pelvic;  6.a. Mesoderm;  b. Endoderm;  c. Ectoderm;
7. superior;  8. distal;  9. proximal;  10. posterior;
11. transverse;  12. inferior;  13. anterior;  14. frontal;
15. circulatory;  16. respiratory;  17. urinary (= excretory);  18. skeletal;  19. endocrine;  20. immune;
21. reproductive;  22. digestive;  23. muscular;  24. nervous;  25. integumentary;  26. G;  27. I;  28. H;  29. B;
30. F;  31. E;  32. C;  33. J;  34. A;  35. D;  36. K.

**33.7. Homeostasis and Systems Control** (pp. 554–555)
1. D;  2. G;  3. B;  4. I;  5. E;  6. C;  7. F;  8. A;  9. H;
10. stable;  11. Homeostasis;  12. control;  13. positive;
14. negative;  15. integrator;  16. glands;  17. receptors;
18. brain;  19. cooling.

**Self-Quiz**
1. d;  2. b ;  3. c;  4. c;  5. d;  6. c;  7. c;  8. d;  9. a;  10. d;
11. d;  12. a;  13. A;  14. H;  15. E. 16. F;  17. K;  18. I;
19. J;  20. G;  21. B;  22. D;  23. C.

# Chapter 34  Information Flow and the Neuron

*TORNADO!* (pp. 558–559)
**34.1. Neurons—The Communication Specialists**
(pp. 560–561)
**34.2. A Closer Look at Action Potentials** (pp. 562–563)
1. neurons;  2. Neuroglial;  3. Sensory;  4. interneurons;
5. motor;  6. cell body;  7. Dendrites;  8. signals (stimuli);  9. axon;  10. cell body;  11. axon;  12. endings;
13. voltage differential;  14. resting membrane potential;
15. action potential (nerve impulse);  16. disturbance;
17. molecules (chemicals);  18. contract;  19. potassium;
20. Na;  21. Channel;  22. Transport;  23. potassium;
24. sodium;  25. negative;  26. millivolts;  27. 30;  28. 10;
29. channel;  30. gates;  31. sodium-potassium pumps;
32. axonal membrane;  33. channel proteins;  34. gates;
35. sodium-potassium pump;  36. lipid bilayer;  37. localized;  38. Graded;  39. duration;  40. trigger zone;
41. action potential;  42. all-or-nothing;  43. threshold;
44. action potential;  45. threshold;  46. resting membrane potential;  47. milliseconds;  48. millivolts.

**34.3. Chemical Synapses** (pp. 564–565)
**34.4. Paths of Information Flow** (pp. 566–567)
**34.5.** *Focus of Health:* **Skewed Information Flow** (p. 568)
1. chemical synapse;  2. neurotransmitters;  3. Acetylcholine (ACh);  4. excitatory;  5. inhibitory;  6. postsynaptic;  7. neuromuscular junction;  8. Serotonin;
9. Endorphins;  10. Synaptic integration;  11. summed;
12. excitatory;  13. hyperpolarizing;  14. myelin;  15. unsheathed node;  16. 120;  17. axon;  18. myelin sheath;
19. blood vessels;  20. axons;  21. unsheathed node;
22. Schwann cell (myelin sheath);  23. sensory;  24. inter;
25. motor;  26. receptor endings or dendrites;  27. unsheathed node (peripheral axon);  28. cell body;  29. axon;
30. axon endings;  31. reflex;  32. stretch reflex;  33. muscle spindles;  34. spinal cord;  35. acetylcholine;  36. H, F;
37. E;  38. K;  39. I;  40. J;  41. B;  42. C;  43. G;  44. A, D.

**Self-Quiz**
1. a;  2. a;  3. c;  4. d;  5. d;  6. a;  7. b;  8. d;  9. c;  10. a.

# Chapter 35   Integration and Control: Nervous Systems

*Why Crack The System?* **(pp. 570–571)**
**35.1. Invertebrate Nervous Systems** (pp. 572–573)
1. nervous;  2. neurons;  3. body;  4. net;  5. directional;
6. radial;  7. net;  8. reflex pathways;  9. sensory;
10. bilateral;  11. midsagittal;  12. two;  13. ganglia;
14. brainlike;  15. sensory;  16. nerve nets;  17. planulas;
18. regulatory;  19. sensory;  20. selection;
21. sensory;  22. Cephalization;  23. invertebrate.

**35.2. Vertebrate Nervous Systems—An Overview**
   (pp. 574–575)
1.a. Midbrain;  b. Forebrain;  c. Hindbrain;  2. B;  3. D;
4. G;  5. H;  6. I;  7. F;  8. C;  9. E;  10. A.

**35.3. The Major Expressways** (pp. 576–577)
1. autonomic;  2. sympathetic;  3. parasympathetic;
4. midbrain;  5. medulla oblongata;  6. cervical;  7. thoracic;  8. lumbar;  9. sacral;  10. spinal cord;  11. ganglion;  12. nerve;  13. vertebra;  14. intervertebral disk;
15. meninges;  16. gray matter;  17. white matter;  18. c;
19. b;  20. d;  21. a;  22. a;  23. c;  24. d;  25. d;  26. b;
27. d;  28. b;  29. b;  30. a;  31. b;  32. d;  33. c;  34. d.

**35.4. Functional Divisions of the Vertebrate Brain**
   (pp. 578–579)
1. brain;  2. information;  3. coordinates;  4. bones;
5. neural tube;  6. stem;  7. reflex;  8. gray;  9. sensory;
10. forebrain;  11. cerebellum;  12. hindbrain;  13. midbrain;  14. tectum;  15. optic;  16. forebrain;  17. reticular;  18. cerebrospinal;  19. blood-brain;  20. cerebral;
21. E;  22. I;  23. C;  24. G;  25. J;  26. B;  27. F;  28. K;
29. H;  30. A;  31. D.

**35.5. A Closer Look at the Human Cerebrum**
   (pp. 580–581)
**35.6.** *Focus on Science:* **Sperry's Split-Brain Experiments**
   (pp. 582–583)

1. cerebrum;  2. hypothalamus;  3. thalamus;  4. pineal;
5. medulla oblongata;  6. pons;  7. cerebellum;  8. midbrain;  9. optic;  10. corpus callosum;  11.a. Limbic system;  b. Cerebral cortex, temporal lobe;  c. Right cerebral
hemisphere;  d. Cerebral cortex, frontal lobe;  e. Corpus
callosum;  f. Cerebral cortex, occipital lobe;  g. Left
cerebral hemisphere;  h. Cerebral cortex, parietal lobe;
i. Cerebral cortex;  12. Dr. Sperry demonstrated that
signals across the corpus callosum coordinate the functioning of the two cerebral hemispheres, each of which
had responded to visual signals from the opposite side of
the body.

**35.7. Memory** (p. 583)
1. touch;  2. hearing;  3. vision;  4. hippocampus;
5. amygdala;  6. smell;  7. prefrontal cortex;  8. basal
ganglia;  9. thalamus and hypothalamus;  10. premotor
cortex;  11. corpus striatum;  12. Memory;  13. Learning;
14. Short-term;  15. long-term;  16. sensory;  17. irrele-
vant;  18. skills;  19. long-term;  20. Skills;  21. input;
22. amygdala (hippocampus);  23. hippocampus (amygdala);  24. amygdala;  25. hippocampus;  26. fact;
27. basal;  28. long-term;  29. Skill;  30. sensory;
31. corpus striatum;  32. cerebellum;  33. Amnesia;
34. skills;  35. Parkinson's;  36. Alzheimer's;  37. hippocampus;  38. information.

**35.8. States of Consciousness** (p. 584)
**35.9.** *Focus on Health:* **Drugging the Brain**
   (pp. 584–585)
1. C;  2. E;  3. B;  4. F;  5. A;  6. D;  7. a;  8. a;  9. a;  10.
c;  11. b;  12. c;  13. d;  14. d;  15. a.

**Self-Quiz**
1. a;  2. d;  3. e;  4. b;  5. d;  6. c;  7. d;  8. e;  9. c;  10. e.

# Chapter 36   Sensory Reception

*Different Stokes for Different Folks* (pp. 588–589)
**36.1. Sensory Receptors and Pathways—An Overview**
   (pp. 590–591)
1. nerve pathways;  2. brain regions;  3. sensation;
4. perception;  5. receptors;  6. stimulus;  7. Chemoreceptors;  8. Mechanoreceptors;  9. photoreceptors;
10. thermoreceptors;  11. action potentials;  12. brain;
13. stimulus intensity;  14. frequency;  15. number;
16. frequency;  17. sensory adaptation;  18. change;
19. Stretch;  20. length;  21. somatic sensations;  22. D;
23. C;  24. A;  25. B;  26. A;  27. E;  28. E;  29. B;  30. D;
31. B;  32. B;  33. A;  34. C;  35. B.

**36.2. Somatic Sensations** (pp. 592–593)
**36.3. Senses of Taste and Smell** (p. 594)
1. somatic sensory cortex;  2. head;  3. ear;  4. cerebral
hemisphere;  5. mouth;  6. hand;  7. touch;  8. cold;
9. skin;  10. skeletal;  11. mechanoreceptors;  12. Free;
13. Pain;  14. nociceptors;  15. action potentials;  16. referred pain;  17. Mechanoreceptors;  18. skin;  19. free
nerve endings (C);  20. Ruffini endings (A);  21. Meissner's corpuscle (D);  22. dermis;  23. epidermis;
24. Pacinian corpuscle (B);  25. Chemo;  26. sensory;
27. taste buds;  28. chemo;  29. nose;  30. five;  31. olfactory;  32. Chemo;  33. smell.

**36.4. Sense of Balance** (p. 595)

**36.5. Sense of Hearing** (pp. 596–597)

1. amplitude; 2. frequency; 3. higher; 4. mechanoreceptors; 5. middle; 6. cochlea; 7. organ of Corti; 8. semicircular canals; 9. middle ear bones (malleus, incus, stapes); 10. cochlea; 11. auditory nerve; 12. tympanic membrane/eardrum; 13. oval window; 14. basilar membrane; 15. tectorial membrane.

**36.6. Sense of Vision** (pp. 598–599)

**36.7. Structure and Function of Vertebrate Eyes** (pp. 600–601)

**36.8. Case Study: From Signaling to Visual Perception** (pp. 602–603)

**36.9.** *Focus on Science*: **Disorders of the Human Eye** (pp. 604–605)

1. photons; 2. Photoreception; 3. Vision; 4. Eyespots; 5. Eyes; 6. cornea; 7. retina; 8. ommatidia; 9. focal point; 10. Accommodation; 11. Farsighted; 12. Cone; 13. fovea; 14. vitreous humor; 15. cornea; 16. iris; 17. lens; 18. aqueous humor; 19. suspensory ligament; 20. retina; 21. fovea; 22. optic nerve; 23. blind spot/optic disk; 24. sclera.

**Self-Quiz**

1. d; 2. c; 3. d; 4. e; 5. d; 6. a; 7. c; 8. b; 9. b; 10. a.

---

# Chapter 37    Endocrine Control

*Hormone Jamboree* (pp. 608–609)

**37.1. The Endocrine System** (pp. 610–611)

1. E; 2. A; 3. B; 4. F; 5. C; 6. D; 7. a. 1, six releasing and inhibiting hormones;  synthesizes ADH, oxytocin; b. 2, ACTH, TSH, FSH, LH, GSH;  c. 2, stores and secretes two hypothalamic hormones, ADH and oxytocin; d. 3, sex hormones of opposite sex, cortisol, aldosterone; e. 3, epinephrine, norepinephrine;  f. 4, estrogens, progesterone;  g. 5, testosterone;  h. 6, melatonin;  i. 7, thyroxine and triiodothyronine;  j. 8, parathyroid hormone (PTH);  k. 9, thymosins;  l. 10, insulin, glucagon, somatostatin.

**37.2. Signaling Mechanisms** (pp. 612–613)

1. a; 2. b; 3. a; 4. b; 5. b; 6. a; 7. b; 8. b; 9. b; 10. b.

**37.3. The Hypothalamus and the Pituitary Gland** (pp. 614–615)

1. A(H); 2. P(G); 3. A(A); 4. A(I); 5. A(D); 6. I(B); 7. P(E); 8. A(C); 9. A(F); 10. hypothalamus; 11. posterior; 12. anterior; 13. intermediate; 14. releasers; 15. inhibitors.

**37.4. Examples of Abnormal Pituitary Output** (p. 616)

1.a. Gigantism;  b. Pituitary dwarfism;  c. Diabetes insipidus;  d. Acromegaly.

**37.5. Sources and Effects of Other Hormones** (p. 617)

1.a. D(c);  b. I(e);  c. F(h);  d. A(g);  e. H(k);  f. E(d); g. C(a);  h. K(j);  i. B(i);  j. J(f);  k. G(b);  l. F (l);  m. A (m);  n. B (o);  o. C (n).

**37.6. Feedback Control of Hormonal Secretions** (pp. 618–619)

1. E; 2. F; 3. D; 4. H. 5. I; 6. J; 7. A; 8. G; 9. B; 10. C; 11. thyroid; 12. metabolic (metabolism); 13. iodine; 14. iodide; 15. TSH; 16. goiter; 17. Hypothyroidism; 18. hypothyroid; 19. Hyperthyroidism; 20. gonads; 21. hormones; 22. testes; 23. ovaries; 24. gametes; 25. hormones; 26. secondary.

**37.7. Responses to Local Chemical Changes** (pp. 620–621)

1. parathyroid; 2. PTH; 3. calcium; 4. calcium; 5. reabsorption; 6. $D_3$; 7. intestinal; 8. rickets; 9. a. glucagon, causes glycogen (a storage polysaccharide) and amino acids to be converted to glucose in the liver (glucagon raises the glucose level);  b. insulin, stimulates glucose uptake by liver, muscle, and adipose cells;  promotes synthesis of proteins and fats, and inhibits protein conversion to glucose (lowers the glucose level); c. somatostatin, helps control digestion; can block secretion of insulin and glucagon;  10. rises; 11. excessive; 12. energy; 13. ketones; 14. insulin; 15. Glucagon; 16. type 1 diabetes; 17. type 1 diabetes; 18. type 2 diabetes; 19. Type 2 diabetes.

**37.8. Hormonal Responses to Environmental Cues** (pp. 622–623)

1. b; 2. a; 3. a; 4. a; 5. b; 6. a; 7. a; 8. a; 9. a; 10. b.

**Self-Quiz**

1. a; 2. e; 3. d; 4. b; 5. e; 6. d; 7. b; 8. a; 9. c; 10. a; 11. H; 12. F; 13. K; 14. L; 15. N; 16. D; 17. E; 18. A; 19. G; 20. B; 21. I; 22. Q; 23. O; 24. C; 25. J; 26. M; 27. P.

# Chapter 38    Protection, Support, and Movement

*Of Men, Women, and Polar Huskies* (pp. 626–627)
**38.1. Integumentary System** (pp. 628–629)
**38.2. A Look at Human Skin** (p. 630)
**38.3.** *Focus on Health:* **Sunlight and Skin** (p. 631)
1. epidermis;  2. dermis;  3. hypodermis;  4. adipose;
5. hair;  6. sensory nerve ending;  7. sebaceous gland;
8. smooth muscle;  9. hair follicle;  10. sweat gland;
11. blood vessel;  12. epidermis;  13. dermis;  14. hypodermis;  15. a;  16. c;  17. d;  18. b;  19. c;  20. c;  21. e;
22. melanin;  23. Oil;  24. acne;  25. Hirsutism;  26. vitamin D;  27. calcium;  28. sunlight;  29. cholesterol;
30. intestine;  31. immune;  32. Langerhans;  33. cold
sores;  34. Granstein.

**38.4. Types of Skeletons** (pp. 632–633)
**38.5. Characteristics of Bone** (pp. 634–635)
**38.6. Human Skeletal System** (pp. 636–637)
1. contract (relax);  2. relax (contract);  3. contractile;
4. antagonistic;  5. hydrostatic;  6. exoskeleton;
7. Bones;  8. muscles;  9. Haversian;  10. Red marrow;
11. calcium (phosphate);  12. phosphate (calcium);
13. osteoblasts;  14. hormonal;  15. osteoporosis;
16. axial;  17. appendicular;  18. Synovial;  19. cartilage;
20. synovial;  21. osteoarthritis;  22. rheumatoid arthritis;  23. nutrient canal;  24. contains yellow marrow;
25. compact bone;  26. spongy bone;  27. connective
tissue covering (periosteum);  28. Haversian system;
29. blood vessel;  30. mineral deposits (calcium phosphate);  31. osteocyte (bone cell);  32. cranium;  33. sternum;  34. clavicle;  35. scapula;  36. radius;  37. carpal
bones;  38. femur;  39. tibia;  40. tarsal bones;
41. metatarsals.

**38.7. Skeletal-Muscular Systems** (pp. 638–639)
**38.8. Muscle Structure and Function** (pp. 640–641)
**38.9. Control of Muscle Contraction** (pp. 642–643)
**38.10. Properties of Whole Muscles** (p. 644)
**38.11.** *Focus on Science:* **Pebbles, Feathers, and Muscle
  Mania** (p. 645)
1. D;  2. C;  3. A;  4. E;  5. B;  6. F;  7. triceps brachii;
8. pectoralis major;  9. external oblique;  10. rectus abdominis;  11. quadriceps femoris;  12. tibialis anterior;
13. gastrocnemius;  14. trapezius;  15. deltoid;  16. biceps brachii;  17. biceps contracts;  18. triceps relaxes;
19. biceps relaxes;  20. triceps contracts;  21. cardiac;
22. skeletal;  23. bones;  24. tendons;  25. Skeletal;
26. bones;  27. joints;  28. biceps brachii;  29. triceps
brachii;  30. excitability;  31. neuron;  32. action potential;  33. motor;  34. myofibrils;  35. sarcomeres;
36. myosin;  37. actin;  38. sliding-filament;  39. myosin;
40. sarcomere;  41. ATP;  42. creatine phosphate;

43. Contraction (Shortening);  44. bone;  45. calcium;
46. sarcoplasmic reticulum;  47. actin;  48. active transport;  49. action potentials;  50. sarcoplasmic reticulum;
51. Its oxygen-requiring reactions provide most of the
ATP needed for muscle contraction during prolonged,
moderate exercise;
52. It provides the energy to make myosin filaments slide
along actin filaments;
53. These are used to clear the actin binding sites of any
obstacles to cross-bridge formation with myosin heads;
54. This supplies phosphate to ADP → ATP, which pow-
ers muscle contraction for a short time because creatine
phosphate supplies are limited;
55. Stored in muscles and in the liver, glucose is stored
by the animal body in this form of starch;
56. An anaerobic pathway in which glucose is broken
down to yield a small amount of ATP; this pathway
operates during intense exercise;
57. This supplies commands (signals) to muscle cells to
contract and relax;
58. Projections from the thick filaments of myosin that
bind to actin sites and form temporary cross-bridges.
Making and breaking these cross-bridges cause myosin
filaments to be pulled to the center of a sarcomere;
59. The repetitive unit of muscle contraction. Many sarcomeres constitute a myofibril. Many myofibrils constitute a muscle cell;
60. This is the endoplasmic reticulum of a muscle cell. It
stores calcium ions and releases them in response to
incoming signals from motor neurons. It uses active
transport to bring the calcium ions back inside;
61. dephosphorylation of creatine phosphate;  62. aerobic respiration;  63. glycolysis alone;  64. ATP;  65. diameter;  66. motor;  67. motor unit;  68. muscle twitch;
69. twitch;  70. Tetanus;  71. weak;  72. frequency (rate);
73. muscle;  74. muscle cell (muscle fiber);  75. myofibril;
76. sarcomere;  77. myosin filament;  78. actin filament;
79. axon of motor neuron serving one motor unit;
80. motor neuron endings;  81. axon of another motor
neuron serving another motor unit;  82. individual muscle cells;  83. time that stimulus is applied;  84. contraction phase;  85. relaxation phase;  86. time (seconds);
87. force;  88. Anabolic steroids;  89. muscle;  90. acne;
91. testes;  92. facial;  93. menstrual cycles;  94. clitoris;
95. aggression.

**Self-Quiz**
1. e;  2. c;  3. c;  4. d;  5. b;  6. d;  7. a;  8. c;  9. d;  10. b;
11. e;  12. d;  13. a;  14. e;  15. b;  16. c;  17. b;  18. d;
19. a;  20. e.

# Chapter 39    Circulation

*Heartworks* (pp. 648–649)

**39.1. Circulatory Systems—An Overview** (pp. 650–651)
1. nutrients (food);  2. wastes;  3. closed;  4. Blood;
5. heart;  6. interstitial fluid;  7. heart;  8. rapidly;
9. capillary;  10. solutes;  11. lymphatic;  12. digestive
(respiratory);  13. respiratory (digestive);  14. respiratory;  15. urinary;  16. heart(s);  17. blood vessels;
18. hearts;  19. open circulatory system;  blood is
pumped into short tubes that open into spaces in the
body's tissues, mingles with tissue fluids, then is
reclaimed by open-ended tubes that lead back to the
heart;  20. closed circulatory system;  blood flow is
confined within blood vessels that have continuously
connected walls and is pumped by 5 pairs of "hearts."

**39.2. Characteristics of Blood** (pp. 652–653)
**39.3. *Focus on Health*: Blood Disorders** (p. 654)
**39.4. Blood Transfusion and Typing** (pp. 654–655)
1. a. Plasma proteins;  b. Red blood cells;
c. Neutrophils;  d. Lymphocytes;  e. Platelets;  2. connective;  3. pH;  4. iron;  5. cell count;  6. 50 to 60;
7. red bone marrow;  8. Stem cells;  9. Neutrophils;
10. Platelets;  11. nucleus;  12. four;  13. nucleus;
14. nine;  15. AB;  16. Agglutination (clumping);
17. stem (F);  18. red blood (E);  19. platelets (C);
20. neutrophils (A);  21. B (D);  22. T (D);  23. monocytes
(A);  24. macrophages (A);  25. A;  26. I;  27. D;  28. G;
29. F;  30. H;  31. B;  32. C;  33. E;  34. E.

**39.5. Human Cardiovascular System** (pp. 656–657)

**39.6. The Heart Is a Lonely Pumper** (pp. 658–659)
1. artery;  2. capillary;  3. endothelial;  4. capillary bed
(diffusion zone);  5. interstitial;  6. Valves;  7. veins
(venules);  8. venules (veins);  9. Arteries;  10. ventricles;  11. Arterioles;  12. pressure;  13. oxygen;  14. oxygen;  15. heart;  16. SA node;  17. ventricles;  18. atrial;
19. atrium;  20. ventricle;  21. systole;  22. diastole;
23. pulmonary;  24. systemic;  25. jugular;  26. superior
vena cava;  27. pulmonary;  28. hepatic;  29. renal;

30. inferior vena cava;  31. iliac;  32. femoral;
33. femoral;  34. iliac;  35. dorsal;  36. renal;
37. brachial;  38. coronary;  39. pulmonary;  40. carotid;
41. aorta;  42. left pulmonary veins;  43. semilunar
valve;  44. left ventricle;  45. inferior vena cava;  46. atrioventricular valve;  47. right pulmonary artery;  48. superior vena cava.

**39.7. Blood Pressure in the Cardiovascular System**
   (pp. 660–661)
**39.8. From Capillary Beds Back to the Heart**
   (pp. 662–663)
**39.9. *Focus on Health*: Cardiovascular Disorders**
   (pp. 664–665)
1. aorta;  2. pressure;  3. resistance;  4. elastic;  5. little;
6. does not drop much;  7. arterioles;  8. nervous;
9. medulla oblongata;  10. beat more slowly;  11. relax;
12. vasodilation;  13. diffusion;  14. bulk flow (blood
pressure);  15. valves;  16. 50–60;  17. vein;  18. artery;
19. arteriole;  20. capillary;  21. smooth muscle, elastic
fibers;  22. valve;  23. F, pressure;  24. F, blood pressure
cannot remain constant because it passes through various kinds of vessels that have varied structures;
25. stroke;  26. coronary occlusion;  27. Plaque;  28. low;
29. thrombus.

**39.10. Hemostasis** (p. 666)
**39.11. Lymphatic System** (pp. 666–667)
1. hemostasis;  2. platelet plug formation;  3. coagulation;  4. collagen;  5. insoluble;  6. d;  7. c;  8. e;  9. b;
10. a;  11. tonsils;  12. thymus;  13. thoracic;  14. spleen;
15. lymph node(s);  16. macrophages, lymphocytes;
17. bone marrow;  18. Lymph;  19. fats;  20. small intestine.

**Self-Quiz**
1. d;  2. e;  3. c;  4. e;  5. e;  6. a;  7. b;  8. a;  9. a;  10. a;
11. I;  12. G;  13. M;  14. H;  15. B;  16. O;  17. F;  18. N;
19. E;  20. L;  21. A;  22. J;  23. D;  24. C;  25. K.

# Chapter 40    Immunity

*Russian Roulette, Immunological Style* (pp. 670–671)
**40.1. Three Lines of Defense** (p. 672)
**40.2. Complement Proteins** (p. 673)
**40.3. Inflammation** (pp. 674–675)
1. mucous;  2. Lysozyme;  3. Gastric;  4. bacterial;
5. phagocytic;  6. clotting;  7. Phagocytic (Macrophage);
8. complement system;  9. histamine;  10. capillaries;
11. nonspecific;  12. specific;  13. lymphocytes;  14. immune;  15. secrete histamine and prostaglandins that
change permeability of blood vessels in damaged or

irritated tissues;  16. attack parasitic worms by secreting
corrosive enzymes;  17. the most abundant white blood
cells;  they quickly phagocytize bacteria and reduce
them to molecules that can be used for other purposes;
18. slow, "big eaters";  engulf and digest foreign agents,
and clean up dead and damaged tissues.

**40.4. The Immune System** (pp. 676–677)
**40.5. Lymphocyte Battlegrounds** (p. 678)

**40.6. Cell-Mediated Responses** (pp. 678–679)
**40.7. Antibody-Mediated Responses** (pp. 680–681)
1. nonspecific; 2. immune system; 3. MHC marker; 4. nonself; 5. lymphocytes; 6. B cell; 7. T cell; 8. thymus; 9. viruses; 10. cancer; 11. identity; 12. antigen; 13. antigen-presenting; 14. effector; 15. helper T; 16. cytotoxic T; 17. cell-mediated; 18. antibodies; 19. antibody-mediated; 20. memory; 21. antigen-presenting cells (macrophages); 22. virgin helper T cells; 23. memory T cells; 24. intracellular; 25. virgin B cells; 26. effector B cells; 27. antibodies; 28. extracellular; 29. a; 30. d; 31. e; 32. b; 33. c; 34. natural killer; 35. perforins; 36. cell -; 37. cytotoxic T; 38. helper T; 39. antigen-MHC; 40. virgin B; 41. antibody; 42. effector B; 43. B cell; 44. primary immune response; 45. memory cells; 46. antigens; 47. Ig A; 48. Ig E; 49. Ig G; 50. Ig M.

**40.8. Focus on Health: Cancer and Immunotherapy**
 (p. 681)
**40.9. Immune Specificity and Memory** (pp. 682–683)
**40.10. Immunity Enhanced, Misdirected, or Compromised** (pp. 684–685)

**40.11. Focus on Health: Aids—The Immune System Compromised** (pp. 686–687)
1. immunotherapy; 2. Cancer; 3. monoclonal antibodies; 4. hybrid; 5. lymphocytes; 6. lymphokine; 7. immunoglobulin; 8. Antibodies; 9. recombination; 10. antibodies; 11. antigen; 12. virgin B; 13. memory cells; 14. effector cells; 15. clonal selection hypothesis; 16. nonself; 17. immunization; 18. active; 19. primary immune response; 20. memory cells; 21. b, d, e, h, i, j; 22. b, d, e, g; 23. a, b, c, f; 24. g, i, (j); 25. g, i, (j); 26. d, f, i; 27. g, h, (i); 28. b, d, e, h, i, j; 29. b, d, e, h, i; 30. Allergy; 31. Autoimmune disease; 32. Myasthenia gravis; 33. Rheumatoid arthritis; 34. human immunodeficiency virus; 35. male homosexuals; 36. retrovirus; 37. reverse transcriptase; 38. body fluids; 39. helper T (antigen-presenting); 40. macrophages; 41. one million; 42. 000 million; 43. 000 million; 44. 000.

**Self Quiz**
1. d; 2. b; 3. b; 4. a; 5. a; 6. e; 7. e; 8. e; 9. a; 10. a; 11. H; 12. D; 13. E; 14. J; 15. B; 16. G; 17. C; 18. I; 19. F; 20. A.

# Chapter 41   Respiration

*Conquering Chomolungma* (pp. 690–691)
**41.1. The Nature of Respiration** (p. 692)
**41.2. Invertebrate Respiration** (p. 693)
**41.3. Vertebrate Respiration** (p. 694)
1. gill; 2. countercurrent flow; 3. tracheas; 4. 21; 5. aerobic metabolism; 6. $O_2$ (oxygen); 7. carbon dioxide ($CO_2$); 8. respiration; 9. pressure gradient; 10. high; 11. lowest; 12. high; 13. lower; 14. surface area; 15. partial pressure; 16. Hemoglobin; 17. lung; 18. airways (blood); 19. blood (airways); 20. water; 21. blood vessel in gill filament; 22. oxygen-poor blood; 23. oxygen-rich blood; 24. water; 25. blood; 26. partial pressure; 27. Diffusion; 28. carbon dioxide; 29. Hypoxia.

**41.4. Human Respiratory System** (pp. 696–697)
**41.5. Breathing—Cyclic Reversals in Air Pressure Gradients** (pp. 698–699)
1. diaphragm; 2. rib cage; 3. increases; 4. drops; 5. ventilating; 6. pleural sac; 7. glottis; 8. larynx; 9. bronchi; 10. bronchioles; 11. alveoli; 12. intercostal muscles; 13. diaphragm; 14. pharynx; 15. epiglottis; 16. vocal cords; 17. trachea; 18. bronchus; 19. bronchioles; 20. thoracic cavity; 21. abdominal cavity; 22. smooth muscle; 23. bronchiole; 24. alveolus, alveoli; 25. capillary; 26. a. alveoli; b. bronchial tree; c. diaphragm; d. larynx; e. pharynx.

**41.6. Gas Exchange and Transport** (pp. 700–701)
**41.7. Focus on Health: When the Lungs Break Down** (pp. 702–703)
**41.8. Respiration in Unusual Environments** (pp. 704–705)
**41.9. Focus on Science: Respiration in Leatherback Sea Turtles** (pp. 706–707)
1. oxygen; 2. aerobic respiration (electron phosphorylation); 3. carbon dioxide; 4. hemoglobin; 5. bicarbonate; 6. oxyhemoglobin (hemoglobin); 7. systemic (low-pressure); 8. partial pressure; 9. medulla oblongata; 10. Emphysema; 11. lung cancer; 12. $N_2$ (nitrogen gas); 13. joints; 14. decompression.

**Self-Quiz**
1. c; 2. e; 3. a; 4. c; 5. a; 6. a; 7. a; 8. a; 9. a; 10. d; 11. I; 12. F; 13. K; 14. L; 15. H; 16. O; 17. G; 18. M; 19. B; 20. D; 21. A; 22. E; 23. C; 24. N; 25. J.

# Chapter 42    Digestion and Human Nutrition

**Lose It—And It Finds Its Way Back** (pp. 710–711)
**42.1. The Nature of Digestive Systems** (pp. 712–713)
**42.2. Overview of the Human Digestive System** (p. 714)
**42.3. Into the Mouth, Down the Tube** (p. 715)
1. Nutrition; 2. carbohydrates; 3. particles; 4. molecules; 5. absorbed; 6. incomplete; 7. circulatory; 8. complete; 9. opening; 10. Motility; 11. secretion; 12. stomach; 13. small intestine; 14. anus; 15. accessory; 16. pancreas; 17. circulatory; 18. respiratory; 19. carbon dioxide $CO_2$; 20. urinary; 21. Salivary amylase; 22. epiglottis; 23. esophagus; 24. degrade (digest); 25. Pepsin; 26. salivary glands; 27. liver; 28. gallbladder; 29. pancreas; 30. anus; 31. large intestine; 32. small intestine; 33. stomach; 34. esophagus; 35. pharynx; 36. mouth.

**42.4. Digestion in the Stomach and Small Intestine** (pp. 716–717)
**42.5. Absorption in the Small Intestine** (pp. 718–719)
**42.6. Disposition of Absorbed Organic Compounds** (p. 720)
**42.7. The Large Intestine** (p. 721)
1. a. Mouth; b. Salivary glands; c. Stomach; d. Small intestine; e. Pancreas; f. Liver; g. Gallbladder; h. Large intestine; i. Rectum; 2. starches; 3. salivary amylase; 4. disaccharide; 5. stomach; 6. small intestine; 7. amylase; 8. small intestine; 9. disaccharidases; 10. stomach; 11. pepsins; 12. small intestine; 13. pancreas; 14. amino acids; 15. Lipase; 16. small intestine; 17. fatty acid; 18. Bile; 19. gallbladder; 20. lipase; 21. small intestine; 22. small intestine; 23. Carboxypeptidase; 24. Pancreatic nucleases; 25. segmentation (peristalsis); 26. T; 27. cellular respiration; 28. T; 29. T; 30. small intestine; 31. constructing hormones, nucleotides, proteins, and enzymes; 32. monosaccharides, free fatty acids, and glycerol; 33. The three uses are (a) to construct components of cells and storage forms (such as glycogen) and specialized derivatives such as steroids and acetylcholine; (b) to convert to amino acids as needed; and (c) to serve as a source of energy;

**42.8. Human Nutritional Requirements** (pp. 722–723)
**42.9. Vitamins and Minerals** (pp. 724–725)
**42.10.** *Focus on Science:* **Tantalizing Answers To Weighty Questions** (pp. 726–727)
1. a. 2,070; b. 2,900; c. 1,230; 2. 18; 3. Complex carbohydrates; 4. 50 to 60; 5. Phospholipids; 6. energy reserves; 7. 30; 8. essential fatty acids; 9. Proteins; 10. essential; 11. milk (soybeans, eggs, meats, wheat germ); 12. soybeans (milk, eggs, meats, wheat germ); 13. Vitamins; 14. Minerals; 15. Consulting Fig. 42.15 (men's column, 6'0'') yields 178 pounds as his ideal weight. 195 – 178 = 17 pounds overweight; 16. Multiply 178 times 10 (see p. 726) to obtain 1780 kilocalories (the daily number of calories that *maintains* weight in the correct size range). The excess 17 pounds should be lost gradually by adopting an everyday exercise program that over many months would gradually eliminate the excess kilocalories that are stored mostly in the form of fat; The smallest range of serving sizes shown in Figure 42.13 will help keep the total caloric intake to about 1600 kcal;
17. a. 6 servings; b. bread, cereal, rice, pasta;
18. a. 2 servings; b. fruits;
19. a. 3 servings; b. vegetables;
20. a. 2 servings; b. milk, yogurt or cheese;
21. a. 2 servings; b. legume, nut, poultry, fish or meats;
22. a. Scarcely any; b. added fats and simple sugars;
23. –39. Choose from Fig. 42.13.
23. food pyramid; 24. carbohydrates; 25. bread; 26. 6–11; 27. vegetable; 28. 3–5; 29. fruit; 30. 2–4; 31. apples; 32. berries; 33. meat; 34. proteins; 35. 2–3; 36. amino acids; 37. milk; 38. 2–3; 39. 0.

**Self-Quiz**
1. b; 2. b; 3. a; 4. e; 5. d; 6. c; 7. d; 8. b; 9. b; 10. c; 11. D; 12. C; 13. H; 14. J; 15. B; 16. A; 17. I; 18. F; 19. E; 20. G.

# Chapter 43  The Internal Environment

*Tale of the Desert Rat* (pp. 730–731)

**43.1. Urinary System of Mammals** (pp. 732–733)
1. metabolism;  2. urine;  3. respiratory surfaces;
4. sweating;  5. Thirst;  6. metabolism;  7. ammonia;
8. urea;  9. uric acid;  10. kidney;  11. ureter;  12. urinary bladder;  13. urethra;  14. cortex;  15. medulla;
16. ureter;  17. kidneys;  18. nephrons;  19. bloodstream;
20. arterioles;  21. Bowman's capsule;  22. glomerular;
23. blood pressure;  24. glomerulus;  25. proximal tubule;  26. peritubular;  27. loop of Henle;  28. distal tubule;  29. collecting duct;  30. ureter;  31. urinary;
32. solutes;  33. extracellular;  34. glomerular capillaries;
35. proximal tubule;  36. Bowman's capsule;  37. distal tubule;  38. peritubular capillaries;  39. collecting duct;
40. loop of Henle.

**43.2. Urine Formation** (pp. 734–735)
**43.3.** *Focus on Health:* **When Kidneys Break Down**
  (p. 736)
**43.4. The Acid-Base Balance** (p. 736)
**43.5. On Fish, Frogs, and Kangaroo Rats** (p. 737)
1. filtration;  2. tubular reabsorption;  3. tubular secretion;  4. glomerular;  5. nephron;  6. peritubular;
7. proximal tubule;  8. osmosis;  9. sodium;  10. dilute;

11. aldosterone;  12. sodium;  13. less;  14. blood pressure;  15. hypertension;  16. table salt (sodium chloride);
17. ADH;  18. inhibited;  19. T;  20. T;  21. Kidneys;
22. $H^+$;  23. 7.43;  24. Acids;  25. bases;  26. lowered;
27. $H^+$;  28. bicarbonate ($HCO_3^-$);  29. urinary;  30. loops of Henle;  31. water;  32. water;  33. solutes;  34. very dilute.

**43.6. Maintaining the Body's Core Temperature**
  (pp. 738–739)

**43.7. Temperature Regulation in Mammals**
  (pp. 740–741)
1. hypothalamus;  2. central (core);  3. pilomotor response;  4. Peripheral vasoconstriction;  5. hypothermia;
6. T;  7. T;  8. T;  9. D;  10. B;  11. A;  12. C;  13. G;
14. F;  15. E.

**Self-Quiz**
1. d;  2. c;  3. e;  4. a;  5. d;  6. d;  7. d;  8. b;  9. e;  10. b.

**Crossword Puzzle**

# Chapter 44    Principles of Reproduction and Development

1. asexual;   2. environmental;   3. variation;   4. reproductive timing;   5. Gamete formation;   6. egg;   7. fertilization;   8. zygote;   9. Cleavage;   10. blastomeres;   11. Gastrulation;   12. nervous;   13. gut;   14. skeleton;   15. Viviparous;   16. ovoviviparous;   17. oviparous;   18. embryonic;   19. Development;   20. F, rich in lipids and proteins;   21. T;   22. T;   23. T;   24. T;   25. T;   26. F;   27. B;   28. C;   29. A;   30. E;   31. D;   32. a. mesoderm; b. ectoderm;   c. endoderm;   d. mesoderm;   e. ectoderm; f. mesoderm;   g. endoderm;   h. mesoderm;   i. mesoderm.

1. F, nuclear;   2. F, by "maternal messages";   3. T;   4. cleavage;   5. blastula;   6. maternal instructions;   7. cytoplasmic localization;   8. animal pole;   9. gray crescent;   10. body axis;   11. gastrulation;   12. germ;   13. gastrula;   14. yolk;   15. three;   16. neural tube;   17. cell differentiation;   18. morphogenesis;   19. active cell migration;   20. Adhesive;   21. microtubules;   22. microfilaments;   23. apoptosis.

1. cell differentiation;   2. genes;   3. restrictions;   4. Morphogenesis;   5. pattern formation;   6. embryonic induction;   7. ectodermal sheets;   8. embryonic induction;   9. Morphogens;   10. larva;   11. metamorphosis;   12. complete;   13. limited division potential. 14. F, internal;   15. T;   16. T.

**Self-Quiz**
1. c;   2. b;   3. b;   4. a;   5. c;   6. e;   7. d;   8. b;   9. e;   10. a.

# Chapter 45    Human Reproduction and Development

1. mitotic (mitosis);   2. seminiferous;   3. meiosis (spermatogenesis);   4. sperm;   5. epididymis;   6. vas deferens;   7. urethra;   8. Seminal vesicles;   9. Prostate gland;   10. Bulbourethral;   11. Leydig;   12. Testosterone;   13. testosterone;   14. anterior;   15. hypothalamus;   16. decrease;   17. LH;   18. Sertoli;   19. increase;   20. hypothalamus;   21. anterior pituitary;   22. Sertoli cells;   23. Leydig cells

1. ovary;   2. oviduct;   3. uterus;   4. cervix;   5. myometrium;   6. endometrium;   7. vagina;   8. labia majora;   9. labia minora;   10. clitoris;   11. urethra;   12. Meiosis;   13. I;   14. 300,000;   15. follicle;   16. hypothalamus;   17. anterior pituitary;   18. estrogens;   19. ovulation;   20. LH;   21. menstruation;   22. endometrial;   23. corpus luteum;   24. progesterone;   25. blastocyst;   26. endometrium;   27. menstrual flow.

1. oviduct;   2. implantation;   3. placenta;   4. eighth;   5. fetus;   6. embryonic disk;   7. amniotic cavity;   8. embryonic disk;   9. amniotic cavity;   10. yolk sac;   11. future brain of embryo;   12. amnion;   13. chorion;   14. yolk sac;   15. umbilical cord;   16. four;   17. pharyngeal arches;   18. somites;   19. six;   20. forelimb;   21. D;   22. B;   23. A;   24. C;   25. human chorionic gonadotropin (HCG);   26. corpus luteum;   27. gastrulation;   28. primitive streak;   29. neural tube;   30. Somites;   31. pharyngeal arches;   32. placenta;   33. umbilical cord;   34. nutrients;   35. wastes;   36. arms;   37. head;   38. muscles;   39. 95.

1. 1,800,000;  2. 1 million;  3. 1,600,000;  4. abstinence;  5. Condoms;  6. diaphragm;  7. estrogens (progesterones);  8. progesterones (estrogens);  9. gonadotropins;  10. tubal ligation;  11. A, F;  12. A, C;  13. D, E;  14. F;  15. F;  16. D;  17. C;  18. F;  19. C;  20. C, F;  21. C;  22. F;  23. D, F;  24. B;  25. A, B, C, D, E, F.

**Self-Quiz**
1. d;  2. e;  3. a;  4. b;  5. c;  6. e;  7. d;  8. a;  9. c;  10. a;  11. b;  12. d.

# Chapter 46    Population Ecology

*Tales of Nightmare Numbers*
**46.1. Characteristics of Populations** (pp. 794–795)
**46.2. Population Size and Exponential Growth**
   (pp. 796–797)
1. K;  2. H;  3. D;  4. I;  5. B;  6. F;  7. A;  8. G;  9. J;  10. L;  11. C;  12. E;  13.a. It increases;  b. It decreases;  c. It must increase;  14. population growth rate;  15. It must decrease;  16. 100,000;  17.a. 100,000;  b. 300,000.

**46.3. Limits on the Growth of Populations** (pp. 798–799)
**46.4. Life History Patterns** (pp. 800–801)
**46.5. *Focus on Science:* Natural Selection and the Guppies of Trinidad** (pp. 802–803)
1. limiting;  2. Carrying capacity;  3. logistic;  4. carrying capacity;  5. increases;  6. decreases;  7. density-dependent;  8. density-independent;  9. dependent;  10. carrying capacity;  11. independent;  12. life history;  13. insurance;  14. cohort;  15. life;  16. "survivorship";  17. Survivorship;  18. III;  19. I;  20. II;  21. I;  22. C;  23. D;  24. E;  25. B;  26. A;  27. F;  28. A, E, H;  29. B, C, D;  30. G.

**46.6. Human Population Growth** (pp. 804–805)
**46.7. Control Through Family Planning** (pp. 806–807)
**47.8. Population Growth and Economic Development**
   (pp. 808–809)
**46.9. Social Impact of No Growth** (p. 809)

1.a. 1962–1963;  b. 2025 or sooner;  c. Depends on the age and optimism of the reader;  2. 5.8 billion;  3. T;  4. short;  5. T;  6. sidestepped;  7. cannot;  8. 10;  9. 11;  10. resources;  11. pollution;  12. Family planning;  13. birth;  14. two;  15. female;  16. fertility;  17. 3;  18. 6.5;  19. baby-boomers;  20. one-third;  21. thirties;  22. China;  23. stabilize;  24. reproductive;  25. D;  26. C;  27. B;  28. A;  29. B;  30. D;  31. A;  32. C;  33. industrial;  34. decreasing;  35. smaller;  36. transitional;  37. transitional;  38. economic;  39. immigration;  40. 16;  41. 4.7;  42. 21;  43. 25;  44. fifty;  45. 25;  46. 1;  47. 3;  48. 3;  49. 12.9;  50. 258;  51. growth;  52. social;  53. older;  54. economic;  55. postponed;  56. cultural;  57. carrying capacity.

**Self-Quiz**
1. d;  2. a;  3. b;  4. a;  5. d;  6. d;  7. a;  8. b;  9. a;  10. a;  11. c;  12. c;  13. b;  14. d;  15. b;  16. a;  17. a;  18. density;  19. size;  20. dispersion;  21. age;  22. pre-reproductive and reproductive;  23. zero population growth;  24. J-shaped;  25. biotic potential;  26. exponentially;  27. decreases.

# Chapter 47    Community Interactions

*No Pigeon Is An Island* (pp. 812–813)
**47.1. Factors That Shape Community Structure** (p. 814)
**47.2. Mutualism** (p. 815)
1. habitat;  2. community;  3. niche;  4. potential;  5. realized;  6. neutral;  7. directly;  8. commensalistic;  9. mutualistic;  10. interspecific;  11. predation;  12. prey;  13. parasitism;  14. host;  15. symbiosis;  16.a. It cannot complete its life cycle in any other plant, and its larvae eat only yucca seeds. b. The yucca moth is the plant's only pollinator.

**47.3. Competitive Interactions** (pp. 816–817)
**47.4. Predation** (pp. 818–819)
**47.5. *Focus on the Environment:* The Coevolutionary Arms Race** (pp. 820–821)

1. Intraspecific;  2. Interspecific;  3. Interspecific;  4. competitive exclusion;  5. resource partitioning;  6. Predators;  7. prey;  8. Parasites;  9. hosts;  10. host;  11. host;  12. Coevolution;  13. camouflage;  14. warning coloration;  15. Mimicry;  16. Moment-of-truth;  17. adaptive;  18. G;  19. K;  20. I;  21. A;  22. D;  23. B;  24. B;  25. C;  26. B;  27. J;  28. D;  29. K;  30. G;  31. F;  32. I;  33. F;  34. B;  35. K;  36. H;  37. K;  38. K;  39. B;  40. G;  41. F;  42. D.

**47.6. Parasitic Interactions** (pp. 822–823)
1. g;  2. a;  3. h;  4. e;  5. g;  6. f;  7. b;  8. c;  9. h;  10. d;  11. In evolutionary terms, killing a host is not good for the parasite's reproductive success. Parasitic infections of longer time durations give the parasite more opportunity to produce more offspring. Usually, death occurs only

when a parasite attacks a normal host. 12. Less than 20 percent of existing selections qualify as effective defenses against pests. Releasing more than one kind of biological control agent in an area may trigger competition between them and lessen their overall effectiveness. There is a chance that the parasites or parasitoids released may attack nontargeted species.

### 47.7. Forces Contributing to Community Stability (pp. 824–825)
### 47.8. Community Instability (pp. 826–827)
### 47.9. *Focus on the Environment:* Nile Perch and Rabbits and Kudzu, Oh My! (pp. 828–829)
1. succession; 2. Pioneer; 3. pioneers; 4. climax; 5. primary; 6. replacement; 7. secondary; 8. facilitate; 9. succession; 10. climax-pattern; 11. community; 12. pioneers; 13. fires; 14. natural; 15. active; 16. b; 17. a; 18. b; 19. b; 20. a; 21. a; 22. b; 23. a; 24. a; 25. B; 26. D; 27. E; 28. A; 29. C.

### 47.10. Patterns of Biodiversity (pp. 830–831)
1.a. Resource availability tends to be higher and more reliable. Tropical latitudes have more sunlight of greater intensity, rainfall amount is higher, and the growing season is longer. Vegetation grows all year long to support diverse herbivores, etc. b. Species diversity might be self-reinforcing. When a greater number of plant species compete and coexist, a greater number of herbivore species evolve because no herbivore can overcome the chemical defenses of all kinds of plants. Then more predators and parasites evolve in response to the diversity of prey and hosts. c. The rates of speciation in the tropics have exceeded those of background extinction. At higher latitudes, biodiversity has been suppressed during times of mass extinction. 2. tropics; 3. Iceland; 4. biodiversity; 5. Iceland; 6. dispersal; 7. distance; 8. area; 9. Larger; 10. diversity; 11. targets; 12. biodiversity; 13. small; 14. small; 15. immigration; 16. extinction; 17. immigration; 18. extinction; 19. Island C.

### Self-Quiz
1. b; 2. b; 3. c; 4. a; 5. b; 6. d; 7. e; 8. b; 9. e; 10. a; 11. d; 12. d; 13. d; 14. a; 15. d.

---

# Chapter 48    Ecosystems

*Crepes for Breakfast, Pancake Ice for Dessert* (pp. 834–835)
### 48.1. The Nature of Ecosystems (pp. 836–837)
### 48.2. Energy Flow Through Ecosystems (pp. 838–839)
### 48.3. *Focus on Science:* Energy Flow at Silver Springs, Florida (p. 840)
1. producers; 2. heterotrophs; 3. herbivores; 4. carnivores; 5. omnivores; 6. parasites; 7. decomposers; 8. detritivores; 9. open; 10. energy; 11. nutrient; 12. energy; 13. nutrients; 14. trophic; 15. food chain; 16. food web; 17. productivity; 18. grazing; 19. detrital; 20. pyramid; 21. producers; 22. biomass; 23. biomass; 24. smallest; 25. energy; 26. trophic; 27. large; 28. 1; 29. 6; 30. 10; 31. low; 32. 4; 33. b; 34. c; 35. b; 36. a; 37. b; 38. b; 39. b; 40. b; 41. b; 42. b; 43. a; 44. b; 45. c; 46. a; 47. b; 48. open; 49. input; 50. cannot; 51. can; 52. have; 53. E; 54. B; 55. D; 56. C; 57. E; 58. D; 59. E.

### 48.4. Biogeochemical Cycles—An Overview (p. 841)
### 48.5. Hydrologic Cycle (pp. 842–843)
### 48.6. Carbon Cycle (pp. 844–845)
### 48.7. *Focus on the Environment:* From Greenhouse Gases to a Warmer Planet? (pp. 846–847)
### 48.8. Nitrogen Cycle (pp. 848–849)
1. Usually as mineral ions such as ammonium ($NH_4^+$); 2. Inputs from the physical environment and the cycling activities of decomposers and detritivores; 3. The amount of a nutrient being cycled through the ecosystem is greater; 4. Common sources are rainfall or snowfall, metabolism (such as nitrogen fixation), and weathering of rocks; 5. Losses of mineral ions occur by runoff; 6.a. oxygen and hydrogen move in the form of water molecules; b. a large portion of the nutrient is in the form of atmospheric gas such as carbon and nitrogen (mainly $CO_2$); c. nutrients are not in gaseous forms; nutrients move from land to the seafloor and only "return" to land through geological uplifting of long duration; phosphorus is an example; 7. F; 8. H; 9. D; 10. B; 11. A (C); 12. E; 13. G (H); 14. C; 15. size; 16. soil; 17. streams; 18. transpiration; 19. vegetation; 20. nutrients; 21. calcium; 22. calcium; 23. biomass; 24. soil; 25. herbicides; 26. six; 27. deforestation; 28. stabilize; 29. minerals; 30. nutrients; 31. C; 32. D; 33. F; 34. B; 35. G; 36. A; 37. E; 38. gaseous; 39. climate; 40. greenhouse effect; 41. heat; 42. CFCs; 43. carbon dioxide; 44. warming; 45. reradiate; 46. energy; 47. atmosphere; 48. greenhouse; 49. global; 50. sea; 51. flood; 52. climate; 53. increasing; 54. d; 55. e; 56. c; 57. a; 58. b; 59. Soil nitrogen compounds are vulnerable to being leached and lost from the soil; some fixed nitrogen is lost to air by denitrification; nitrogen fixation comes at high metabolic cost to plants that are symbionts of nitrogen-fixing bacteria; losses of nitrogen are enormous in agricultural regions through the tissues of harvested plants, soil erosion, and leaching processes.

### 48.9. Sedimentary Cycles (p. 850)

**48.10. Predicting the Impact of Change in Ecosystems**
   (p. 851)
1. reservoir; 2. sedimentary; 3. phosphates; 4. ocean; 5. continental; 6. geochemical; 7. ecosystem; 8. organisms; 9. ionized; 10. plants; 11. herbivores; 12. Decomposition; 13. ecosystem; 14. Fertilizers; 15. lakes; 16. growth; 17. nitrogen; 18. potassium; 19. sediments; 20. phosphorus; 21. algal; 22. eutrophication; 23. Ecosystem modeling; 24. mosquitoes; 25. malaria; 26. biological magnification; 27. food webs; 28. predators; 29. top; 30. biological magnification; 31. Disturbances.

**Self-Quiz**
1. c; 2. c; 3. d; 4. b; 5. a; 6. a; 7. b; 8. a; 9. a; 10. b; 11. b; 12. d.

# Chapter 49   The Biosphere

*Does a Cactus Grow in Brooklyn?* (pp. 854–855)
**49.1. Air Circulation Patterns and Regional Climates**
   (pp. 856–857)
**49.2. The Ocean, Land Forms, and Regional Climates**
   (pp. 858–859)
1. D; 2. H; 3. K; 4. B; 5. J; 6. F; 7. C; 8. A; 9. L; 10. E; 11. I; 12. M; 13. G; 14. equatorial; 15. warm; 16. rises or ascends; 17. moisture; 18. descends; 19. moisture; 20. ascends; 21. moisture; 22. descends; 23. east (easterlies); 24. west (westerlies); 25. tropical; 26. warm; 27. cool; 28. cold; 29. solar (sun's); 30. rotation; 31. F; 32. B; 33. E; 34. D; 35. G; 36. C; 37. A.

**49.3. The World's Biomes** (pp. 860–861)
**49.4. Soils of Major Biomes** (p. 862)
**49.5. Deserts** (p. 863)
**49.6. Dry Shrublands, Dry Woodlands, and Grasslands**
   (pp. 864–865)
1. F; 2. I; 3. G; 4. A; 5. H; 6. C; 7. B; 8. K; 9. E; 10. D; 11. J; 12. L; 13. E; 14. D; 15. E; 16. F; 17. B; 18. C; 19. H; 20. D; 21. d; 22. a; 23. d; 24. b; 25. c; 26. a; 27. e; 28. d; 29. e; 30. b.

**49.7. Tropical Rain Forests and Other Broadleaf Forests**
   (pp. 866–867)
**49.8. Coniferous Forests** (p. 868)
**49.9. Tundra** (p. 869)
1. d; 2. c; 3. c; 4. a; 5. b; 6. b; 7. a; 8. d; 9. c, d; 10. b; 11. tropical rain; 12. 25; 13. humidity; 14. decomposition; 15. tropical deciduous; 16. temperate deciduous; 17. coniferous; 18. needle; 19. stomata; 20. adaptations; 21. Spruce; 22. taiga; 23. montane; 24. Spruce; 25. pines; 26. pine barrens; 27. tundra; 28. permafrost; 29. alpine; 30. permafrost.

**49.10. Freshwater Provinces** (pp. 870–871)
**49.11. The Ocean Provinces** (pp. 872–873)
1. lake; 2. littoral; 3. limnetic; 4. plankton; 5. profundal; 6. overturns; 7. 4; 8. spring; 9. thermocline; 10. cools; 11. fall; 12. down; 13. up; 14. higher; 15. short; 16. Oligotrophic; 17. eutrophic; 18. eutrophication; 19. Streams; 20. runs; 21. benthic (C); 22. pelagic (A); 23. neritic (B); 24. oceanic (D); 25. a; 26. c; 27. c; 28. b; 29. c; 30. a; 31. b; 32. a; 33. b; 34. c; 35. a; 36. b; 37. a; 38. c; 39. b; 40. c; 41. a; 42. b; 43. b; 44. c.

**49.12. Coral Reefs and Banks** (pp. 874–875)
**49.13. Life Along the Coasts** (pp. 876–877)
**49.14.** *Focus on Science:* **El Niño and Seesaws in the World's Climates** (pp. 878–879)
1.a. Atolls are ring-shaped coral reefs that enclose or almost enclose a shallow lagoon; b. Fringing reefs form next to the land's edge in regions of limited rainfall, as on the leeward side of tropical islands; c. Barrier reefs form around islands or parallel with the shore of a continent. A calm lagoon forms behind them; 2. d; 3. a; 4. c; 5. e; 6. c; 7. a; 8. b; 9. d; 10. b; 11. a; 12. b; 13. e; 14. d; 15. a; 16. d; 17. e; 18. a; 19. c; 20. e; 21. c; 22. a.

**Self-Quiz**
1. d; 2. d; 3. b; 4. d; 5. a; 6. b; 7. c; 8. c; 9. d; 10. c; 11. c; 12. b.

# Chapter 50   Human Impact on the Biosphere

*An Indifference of Mythic Proportions* (pp. 882–883)
**50.1. Air Pollution—Prime Examples** (pp. 884–885)
**50.2. Ozone Thinning—Global Legacy of Air Pollution**
(p. 886)
1. Pollutants;  2. thermal inversion;  3. winters;  4. industrial smog;  5. photochemical smog;  6. oxygen;  7. nitrogen dioxide;  8. photochemical;  9. gasoline;  10. PANs;  11. sulfur (nitrogen);  12. nitrogen (sulfur);  13. sulfur;  14. nitrogen;  15. nitrogen;  16. acid deposition;  17. sulfuric (nitric);  18. nitric (sulfuric);  19. acid rain;  20. 5;  21. Chlorofluorocarbons;  22. c;  23. d;  24. e;  25. b;  26. a;  27. b;  28. f;  29. f;  30. a;  31. b;  32. b;  33. d;  34. a (b);  35. d;  36. f;  37. c;  38. d;  39. e;  40. f;  41. e;  42. c;  43. d.

**50.3. Where to Put Solid Wastes, Where to Produce Food** (p. 887)
**50.4. Deforestation—Concerted Assaults on Finite Resources** (pp. 888–889)
**50.5. *Focus on Bioethics:* You and the Tropical Rain Forest** (p. 890)
1. F;  2. H (G);  3. I;  4. E;  5. J;  6. B;  7. C;  8. A;  9. D;  10. G.

**50.6. Trading Grasslands for Deserts** (p. 891)
**50.7. A Global Water Crisis** (pp. 892–893)
1. desertification;  2. overgrazing on marginal lands;  3. domestic cattle;  4. native wild herbivores;  5. deserts;

6. desalination;  7. energy;  8. agriculture;  9. salination;  10. water table;  11. saline;  12. 20;  13. pollution;  14. wastewater;  15. tertiary;  16. 55–66;  17. oil;  18. a. Screens and settling tanks remove sludge, which is dried, burned, dumped in landfills, or treated further;  chlorine is often used to kill pathogens in water, but does not kill them all. b. Microbial populations are used to break down organic matter after primary treatment but before chlorination. c. It removes nitrogen, phosphorus, and toxic substances, including heavy metals, pesticides, and industrial chemicals;  it is largely experimental and expensive.

**50.8. A Question of Energy Inputs** (pp. 894–895)
**50.9. Alternative Energy Sources** (p. 896)
**50.10. *Focus on Bioethics:* Biological Principles and the Human Imperative** (p. 897)
1. J-shaped;  2. increased numbers of energy users and to extravagant consumption and waste; 3. Net energy;  4. plants;  5. next;  6. decreases;  7. acid deposition;  8. less;  9. meltdown;  10. have not;  11. solar-hydrogen energy;  12. wind farms;  13. Fusion power;  14. C (N);  15. B (N);  16. F (N);  17. A (R);  18. D (N);  19. E (R).

**Self-Test**
1. c;  2. b;  3. a;  4. b;  5. d;  6. d;  7. c;  8. c;  9. d;  10. b.

---

# Chapter 51   An Evolutionary View of Behavior

*Deck the Nest With Sprigs of Green Stuff* (pp. 900–901)
**51.1. The Heritable Basis of Behavior** (pp. 902–903)
**51.2. Learned Behavior** (p. 904)
1. C;  2. F;  3. E;  4. B;  5. G;  6. A;  7. D;  8. the banana slug;  9. ate;  10. inland;  11. were not;  12. genetic;  13. pineal gland;  14. directly;  15. estrogen;  16. testosterone;  17. Hormones;  18.a. Cuckoo birds are social parasites in that adult females lay eggs in the nests of other bird species;  young cuckoos instinctively eliminate the natural-born offspring (eggs are maneuvered onto their backs and pushed out of the nest) and then receive the undivided attention of their unsuspecting foster parents. b. Young toads instinctively capture edible insects with sticky tongues;  if a bumblebee is captured and then stings the tongue, the toad learns to leave bumblebees alone;  19. E;  20. C;  21; F;  22. A;  23. D;  24. B

**51.3. The Adaptive Value of Behavior** (p. 905)
1. Natural selection;  2. alleles;  3. evolution;  4. adaptive;  5. reproductive;  6. individual;  7. a. Consideration of individual survival and production of offspring. b.

Any behavior that promotes propagation of an individual's genes and tends to occur at increased frequency in future generations. c. Cooperative, interdependent relationships among individuals of the same species. d. Within a population, any behavior that increases an individual's chances to produce or protect offspring of its own, regardless of the consequences for the population. e. Within a population, a self-sacrificing behavior. The individual behaves in a way that helps others but that decreases its own chances to produce offspring.

**51.4. Communication Signals** (pp. 906–907)
1. i;  2. c;  3. e;  4. a;  5. g;  6. j;  7. f;  8. b;  9. h;  10. c;  11. g;  12. d.
**51.5. Mating, Parenting, and Individual Reproductive Success** (pp. 908–909)
1.a. Hangingflies;  b. Sage grouse;  c. Lions, sheep, elk, and bison;  d. Caspian terns

**51.6. Benefits of Living in Social Groups** (pp. 910–911)
**51.7. Costs of Living in Social Groups** (p. 912)

**51.8. Evolution of Altruism** (p. 913)

**51.9.** *Focus on the Social Environment:* **About Those Self-Sacrificing Insects** (pp. 914–915)

**51.10.** *Focus on Science:* **About Those Naked Mole-Rats** (p. 916)

1. C;  2. A;  3. E;  4. D;  5. B;  6.a. Royal penguins, herring gulls, cliff swallows, and prairie dogs live in huge colonies and must compete for a share of the same food resources. b. Under crowded living conditions, the individual and its offspring are more likely to be weakened by pathogens and parasites that are more readily transmitted from host to host in crowded groups. Plagues spread like wildfire through densely crowded human populations;  this is especially the case in settlements and cities with chronic infestations of rats and fleas, and with inadequate or nonexistent sewage treatment and medical care. c. Breeding pairs of herring gulls will quickly cannibalize the eggs or young chicks of the neighbors in an instant;  long-lived lions compete for permanent hunting territories even though they can live for an extended time between kills;  a lion pride of three or more actually eat less well than one or a pair of lions; male lions intent on taking over a pride will show infanticidal behavior and will kill the cubs;  group living is costly for lionesses in terms of food intake and reproductive success;  they stick together to defend territories against smaller groups of rivals;  aggressive males almost always kill the cubs of a single lioness, but occasionally a group of two or more lionesses can save some of the cubs. 7. subordinate;  8. altruistic;  9. reproductive;  10. dominant;  11. nonbreeding;  12. altruistic; 13. insect;  14. worker;  15. genes;  16. indirect selection; 17. parenting;  18. indirect;  19. reproduce;  20. altruistic;  21. genes;  22. vertebrates;  23. nonbreeding; 24. perpetuate;  25. genes (alleles).

**51.11. An Evolutionary View of Human Social Behavior** (p. 917)

1. human;  2. a trait valuable in gene transmission; 3. adopt;  4. strangers;  5. can;  6. are;  7. is not;  8. is; 9. are;  10. nonrelative.

**Self-Quiz**

1. c;  2. d;  3. b;  4. d;  5. a;  6. b;  7. b;  8. c;  9. a;  10. c; 11. b;  12. b;  13. d;  14. c;  15. a